CLIMATE-RESILIENT AGRICULTURE

A Molecular Perspective

CLIMATE-RESILIENT AGRICULTURE

A Molecular Perspective

Edited by

Ashutosh Singh, PhD

Saurabh Pandey, PhD

APPLE ACADEMIC PRESS

First edition published 2025

Apple Academic Press Inc.
1265 Goldenrod Circle, NE,
Palm Bay, FL 32905 USA

760 Laurentian Drive, Unit 19,
Burlington, ON L7N 0A4, CANADA

CRC Press
2385 NW Executive Center Drive,
Suite 320, Boca Raton FL 33431

4 Park Square, Milton Park,
Abingdon, Oxon, OX14 4RN UK

Library and Archives Canada Cataloguing in Publication

Title: Climate-resilient agriculture : a molecular perspective / edited by Ashutosh Singh, PhD, Saurabh Pandey, PhD.
Names: Singh, Ashutosh (Assistant professor at the Centre for Advanced Studies on Climate Change), editor. | Pandey, Saurabh (Lecturer in agriculture), editor.
Description: First edition. | Includes bibliographical references and index.
Identifiers: Canadiana (print) 20240281772 | Canadiana (ebook) 20240281802 | ISBN 9781774915066 (hardcover) | ISBN 9781774915073 (softcover) | ISBN 9781003455271 (ebook)
Subjects: LCSH: Crops—Effect of stress on. | LCSH: Crops and climate. | LCSH: Plant breeding. | LCSH: Agricultural innovations.
Classification: LCC SB112.5 .C55 2024 | DDC 632/.1—dc23

Library of Congress Cataloging-in-Publication Data

CIP data on file with US Library of Congress

ISBN: 978-1-77491-506-6 (hbk)
ISBN: 978-1-77491-507-3 (pbk)
ISBN: 978-1-00345-527-1 (ebk)

About the Editors

Ashutosh Singh, PhD
Assistant Professor, Centre for Advanced Studies on Climate Change, Dr. Rajendra Prasad Central Agricultural University, Pusa, Samastipur, Bihar, India

Ashutosh Singh, PhD, is working as an Assistant Professor at the Centre for Advanced Studies on Climate Change at Dr. Rajendra Prasad Central Agricultural University, Pusa, Samastipur, Bihar, India. Dr. Singh has been awarded a PhD degree from the University of Agricultural Science Bangalore, GKVK, Karnataka (Plant Biotechnology) in 2018. He has more than eight years of experience in different areas of plant biotechnology aspects such as plant stress biology, pollen-based selection, molecular marker studies, and so on. Currently, he is working on transcriptomic profiling and the development of gene-based markers for pollen selection mediated reproductive stage heat stress tolerance in maize. He has published more than a dozen research papers in various journals of international repute. Dr. Singh has published many book chapters and more than 20 popular articles.

Saurabh Pandey, PhD
Assistant Professor, Department of Agriculture, Guru Nanak Dev University, Amritsar, Punjab, India

Saurabh Pandey, PhD, is currently working as an Assistant Professor, Department of Agriculture at Guru Nanak Dev University, Amritsar, Punjab, India. He did his MSc (Agriculture), specializing in Plant Biotechnology, at the University of Agricultural Sciences, Bangalore, and was conferred his PhD (Plant Virology) from the National Institute of Plant Genome Research, New Delhi. He has eight years of experience in different areas of plant biotechnology aspects such as plant stress biology, plant virus interactions, abiotic stresses, biofortification, molecular marker studies, and metabolism related to the tomato and rice crop. Currently, he is working on the introduction and biotic stress management of unconventional crops like banana and apple in the Punjab region, India. He has published about half a dozen research papers in various journals of international repute, such as *Frontiers in Plant Science, Environmental and Experimental Botany, Plant Science, Journal of Plant Growth and Regulation, Advances in Genetics, Physiology,* and *Molecular Biology of Plants and Plant Stress*. He has also contributed articles of varied interest in many magazines and periodicals. Dr. Pandey has authored many book chapters on molecular markers, genomics, plant breeding, and plant molecular virology, catering to the needs of modern agriculture research. He has been an active member of the Indian Virological Society, the Indian Academy of Horticultural Sciences, and the Society of Plant Biochemistry and Biotechnology. Dr. Pandey is the recipient of a Gold Medal for his MSc degree from UAS, Bangalore, in 2016. He has also received the Best PG Thesis Award for his MSc thesis and various other awards for his research work. Also, Dr. Pandey has received NTS, JRF, and SRF during his degree program, awarded by the Indian Council of Agriculture Research (ICAR), India, and Department of Biotechnology, India.

Contents

Contributors

Sneha Adhikari
ICAR-IIWBR, Regional Station Flowerdale, Shimla, Himachal Pradesh, India

Ayaz Ahamad
Biosciences Group, Homi Bhabha National Institute (HBNI), Bhabha Atomic Research Centre, Mumbai, Maharashtra, India

S. K. MD. Ajaharuddin
Department of Entomology, Bidhan Chandra Krishi Viswavidyalaya (BCKV), Mohanpur, Nadia, West Bengal, India

Mohd Aftab Alam
Department of Plant Pathology, Swami Keshwanand Rajasthan Agriculture University (SKRAU), Rajasthan, India

Suresh Antre
University of Agricultural Sciences, GKVK Campus, Bangalore, India

Aditi Arya
Department of Biotechnology, Deenbandhu Chottu Ram University of Science and Technology, Murthal, Haryana, India

M. Ashajyothi
ICAR- Central Agroforestry Research Institute, Jhansi, Uttar Pradesh, India

Kousik Atta
Department of Plant Physiology, Bidhan Chandra Krishi Viswavidyalaya (BCKV), Mohanpur, Nadia, West Bengal, India

Partha Pratim Behera
Department of Plant Breeding and Genetics, Assam Agricultural University, Jorhat, Assam, India

S. C. Bhardwaj
ICAR-IIWBR, Regional Station Flowerdale, Shimla, Himachal Pradesh, India

Tanmaya Kumar Bhoi
Forest Protection Division, ICFRE-Arid Forest Research Institute (AFRI), Jodhpur, Rajasthan, India

Alka Bharati
ICAR-Central Agroforestry Research Institute, Jhansi, Uttar Pradesh, India

Ajay Kumar Chandra
ICAR-Indian Agricultural Research Institute, New Delhi, India

Divya Chauhan
Banasthali Vidyapeeth, Jaipur, Rajasthan, India

Sharani Choudhury
ICAR-Indian Agricultural Research Institute, New Delhi, India

Vinay Kumar Dhiman
Dr. YSP University of Horticulture and Forestry, Solan, Himachal Pradesh, India

Vivek Kumar Dhiman
Himachal Pradesh University Shimla, Himachal Pradesh, India

Vishal Dinkar
ICAR-Central Institute of Temperate Horticulture, Srinagar, Jammu and Kashmir, India

Suresha P. G.
College of Agriculture, KSNUAHS, Shivamogga, Karnataka, India

O. P. Gangwar
ICAR-IIWBR, Regional Station Flowerdale, Shimla, Himachal Pradesh, India

Saidul Islam
Nadia KVK, Goyeshpur, Kalyani, Nadia, West Bengal, India

Anjali Joshi
Genetics and Tree Improvement Division, Arid Forest Research Institute, Jodhpur, Rajasthan, India

Bhavyasree R. K.
Punjab Agricultural University, Regional Research Station, Gurdaspur, Punjab, India

H. G. Kencharaddi
College of Agriculture, CAU(I), Kyrdemkulai, Meghalaya, India

Amarjeet Kumar
Department of Genetics and Plant Breeding, MTTC & VTC, Selesih, CAU, Imphal, India

Amresh Kumar
ICAR-National Institute for Plant Biotechnology, Pusa Campus, New Delhi, India

Love Kumar
Centre for Protected Cultivation Technology, IARI, New Delhi, India

Subodh Kumar
ICAR-IIWBR, Regional Station Flowerdale, Shimla, Himachal Pradesh, India

Saumya Kumari
University of Agricultural Sciences, GKVK Campus, Bangalore, India

Santanu Kundu
Department of Agronomy, Professor Jayashankar Telangana State Agricultural University, Hyderabad, Telangana, India

Amar Kant Kushwaha
ICAR-Central Institute of Sub-tropical Horticulture, Lucknow, Uttar Pradesh, India

Charu Lata
ICAR-IIWBR, Regional Station Flowerdale, Shimla, Himachal Pradesh, India

Devraj Lenka
Department of Plant Breeding and Genetics, College of Agriculture, Odisha University of Agriculture and Technology (OUAT), Bhubaneswar, Odisha, India

Chirag Maheshwari
Division of Biochemistry, ICAR-Indian Agricultural Research Institute, New Delhi, India

Prasanta Kumar Majhi
Department of Plant Breeding and Genetics, Regional Research and Technology Transfer Station (RRTTS), Odisha University of Agriculture and Technology (OUAT), Keonjhar, Odisha, India

P. K. Mandal
ICAR-National Institute of Plant Biotechnology, Pusa, New Delhi, India

Apurba Pal
Horticulture College, Khuntpani, Birsa Agriculture University, Ranchi, Jharkhand, India

Chaitanya B. Pande
Department of Remote Sensing, Primus Partners Private Limited, Maharashtra, India

Himanshu Pandey
Dr. YSP University of Horticulture and Forestry, Solan, Himachal Pradesh, India

Saurabh Pandey
Department of Agriculture, Guru Nanak Dev University, Amritsar, Punjab, India

Biswaranjan Patra
Center for Biotechnology, Siksha 'O' Anusandhan Deemed to be University, Bhubaneswar, Odisha, India

Sachin Phogat
ICAR-Indian Agricultural Research Institute, New Delhi, India

Ankur Poudel
ICAR-Indian Agricultural Research Institute, New Delhi, India

Kalipada Pramanik
Department of Agronomy, Palli Siksha Bhavana, Institute of Agriculture (Visva-Bharati), Sriniketan, Birbhum, West Bengal, India

Geeta Prasad
ICAR-National Bureau of Plant Genetic Resources, New Delhi, India

Pramod Prasad
ICAR-IIWBR, Regional Station Flowerdale, Shimla, Himachal Pradesh, India

Ashutosh Purohit
Department of Plant Breeding and Genetics, College of Agriculture, Odisha University of Agriculture and Technology (OUAT), Bhubaneswar, Odisha, India

Manjesh Saakre
ICAR-Indian Agricultural Research Institute, New Delhi, India

Kuleshwar Prasad Sahu
School of Agricultural Science, GD Goenka University, Gurugram, Haryana, India

Sayantan Sarkar
Blackland Research and Extension Center, Texas A&M Agrilife Research, Temple, TX, USA

Ranjit Saroj
Division of Genetics, Indian Agricultural Research Institute, New Delhi, India

Maksud Hasan Shah
Department of Agronomy, Bidhan Chandra Krishi Viswavidyalaya (BCKV), Mohanpur, Nadia, West Bengal, India

Aalok Shiv
Division of Crop Improvement, ICAR-Indian Institute of Sugarcane Research (ICAR-IISR), Lucknow, Uttar Pradesh, India

Ashish Kumar Singh
ICAR-Vivekananda Parvatiya Krishi Anusandhan Sansthan, Almora, Uttarakhand, India

Ashutosh Singh
Centre for Advanced Studies on Climate Change, RPCAU, Pusa, Samastipur, Bihar, India

Badal Singh
Division of Conservation, ICAR-National Bureau of Plant Genetic Resources, New Delhi, India

Devendra Singh
B. N. College of Engineering and Technology, Lucknow, Uttar Pradesh, India

Subodh Kumar Sinha
ICAR-National Institute for Plant Biotechnology, Pusa Campus, New Delhi, India

C. D. Soregaon
College of Agriculture, UASD, Vijayapura, Karnataka, India

A. E. Sunil Subramanya
University of Agricultural Sciences, GKVK Campus, Bangalore, India

Jyotsana Tilgam
ICAR-National Bureau of Agriculturally Important Micro-organisms, Maunath Bhanjan, Uttar Pradesh, India

Abbreviations

ABA	abscisic acid
AFLP	amplified fragment length polymorphism
AI	artificial intelligence
ALS	acetolactate synthase
AM	association mapping
ANN	artificial neural network
AOX	alternative oxidase
APOX	ascorbate peroxidase
AQY	apparent quantum yield
ARF	auxin response factors
ASI	anthesis-silking interval
BABA	beta-aminobutyric acid
BC	backcrosses
BILs	backcross inbred lines
BLB	bacterial leaf blight
BLSB	banded leaf and sheath blight
BNF	biological nitrogen fixation
BRs	brassinosteroids
BTH	benzothiadiazole
CAPS	cleaved amplified polymorphic sequences
CAT	catalase
CAX	cation exchanger
CBP	CREB-binding protein
CBSD	cassava brown streak disease
CCN	cereal cyst nematode
CDF	cation diffusion facilitator
CDPK	calcium-dependent protein kinase
CDPK	cyclin-dependent protein kinases
CE	carboxylation efficiency
CMS	cytoplasmic male sterility
CMTS	cell membrane thermostability

CMV	cucumber mosaic virus
Co	cobalt
CO_2	carbon dioxide
Cr	chromium
CRA	climate resilient agriculture
CRISPR	clustered regularly interspaced short palindromic repeat
CRISPR-Cas9	clustered regularly interspaced short palindromic repeats and CRISPR-associated protein 9
CSA	climate-smart agriculture
CSSLs	chromosome segment substitution lines
CTD	canopy temperature depression
Cu	copper
CUC	cup-shaped cotyledon
CVYV	cucumber vein yellowing virus
CWRs	crop wild relatives
CymMV	Cymbidium mosaic virus
DAMP	damage-associated molecular pattern
DBM	diamondback moth
DEGs	differentially expressed genes
DGK	diacylglycerol kinase
DH	doubled haploids
DHAR	dehydroascorbate reductase
DNN	deep neural network
DSB	double stranded breaks
dsRNA	double-stranded RNA
ELM	extreme learning machine
EN	engineered nucleases
ER	endoplasmic reticulum
ERFs	ethylene-responsive factors
ETC	electron transport chain
ETI	effector-triggered immunity
flg22	flagellin 22
FMs	functional markers
Fo	fluorescence
FS	full spectrum
GABA	gamma-aminobutyric acid
GAT	glyphosate acetyltransferase

GB	glycine betaine
GBS	genotyping-by-sequencing
GE	gene editing
GEBV	genome estimated breeding values
GFD	grain-filling duration
GHG	greenhouse gases
GMMs	genic molecular markers
GOX	glyphosate oxidoreductase
GPx	glutathione peroxidase
GR	glutathione reductase
GS	genomic selection
GSH	glutathione
GWAS	genome-wide association study
H_2O_2	hydrogen peroxide
HDR	homology-dependent repair
HDR	homology-directed recombination
HDR	homology-directed repair
HI	harvest index
HIGS	host-induced gene silencing
HMAs	heavy metal ATPase
HR	hypersensitive response
HSF	heat shock factors
Hsfs	heat stress transcription factor
HSG	heat stress granule
HSP	heat shock protein
HT	heat stress
IBD	isolation by distance
IBE	isolation by environment
INDELs	insertion/deletion
IoT	Internet of Things
IPCC	intergovernmental panel on climate change
ISSR	inter simple sequence repeat
JA	jasmonic acid
KASP	Kompetitive allele-specific PCR
LBD	lateral organ boundaries domain
LD	linkage disequilibrium
LD	long day

LE	linkage equilibrium
LEA	late embryogenesis abundant
LED	light-emitting diode
LRK	receptor-like kinase
LRR	leucine-rich repeats
LRs	landraces
MAB	marker-assisted backcrossing
MAB	marker-assisted breeding
MABB	marker-assisted backcross breeding
MABC	marker-assisted backcross
MAF	minor allele frequency
MAGIC	multiparent advanced generation intercross
MAPK	mitogen-associated protein kinases
MAPKs	mitogen-activated protein kinases
MARS	marker-assisted recurrent selection
MAS	marker-assisted selection
MDA	malondialdehyde
MDHAR	monodehydroascorbate reductase
MEP	methyl erythritol phosphate
ML	machine learning
MNs	meganucleases
MNSV	melon necrotic spot virus
MS	mass spectrometry
MSV	maize streak virus
MTA	marker trait association
MTP	metal transporter
MVs	maize varieties
MWMV	Moroccan watermelon mosaic virus
NAM	nested association mapping
NAM	no apical meristem
NAR	net assimilation rate
NAS	nicotianamine synthase
NBS-LRR	nucleotide-binding sequences-leucine-rich repeat
NBT	new breeding techniques
NHEJ	nonhomologous end joining
Ni	nickel
NICRA	National Innovations in Climate Resilient Agriculture

NIL	near-isogenic lines
OA	osmotic adjustment
ORSV	Odontoglossum ringspot virus
OsMPK5	overexpression of rice MPK5
O3	ozone
PAL	phenylalanine ammonia-lyase
PAMP	pathogen-associated molecular pattern
PAR	photosynthetic active radiation
PBS	pollen-based selection
PCD	programmed cell death
PEG	polyethylene glycol
PGR	plant growth regulators
PHR	phosphate starvation response
PHTs	phosphate transporters
PIF3	phytochrome-interacting factor 3
PIL	PIN-like proteins
POD	peroxidase
PPNs	plant-parasitic nematodes
PPOs	polyphenol oxidases
PRRs	plant-based recognition receptors
PRX	peroxiredoxin
PSII	photosystem II
PTI	PAMP-triggered immunity
PVY	potato virus Y
QTL	quantitative trait loci
RAPD	random amplified polymorphic DNA
RB	red and blue light
RBOHD	respiratory burst oxidase homologue D
RBOHs	respiratory burst oxidase homologs
Rca	Rubisco Activase
RDMs	random DNA markers
RdRp	RNA-dependent RNA polymerase
RF	random forest
RFLP	restriction fragment length polymorphism
RGA	rapid generation advance
RGR	relative growth rate
RILs	recombinant inbred lines

RISC	RNA-induced silencing complex
RLK	receptor-like kinase
RMR	root mass ratio
RNAi	RNA interference
ROS	reactive oxygen species
RSA	root system architecture
RTD	rice tungro disease
Rubisco	ribulose bisphosphate carboxylase/oxygenase
RuBP	ribulose bisphosphate
RUE	radiation use efficiency
RWA	Russian wheat aphid
RWC	relative water content
SA	salicylic acid
SAR	systemic acquired resistance
SB	speed breeding
SBR	sheath blight of rice
SCAR	sequence characterized amplified region
SCMV	sugarcane mosaic virus
SCN	soybean cyst nematode
sgRNA	single guide RNA
SLs	strigolactones
SMRT	single-molecule real-time
SNP	single-nucleotide polymorphism
SO_2	sulfur dioxide
SOD	superoxide dismutase
SPAD	soil plant analysis development
SPD	single-pod descent
SPS	single-plant selection
SPS	sucrose phosphate synthase
SRAP	sequence-related amplified polymorphism
SSC	single-seed chipping
SSD	single-seed descent
SSRs	simple sequence repeats
STMS	sequence tagged microsatellite
SVL	sodium vapor lamps
SVM	support vector machine
SVR	support vector regression

SVs	structural variations
TALENs	transcription activator-like effector nucleases
TEs	transposable elements
TFB	transcription factor binding
TFs	transcription factors
TMV	tobacco mosaic virus
TOR	target of rapamycin kinase
TPE	target population environment
TPR	tetratricopeptide repeat
TSWV	tomato spotted wilt virus
UAVs	unmanned aerial vehicles
UPR	unfolded protein response
UV-B	ultraviolet-B
VIT	vacuolar iron transporter
VPD	vapor pressure deficit
VSC	vacuolar sequestration capacity
WRs	wild relatives/cousins
WSMV	wheat streak mosaic virus
ZFNs	zinc-finger nucleases
ZIF	zinc-induced facilitator
ZYMV	zucchini yellow mosaic virus

Foreword

The world population is steadily growing and is expected to reach over 10 billion by 2050. Consequently, the increasing pressure to feed a growing population in the future would need additional production of major crops. On the contrary, the rapidly changing climate is imposing a heavy toll on agriculture, threatening food, and nutritional security. This has warranted that researchers to deploy classical and advanced tools to improve the climate resilience of cultivated species. In this context, this book, *Climate Resilient Agriculture: A Molecular Perspective,* is a timely attempt to collate the efforts and outcomes of research on climate-resilient agriculture.

The book has four parts, where part A introduces the readers to climate change, its effect on plant growth, development, reproduction, and productivity, followed by the issues of increased biotic and abiotic stresses. Part B deals with breeding and molecular approaches used to improve crop species' climate resilience. Each chapter deals with stress and the nine chapters under this part give a detailed insight into how different tools and techniques have enhanced the climate resilience of cultivated species. Part C summarizes the advanced tools, including speed breeding and genome editing, which have direct implications for expediting the research on climate-resilient agriculture. Part D has three chapters on the success stories of climate-resilient agriculture.

Altogether, the book is an outstanding effort to bring together the knowledge on climate-resilient agriculture, which will serve as a one-stop solution for researchers as well as students. Also, the book will serve as a guide for researchers aspiring to work in the area of climate-resilient agriculture. Reliability and recent knowledge integration are critical aspects of any book. The effort of the editors to bring out this book is very much appreciated, and the contributions of authors from different subject domains is also acknowledged. The book evidences the conclusions from various issues and volumes to ensure timely collation of available knowledge in the given field. I believe this book is essential for students, early career researchers, and scientists working in molecular biology, biotechnology, molecular breeding, and plant physiology. Suggestions for improvement are always welcome.

—**Manoj Prasad**
Professor,
Delhi University, South campus, India

Preface

Today, in a world of around 7.95 billion people, agriculture crops are facing tremendous challenges. Food security faces constraints with regards to number of produce as well as in quality standards. Climate change has induced alterations in weather and increased the frequency of natural disasters causing a worsening scenario directly impacting global crop productivity. Currently, the agricultural sector and the economy faces huge challenges such as growing demand for food, reduced crop productivity due to unpredictable rainfall, water deficit, high salinity, and temperature fluctuations.

Plant stress response has been explored in the recent years as a field of study, however, it is still largely unexplored when it comes to delivering effective strategies for improving agriculture. Agriculture is highly vulnerable to climate and even a slight change in climatic factors such as drought and temperature causes tremendous losses in yield potential. Food crops are crucial for global food security and environmental sustainability and exposed to various environmental encounters in the field.

In relation to that, crop research is still lagging in addressing its problems causing a substantial gap in demand and productivity. Further, this gap is directly associated with a decreased crop yield worldwide. Climate change brings various stresses affecting the crucial plant processes such as plant hormone balance, seedling vigor, seed germination, leaf area development, mineral nutrition, water relation, functional genes, metabolite productions, photosynthesis, and many more reducing the final yield. Abiotic stresses are persistent in modern crops grown and cultivated in less productive soil and harsh conditions. Therefore, developing climate-smart crops is vital to secure global food security.

Keeping this in mind, Climate-Resilient Agriculture: A Molecular Perspective compiles the current aspects of climate-resilient agriculture, including the crucial physiological, biochemical, and molecular aspects of cultivated crops under stress conditions, which could play a pivotal role in developing climate-smart crops. This book discusses climate change's current status and effect on crop diversity and germplasm. The text's primary focus is on advancing technologies and approaches such as breeding, omics, transgenic, bioengineering of metabolic pathways in unique ways, and artificial intelligence which helps develop climate-resilient crops. Each has been laid out with systematic approaches to developing abiotic stress-tolerant genotypes using biotechnological tools.

This book covers all the crucial advanced technologies which help mitigate the plant abiotic stresses in cultivated crops and will familiarize the readers with the state-of-the-art advances in climate-resilience agriculture. This book will assist students and practitioners as a valuable tool for dealing with the complexities in this area, and researchers will find it an assistive instrument in their work as well.

Introduction

Global population is progressively growing and is projected to touch over 9.8 billion by 2050. Consequently, an increasing pressure to feed a growing population in the future would need additional production of major crops. On the contrary, the limiting resources and challenges associated with changing climate and unpredictability pose a severe threat to global food security. The abiotic stresses include drought, heat, cold, freezing, chilling, salinity, nutrient, ozone (O3), high light intensity, and anaerobic stress. Depending on different geographical locations, abiotic stresses cause a yield loss of around 50%. A significant portion of loss under abiotic stresses is due to drought and heat stress. These stresses are complex and affect plant physiology by affecting processes such as grain filling, maturation, and plant flowering. Understanding the molecular mechanism, plant responses to various stresses and inter-related cross talk is vital for providing future prospects to develop multiple stress-tolerant crops. Therefore, it is imperative to recognize the abiotic stress tolerance mechanisms importance under changing climatic conditions.

This book systematically studies effects of climate change on crop plants. The book has four parts, where Part A introduces the readers to climate change, its effect on plant growth, development, reproduction, and productivity, followed by the issues of increased biotic and abiotic stresses. Part B deals with breeding and molecular approaches used to improve crop species' climate resilience. Each chapter deals with stress and the nine chapters under this part give a detailed insight into how different tools and techniques have enhanced the climate resilience of cultivated species. Part C summarizes the advanced tools, including speed breeding and genome editing, which have direct implications for expediting the research on climate-resilient agriculture. Part D has three chapters on the success stories of climate-resilient agriculture.

Part A, "Climate Resilient Agriculture: Introduction" introduces the readers to climate change effect crops at phenotypic and molecular levels. Further, Part B, "Climate Resilient Agriculture: Breeding and Molecular Approaches" discusses the breeding and molecular mechanism and approaches utilized by crops for tackling these abiotic stresses In crop-breeding techniques, identifying and characterizing different genes and associated specific regions with the influence on quantitative and qualitative traits have been a significant challenge. Crop scientists have always tried identifying breeding strategies to make their selections more efficient and reliable. It would facilitate the utilization of unadopted germplasm with the minimum risk of linkage drag. One potential option pursued with significant success is using inherited traits for an indirect selection for complex abiotic stress-tolerant traits. Indeed, the development of DNA-based molecular markers has revolutionized crop improvement programs by allowing the exploration of the widespread polymorphism in natural plant populations. Moreover, molecular markers will enable the mapping of potential QTLs and genes in the chromosomes. However, a clear and thorough understanding is needed to ensure a reliable procedure for screening and characterizing crop plants for desired traits. Over the 21st century, a high-throughput marker-assisted selection has been extensively used for enhancing the selection efficiency and precision. The advancement of the marker system provides many other faithful techniques, such as genome-wide association study, genomic selection, and others that accelerate crop improvement program. It provides a platform for developing advanced high-yielding, drought-, and heat-stress-resilient varieties. We have discussed such studies throughout Part B of the book.

In Part C, "Climate Resilient Agriculture: Advanced Techniques," the advancement in modern biology and new biotechnological tools that deals with abiotic stresses harmful effect on the agricultural economy are discussed. Advent of new generation bioinformatics and integrating omics technologies facilitate enormous data interpretation to envisage prospective approaches for improving abiotic stress tolerance in cereal crops. The multiomics approaches advances our knowledge in the area of abiotic stress tolerance mechanisms through the use of functional genomics, transgenics, comparative gene expression, epigenomics, protein or metabolite profile(s), and their advanced plant phenotyping tools. Next-generation sequencing enables the mutagenesis-related generation of mutants involved in metabolite biosynthesis. Technological advances in the genomics have produced enormous data set from pan genome-wide studies at different omics levels. Significant progress has been made with extracting the relevant data from NGS and mutagenesis to explore functional pathways for manipulation of drought and heat resistance. Recently, genetic engineering technologies were used for manipulating specialized metabolic pathways in plants for improving the abiotic stress tolerance. The clustered regularly interspaced short palindromic repeat (CRISPR)/CRISPR-associated protein 9 (Cas9) tools have been employed in plant genome editing. CRISPR/Cas9 technique could be the potential technique to tackle complex traits. Lastly, in Part D "Climate Resilient Agriculture: Success Stories," success stories are summarized in book chapters for translating the application of molecular, physiological, and biochemical knowledge gained through modern breeding and biotechnological tools.

The compiled book "Climate Resilient Agriculture: A Molecular Perspective" deals with climate change facilitated stresses affecting different crops. The book's chapters are focused on understanding the climate change impact on crops' physiological, molecular, and biochemical levels and the strategies for generating climate resilience crops. The compiled chapters are divided into different parts for better understanding and then substantiated with figures, and table for comprehensive thoughtful understanding of the subject matter. Each chapter also provides appropriate and current references for additional readings. The material provided on different aspect of climate resilient agriculture will be useful for academician and researchers along with undergraduate and postgraduate students in colleges and universities.

—EDITORS

PART I

Climate-Resilient Agriculture: Introduction

CHAPTER 1

Climate Change: Effects and Need of Climate-Resilient Agriculture

SAURABH PANDEY[1] and ASHUTOSH SINGH[2]

[1]*Department of Agriculture, Guru Nanak Dev University, Amritsar, Punjab, India*

[2]*Centre for Advance Studies on Climate Change, Rajendra Prasad Central Agricultural University, PUSA, Samastipur, Bihar, India*

ABSTRACT

Climate changes in the global warming scenarios are constantly threatening the agricultural production. Existing agricultural and crop systems are extremely vulnerable to slightest change in temperature, precipitation amount and pattern, with frequent natural disasters. These vulnerabilities are ultimately resulting in huge yield losses and hampering the global food supply chain. This situation warrants a sustainable strategy for transforming and reorienting agricultural system toward realities of the changing climatic conditions. Climate resilience agriculture is one such approach where we enforce the agriculture with improved cultivars adaptable for climate change. These cultivars will boost the crop productivity by enhancing the resource use efficiency and can be developed by modern breeding and biotechnological approaches. In this chapter, we will discuss the impact of climate change on agricultural system, need of development of climate-resilient cultivars, and how modern technological inventions through molecular approaches will help us to achieve this.

Climate-Resilient Agriculture: A Molecular Perspective. Ashutosh Singh, Saurabh Pandey (Eds.)

1.1 INTRODUCTION

1.1.1 CLIMATE CHANGE: CHALLENGES TO AGRICULTURE

Long-lasting change in the weather patterns over a period of time known as climate change which threatens the global ecology, environment, and food security. One of the chief contributors of climate change is greenhouse gases (GHG). Interestingly, agriculture is the producer as well as victim of these emissions (FAO, 2013). Climate change can be caused by both natural (volcanoes, ocean current, earths tilt, etc.) and anthropogenic activities (deforestation, fossil fuel burning, industrialization, urbanization, etc.). Both types have to mitigate with adaptive technologies synchronized with the pace of change.

With the ever growing human population and constantly changing food habits, demands larger supply of food. Decreasing health of natural resources put enormous pressure to keep up the crop production at sustainable level. With the 690 million people hungry worldwide it is estimated that we need to produce 70% more food by 2050 to fulfill the global demand (FAO, 2013; Taiz, 2013; Zahid et al., 2016). The major challenge in achieving the sustainable food production is the extreme vulnerability of agriculture system to climate change. Climatic factors such as extreme temperature range, variable rainfall, and frequency of disasters leads to disturbance in agro-ecology, increases the incidence of various pathogens and pest ultimately reducing the crop yield (Elad and Pertot, 2014; Post, 2013). Improvements in the existing agro-ecosystem could be achieved by adaptation of improved cultivar, which are resource use efficient and ensures the quality and quantity of food produce.

1.1.2 CLIMATE RESILIENCE AGRICULTURE (CRA): NEED OF THE PRESENT AGRICULTURE

Vulnerability of agriculture components to climate change requires implementation of management strategies to minimize the risk associated with crop production. Climate resilience is one such strategy which prepares the agriculture system to anticipate, adopt, absorb, and recover from the adverse effects of extreme weather and climate change (IPCC, 2012). Resilience in agriculture needs the implementation of both short- and long-term mitigation strategies based upon onset and intensity of hazard including the adaptation of improved cultivars, resource management, and flexible technologies suitable for large agro-climatic zones. CRA provides the risk assessment on long- and short-term hazards to crop system and help us to devise strategies accordingly. Similarly, forecasting is the crucial component in determining the mitigation strategies against biotic and abiotic factors. It will help to predict the pest and pathogen attacks based upon the weather data and apply the preventive measures. These strategies are already present in the existing farming systems. However, the shear pace of the climate change needs fine tuning and synchronization of all management practices in an integrated manner.

1.2 ROLE OF MOLECULAR TOOLS AND BIOTECHNOLOGY IN CLIMATE RESILIENT AGRICULTURE

With the associated threats of climate change, it is imperative to integrate cutting edge technologies from various plant science disciplines to develop adaptive climate smart cultivars and varieties that can sustain the extreme stresses (Espeland and Kettenring, 2018; Tesfahun, 2018; Varshney et al., 2018). Conventional breeding methods are time consuming and could not sustain the pace and requirements of climate change-driven multiple stresses (Halterman et al., 2016). However, requirement of pure and desirable characters for transfer often needs these cumbersome breeding methods (Miedaner and Korzun, 2012). Considering the rapidness of climatic variation limited time is a major constraint, which requires exploring other options to fast pace the cultivar development. For example, we can utilize the variation of 7.2 million deposited accessions throughout the 1700 seed banks (Crossa et al., 2017; Venkateswarlu and Shanker, 2012; Wambugu et al., 2018). These variations could be mined through advanced biotechnological and molecular tools for crop improvement.

Biotechnological tools could be employed to mine superior alleles through high-throughput phenotyping and genotyping and develop the climate-resilient cultivars (Kole, 2013; Scheben et al., 2016). High throughput marker systems such as single nucleotide polymorphisms (SNPs) make these genotyping much more accessible (Wu et al., 2018). Other exciting advancements in this technology are genome editing tools. New generation tools such as CRISPR/Cas9 offer precise alteration in the desired allele and have enormous potential to produce climate-resistant cultivars in various crops (Arora and Narula, 2017; Bertier et al., 2018). Although the technology is new its application has been started in Arabidopsis, tobacco, cereals, legume, and fruit crops for various traits of stress improvement (Fauser et al., 2014; Mushtaq et al., 2018; Osakabe et al., 2016). With clearance of genome edited crops in USA and other countries it provides a future option for plant scientists to mitigate the adverse effects of plant stress through precise genetic manipulation.

Integration of interdisciplinary approaches such as genomic selection, along with NGS technologies through genome editing will provide the beneficial alleles important for climate resilient agriculture. Further multi-stress scenarios make it inevitable to use a multidisciplinary approach for developing crop varieties through breeding and biotechnological interventions (Lenaerts et al., 2018). This requires proper understanding of the genetic variation that could be utilized in a time efficient manner (Taranto et al., 2018; Wakelin et al., 2018). Subsequently, these variations through marker-assisted selection and molecular marker systems could be transferred into popular cultivars (Aruna kumari et al., 2016; Scheben et al., 2016). Diverse plant genotypes often offer various QTLs identification and mapping for stress-resistant genes across the population. These identified regions can be transferred through gene pyramiding and other techniques for breeding superior varieties (Elkot et al., 2015). However, applicability of these methods sometimes is complicated, labor-intensive, and random. Therefore, integration of genome editing technologies will make them more efficient and time-effective as the need of climate resilient agriculture.

Therefore, these molecular and biotechnological tools offer enormous potential in developing the climate resilient crops for future generations. However, critical consideration of gene and allele identification, their precise mapping, advancement in transfer technologies, etc., need constant innovation and varies from crop to crop. More efforts in direction will definitely lead us to more sustainable and climate-smart agricultural system.

KEYWORDS

- **climate change**
- **climate resilient agriculture**
- **molecular approaches**
- **breeding**
- **biotechnological interventions**

REFERENCES

Arora, L.; Narula, A. Gene Editing and Crop Improvement Using CRISPR-Cas9 system. *Front. Plant Sci.* **2017,** *8*, 1932.

Arunakumari, K. et al. Marker-Assisted Pyramiding of Genes Conferring Resistance Against Bacterial Blight and Blast Diseases into Indian Rice Variety MTU1010. *Ric. Sci.* **2016,** *23*, 306–316. https://doi.org/10.1016/j.rsci.2016.04.005

Bertier, L. D.; Ron, M.; Huo, H.; Bradford, K. J.; Britt, A. B.; Michelmore, R. W. High-Resolution Analysis of the Efficiency, Heritability, and Editing Outcomes of CRISPR/Cas9-Induced Modifications of NCED4 in Lettuce (*Lactuca sativa*). *G3* **2018,** *8*, 1513–1521. https://doi.org/10.1534/g3.117.300396

Crossa, J. et al. Genomic Selection in Plant Breeding: Methods, Models, and Perspectives. *Trends Plant Sci.* **2017,** *22*, 961–975. https://doi.org/10.1016/j.tplants.2017.08.011

Elad, Y.; Pertot, I. Climate Change Impacts on Plant Pathogens and Plant Diseases. *J. Crop Improv.* **2014,** *28*, 99–139. https://doi.org/10.1080/15427528.2014.865412

Elkot, A. F. A.; Chhuneja, P.; Kaur, S.; Saluja, M.; Keller, B.; Singh, K. Marker Assisted Transfer of Two Powdery Mildew Resistance Genes PmTb7A.1 and PmTb7A.2 from *Triticum boeoticum* (Boiss.) to *Triticum aestivum* (L). *PLoS One* **2015,** *10*, e0128297. https://doi.org/10.1371/journal.pone.0128297

Espeland, E. K.; Kettenring, K. M. Strategic Plant Choices Can Alleviate Climate Change Impacts: A Review. *J. Environ. Manage* **2018,** *222*, 316–324. https://doi.org/10.1016/j.jenvman.2018.05.042.

FAO. *FAO Statistical Yearbook: World Food and Agriculture*; FAO: Rome, 2013.

Fauser, F.; Schiml, S.; Puchta, H. Both CRISPR/C as-Based Nucleases and Nickases Can Be Used Efficiently for Genome Engineering in *Arabidopsis thaliana. Plant J.* **2014,** *79*, 348–359.

Halterman, D.; Guenthner, J.; Collinge, S.; Butler, N.; Douches, D. Biotech Potatoes in the 21st Century: 20 Years Since the First Biotech Potato. *Am. J. Potato Res.* **2016,** *93*, 1–20. https://doi.org/10.1007/s12230-015-9485-1

IPCC. Managing the Risks of Extreme Events and Disasters to Advance Climate Change Adaptation: A Special Report of Working Groups I and II of the Inter-Governmental Panel on Climate Change, 2012; p. 582.

Kole, C. *Genomics and Breeding for Climate-Resilient Crops*; Springer-Verlag: Berlin, 2013. https://doi.org/10.1007/978-3-642-37045-8.

Lenaerts, B.; de Mey, Y.; Demont, M. Global Impact of Accelerated Plant Breeding: Evidence from a Meta-Analysis on Rice Breeding. *PLoS One* **2018,** *13*, e0199016. https://doi.org/10.1371/journal.pone.0199016

Miedaner, T.; Korzun, V. Marker-Assisted Selection for Disease Resistance in Wheat and Barley Breeding. *Phytopathology* **2012,** *102*, 560–566. https://doi.org/10.1094/PHYTO-05-11-0157.

Mushtaq, M. et al. CRISPR/Cas Approach: A New Way of Looking at Plant-Abiotic Interactions. *J. Plant Physiol.* **2018,** *224–225*, 156–162. https://doi.org/10.1016/j.jplph.2018.04.001

Osakabe, Y.; Watanabe, T.; Sugano, S. S.; Ueta, R.; Ishihara, R.; Shinozaki, K.; Osakabe, K. Optimization of CRISPR/Cas9 Genome Editing to Modify Abiotic Stress Responses in Plants. *Sci. Rep.* **2016,** *6*, 26685. https://doi.org/10.1038/srep26685

Post, E. Erosion of Community Diversity and Stability by Herbivore Removal Under Warming. *Proc. R. Soc. B Biol. Sci.* **2013,** *280*, 20122722. https://doi.org/10.1098/rspb.2012.2722

Scheben, A.; Yuan, Y.; Edwards, D. Advances in Genomics for Adapting Crops to Climate Change. *Curr. Plant Biol.* **2016,** *6*, 2–10. https://doi.org/10.1016/j.cpb.2016.09.001

Taiz, L. Agriculture, Plant Physiology, and Human Population Growth: Past, Present, and Future. *Theor. Exp. Plant Physiol.* **2013,** *25*, 167–181.

Taranto, F.; Nicolia, A.; Pavan, S.; De Vita, P.; D'Agostino, N. Biotechnological and Digital Revolution for Climate-Smart Plant Breeding. *Agronomy* **2018,** *8*, 277.

Tesfahun, W. Climate Change Mitigation and Adaptation Through Biotechnology Approaches: A Review. *Cogent Food Agric.* **2018,** *4*, 1512837. https://doi.org/10.1080/23311932.2018.1512837

Varshney, R. K.; Singh, V. K.; Kumar, A.; Powell, W.; Sorrells, M. E. Can Genomics Deliver Climate-Change Ready Crops? *Curr. Opin. Plant Biol.* **2018,** *45*, 205–211. https://doi.org/10.1016/j.pbi.2018.03.007

Venkateswarlu, B.; Shanker, A. K. Dryland Agriculture: Bringing Resilience to Crop Production Under Changing Climate. In: *Crop Stress and Its Management: Perspectives and Strategies*, 2012; pp 19–44. https://doi.org/10.1007/978-94-007-2220-0_2

Wakelin, S. A.; Gomez-Gallego, M.; Jones, E.; Smaill, S.; Lear, G.; Lambie, S. Climate Change Induced Drought Impacts on Plant Diseases in New Zealand. *Aust. Plant Pathol.* **2018,** *47*, 101–114. https://doi.org/10.1007/s13313-018-0541-4

Wambugu, P. W.; Ndjiondjop, M. N.; Henry, R. J. Role of Genomics in Promoting the Utilization of Plant Genetic Resources in Genebanks. *Brief Funct. Genom.* **2018,** *17*, 198–206. https://doi.org/10.1093/bfgp/ely014

Wu, P. et al. Development of Molecular Markers Linked to Powdery Mildew Resistance Gene Pm4b by Combining SNP Discovery from Transcriptome Sequencing Data with Bulked Segregant Analysis (BSR-Seq) in Wheat. *Front. Plant Sci.* **2018,** *9*, 95–95. https://doi.org/10.3389/fpls.2018.00095

Zahid, H. J.; Robinson, E.; Kelly, R. L. Agriculture, Population Growth, and Statistical Analysis of the Radiocarbon Record. *Proc. Natl. Acad. Sci.* **2016,** 113, 931–935. https://doi.org/10.1073/ pnas.1517650112

CHAPTER 2

Effects of Climate Change on Plant Growth and Development

MAKSUD HASAN SHAH[1], KALIPADA PRAMANIK[2], S. K. MD. AJAHARUDDIN[3], SANTANU KUNDU[4], MOHD AFTAB ALAM[5], SAIDUL ISLAM[6], APURBA PAL[7], KOUSIK ATTA[8], and CHAITANYA B. PANDE[9]

[1]*Department of Agronomy, Bidhan Chandra Krishi Viswavidyalaya (BCKV), Mohanpur, Nadia, West Bengal, India*

[2]*Department of Agronomy, Palli Siksha Bhavana, Institute of Agriculture (Visva-Bharati), Sriniketan, Birbhum, West Bengal, India*

[3]*Department of Entomology, Bidhan Chandra Krishi Viswavidyalaya (BCKV), Mohanpur, Nadia, West Bengal, India*

[4]*Department of Agronomy, Professor Jayashankar Telangana State Agricultural University, Hyderabad, Telangana, India*

[5]*Department of Plant Pathology, Swami Keshwanand Rajasthan Agriculture University (SKRAU), Rajasthan, India*

[6]*Programme Assistant, Nadia KVK, Goyeshpur, Kalyani, Nadia, West Bengal, India*

[7]*Horticulture College, Khuntpani, Birsa Agriculture University, Ranchi, Jharkhand, India*

[8]*Department of Plant Physiology, Bidhan Chandra Krishi Viswavidyalaya (BCKV), Mohanpur, Nadia, West Bengal, India*

[9]*Department of Remote Sensing, Primus Partners Private Limited, Maharashtra, India*

Climate-Resilient Agriculture: A Molecular Perspective. Ashutosh Singh, Saurabh Pandey (Eds.)

ABSTRACT

Because numerous interrelated physical, biological, and chemical progressions are continuing on earth and human systems, measuring the implications of climate change will be a critical challenge in both developed and developing nations. Climate change has the potential to disrupt these processes, altering natural resources such as water, forestry products, and so on, biodiversity, ecosystem services, and plant growth and development in general. Climate change affects a number of aspects which influence plant growth potential. Based on satellite data and weather forecasts, a 7% reduction in the mean number of cold days will normally help in plant development, according to the observation. Plant survival will be made more difficult by extreme temperatures, low water availability, and changes in soil environment. Plants have a mechanism in place that allows them to withstand a particular level of temperature change. When soil temperature rises, the rate of organic matter decomposition increases, as does nutrient mineralization and availability for plant uptake, assuming all other factors stay constant. As a result, soil properties, water, mineral, and nutrient availability, among other things, impact plant response in different environments, as well as climatic variability. Plant growth is anticipated to be stifled as a result of climate change. Plant growth would stall, destroying forests, and significantly changing the ecosystems in which many species live. Furthermore, if circumstances deteriorate significantly, trees may begin to produce carbon rather than remove it from the atmosphere, worsening the primary driver of climate change.

2.1 INFLUENCE OF TEMPERATURE ON PLANT GROWTH AND DEVELOPMENT

Temperature is an important meteorological element in plant growth and function. Temperature affects enzyme function in leaves (Walker et al., 2013; Florian et al., 2014; Kumarathunge et al., 2019; Timm et al., 2019) and causes changes in the developmental stages of growth associated with plant productivity (Ruiz-Vera et al., 2018; Zhu et al., 2018). In addition, when temperature enhances, the quantity of water vapor that is present in the aerosphere rises exponentially at saturation, increasing the vapor pressure deficit (VPD) and potential loss of water from floras (Novick et al., 2016; Grossiord et al., 2020). Based on these wide plants' physiological responses to temperature, changes in long-term average yearly temperature, and the occurrence of high temperatures are likely to have significant influences on crop productivity in major food- and fuel-producing tracts of the earth.

2.2 EFFECTS OF HEAT STRESS ON PLANTS

Heat stress (HT) has different impacts on plants dependent on temperature, duration, and plant species. Extreme HT can cause cell damage or death within minutes, resulting in catastrophic breakdown of cell structure (Ahuja et al., 2010). Heat stress influences all aspects growth of plant, development, reproduction, and yield (Hasanuzzaman et al., 2013; Mittler and Blumwald, 2010; Lobell et al., 2011; McClung and Davis, 2010). Heat stress has different effects on the stability of various proteins, membranes, RNA species, and cytoskeletal structures, as well as on the efficiency of enzymatic reactions in the cell, hindering metabolism, and causing imbalance (Ruelland and Zachowski, 2010; Suzzuki et al., 2011; Suzuki et al., 2012; Pagamas and Nawata, 2008). Table 2.1 provides an overview of some of the most common effects of heat stress.

2.2.1 EFFECTS ON PLANT GROWTH

The first step in plant development is germination, to be affected by heat stress. Heat stress has a negative effect on seed germination in a wide range of plants, and the temperature ranges vary widely among plant species (Johkan et al., 2011; Kumar et al., 2011). Heat stress has been revealed to affect percentage of germination, emergence of plant, seedlings abnormality, weak seedling growth performance, and reduced the development of radicle and plumule in geminated seedlings in several crop species (Kumar et al., 2011; Piramil et al., 2012; Toh et al., 2008). Inhibition of seed germination has also been reported at HT, often caused by ABA induction (Essemine et al., 2010). Germination of wheat was severely suppressed at very high temperatures (45°C), resulting in cell and embryo death and a reduction in seedling establishment rate (Cheng et al., 2009). The plant height, number of shoots, and total biomass of paddy were decreased in response to HT (Mitra et al., 2008). High temperatures favor a decrease in cell turgidity, which reduces cell size and eventually growth (Rodriguez et al., 2005; Ashraf and Hafeez, 2004). Decreased relative growth rate (RGR) at HT has also been associated with a decrease in net assimilation rate (NAR), as shown in maize and millet (Wahid, 2007) and sugarcane (Srivastava et al., 2012). Burning and sunburn of young leaves and twigs, branches and shoots, senescence and dropping of leaf, inhibition of shoot, and root development, discoloration of fruits as well as damage are all morphological signs of heat stress (Rodriguez et al.,

TABLE 2.1 Effects of High Temperature Stress in Different Crop Species.

Crops	Thermal treatment	Phenological stage	Major effects	References
Capsicum annuum	38/30°C (day/night)	Reproductive, maturity and harvesting stage	Fruit girth and mass were decreased, but the proportion of aberrant seeds per fruit increased	Cao et al. (2009)
Oryza sativa	More than 33°C, 10 days	Heading stage	Pollen and spikelet fertility rates were reduced	Hurkman et al. (2009)
Triticum aestivum	37/28°C (day/night), 20 days	Grain filling and maturity stage	Grain filling and maturity periods are shorter, and kernel weight and yield are lower	Rahman et al. (2009)
Triticum aestivum	30/25°C day/night	From 60 DAS upto maturity stage	Decreased size of leaf, shorter times to reproductive stages like booting, heading, anthesis, and maturity, drastically reduced number of grains per spike and reduced size of grain, and lower productivity	Djanaguiraman et al. (2010)
Hordeum vulgare	40/30°C (day/night)	65 DAS to maturity stage	Reduced yield due to decreased chlorophyll (chl) concentration, fluorescence, photosystem II (PSII) photochemistry, Pn and antioxidant enzyme activity, and enhanced reactive oxygen species (ROS) content, as well as damage of thylakoid membrane	Mohammed and Tarpley (2010)
Oryza sativa	32°C (night temperature)	Reproductive stage	Reduced length of grain, breadth, and weight, as well as increased sterility of spikelet	Suwa et al. (2010)
Zea mays	35/27°C (day/night), 14 days	Post vegetative stage	Decreased ear development, notably inhibition of cob extensibility, due to reduced photosynthate availability affecting hemicellulose and cellulose production	Yin et al. (2010)
Oryza sativa	25–42.5°C	Pre reproductive stage	The rate of CO_2 assimilation has decreased	Djanaguiraman et al. (2011)

TABLE 2.1 *(Continued)*

Crops	Thermal treatment	Phenological stage	Major effects	References
Glycine max	38/28°C (day/night), 14 days	Blooming stage	Leaf Pn and conductance of stomata were reduced. (*gs*), Damaging membranes of plasma, chloroplast, and thylakoid; deformation of membranes of mitochondrial, cristae, and matrix, increased palisade and spongy layer thicknesses	Tan et al. (2011)
Nicotiana tabacum	43°C, 2 h	Initial growth stage	Reduction in net photosynthetic rate (Pn), conductance of stomata, apparent quantum yield (AQY), and carboxylation efficiency (CE) of photosynthesis. Antioxidant enzyme activity is reduced	Gunawardhana and de Silva (2011)
Abelmoschus esculentus	32 and 34°C	Entire growing period	Lower productivity as well as impairment of pod quality traits like fibre content and Ca-pectate degradation	Edreira and Otegui (2012)
Zea mays	33–40°C, 15 days	At Pre-anthesis and silking onward	Plant and growth rate of ear are severely affected	Zhang et al. (2013)
Triticum aestivum	38°C, 24 and 48 h	Seedling stage	Antioxidative ability is reduced due to reduced chl and relative water content (RWC)	Hasanuzzaman et al. (2013)
Triticum aestivum	32/24°C (day/night), 24 h	During end of spikelet initiation stage	Decreased grain production due to spikelet sterility	Saitoh (2008)

2005). HT stress also causes damage to the leaf tip and margins as well as leaf curling and desiccation, necrosis in sugarcane (Omae et al., 2012). Heat stress significantly affects the morphophysiological features of *Phaseolus vulgaris*, such as phenology, distribution, plant–water interactions, and shoot development and spread (Koini et al., 2009). Growth under heat stress (28/29°C) results in extended stems and inverted (intended) leaves (hyponasty) and a decrease in total dry matter accumulation in some plant species (Patel and Franklin, 2009; Savin et al., 1997). Under heat stress, wheat plants formed fewer plantlets with increased shoot elongation (Kumar et al., 2011). At HT (30/25°C, day/night), area of green leaf and ear bearing shoots per plant were significantly decreased in wheat (Djanaguiraman et al., 2010). High temperatures can shorten the entire phenological period by shortening the life span. Temperature increases of 1–2°C above optimal result in lesser grain filling duration and negatively affects wheat yield components (Zhang et al., 2006; Nahar et al., 2010). Extreme heat stress leads to programmed cell death in certain cells or tissues within a short period, that is, minutes or even seconds through proteins denaturation or aggregation, whereas moderate heat stress leads to gradual death; both types of injury or death can result in leaf drop, death of flower, and fruit, or even the entire death of plant (Hasanuzzaman et al., 2010; Rodriguez et al., 2005).

2.2.2 EFFECTS OF TEMPERATURE ON PHOTOSYNTHESIS WITHIN THE LEAF

Among the most important physiological processes in plants is photosynthesis which is heat-sensitive (Crafts-Brandner and Salvucci, 2002; Pokharel et al., 2020). High temperatures have a significant effect on the plant's photosynthetic capacity, particularly in C3 plants over C4 plants (Yang et al., 2006; Ogren and Bowes, 1971). The primary sites in the chloroplast are carbon metabolism in the stroma and photochemical reactions in the thylakoid lamellae are severely damaged by HTs (Wang et al., 2009; Marchand et al., 2005). The thylakoid membrane is extremely vulnerable to HT. Under heat stress, significant changes occur in chloroplasts, such as changed structural organization of thylakoids, losing of grana stacking, and grana swelling (Ashraf and Hafeez, 2004; Essemine et al., 2010). Under heat stress, photosystem II (PSII) activity is significantly decreased or even stops altogether (Morales et al., 2003). It is found that the number of photosynthetic pigments is decreased by heat shock (Marchand et al., 2005). In fact, most biological temperature reactions increase exponentially with temperature until they reach a thermal optimum (Topt), beyond which they begin to decrease at higher temperatures because of enzyme deactivation and denaturation. Many components of photosynthetic metabolism are temperature sensitive. Therefore, when evaluating the impacts of temperature on plant photosynthesis, it is useful to start with the photosynthetic machinery within a leaf. Rubisco efficiency and activation, and ribulose bisphosphate (RuBP) regeneration play important roles in photosynthetic net carbon assimilation (A) (Farquhar et al., 1980). The most important determinant changes with chloroplastic [CO_2]; RuBP regeneration limits A at high [CO_2], whereas Rubisco production limits A at low [CO_2]. At high temperatures, enzyme degradation can interruption PSII activity, reduce electron transport rates, block Rubisco activase (Rca), and reduce chlorophyll concentration (Salvucci et al., 2001; Guo et al., 2006; Allakhverdiev et al., 2008; Prasad and Djanaguiraman, 2011).

Increased temperatures can also cause membrane permeability, which can lead to direct damage to the thylakoid membranes of chloroplasts, reducing light collection, electron transport rates, and ATP production (Schrader et al., 2004; Prasad et al., 2008; Djanaguiraman et al., 2013; Pokharel et al., 2020). With increasing temperature, the fraction of the enzyme capable of reaching or exceeding the Ea required for catalysis increases, resulting in increased carboxylation activity of Rubisco. Rubisco, on the other hand, is a bifunctional enzyme that also catalyzes the oxygenation of RuBP (Ogren and Bowes, 1971; Tcherkez, 2016; Bathellier et al., 2020; von Caemmerer, 2020). As temperature increases, the specificity of Rubisco for CO_2 over O_2 (SC/O) decreases, lowering the ratio of carboxylation to oxygenation *in vivo*. At higher temperatures, this greater propensity for Rubisco oxygenation generates more 2-phosphoglycolate that must be converted via the photo respiratory pathway, resulting in the loss of previously fixed carbon and incurring high energy costs (Walker et al., 2016).

In C4 photosynthesis, CO_2 is localized to Rubisco in the chloroplasts of the bundle sheath. Therefore, elevated temperatures have no effect on photorespiration, and A has a greater effect in C4 plants over C3 plants (Sage and Kubien, 2007). C4 photosynthesis can also be reduced by inactivation of Rubisco (Crafts-Brandner and Salvucci, 2002) or rates of other C4 bundle sheath enzymes (Boyd et al., 2015), which are species-specific temperature sensitive (Sonawane et al., 2017), above the top level. This effect can be observed in field-grown maize, where A and yield at the leaf level decrease with increasing temperature, even when CO_2 levels are high (Ruiz-Vera et al., 2015). The period and severity of future warming events are expected to shift (Hoegh-Guldberg et al., 2018), with significant implications for potential thermal acclimation of A (Kattge and Knorr, 2007; Vico et al.,

2019). Long-term growth at higher temperatures may lead to photosynthetic acclimation through enhanced electron transport capacity, differential expression of Rca isoforms, and expression of heat shock proteins in sunny leaves near the tip of the leaf canopy (Yamori et al., 2014). Short-term increases in temperature, on the other hand, can boost leaf respiration, resulting in lower A than at ambient temperature, and a robust and relatively rapid acclimation response whose effects diminish with sustained higher temperatures (Way and Yamori, 2014; Kumarathunge et al., 2019). Acclimation responses may be too slow or too weak to provide a discernible advantage during heat waves or acute heat stress, defined as an abrupt rise in temperature (Smith and Dukes, 2017) with large but reversible effects on photosynthesis (Siebers et al., 2015; Thomey et al., 2019). Energy balances change in these scenarios as photosynthetic rates fall below the Topt and respiration rates increase. Therefore, in a warmer environment, most of the potential for increasing agricultural yields is focused on increasing photosynthetic carbon gain above Topt.

2.2.3 EFFECTS OF TEMPERATURE ON STOMATAL BEHAVIOR

Heat tolerance is directly related to a plant's capability to maintain leaf gas exchange and assimilation rates of CO_2 under heat stress (Kumar et al., 2005). Leaf hydration status, stomatal conductance (gs), and intercellular CO_2 concentration are all affected by heat (Greeer and Weedon, 2012). Another cause of reduced photosynthesis at HT is stomatal closure, which affects intercellular CO_2 (Essemine et al., 2010). Heat stress (day/night, 40/30°C) promotes lipid peroxidation of chloroplast and thylakoid membranes, leading to a decrease in Chl pigment (Mohammed and Tarpley, 2010). Under the same stress situation, the photochemistry of photosystem II (Fv/Fm ratio) and gs was also decreased. Compared to OT in sorghum, all these events significantly reduced photosynthesis (Mohammed and Tarpley, 2010). Heat stress (38/28°C) significantly reduced total Chl content (18%), Chla content (7%), Chl a/b ratio (3%), Fv/Fm ratio (5%), Pn (20%), and gs content in soybean (16%). The result was a decrease in sucrose content (9%) and an increase in reducing sugars (47%) and soluble sugars content in leaf (36%) (Tan et al., 2011). HT (33°C, 5 days) reduced the photosynthetic rate of rice plants by 16% in Shuanggui 1 and 15% in T219 [19]. Greer and Weedon (Kepova et al., 2005) found that when temperature increased from 25 to 45°C, the average photosynthetic rate of Vitis vinifera leaves reduced by 60%. Stomatal closure by 15–30% was found to be responsible for the loss of photosynthesis.

Stomatal response to high temperature is complicated not only by the disruption of a number of metabolic processes, including those occur in guard cells, but also by the fact that temperature influences photosynthesis, VPD, transpiration, and water balance in the plant, all of which affect stomatal behavior (Urban et al., 2017). Temperature changes affect VPD (see Scaling of Plants to the Ecosystem), which in turn affects transpiration as stomata adapt to changes in atmospheric dryness (Brodribb and McAdam, 2011). Higher VPD causes an increase in the diffusion gradient between the leaf and the atmosphere, resulting in higher loss of water and stomata closure to sustain the plant's water status (Mott and Peak, 2013). Apart from a general categorization into two hydraulic responses: active and passive, the exact mechanisms for stomatal responses to VPD are still unknown (Xie et al., 2006; Chater et al., 2011; Bauer et al., 2013; Franks, 2013; McAdam and Brodribb, 2014). Studies of stomatal responses to temperature have received fewer attention than studies of other environmental parameters (Way, 2011; Teskey et al., 2015), and results vary widely among species (Sage and Kubien, 2007; Matthews and Lawson, 2019). At elevated temperature, gs has a mixed response across plant species (Schulze et al., 1975; Lu et al., 2000; von Caemmerer and Evans, 2015; Urban et al., 2017), with a 163% increase in maize (Zheng et al., 2013) but a decrease (Raven et al., 2005) or no effect on gs in other plants (Sage and Sharkey, 1987; Aphalo and Jarvis, 1991; von Caemmerer and Evans, 2015).

Interactions between temperature and VPD make generalization of stomatal response to changes in leaf temperature difficult, as does the nonlinearity of the responses, which are usually characterized as bell-shaped (Way, 2011; Schulze et al., 1975). The gs increases with temperature until it reaches a threshold (Way, 2011; Tricker et al., 2018), then decreases rapidly at higher temperatures (Santrucek and Sage, 1991), and may increase again when stomata reopen at extremely high temperatures. The temperature at which stomata begin to close in the bell-shaped reaction is species-specific and depends on growth temperatures (Sage and Sharkey, 1987). Differences in hydraulic conductivity and the effects of temperature on viscosity (Mittler and Blumwald, 2010), as well as photosynthetic demand (Sage and Sharkey, 1987), likely explain this variance. In gs, heat stress leads to genotypically different responses (Zhou et al., 2017; Ferguson et al., 2020); however, it is unclear whether this variance in gs is related to the degree of heat sensitivity. Plants may also adapt to various growth temperatures, resulting in lower stomatal sensitivity to short-term (i.e., minute) fluctuations in ambient temperature (Santrucek and Sage, 1996). The gs responses of plants can have a similar form at different growth temperatures, but the magnitude can vary considerably (Yamori et al., 2006; Way, 2011).

Higher gs values at advanced temperatures improve plant performance by reducing CO_2 diffusion into the leaf, and the subsequent increase in intercellular CO_2 helps to mitigate the adverse effects of higher leaf temperatures on photorespiration (see previous section). In addition, higher gs allows for increased transpiration and evaporative cooling, so that leaf temperature can be maintained closer to the maximum value for photosynthesis while photorespiration activities decrease (Urban et al., 2017). However, higher gs can affect plant water content, which can adversely affect plant performance and development depending on the degree of water stress (Schulze et al., 1975). Because high air temperatures are often associated with decreased water availability, stomatal temperature responses are closely linked not only to VPD but also to drought and water potential (Urban et al., 2017). Even when photosynthetic demand is high, stomata close when water becomes scarce to prevent catastrophic water loss. Leaf temperature increases even more when gs decreases with increasing temperature and/or limited water supply, resulting in metabolic disturbances (Perdomo et al., 2017) and reduced photosynthesis due to restricted CO_2 diffusion (Chaves et al., 2003).

2.2.4 DEVELOPMENTAL OF ROOT RESPONSES TO ELEVATED TEMPERATURE

Global climate change is occurring that can have significant influences on plants, such as changes in the availability of certain nutrients. Compared to shoots, less is known about how roots, particularly root system architecture (RSA), adapt to increased temperatures. In addition, there are few data on the particular effects of high temperatures on the growth of the root system of invasive plants. The root system is illuminated to a heterogeneous soil matrix under a large extent of ambience situations that include a soil temperature gradient (Lundholm, 2009; Onwuka and Brown, 2018; Ulrich et al., 2014). On the other hand, the root enlargement process exhibits a spatial arrangement in which roots develop and respond progressively to alter in the environment, such as temperature differences (Bardgett et al., 2014; Zhu et al., 2011). Reductions in primary root length and seminal root density, as well as alters in root development angle, occur when plants are exposed to stressful high temperatures (>29°C) (Mc Michael and Qiusenberry, 1993; Nagel et al., 2009; Seiler, 1998). This decline has detrimental consequences for nutrient and water uptake, beneath ground interactions with other seedlings, and resistance to other stressors such as drought and pests (Luo et al., 2020; Roman-Aviles, 2004). In addition, the temperature does not affect all species in the same manner (Gray and Brady, 2016) found that different species have different optimal temperatures for root system architecture (Gray and Brady, 2016; Walter et al., 2009). Root temperature response profiles are strikingly similar to the wide optimal temperature ranges for the growth of aboveground parts of numerous species from various (Alsajri et al., 2019; Ayenan et al., 2019; Begcy et al., 2019; Chavan et al., 2019; Draeger et al., 2020) areas.

In reality, despite the fact that certain tundra plants may thrive in extremely cold soils temperature (1–3°C), optimal root enlargement occurs between 12 and 20°C (Bell and Bliss, 1978), which is significantly closer to the temperature of plants found in different biomes. Because of the root's concealed, buried nature, temperature-based root investigations are particularly difficult, as most tests focus on conventional parameters like biomass and length (Jarvi and Burton, 2020). Non-destructive observations, such as the use of a clear gel growth system, provide an alternative to soil testing and yield similar observations (Walter et al., 2019). At 36°C, for example, subtropical Agastache rugosa plants exposed to constant air temperature but increasing root temperature exhibit significantly changes root architecture (Lam et al., 2020). Gladish and Rost (1993) observed a downward trend in primary root growth rate and inhibition of seminal root development in garden pea when temperature was increased from low (15°C) to high (32°C) (Gladish and Rost, 1993). When seedlings were rise to 32°C for 17 days and exhibited suppression of primary root elongation, they were able to recover normal root development when transplanted to 25°C (Gladish and Rost, 1993). Soybean plants exposed to high temperatures (40/32°C) showed a significant decline in numerous metrics related to overall root growth (e.g., root density, cumulative root length, and root volume) compared with the control (30/22°C) (Alsajri et al., 2019). Surprisingly, not all traits studied revealed a diminishing trend. In the same study, secondary developmental traits of soybean roots viz. number of root tips, and root forks were increased at high temperatures (Alsajri et al., 2019). Compared with the control at 25/20°C, 21 days old wheat plants exposed to 36°C/28°C (diurnal regime) revealed a significant decline in various root metrics, including biomass of root, shoot-to-root ratio, length of primary root, root density, and root volume (Rehman et al., 2019).

The ideal root temperature of plants depends, of course, on the species. Higher temperatures are often allied with altered root to shoot ratios within this range, but further increases in temperature would inhibit root growth and cause a change in RSA, decreasing the root-to-shoot ratio (Rebeiro et al., 2014). However, at high temperatures, more extensive root systems grow in certain plants. For example, in mature corn plants, temperature increases reduce seminal root growth and stimulates the formation of long-axial roots to reach deeper soil layers for water uptake (Hund et al., 2008; Calleja-Cabrera et al., 2020). In potatoes, on the other hand, higher temperature inhibits the initiation and expansion of adventitious and seminal roots. Another bad effect of high soil temperature is the development of the root cap meristem and bending of the root tip of potato crop. The lower cell division rate in these plants seems to affect root development (Sattlemacher et al., 1990; Joshi et al., 2016).

Apart from the effects on lateral root branching, seminal root branching enhanced in response to elevation of temperature (up to the ideal temperature) and was further sensitive to temperature increases than taproot development. Temperature also affected the angle of lateral roots, with branching angles being significantly higher in plants growing at 20°C than in plants growing at 10°C (Nagel et al., 2009). Because soil temperature is affected by different factors, such as soil texture and moisture parameters, degree of insulation by snow cover or vegetation, latitude, and season, predictions for increases in soil temperature related with climate change are more complicated than those for corresponding changes in air temperature (Jungvist et al., 2014). Among the minerals taken up by plant roots, nitrogen is one of, if not the, most important macronutrient for plant growth, development, and response to biotic and abiotic stress. Although the development of techniques to synthesize and deliver ammonia and other nitrogenous compounds is an important component of the green revolution, soil temperature affects (1) the availability of fertilizers in the soil and (2) plant preferences for chemical forms (Zheng et al., 1993; IPCC 2014). Because soil temperature is directly proportional to and dependent on-air temperature (Zheng et al., 1993), the 1.0 to 3.7°C increase in earth average surface temperature expected this era will lead to higher soil temperatures (IPCC 2014).

Anhydrous ammonia, urea, ammonium sulfate, and ammonium nitrate are among the most commonly used modern fertilizers at the soil level. Because NH_3 converts to nitrate most slowly in soil, it is less vulnerable to denitrification (as opposed to soil N fixation) and leaching than the others. However, within the temperature range of 15–35°C, the actual rate of this reaction is stimulated as soil temperature increases, doubling every 10°C. Similarly, at different soil temperatures, certain plant species show a preference for certain forms of nitrogen over others. When both cationic (NH_4^+) and anionic (NO_3^-) forms of nitrogen are present in the soil, NH_4^+ absorption increases with decreasing temperature (Sato et al., 2006). It is thought that the lesser energy consumption for NH_4^+ assimilation compared to the higher biochemical processes involved in NO_3^- assimilation is the reason. Plant tolerance to NH_4 decreases as root zone temperature increases (>25°C). Root respiration changes within a temperature range in addition to mineral uptake. During the period of root respiration, cells extract O_2 from the air in the crevices between soil particles. This mechanism is most visible at the root hair level, where O_2 and carbon dioxide diffuse among various gases. With each 10°C increase in temperature of soil, root respiration, and oxygen uptake double, although O_2 solubility is inversely related to temperature. Root hypoxia could be due to increased oxygen demand and decreased oxygen supply (Lam et al., 2020).

2.2.5 REPRODUCTIVE DEVELOPMENT TO ELEVATED TEMPERATURE

Although HT stress influences all plant tissues at virtually all stages of growth and development, reproductive tissues are the most susceptible, and a temperature increase of a few degrees at flowering time can result in the loss of complete grain crops (Mittler and Blumwald, 2010). A brief period of HT stress during reproduction can significantly reduction of flower bud and flower death, although sensitivity varies widely within and among the plant species and cultivars (Sato et al., 2006). Even during a heat wave, the plant may fail to produce flowers or the flowers may fail to produce fruit or seeds (Maheswari et al., 2012; Foolad, 2005). Impaired meiosis in both male and female organs, impaired emergence of pollen, and development of pollen tubes, declined viability of ovules, abnormalities in the location of the stigma and pistil, decreased no. of pollen grains retained by the stigma, disturbed fertilization processes, impeded endosperm growth, proembryo, and unfertilized embryo are the cause of increased sterility under abiotic stress conditions, including HT (Cao et al., 2008).

The susceptibility of plants to HT stress depends on their developmental stage, the duration of heat exposure, and the severity of the stress (Prasad et al., 2015; Wahid et al., 2007). Heat stress causes sterility in several plant species, especially legumes, just before or during anthesis (Siddique, 1999). During flowering, field pea (Pisum sativum) is particularly susceptible to HT stress. Flowering begins after 10 or more vegetative nodes, and these flowering nodes are added consecutively until the reproductive node and cessation of plant growth, which occurs many weeks later. Guilioni et al. (2003) discovered that stress affects all phases of flowering from early blooming to seed filling at the last seed-bearing ovary. Many crops, including chilli peppers (*Capsicum annuum* and *Capsicum baccatum*), coyote melon (*Cucurbita palmata*), tepary bean (*Phaseolus acutifolius*), pole bean (*Phaseolus coccineus*), maize (*Zea mays*), and other staple crops, are very sensitive to elevated temperatures during pollen production (Dolferus et al., 2011). In sorghum (*Sorghumbicolor*), 14 days of heat exposure resulted in a much greater reduction in flower fertility than 7 days of heat exposure (Prasad et al., 2015). In field pea, heat stress affected pollen viability, pollen germination *in vitro*, and pollen tube length (Jiang et al., 2018; Jiang et al., 2019a, b). When pollen grains in the early panicle differentiation stage of barley (*Hordeum vulgare*) were exposed to diurnal temperatures of 30/25°C for 5 days, they produced normal exines but little or no cytoplasm, whereas the similar high temperature regime at the time of heat exposure in the pre-meiotic stage of pollen mother cells resulting in the formation of small anthers devoid of pollen grains (Sakata et al., 2000). When panicles were exposed to heat at the headstand stage, barley seeds were virtually barren (Sakata et al., 2000). Pollination is reduced and flower buds, flowers, and pods are lost as a result of high temperatures during reproduction (Devasirvatham et al., 2012; Guilioni et al., 1997). Several research groups have observed that heat stress leads to a decline in the number of seeds, reproductive nodes, and seed production in the field pea (Bueckert et al., 2015; Guilioni et al., 2003; Jeuffroy et al., 1990; Sadras et al., 2012). Pea seed development was particularly susceptible to high temperatures 6–12 days after anthesis; and HT decreases the no. of seeds per pod, resulting in loss of seed production (Jeufroy et al., 1990). Heat stress treatment (>33°C) at flowering diminish anther dehiscence and pollen fertility, resulting in a decrease in the no. of pollens on the stigma. These were the reasons of reduced pollination, consequent spikelet fertility, and sterile seeds in rice (Hurkman et al., 2009; Ahamadet al., 2010), with sensitive cultivars being more susceptible to this event than tolerant cultivars (Hurkman et al., 2009). Sterility of spikelets of rice increased (by 61% compared to the control) as pollen germination (36%) was reduced due to high night temperatures (32°C) (Suwa et al., 2010). It is hypothesized that ethylene (Eth) at excessive temperatures disrupts critical enzymes of sugar–starch metabolism, weakens sink strength, and restricts grain filling, resulting in sterile grains. Length of ear, number of spikelet main shoots, and number of viable flowering main shoots were significantly reduced in wheat plants as a result of heat stress at late sowing, resulting in lower grain production (Prasad et al., 2008). According to (Edreira and Otegui, 2012; Zhang et al., 2013), heat stress throughout the flowering period, especially during the pre-silking and post silking stages, as a result in a greater yield reduction than heat stress during the grain development stage of maize. High temperature stress resulted in death of flowers, early pods, and developing seeds, leading to lower seed number in soybean (Tubiello et al., 2007). High temperatures during flowering have been reported to reduce pollen viability in soybean (Tan et al., 2011). Although there are several studies on the effects of high temperature on reproduction in field crops, major developmental phases, stress duration, and pollen and ovule susceptibility to heat stress have not been fully defined and described. We expected that certain reproductive stages during flowering would be more susceptible to heat stress than others and that prolonged heat exposure would affect reproductive development more than shorter heat exposure. In addition, male floral organs are more susceptible to heat stress than female floral organs. As a result, the goal of this research was to assess the impacts of heat stress on pollen and ovule function, as well as to look into the effects of varying the length of heat exposure on different phases of flower development.

2.2.6 YIELD RESPONSES TO HEAT STRESS

Crop production and food security are threatened by rising temperatures (Srivastava et al., 2012). It has such a devastating impact that even a small increase in temperature (1.5°C) has major adverse influences on crop production (Warland et al., 2006). Higher temperatures affect grain yield by altering phenological development stages. Heat-related yield losses have been observed in various crops, including cereals (e.g., rice, wheat, barley, sorghum, and corn), legumes (e.g., chickpea and cowpea), oil-producing crops (mustard and canola), and others (Zhang et al., 2013; Jeuffroy et al., 1990; Ahamad et al., 2010; Tubiello et al., 2007; kalra et al., 2008; Hatfield et al., 2011). A 1°C increase in seasonal average temperature has been found to reduce wheat yield by 4.1–10% (Wang et al., 2012). Heat stress affects sensitive plant species more than tolerant plant species. The, 1000-grain weight of sensitive Shuanggui 1 and tolerant Huanghuazhan rice decreased by 7.0–7.9% and 3.4–4.4%, respectively, at 35–40°C heat stress. A comparison between the heat-sensitive rice variety Shuanggui 1 (35.3–39.5% yield reduction) and the heat-tolerant rice variety Huanghuazhan (21.7–24.5%

yield reduction) was also published (Ahamad et al., 2010). In O. High night temperatures (32°C) reduced grain length, width, and weight by 2%, while spikelet sterility increased by 2% in Sativa (about 61%). In addition, grain nitrogen (N) content increased by 44%, which was inversely proportional to grain weight. All these variables resulted in a 90% yield reduction (Suwa et al., 2010). Heat stress alters the early dough and maturity stages of wheat, shortens the duration of grain desiccation, and reduces grain yield (Hasanuzzaman et al., 2013). Heat also lowers the weight of the individual grain, which contributes significantly to yield loss (Kutcher et al., 2010). Heat stress caused by late sowing (28–30°C) resulted in significant yield losses in several wheat cultivars: 70% in "Sourav," 58% in "Pradip," 73% in "Sufi," 55% in "Shatabdi," and 53% in "Bijoy" (Koini et al., 2009). Heat stress reduced filled seed weight and seed size in sorghum by 53 and 51%, respectively, which reduced yield (Mohammed and Tarpley, 2010). Seed yield on the main stem of oilseed rape (*Brassica* spp.) was reduced by 89%, although all branches contributed to an overall loss of 52% at HTs of >30°C. Reduced photosynthesis due to altered membrane stability (Zhang et al., 2006) and increased cost of maintenance respiration (Reynolds et al., 2007), as well as a decrease in radiation efficiency, are the main causes of productivity loss under heat stress (RUE, biomass production per unit of light intercepted by the canopy). These phenomena have been demonstrated in both wheat (Cicchino et al., 2010) and maize (Hogy et al., 2013). High temperatures (33–40°C) affected light yield, RUE, biomass and yield, and harvest index of maize, with heat during the flowering stage reducing yields more than heat during the grain filling stage (Zhang et al., 2013). Under heat stress, barley quality traits changed dramatically. Numerous proteinogenic amino acids and maltose content enhanced in barley grain, but total nonstructural carbohydrates, starch, fructose and raffinose, lipids, and aluminum concentration decreased (Vasseur et al., 2011). HT in okra (Abelmoschus esculentus) and stress-affected pod quality, such as fiber content and Ca-pectate degradation (Edreira and Otegui, 2012).

2.2.7 EFFECTS OF OXIDATIVE STRESS ON PLANT

Enzymes that respond to varying degrees to heat stress are required for various metabolic processes. It has been suggested that heat stress, like other abiotic stressors, uncouples enzymes, and metabolic pathways, leading to the formation of undesirable and defective reactive oxygen species (ROS). The oxidative stress is caused by singlet oxygen ($1O_2$), superoxide radicals

(O_2^-), hydrogen peroxide (H_2O_2), and hydroxyl radicals (OH^-) (Asada, 2006). Although ROS is also formed another organelles like those mitochondria and peroxisome, the center of reaction of PSI and PSII in chloroplasts are the primary location of formation of ROS (Soliman et al., 2011). The maximum potentiality of PSII and the accumulated ROS are linearly related. It has been proposed that fewer photons are absorbed under such HTs due to thermal damage to the photosystems (Halliwell, 2006). Excess electrons are generated when PSI and PSII both exploit photon intensity in such a stress condition, and they function as a generator of ROS (Halliwell, 2006). Among the ROS that generate O_2 are photo oxidation mechanisms (flavoproteins, redox cycles), the Mehler reaction in chloroplasts, during mitochondrial ETCs and glyoxisomal photorespiration, NADPH oxidase in plasma membranes, xanthine oxidase, and membrane polypeptides. Hydroxyl radicals are produced via the interactions of hydrogen peroxide with oxygen (Haber-Weiss reaction), hydrogen peroxide with Fe^{2+} (Fenton reaction), and the degradation of ozone in the apoplastic region (Moller et al., 2007; Karuppanapandian et al., 2011). O_2 is generated in chloroplasts at the timing of photoinhibition and PS II electron transport pathways (Karuppanapandian et al., 2011b, huang and Xu, 2008). Despite the fact that the reactions' products might trigger signaling responses, the hydroxyl radical is not thought to have a signaling role, and the cells successfully sequester the catalytic metals to metallochaperones, avoiding hydroxyl radical (Moller et al., 2007; Karuppanapandian et al., 2011).

Because of these varying degrees of heat stress, plants experience a variety of physiological abnormalities (Halliwell, 2006). OH• have the capability to reaction with essentially all biomolecules, including pigments, proteins, lipids, and DNA, as well as almost all cell components (Moller et al., 2007; Karuppanapandian et al., 2011). Singlet O_2 can oxidise protein, polyunsaturated fatty acids, and DNA right away (Karuppanapandian et al., 2011b, Huang and Xu, 2008). Thermal stress can lead to oxidative stress by generating membrane lipid peroxidation and affecting cell membrane integrity through protein denaturation (Rodriguez et al., 2005; Camejo et al., 2006). Even at low HTs, a functional decrease in photosynthetic light response has been demonstrated to enhance oxidative stress via enhanced electron loss from the thylakoid membrane, which drives to the generation of ROS (Halliwell, 2006; Bavita et al., 2012). In rice plant leaves, high temperature enhanced leaf temperature, decreased antioxidant enzyme activity, and enhance malondialdehyde (MDA) content (Hurkman et al., 2009). Heat exposure (33°C) caused oxidative stress in wheat, which resulted in membrane damage, protein breakdown, and enzyme deactivation, all of which lowered cell viability significantly. Heat stress-influenced oxidative stress enhanced membrane peroxidation by 28% and decreased membrane thermostability by 54%, resulting in unexpectedly increased electrolyte loss in wheat (Savita and Skute, 2010). Oxidative stress was evident in populations of perennial ryegrass (*Lolium perenne* L.) illuminate to modest (36°C) and acute (40°C) HT stress, as evidenced by higher hydrogen peroxide levels, and it was also the cause of significant physiological harm to PS II's maximum effectiveness, cell membrane stability, and lipid peroxidation (Halliwell, 2006). HT stress caused membrane lipid peroxidation and provoked membrane damage in cotton, sorghum, and soybean (Mohammed and Tarpley, 2010; Tan et al., 2011; Young et al., 2004). Heat stress increased damage of membrane and MDA content in sorghum by 110 and 75%, respectively, as well as hydrogen peroxide and singlet oxygen concentration (124 and 43%, respectively) (Mohammed and Tarpley, 2010). Furthermore, the ROS produced by high temperature stress are involved in protein proteolysis or the breakdown of polymeric proteins into simple soluble forms, both of which are linked to early leaf senescence in cotton (Young et al., 2004). After 2 days of temperature exposure, root development was suppressed in wheat, which was connected to severe oxidative stress as evidenced by a significant enhance (68%) in singlet oxygen generation in root cells. The MDA concentration enhanced by 27% in the first leaf 2 days after exposure in the early stages of seedling growth, and this tendency (by 58%) continued in the later stages of development (Miller et al., 2009).

High temperature stress causes an accumulation of ROS on the plasma membrane's outer surface, which can lead to membrane depolarization and the activation of Ca-induced RBOHD (Qi et al., 2010) (the ROS-producing enzyme RBOHD located at the plasma membrane). In such severe cases, ROS build up in cells might lead to programmed cell death (Qi et al., 2010). Although ROS can be harmful to plant metabolic systems, they are also hypothesized to have signaling behaviors that cause plants to gain heat tolerance by activating heat shock responses (Asada, 2006).

2.3 PLANT RESPONSES TO CO_2

2.3.1 RESPONSES OF LEAF GROWTH TO INCREASED CO_2 LEVELS

Elevated CO_2 stimulates above ground biomass which is associated with larger leaf area in poplar and soyabean (Dermody et al., 2006; Taylor et al., 2003). Increased cell formation and/or enlargement may result in increased leaf size in the presence of increasing CO_2, and both activities result in high leaf size in researched species. There are interesting differences among cell types in these responses (Taylor et al., 2003). Taylor observed that increasing CO_2 enhanced the numerals epidermal cells in growing leaves but not mature leaves of the hybrid Populus euramericana (Populus deltoides, P. nigra, clone I-214), whereas a result of increased CO_2 level, volume of spongy cells and palisade cells of the mesophyll increased in young and old leaves (Masle, 2000). Elevated CO_2 has an effect on cell type or morphology of wheat leaf, including formation of additional cell strata and spongy mesophyll cells become more intercellular, but there is no specific effect on epidermal structure. This phenomenon can be observed in poplar leaves, although it is age-dependent (Taylor et al., 2003; Ranasinghe and Taylor, 1996; Ferris et al., 2001). These data indicate that elevated CO_2 in leaves induces cell growth responses that are geographic, temporal, and species specific. Within the leaf, different cell types or cell type specifications respond differently to increase CO_2. In some species, including several Arabidopsis cultivars, increased CO_2 declines the stomatal index (the ratio of stomata to epidermal cells) (Woodward and Kelly, 1995). Arabidopsis stomatal maturation is inhibited by a CO_2-responsive down regulation by elevated CO_2. It was identified as a regulator of stomatal index (Gray et al., 2000). In reaction to high CO_2, the hic mutant displayed 18–28% rise in stomatal index, indicating that HIC, a potential 3-keto-acyl coenzyme A synthase involved in cell wall wax manufacture, is critical for the negative control of stomatal growth (Gray et al., 2000). Further research revealed that an extracellular signaling pathway involves carbonic anhydrases that regulates stomatal growth in response to CO_2 (Engineer et al., 2014). Ferris et al. (2002) used a mapping population of P. trichocarpa and P. deltoides to find QTL in stomatal sensitivity to high CO_2. They researched that elevated CO_2 lowered stomatal thickness and stomatal index in P. deltoides, but had no influence on these features in P. trichocarpa's adaxial leaf surface. The researchers revealed QTL which is used for stomatal trait responses to elevated CO_2, but candidate genes for these species yet to be discovered (Ferris et al., 2002). Xu et al. investigated potential stomatal growth response controllers to high CO_2 (2016). CO_2 raised the transcription randomness of genes involved in starch, sugar, glycolysis, the tricarboxylic acid cycle, and mitochondrial electron transport in soybeans, according to (Leakey et al., 2009). Increased photosynthetic carbon absorption and night-time respiration were linked to these transcriptional alterations (Leakey et al., 2009). Rice showed similar transcriptional responses to high CO_2, with increased gene indication involved in sucrose production, glycolysis, and the

TCA cycle (Fukayama et al., 2011). CO_2 boosts the flow of carbohydrates and others metabolites, which are crucial for increase and development as well as the activity of glucose as a signaling molecule that alters plant development. Glucose sensor hexokinase1 (HXK1), kinases KIN10, or KIN11, and target-of-rapamycin kinase (TOR) have been identified as glucose-modulated master regulators in Arabidopsis (Sheen, 2014). HXK1 (223) regulates glucose-dependent transcription and plant growth, KIN10 (Baena-Gonzalez et al., 2007) regulates the transition from vegetative to reproductive phase, and TOR-kinase (reviewed in Shen, 2014) regulates primary and secondary metabolism, transcription, and translation. It's still a mystery if these major regulators are to blame for changed transcriptional regulation of metabolic genes in rice, soybean, and Arabidopsis when CO_2 levels rise. Thus, for leaves of a variety of species, common transcriptional responses to elevated CO_2 have been observed, potentially leading to alterations in primary carbon metabolism. It is unknown whether these changes in transcription occur before or after the observed changes in leaf structure. Furthermore, although the role of glucose as a signaling molecule influencing regulatory and developmental reaction to rising CO_2 is feasible, it remains unclear.

2.3.2 THE EFFECTS OF INCREASED CO_2 ON ROOT DEVELOPMENT

It's still unknown whether these key controllers are accountable for the changed transcriptional control of metabolic genes in paddy, soybeans, and Arabidopsis when CO_2 levels rise. Thus, in the leaves of a range of species, conserved transcriptional responses to high CO_2 have been reported, likely leading to alterations in primary carbon metabolism. It is unknown whether these transcriptional alterations occur before or after leaf shape changes. Moreover, the significance of glucose as a signaling molecule regulating transcriptional and developmental responses to high CO_2 is highly speculative. To clarify how this rise in biomass may occur, more detailed studies reveal alterations in root system architecture and changes in cell anatomy. Furthermore, no systematic research of the maintenance of all of these interactions between and within plant species has been conducted. Minirhizotron tests on *Glycine max* have shown that high level CO_2 enhanced root length, especially at low and subsoil depths, and that high CO_2 linked with reducing rainfall enhances the amount and volume of root nodules harboring nitrogen-fixing Bradyrhizobia bacteria (Gray et al., 2013; Gray et al., 2016). Increased root length, improved root branching, and increased root width have been found to increase biomass of root in agricultural and forest plants (reviewed in Madhu and Hatfiled, 2013). Elevated CO_2 leads to increased primary root length and growth rate and increased lateral root development and elongation in Arabidopsis (Crookshanks et al., 1998). Root system structure under high carbon dioxide may impact root system efficacy in water absorption by changing the root length dispersion in relation to water availability (Gray et al., 2013). Changes in root depth distribution can also affect the ability of plant to accumulate nutrient availability that are unevenly spreaded across the soil depth gradient (Lynch, 2015). Increased branching and increase of adjacent roots in spring wheat, winter wheat, cotton, and sorghum contributes to root dispersion in shallow layers rather than root system growth in deeper soil layers (Revised in Pritchard and Rogers, 2000). Cotton exposed to increased CO_2 showed increased stem and cortical diameters, anatomically, increasing plant length of the root in the root development zone enhanced overall root system density (Rogers et al., 1992). Increased growth of cortical cells and greater extensibility of cell walls were also observed in Arabidopsis (Crookshanks et al., 1998). In a future climate with high CO_2, alteration in the structure, root function may be influenced by cellular architecture and interactions between root systems and soil microbes. *Populus deltoides* and *Populus strichocarpa* have shown interspecific variation of gene in root density response to increased CO_2, with P. trichocarpa exhibiting greater response of root density to elevated CO_2 (Rae et al., 2007). Using 285 individuals from an F2 population of *P. deltoides, P. trichocarpa* hybrids, three QTL were identified that were attached with rate of root development and the main root density feedback to high CO_2 (Rae et al., 2007). These QTL were in turn detected by rhizotron assays. Because the root growth sequence is obscured by the soil environment around a rhizotron, and quantification restrictions contribute to phenotyping inaccuracies, rhizotrons are unlikely to capture the full extent of loci affecting root carbon dioxide feedback. Candidate genes discovered from QTL of CO_2-exposed roots of plant should be further investigated. A small number of experiments (Plett et al., 2015) looked at changes in root tissue gene expression in response to elevated CO_2, therefore, our understanding of the molecular processes and genetic variability of root responses to climate change is still limited.

2.3.3 THE EFFECTS OF INCREASED CO_2 ON REPRODUCTIVE DEVELOPMENT AND PHENOLOGY

Increased CO_2 levels boost seed production in a wide range of crop plants, but it lowers grain nutrient quality because of changes ion profiles, especially lower

Fe and Zn concentrations (Loladze, 2014; Myers et al., 2014). Increasing CO_2 lowers the "N" and protein content of non-husk seeds (Jablonski et al., 2002; Myers et al., 2014). A meta-analysis of 79 species that flourished under raised CO_2 demonstrated similar impacts on reproductive performance: increasing CO_2 enhanced flower, fruit set, and seed mass on an average of 16–19%, total seed mass by 25%, and individual seed mass by just 4% (Jablonski et al., 2002). This is confirmed by a study on soybean, which found that increasing the number of pods or the quantity of seeds in each pod, rather than increasing the weight of each seeds, increased seed production (Morgan et al., 2005). The chemical mechanism causing this increase in seed number in all these species is unclear. Bishop et al. (2014) found that increased level of CO_2 increased the production of seed by an average of 9% in 18 soybean genotypes over several seasons of growth, while the partition coefficient declined by 11%. The extent to which high CO_2 stimulated seed yield varied with environment, with seed yield stimulation decreasing to zero under hot and dry conditions (Ruiz-Vera et al., 2013; Gray et al., 2013). While agricultural and in the wild species, there was no response, total biomass to increased carbon dioxide, (Jablonski et al., 2002) discovered fruit production increased an average of 28% in crop species but only 4% in wild species, suggesting that crop species were artificially selected for better allocation of carbon to fruit and seed development (Jablonski et al., 2002). The response of wild species' fruit and seed output to increase CO_2 was likewise more variable than that of domesticated species (Jablonski et al., 2002).

2.3.4 ELEVATED CO_2 CAUSES MAJOR PHYSIOLOGICAL AND BIOCHEMICAL CHANGES IN PLANTS

High level of CO_2 has been shown to enhance plant development in nitrogen-rich environments, but it has been shown to reduce plant development in "N" poor environments by altering primary metabolism (Takatani et al., 2014). Climate scientists are intrigued by the structural variations among C3, and C4 plants and associated with carbon sequestration strategies utilizing 3C and 4C compounds, respectively. Initially, it was believed that the benefits of increased CO_2 would outweigh those of C4 plants, but Reich et al. (Reich et al., 2018) reported a 20 years study that found a primary enhancement in biomass in C3 grasslands for 8 years, after that the trend overset. The period of a trial also affects the response of increasing CO_2 (Hovenden and newton, 2018). Plants exposed to high CO_2 undergo physiological and metabolic changes. Elevated CO_2 has an impact on nutritional quality, nodulation, rhizosphere alterations, and biological nitrogen fixation (BNF) in legumes. The effects of increasing CO_2 levels on key physiological traits of legumes have been demonstrated for several crops. Elevated CO_2 alters important phenotypic and biochemical markers in legumes. Thus, these are the succeeding changes in the root zones under conditions of C excess (as a result of elevated CO_2). The effects of high CO_2 exposure on physiological parameters (Table 2.1) are discussed in detail in the following sections.

2.3.5 ELEVATED CO_2 ALTERS THE NITROGEN BALANCE AND AFFECTING THE NUTRITIONAL QUALITY OF LEGUMES

Generally, poor crop nutrient content leads to severe economic losses, as indicated by reduced protein content in pre reproductive tissue and grains. Nitrogen fixation is increased along with production; however, the symbiotic nitrogen fixation capacity of legumes serves to maintain the carbon–nitrogen balance in check (Rogers et al., 2009). Legumes can further mitigate the effects of photosynthetic acclimation under elevated CO_2 by providing more photosynthates for root micro-organism and maintaining nitrogen levels through symbiotic "N" fixation (Grossiord et al., 2020). In *Glycine max*, elevated CO_2 reduced total protein content, flavonoid content, and free amino acid at the mature stage of the plant, but had no effect on the edible stage of the plant. Sulfur, phosphorus, and calcium contents increased in a specific stage, while Fe and Zn concentrations declined. Due to the small sample size, they found a decrease in the nutrient quality of soybeans. This must be considered before drawing conclusions about possible major alteration in these studies (Li et al., 2017). Dietterich et al. (2015) compared 18 rice and wheat, seven soybeans, five field pea, and one sorghum cultivars under ambient (364–386 ppm) and high (546–586 ppm) CO_2 conditions over a 3-year tenure (1998–2001). The study discovered decreased nutritional content, notably in the levels of Zn, Fe, and protein in these plants, due to its functional type (C3 or C4 photosynthetic pathways) and crop-specific responses. C3 grasses & legumes, on the other hand, were consistently affected, although C4 plants were not (Dietterich et al., 2015).

2.3.6 INCREASED CO_2 CAUSES CHANGES IN PLANT NODULATION AND RHIZOSPHERE

Stimulation of nitrogen fixation is characterized by an increase in nodule size, number, or activity (amount of nitrogen fixed per unit mass) (Rogers et al., 2009). A modern study of formation of nodulation in root and plant development in Medicago sativa found that increasing temperature can decline the beneficial impact of increased CO_2, although silicon supplementation promoted growth at various levels of high level of CO_2 and temperature (Johnson et al., 2018). Chickpea and other legumes produce more nodules and biomass when CO_2 levels are higher, according to several researches. Nodule size and quantity, as well as plant nitrogen content, rose in a variety of legume species, including white clover (Trifolium repens), lupin (Lupinus albus), pea, and soybean, resulting in an increase in biomass/seed output (Butterly et al., 2015; Saha et al., 2013; Zanetti et al., 1997). According to a meta-analysis, legumes fix 38% more nitrogen under increasing CO_2 because they have 33% more nodules, 39% more nodule biomass, and 37% more nitrogenase activity (Lam et al., 2012). A 2-year experiment with lentils (Lens culinaris) based on FACE showed greater stimulation of N fixation under aerobic situation and increased CO_2 (400 and 550 ppm, respectively). Under well-watered conditions, this was reflected in a higher number of nodules as 27%, mass 18%, and specific fixation activity +17% than during the low rainfall/dry season (Parvin et al., 2018). As a result, the benefits of increased CO_2 were found to occur only when other non-living factors, such as availability of plant water, were not restricted during the grain formation period (Parvin et al., 2018). The immediate CO_2-mediated activation of biological nitrogen fixation in legumes is highest (Lam et al., 2012), although it is weaker under nutrient-rich conditions (Rogers et al., 2009). Increased numbers of N_2-fixing bacteria in the rhizosphere, increased numbers of nodules in which N-fixing bacteria nest, and increased nitrogenase activity are likely processes (Aranjuelo et al., 2014). Increased carbon supply to the root has been shown to increase activity in the rhizosphere BNF (Butterly et al., 2015; Arajeulo et al., 2014). Plant nutrient uptake has been found to be enhanced by rhizobial diversity and root morphological changes (Lam et al., 2012; Beidler et al., 2015). Increased P dosages and cyanobacterial inoculation increased root growth, N fixation, and soil nitrogen availability under enhanced CO_2 (Dey et al., 2017). Temperature affects root nodule symbiosis, with the ideal temperature for legumes being 15–25°C. Consequently, an increase of as little as 2°C would have an expected effect on root nodule growth and functionality. Plant-mediated processes such as reduced root hair production, fewer nodulation sites, and reduced bacterial adhesion to root hairs would accelerate nodule senescence (Arajeulo et al., 2014).

2.3.7 THE INTERPLAY OF INCREASING CO2 WITH OTHER ABIOTIC STRESSORS HAS A MASSIVE EFFECT

The impact of increased level of CO_2 on a plant is regulated by various elements such as unavailability of water, increased temperature, etc. For example, it was reported (Kimball, 2016) that increased CO_2 caused a 10% reduction in evapotranspiration rate in C3 and C4 plants. This resulted in a 0.7°C increase in canopy temperature and a 19% increase in yield for C3 plants. Increasing CO_2 has also been shown to reduce the effects of oxidative stress (AbdElgawad et al., 2015). Drought-induced restrictions of plant growth, photosynthesis, and stomatal conductance were reported in a recent comparative investigation of the impacts of drought, increased temperature, and elevated CO_2 in grasses and legumes. Elevated CO_2 had a detrimental effect on osmolytes and antioxidants in this situation. In addition, oxidative stress indices were lower in legumes, although photosynthesis and chlorophyll content were higher in grasses. According to the study (AbdElgawad et al., 2015), the effects of the reduction in drought stress mediated by high CO_2 are stronger in legumes than in grass species. In this field, the combination of increasing CO_2 with water deficiency stress and temperature changes has been thoroughly addressed. Different theories exist on how rising CO_2 levels alter water balance and drought tolerance systems. This is because the condition of soil water affects whether or not higher CO_2 levels are beneficial to plant response. While some studies have shown declined transpiration rate under increased CO_2 (Miranda-Apodaca et al., 2018), others have found no change (Manderscheid et al., 1995), while still others have found a negative effect (Manderscheid et al., 1995). Drought-induced osmotic adjustment under high CO_2 levels is another interesting issue. Faster leaf growth decreases solute concentration, resulting in less osmotic adjustment, according to some study (Tschaplinski et al., 1995). Under increasing CO_2, the root–shoot ratio increased because to enhanced drought tolerance and fine root development (Xu et al., 2013). Under water shortage situations, elevated CO_2 may decrease or even prevent plant growth stimulation. As a result, agricultural productivity in many dry and semi-arid locations throughout the world may drop under forecasted future climatic conditions. Increasing temperatures or other extreme climate variables, in combination with increased CO_2, are key climate change elements that might alter plant fitness and flowering-related events, resulting in reduced agricultural productivity (Jagadish et al., 2016). Groundnut was the subject of the first studies on the effects of prolonged CO_2 exposure time in tropical legumes in semi-arid environments. Although it is a C3 plant, under optimal conditions it had photosynthetic rates comparable to those of a C4 plant (Pallas and Samish, 1974). The impact of high level CO_2 and the physiological feedback on the plants mitigated the effects of drought stress in soybean (Li et al., 2017). Drought mitigation is species-specific when CO_2 levels are increased. High CO_2 can counteract the negative effects of dryness on turgor potential in two ways, according to research on two grass species (a legume and a grass in white clover (legume), there was osmotic adaptation and root to shoot ratio, while in common knapweed (Agrostis capillaris), it was increased relative leaf water content created by hydraulic conductivity (grass). Drought effects were not mitigated in either species despite increased soil water maintenance (Miranda-Apodaca et al., 2018). In a current experiment, the legume common bean (*Vicia faba*) was cultivated under well-watered (80% of FC) and dry (30% FC) conditions at general (400 ppm) and high CO_2 (550 ppm) levels. Nitrogen fixation was inhibited, as evidenced by lower carbohydrate and higher amino acid content in nodules. Furthermore, both elevated CO_2 and drought circumstances resulted in lower N concentrations in seeds (Parvin et al., 2019). Elevated CO_2 increases heat tolerance in relation to vegetative development in legumes like peanut (Prasad et al., 2003) and cereals like rice (Lai et al., 2015), wheat (Alonso et al., 2009), and maize (Abebe et al., 2016). Increased activity of sucrose phosphate synthase (SPS) and adenosine 5'-diphosphoglucose pyrophosphorylase (AGP) were discovered in kidney beans cultivated at high temperatures (Prasad et al., 2004).

2.3.8 PATHOGEN AGGRAVATION AND VIRULENCE ARE MODULATED BY INCREASED CO_2 AND ITS INTERACTION WITH BIOTIC STRESS

One of the most serious repercussions of climate change is the impact on biological stress and the plant environment (Garrett et al., 2006). Pest-crop interactions have been altered by climate change, resulting in more frequent outbreaks and a shift in pest distribution, putting legumes in jeopardy (Sharma et al., 2015). For instance, increasing CO_2 enhanced the density and area of the soybean canopy, increasing the incidence of pathogenic fungi such downy mildew, brown spots, and early mortality syndrome (Gautam et al., 2013). Other abiotic stressors, such as drought, have been shown to promote fungal virulence in the drought-tolerant peanut and Aspergillus relationship (Luo et al., 2005). On the other hand, a combination interaction isn't always additive. There have been a variety of responses (Pande and Sharma, 2010), both unexpected and common. Increased CO_2 encourages higher photosynthate availability after a pest incursion.

2.3.9 PLANT COMMUNITY INTERACTIONS UNDER ELEVATED CO_2

Several studies have demonstrated that some plant species that grow well in high CO_2 environments when cultivated alone grow less well in mixed plant communities (Poorter, 2002). This impact occurs because the direct benefits of increased CO_2 are offset by the negative consequences of stimulating rival development. The CO_2 levels in the atmosphere are growing, because certain species benefit more than others from increased CO_2, this might have an impact on the composition of plant communities. In mixed-species experiments with more fertile, C4 plants drop as a segment of the biomass of crop communities. Similarly, as CO_2 levels rise, legumes expand as a fraction of plant community biomass in low fertility situations (Poorter, 2002).

2.3.10 INCREASED CO_2 AFFECTS SOLUBLE SUGAR AND ACIDITY LEVELS

Increased CO_2 encourages the buildup of soluble sugars in the eatable portions of leafy vegetables. Enhanced fixation of CO_2 in the presence of eCO_2 stimulates the formation of triose phosphate in leaves (Long et al., 2004), which can subsequently be turned into glucose, fructose, and sucrose. In terms of all vegetables, our meta-analysis found that elevated CO_2 enhanced glucose level by 13.2%, fructose concentrations by 14.2%, sucrose concentrations by 3.7% (at $p = 0.07$), and concentrations of total soluble sugar by 17.5%. Under eCO_2, the increase in total soluble sugar in the leaf (a carbohydrate-synthesis organ) was the highest (36.2%) of all the vegetable classes. In Chinese cabbage leaves, the increase can be as high as 38–188%, while in oily sowthistle leaves, it can be as high as 16–53% (Jin et al., 2009). In comparison to leafy vegetables, the increases are smaller.

2.3.11 THE IMPACT OF INCREASED CO_2 ON NITROGENOUS COMPOUNDS

The eCO_2 reduced the protein content of vegetables by 9.5%, with fruit vegetables losing 10.5%, stem vegetables 12.6%, and root vegetables 20.5%. On the other hand, leafy veggies had no effect. Because leafy plants have a higher nitrate content, eCO_2 may aid in the absorption of nitrogen by the leaves (Stitt and Krapp, 1999). The amount of nitrogen in lettuce increased as CO_2 levels rose. More than the outer leaves of "Batavia Rubia Mungua" that hadn't been treated with arbuscular mycorrhizal fungi. Furthermore, eCO_2 inhibits vegetable nitrogen uptake and synthesis to a lesser extent than other crops (primarily grain crops) (9.5% vs. 10–15%) (Taub et al., 2008; Loladze, 2014), possibly due to the fact that soil nitrogen shortages are more common in grain crop production than in vegetable cultivation (Taub et al., 2008; Loladze, 2014).

2.3.12 *ANTIOXIDANT RESPONSE TO INCREASED CO_2*

In general, elevated CO_2 encourages plants to accumulate antioxidants, resulting in high-quality food. With eCO_2, total antioxidant capacity, total phenols, total flavonoids, ascorbic acid, and chlorophyll b increased by 59.0, 8.9, 45.5, 9.5, and 42.5%, respectively. Leafy vegetables had the highest total antioxidant capacity (72.5%) and ascorbic acid content of all the veggies studied (15.3%). Increasing soluble sugar as a precursor, according to (Wang et al., 2003; Jaafar et al., 2012; Becker and Klaring, 2016), can boost anti-oxidant production and accumulation. For example, 1000 mmol mol^{-1} CO_2 is expected to increase accumulation of sugar and, as a result, synthesis of phenol in lettuce leaves when compared to a control of 200 mmol mol^{-1} CO_2. Increased CO_2 is thought to have a bad impact on the environment.

2.3.13 *INFLUENCED OF CROP QUALITY BY INCREASED CO_2*

The influence of elevated CO_2 on lettuce quality and its interactions with other variables has gotten more attention than that of other crops (Pérez-López et al., 2003; Pérez-López et al., 2015a,b, Pérez-López et al., 2018). According to numerous research, increased CO_2 improves lettuce flavor by increasing soluble sugar accumulation by 27.1% ($n = 18$, meta-analysis) (Baslam et al., 2012; Pérez-López et al., 2015b). Generally, lettuce crop is highly benefited from eCO_2, and eCO_2 increased the buildup of antioxidant molecules in cv. "Oak Leaf" to a greater extent than cv. "Blonde of Paris Batavia" under salt stress or high light intensity (Pérez-López et al., 2013; Pérez-López et al., 2015a, b). Otherwise, the concentration of reducing sugars can be reduced (Kumari and Agarwal, 2014). The effect of eCO_2 on the quality of tomato fruits has also been investigated. According to research, increasing the amount of fructose, glucose, and total soluble sugars in tomato fruits increases quality and flavor (14.7%, $n = 24$) (Islam et al., 1996; Behboudian and Tod, 1995; LI et al., 2017; Zhang et al., 2014). For fruits from the first few harvests (Islam et al., 1996; Zhang et al., 2014) and under moderate N supply, this effect was minimized when less number of carbohydrates was stored (Helyes et al., 2012). Due to less sugar modification, total TSS and different organic acids were raised to a lesser extent than soluble sugar (Behboudian and Tod, 1995; Zhang et al., 2014).

2.4 EFFECT OF DROUGHT STRESS ON PLANTS

Plants may be subjected to a variety of stress based on the dynamics of the external environment (Bukhari et al., 2019; Battaglia et al., 2019). To help the plant to live in these conditions, certain metabolic changes and gene expressions occur (Mostofa et al., 2018; Ahanger et al., 2017). Drought stress, which is recognized as agriculture's most limiting factor, might have a significant impact on grain quality and productivity. Therefore, the study of the capability of plants to cope with water scarcity is extremely important and should be pursued in coming days, especially in arid and semiarid areas (Sobhanian et al., 2020). Basic crops are now being extensively studied to uncover mechanisms that respond to drought to maximize grain yield and quality, but future research should concentrate on the combined effects of heat and water scarcity on the reproductive stages of major cereal crops (Seleiman and Kheir, 2018). Plants need the right amount of water to grow and develop, and too much difference in soil moisture can have adverse effect on grain yield and quality. On the contrary, insufficient water availability in the rhizosphere inhibits plant development and thus nutrient uptake (Elemike et al., 2019). Due to global warming and climatic uncertainties, the latter has recently led to massive yield losses in cereal crops, and this is foreseeable to worsen (Bal and Minhas, 2017; Hafez and Seleiman, 2017).

Water scarcity occurs when there is no or insufficient rainfall, as a result of a decrease in soil moisture content and decreased water potential in aboveground plant parts such as leaves and stems (Ristvey et al., 2019). In dry situations, water loss through leaf transpiration exceeds water uptake by roots (Goche et al., 2020). When water is scarce, roots try to take up more water by expanding, which eventually leads plants to limit stomatal water loss (Martínez-Vilalta and Garcia-Forner, 2017). Leaf rolling, stunned plants, yellowed leaves, blistering, and persistent wilting are common signs of drought stress in plants (Corso et al., 2020), and plant reaction on water scarcity is highly promoted by the frequency and severity of previous water stress episodes (Osakabe et al., 2014; Hafez et al., 2015; Battaglia et al.,

2020) as well as the presence of other stressors (Thomason and Battaglia). Inspite of the negative effects of water scarcity on crop performance, plants can adapt to varying degrees of water scarcity. Water shortage has a greater effect on cell expansion than on cell division; therefore, there is a significant relationship between plant growth and water availability (Humplik et al., 2017). Under these circumstances, plant development is hindered by a loss of cell wall extensibility and turgor (Seleiman et al., 2019). To compensate for the lack of water, stressed plants maintain their osmotic adjustment by increasing the sugar content of their roots and leaves, and plants with a history of drought stress have been shown to develop roots faster than shoots (Miranda et al., 2020; Ullahetal. 2017). Duration, severity, and frequency of drought, as well as soil properties, growth conditions and stages, and plant species, all have a significant influence on the amount and duration of drought-related symptoms in plants (Zoghi et al., 2019). Increased leaf senescence and drooping, scorched and drooping leaves, leaf curling and brittleness, closed flowers and drooping flowers, etiolation, wilting, swelling, premature leaf drop, senescence, and yellowing of leaves are among the most common indicators of drought stress in plants (Ruehr et al., 2019; Khan et al., 2018). Other symptoms presented by plants in drought circumstances include twig cracks, branch dieback, necrosis, stunted growth, bark crack, shrub canopy, and tree thinning (Toscano et al., 2019). Plants may succumb to acute drought stress in some instances. Water scarcity has a significant influence on plant growth and development, whereas water excess has a negative impact on plant performance and slows growth and yield (Zargar et al., 2017). Increase in water stress causes soft succulent leaves, leaves with rotting spots, plant parts affected by fungi and moulds.

Drought stress is also recognized as a limiting factor that affects plant growth and development in a variety of ways. The most important factors for plant development are seed germination, health, and coleoptile length (Sourour et al., 2017). Drought stress affects seed germination, which is the most important part of growth. A variety of plant species, in addition to some of the most repeated farmed crops including Zea mays (Queiroz et al., 2019), Sorghum bicolor, and Triticum aestivum (Quayyum et al., 2011), have shown notable changes in seed germination. Leaf withering, loss in plant height, and stoppage in the development of buds and flowers are visible indications of plants subjected to water scarcity in the early vegetative stage (Bhatt and Rao, 2005). Water scarcity conditions also decline plant nutrient uptake owing to low soil moisture, resulting in shorter stem length (Razmjoo et al., 2008). In Lathyrus sativus L., shoot length was likewise decreased when water was scarce (Gheidary et al., 2017).

Plants enhance their root architecture to take water from deeper soil layers when there is a water shortage (Asadi et al., 2012). Furthermore, roots are the first to detect water availability, which influences development and organization traits like length of the root, spread, and the quantity and lateral roots length (Salazar et al., 2015). Roots are important for a variety of biological processes and plant productivity, including nutrient buildup and water absorption, as well as rhizosphere symbiotic relationships with other microbes. In Crocus sativus L., drought stress increased root length (Maleki et al., 2011). Thus a healthy root system benefits the maintenance of plant growth escalation, particularly during the basic plant development phase (Smith and De Smet, 2012). Root length escalation has been identified as a beneficial method for increasing soil water holding capacity and nutrient procurement to improve plant biomass output (Zulfiqar et al., 2020). When there is a water shortage, the plant root to shoot ratio enhances, and the plant biomass drops significantly (Akhtar and Nazir, 2013).

2.4.1 DROUGHT STRESS HAS AN IMPACT ON PLANT MORPHOLOGICAL, ANATOMICAL, AND BIOCHEMICAL CHANGES PERFORMANCES

Water shortages trigger a variety of plant responses, including morphological, physiological, biochemical, and molecular changes, all of which disrupt plant function (Zlatev and Lidon, 2012). Drought occurrences reduce plant performance at various phases of growth. The germination rate and growth of young plants can both be hampered by a lack of water (Yigit et al., 2016). Drought affects the water balance of plants, leading to severe disturbances in plant metabolism depending on the magnitude and duration of the stress (Sharma et al., 2020; Toscano et al., 2019). Water deficit affects a wide range of plant activities, but one of the most striking is the decline in photosynthetic activity (Bota et al., 2004; Flexas et al., 2006) and eventually plant production (Vurayai et al., 2011; Mafakheri et al., 2010). In drought stress, oxidative stress is one of the most important factors in plant response, resulting in cell membrane damage, altered membrane integrity, physiological and biochemical abnormalities, and finally altered plant production (Wang et al., 2019; Sharma and Zheng, 2019).

2.4.2 BIOCHEMICAL AND PHYSIOLOGICAL RESPONSES

Gaseous movement, leaf water potential, and relative water content in wheat are all affected by drought stress (Nezhadahmadi et al., 2013; Huseynova et al., 2016). Under water deficit conditions, the root–shoot ratio is also adjusted by abscisic acid concentration in order to improve water absorption (Mahdid et al., 2011; Nezhadahmadi et al., 2013). Abscisic acid (ABA) affects leaf and root development, which regulates plant growth (Farooq et al., 2013; Reddy et al., 2014). Plants with low water status close their stomata and reduce their photorespiration rate, improving their water efficiency (Keyvan, 2010). When leave are exposed to water stress, their chlorophyll content decreases, but the extent of the decrease is determined by the age of the leaves. Because enzymes are activated in younger wheat leaves, they have more chlorophyll, but older leaves have less (13–15%) due to enzyme deactivation (Nikolaeva et al., 2010). In stress conditions, osmolytes concentrate in cells to maintain their stability and prevent cell desiccation (Loutfy et al., 2012). Under water deficit situations, wheat accumulates a number of inorganic (K, Ca, Si, and salicylic acid) and organic (glycine-betaine, proline, soluble carbohydrates, and proteins) solutes in its cytoplasm to maintain cell turgor by decline its osmotic potential (Kang et al., 2013; Gou et al., 2015). The increased formation of ROS in wheat plants alters cell structure and harmful affects on photosynthesis. They also lead to enzyme inactivation and oxidative damage (Huseynova et al., 2016). The wheat plant creates several antioxidant enzymes in response to water stress, including catalase (CAT), superoxide dismutase (SOD), peroxidase (POD), ascorbate peroxidase (APX), and glutathione peroxidase (GPx), all of which suppress the formation of ROS.

2.4.3 IMPACTS ON REPRODUCTIVE STAGE UNDER DROUGHT STRESS SITUATION

In a dry climate, flowering time can be a significant predictor of a plant's reproductive performance, with drought having the greatest effect on reproductive development (reviewed in Saini and Westgate, 1999). Drought has a particularly negative effect on the development of male floral organs and gametophytes. Lack of water during inflorescence development in cereal plants can delay inflorescence development or cause partial or complete flower suppression (Saini and Westgate, 1999). In Arabidopsis, reproductive organ development is generally affected by early interruption of flower development during drought (Su et al., 2013). In all crop species and model species, loss of pollen hardiness is the most common response to water scarcity during gametophyte development, resulting in reduced seed filling or grain production in cereal crops (Saini and Westgate, 1999; Su et al., 2013). Compared with the development of male reproductive structures, drought stress has a lesser impact on female floral structures and gametophytes. According to Su et al. (2013), pollen is abundant and small compared to ovules, so protecting female reproduction from male reproductive development is an efficient way to maintain fertility in a stressed environment. Su et al. (2013) discovered that water deficit from flowering to seed maturity enhances the ovule abortion rate in Arabidopsis. Water deficit declines the no. of branches, flowers, siliques, and seeds in Arabidopsis (ma et al., 2014). When acute drought stress was compared with mild drought stress in inflorescence tissue, different transcriptional signatures were found. In moderate drought, genes involved in abscisic acid and response to water deprivation were up-regulated, as was pollen tube growth; in severe drought, a higher no. of genes were differentially expressed, including nuclear factor Y, which is known to have a role in drought tolerance and flowering control (Ma et al., 2014).

2.4.4 DROUGHT STRESS AND PHOTOSYNTHESIS

Reduction or inhibition of photosynthesis is one of the most important consequences of water deficiency in plants (Nezhadahmadi et al., 2013). Osmotic stress is enhanced by lower leaf area, greater stomata closure, and hence decreases leaf cooling by evapotranspiration (Zare et al., 2011; Bhargava and Sawant, 2011). The loss of photosynthetic process in drought-stressed plants is mainly due to a reduction in CO_2 conductance across stomata and mesophyll boundaries (Singh and Thakur, 2018). Loss of CO_2 absorption (Deepak et al., 2019) triggers a decline in photosynthesis, which has been proven to have an impact on Rubisco activity and reduce nitrate reductase and sucrose phosphate synthase function, as well as the ability to produce ribulose bisphosphate (RuBP). CO_2 enrichment, on the other hand, reduced several early drought-related responses in maize metabolites and transcripts (Sicher and Barnaby, 2012).

Water deficit also led to reduced leaf area per shoot, resulting in a change in canopy architecture that can affect gas exchange, water relations, seedling growth, vegetative growth, reproductive growth, and sink development (e.g., fruits or grains) (Rahmati et al., 2018), which affected, for example, sugar concentration in grape berries (Zsófi et al., 2011) and biomass distribution in maize (i.e., number of grains and dry weight of 100 grains decreased with increasing duration of water stress) (Ge et al., 2012).

Chlorophyll concentration, which is troublesome for photosynthesis (Rahdari et al., 2012), is another photosynthetic property that is altered by water deficit and has been identified as a clear indicator of chlorophyll oxidation and degradation (Anjum et al., 2011). Water deficit, for example, affects leaf chlorophyll production and chlorophyll a/b ratio in soybean (Chowdhury et al., 2017). Plant production was reduced due to a decrease in photosynthetic activity, chlorophyll content, photochemical efficiency of the photosystem II, disruption of stomatal movement, and disruption of water status (Xiang et al., 2013). Drought-induced O_2 and H_2O_2 efflux, for example, is a major source of chlorophyll loss due to drought stress, leading to lipid peroxidation and eventual chlorophyll degradation (Karimpour, 2019). In the case of water stress, a decrease in plant growth and production is often associated with a decrease in photosynthetic activity and impaired chlorophyll content (Abid et al., 2017). Water deficiency alters the function of photosynthetic moieties and chlorophyll pigments, resulting in decreased photosynthetic activity in moong crops (Gurumurthy et al., 2019). Water deficiency leads to a decrease in net photosynthesis and alteration in carbon partitioning and metabolism in plants, resulting in energy losses and lower yields (Cuellar-Ortizetal. 2008). For example, water stress led to physiological metabolic abnormalities in soybean by inhibiting photosynthetic product synthesis and disturbing carbon balance. Drought stress reduced the levels of a number of Calvin cycle proteins, including downregulation of Rubisco in olives (Abdallah et al., 2018). Severe drought stress causes damage to the Rubisco enzyme and other enzymes related to photosynthesis, resulting in a decrease in photosynthetic pigment content (Brito et al., 2019).

2.4.5 THE ANTIOXIDANT DEFENSE MECHANISM RESPONSES TO DROUGHT STRESS

The majority of plant defense mechanisms are designed to counteract the negative effects of drought ROS. In this environment, a fast, strong, and effective antioxidant system is crucial for drought tolerance (Hussain et al., 2019). This machinery consists of enzymatic and non-enzymatic detoxification units that reduce and repair ROS-induced damage. The antioxidant system contributes to ROS scavenging, which reduces electrolyte loss and lipid peroxidation and preserves the strength and integrity of organelles and cell membranes (Gharibi et al., 2016). Drought causes oxidative stress by producing ROS, such as O_2, hydroxyl radicals (OH), singlet oxygen ($1O_2$), and H_2O_2 (Impa et al., 2012). When ROS levels surpass physiological levels, the proportion of ROS formation and antioxidant enzyme activity governs the cell redox state, resulting in ROS control or cell damage and death (Moller et al., 2007). CAT, SOD, POD, and APX activities were reported to be increased in a number of experiments done under water deficiency circumstances (Cao et al., 2017). In general, tolerant species/varieties/genotypes have higher antioxidant enzyme activity than non-tolerant plants, suggesting that antioxidant enzymes play an important role in drought resistance, especially in controlling H_2O_2 and O_2 production and diffusion in leaf tissue (Moussa and Abdel-Aziz, 2008). The activity of superoxide dismutase (SOD), peroxidase (POX), and catalase (CAT), which was increased in drought-resistant potato genotypes (Shi et al., 2015; Kumarathunge et al., 2019), reduced the production of O_2 and H_2O_2. Excess ROS is scavenged by ascorbate peroxidase (APX), which uses ascorbate as a substrate to accelerate the conversion of H_2O_2 to H_2O and whose activity is normally increased under stress conditions (Hossain et al., 2012). Since APX is largely present in chloroplasts and cytoplasm and is a key enzyme for scavenging H_2O_2 in chloroplasts, changes in the activity of APX are more dramatic in leaves than in fibrous roots (Caverzan et al., 2016). The activities of POD, CAT, and APX were altered, and they played an important role in protecting peonies from severe water deprivation (Wang et al., 2019). In Coleus plectranthus suffering from drought, both the levels of non-enzymatic antioxidants (ascorbic acid, reduced glutathione, and tocopherol) and antioxidant enzymes (SOD, CAT, and APX) were increased (Prathyusha and Chaitanya, 2019). Limited water availability increased the activity of the enzymes SOD, CAT, and POX in Vicia faba (Abid et al., 2019). Compared to drought-sensitive field bean plants, drought-tolerant genotypes showed increased SOD, POX, and CAT activity (Siddiqui et al., 2015). Under mild and moderate water deficit, the content of enzymatic and non-enzymatic antioxidants increased in drought-resistant plants (Flexas et al., 2012). Drought increased the activity of CAT, SOD, POD, and APX in Adonis amurensis and Adonis pseudoamurensis, showing that better enzyme performance helps to decrease the levels of ROS and reduce drought-induced oxidative stress (Gao et al., 2020). Water scarcity increased the levels of SOD and POD in *Vigna mungo*, and the authors concluded that higher levels of these enzymes influence drought resistance and are necessary to mitigate the negative impacts of drought (Gurumurthy et al., 2019). Water scarcity increased the levels of CAT, POX, and SOD in *Glycyrrhiza glabra* L. leaves, thus preventing the spread of H_2O_2 (Hosseini et al., 2018).

2.4.6 SECONDARY METABOLITES AND DROUGHT STRESS

Plants create secondary metabolites in order to adapt to varied environmental challenges (Kleinwächter and Selmar, 2015; Sharma et al., 2019). The production of secondary metabolites is known to be influenced by environmental variables such as temperature, light regime, and nutrition availability (Ballhom et al., 2011). Under water deficiency conditions, improved production of secondary metabolites is more frequently observed, driven by a decrease in biomass formation and the destination of absorbed CO_2 to C-based secondary metabolites to inhibit sugar-influenced photosynthetic feedback (Selmar and Kleinwächter, 2013).

The quantity of light energy captured by photosynthetic equipment is significantly greater than the energy necessary to fix CO_2. To dissipate energy, non-photochemical quenching and re-oxidation of $NADPH^+$ H^+, that is, via the xanthophyll cycle and C2 cycle, are utilized. Because of the increased diffusion resistance produced by stomata closure, endogenous CO_2 levels are low. As a result, the C3 cycle uses less $NADPH^+$ H^+ for CO_2 fixation and reduction, resulting in a larger quantity of energy being squandered. Feedback processes promote defensive functions such non-photochemical quenching, the C2 cycle, and the xanthophyll cycle; a large amount of e is transferred to O_2 (Mehler reaction). The production of O_2 ions generates a variety of ROS. Detoxification of O_2 ions occurs as a result of stress-related activation of SOD and APX, resulting in a decrease in ROS production. Increased reduction potential, or the ratio of $NADPH^+$ H^+ to $NADP^+$, boosts secondary metabolite production in plants (adapted from Kleinwächter and Selmar, 2015).

When Hypericum brasiliense is cultivated in water shortage circumstances, the content of phenolic acids increases dramatically (Nacif de Abreu and Mazzafera, 2005). Water restriction increased the content of sesquiterpenes (E)-caryophyllene in two native subspecies of Oregano plant, namely subsp. gracile and subsp. virens (Morshedloo et al., 2017). The content of oleanolic acid and betulin in Asian White birch (Yin et al., 2015) and the amount of the triterpenoid glycyrrhizin in Licorice plant (Nasrollahi et al.,

2014) increased under mild and mild/severe drought conditions. Under drought conditions, the lignin content of Bermuda grass Tifton-85, a cultivar of *Cynodon dactylon L.*, increased (Saha et al., 2016). In stress condition, the concentration of flavonoids increased, and under severe drought conditions, the therapeutic abilities of Labisia pumila were enhanced (Jaafar et al., 2012). In the Lamiaceae family, the essential oil content of *Lavandula latifolia* and *Salvia sclarea* are diminished, while the essential oil content of *Mentha piperita, Salvia lavandulifolia, Thymus capitatus*, and *Thymus mastichina* is increased under water scarcity conditions, the increase being due to maximum concentration of oil glands due to a reduced leaf area (García-Caparrós et al., 2019). Drought stress increased the amount of phenolics and flavonoids in Achillea species (Hussain et al., 2019). In Achillea pachycephala, the amount of phenolic acids increased, while the content of flavonoids decreased (Gharibi et al., 2019).

2.4.7 DROUGHT STRESS RESPONSE TO MOLECULAR RESEARCH

A plant's react to water scarcity situation involves a number of genes and signaling mechanism. Several water stress-responsive genes have been discovered by high-throughput molecular research, and their regulation is split into two groups: (1) Late embryogenesis abundant (LEA) proteins and antioxidants are two functional proteins that protect plants from the effects of dehydration. Aquaporins and sugar transporters are involved in the uptake and transfer of water. Signaling cascades involving regulatory proteins (kinase, phosphatase, and transcription regulatory factors-APETALA2/ethylene-responsive element-binding proteins [AP2/EREBP], zinc finger protein [bZIP], and myeloblastosis oncogene protein [MYB]) are involved in the induction of transcription factors and activation of target genes (Zahoor et al., 2020). Through their functions as osmoprotectants, osmotic regulators, antioxidant defense systems, and oxygen radical scavengers, these functional and regulatory proteins play important roles in prevention and tolerance systems in wheat plants (Yang et al., 2010; Ashraf, 2010). There are two types of transcriptional regulatory pathways: abscisic acid-dependent and abscisic acid-independent (Budak et al., 2013). Stimulation of two transcription factors (TFs) by ABA-dependent pathways-mediated stress adaptation: (1) zinc finger proteins (bZIP) (2) MYC (myelocytomatosis oncogene)/MYB (myeloblastosis oncogene). CBF/DREB (cold-binding factor/dehydration-responsive element binding), NAC, and ZF-HD (zinc finger homeodomain) are examples of ABA-independent regulatory pathways (Lata and prasad, 2011). Several transcriptional factors controlled both pathways (TFs). Under stressful situations, transcriptional factors known as WRKY (TFs) control a variety of functions and are regulated by the ABA hormone. Under water stress, the transcriptional factors WRKY1 and WRKY2 were increased in wheat (Proietti et al., 2011). Drought-tolerant wheat also has 35 WRKY transcripts from 10 TaWKRY genes, and their expression was up-regulated in leaf and root–shoot tissues (Okay et al., 2014). In response to abiotic stress, NACs (TFs) have a comparable effect on plant growth (Nakashima et al., 2014). In wheat, NAC (TFs) from the TaNAC69 gene have been discovered, and their overexpression can improve stress tolerance (David et al., 2016). AREB transcription factors are also known as bZIP transcription factors (ABA-responsive element-binding protein). WLIP 19, TaOBF1, and TaABF1 (bZIP) TFs were also discovered to be up-regulated in response to ABA in wheat in response to water stress (Rahaie et al., 2013).

2.4.8 DROUGHT STRESS RESPONSE TO ROOT DEVELOPMENT

Root enlargement, unlike shoot development, is often maintained or even promoted in response to water scarcity. In maize, soybean, cotton, and cucurbit, a smaller reduction in root growth rate was observed compared with shoot growth rate (Spollen et al., 1993). Under water stress, greater root development and deeper root depth distribution were observed in numerous species, indicating a dynamic flexibility of root stress response throughout soil depth and the developmental trajectory of both the root and the whole plant. Drought stress increased root growth rates in field-grown soybeans and the enormity of this effect varied with plant developmental stage and soil depth. Root growth rate acceleration was highest at end of vegetative stage and early maturity stages of development and at soil depths greater than 0.6 meters (Hoogenboom et al., 1987). Due to a lack of water, root system growth accelerated and lateral root tips were directed downward, according to (Rellán-Álvarez et al., 2015). They also discovered that mutants of the auxin receptor TIR1 (Transport Inhibitor Response 1) did not adjust their root position in response to water scarcity, suggesting that auxin plays an important role in redirecting root development angles downward during drought stress. Deeper Rooting 1 (DRO1), a gene that affects rooting depth of rice via erect root development angle of nodal roots, was cloned by (Uga et al., 2013). They discovered that whereas this genotype had no effect on root or shoot biomass or grain output in well-watered settings, it did dramatically improve grain production in moderate or severe water stress (Uga et al., 2013). Auxin produces polar expression patterns of DRO1 across the root in response to a gravity challenge, with a rise near the outer margin of the elongation zone, likely encouraging cell elongation on the upward side of the root (Uga et al., 2013). These findings show that one specific root morphological characteristic is regulated in a unique way by genetics, as well as the fact that root attributes can directly affect seed yield under stressful situations.

2.4.9 DROUGHT STRESS AND MINERAL NUTRITION

Water shortages diminish total soil nutrient unavailability, root nutrient displacement and as a result, the ion concentration in different plant tissues (Kheradmand et al., 2014). Plant potassium (K) uptake was reduced when water was scarce (Hu and Schmidhalter, 2005). K mobility is reduced, and the rate of transpiration is reduced, and weaker root membrane transporter function were all factors in the decrease in K (Hu and Schmidhalter, 2005; Qi et al., 2019). Drought-stressed *Malus hupehensis* plants had lower K levels as well (Qi et al., 2019). The maximum potassium (K) levels were found in resistant Triticum durum genotypes, while the maxium sodium (Na) values were found in susceptible genotypes (Karimpour, 2019; von Caemmerer, 2020). Internal K channels are triggered by a protein kinase, CIPK23, which associates with calcineurin B-like calcium sensors, and genes expressing K transporters are blocked by water deprivation (Li et al., 2009). The K channel in grapevine roots was inhibited, whereas it was activated in leaves (Cuéllar et al., 2010). In water-stressed plants like, *Salvia lavandulifolia, Salvia sclarea, Mentha piperita,* and *Thymus capitatus,* etc. leaf nitrogen (N) did not change, whereas N content decreased in *Lavandula latifolia* and Thymus mastichina plants, while leaf phosphorus (P) content decreased in all plant species without S. sclarea, in which there was no change (García-Caparrós et al., 2019). This decrease in N content was considered to be the main cause of the decline in photosynthesis and leaf senescence (Da Silva et al., 2011). Under water deficit, both the amount of P in the leaves of African basil (Osuagwu et al., 2010) and the amount of K in Thymus daenensis reduced significantly (Bahreininejad et al., 2013). In Sweet basil and American basil, K content decreased significantly when water availability was restricted (Khalid, 2006). Water stress causes stomatal activity and turgidity of guard cell to be disrupted, resulting in reduced photosynthesis and plant biomass output (Sarani et al., 2014). Drought stress enhanced the micronutrient like Mn, Mo, Cu, and Zn and macronutrient like P, K, and Ca build up in soybean (Samarah et al., 2004).

2.4.10 IMPACT OF WATER STRESS ON YIELD

Water stress affects many physiological processes in plants that determine yield. Yield intricately connects many of these physiological processes. The intensity, frequency, and duration of water scarcity, as well as plant responses after stress retreat and the relationship between stress and other parameters, are all significant factors (Plaut, 2003). Pre-anthesis water stress reduced the duration of anthesis in triticale genotypes, while post-anthesis water deficit shortened the time of grain filling (Estrada-Campuzano et al., 2008). Drought stress affects the economic yield of a variety of field crops. In barley (Estrada-Campuzano et al., 2008) (*Hordeum vulgare*), grain yield is reduced to drought stress. Drought stress after anthesis was deleterious to grain yield regardless of intensity (Samarah et al., 2005). Many crop species have experienced yield reductions due to drought, which varies depending on the harshness and timing of the anxiety period (Table 2.2). Water stress in maize diminishes output by late silking and hence enhances the interval between anthesis and silking. Drought had no influence on the rate of kernel fulfilment in wheat after heading, but it did lower the dry weight at maturity and shorten the duration (from fertilization to maturity) (Wardlaw and Willenbrink, 2000). In soybeans, drought reduced overall seed production and branch seed output (Frederick et al., 2001). Increased drought tolerance induces greater storage of dry matter in grain than stover in peral millet (*Pennisetum glaucum*) (Yadav et al., 2004). Prolonged water stress in flowering period often leads to barrenness. A decrease in assimilate flow into the developing ear below the threshold required for good grain growth was a major cause, although not the only one (Yadav et al., 2004). Moisture deprivation lowered lint output in cotton (*Gossypium hirsutum*), and the timing, length, intensity, and speed of development all likely had a part in how the plant responded: When stress intensity was higher throughout the reproductive growth stage, lint yield was generally reduced by poorer pollen production due to less flowers and higher pollen apertures (Pettigrew, 2004).

TABLE 2.2 Drought Stress Affects the Economic Yield of a Variety of Field Crops.

Scientific name of crops	Growth period	Yield decrease	References
Hordeum vulgare	Seed filling	49–57%	Samrah (2005)
Zea mays	Grain filling	79–81%	Monneveux et al. (2006)
Zea mays	Reproductive	63–87%	Kamara et al. (2003)
Zea mays	Reproductive	47–70%	Chapman and Edmeades (1999)
Zea mays	Vegetative	25–60%	Atteya (2003)
Zea mays	Reproductive	32–92%	Atteya (2003)
Oryza sativa	Reproductive (mild stress)	53–92%	Lafitte et al. (2007)
Oryza sativa	Reproductive (severe stress)	48–94%	Lafitte et al. (2007)
Oryza sativa	Grain filling (mild stress)	30–55%	Basnayake et al. (2006)
Oryza sativa	Grain filling (acute stress)	60%	Basnayake et al. (2006)
Oryza sativa	Reproductive	24–84%	Venuprasad et al. (2007)
Cicer arietinum	Reproductive	45–69%	Nayyar et al. (2006)
Cajanus cajan	Reproductive	40–55%	Nam et al. (2001)
Lens esculenta	Reproductive	58–87%	Martinez et al. (2007)
Glycine max	Reproductive	46–71%	Samarah et al. (2006)
Vigna sinensis	Reproductive	60–11%	Ogbonnaya et al. (2003)
Helianthus annuus	Reproductive	60%	Mazahery-Laghab et al. (2003)
Brassica napus	Reproductive	30%	Sinaki et al. (2007)
Solanum tuberosum	Flowering	13%	Kawakami et al. (2006)

The biosynthesis of starch from simple carbohydrates serves to fill the grains in cereals. Sucrose synthase, adenosine diphosphate, glucose pyrophosphorylase, starch synthase, and starch branching enzyme are thought to play significant roles in this system (Taiz and Zeiger, 2006). Decreased sucrose synthase activity resulted in reduced grain growth, whereas deactivation of adenosine diphosphate glucose pyrophosphorylase resulted in growth arrest in water-saturated crop like *Triticum aestivum* (Ahmadi and Baker, 2001). Water stress during pollination promotes defect grain in maize (*Zea mays*); water scarcity increases abscisic acid content, which negatively affects grain formation and development of wheat crop, and reduces endosperm cell division in maize (ober et al., 1991); lack of adequate water during flowering in pigeon pea reduces seed production by 40–45%. Water stress leads to remobilization of pre-existing carbon in grains, which contributes to proper grain filling in rice (Yang et al., 2001). In summary, the current drought inhibits plant growth and development, resulting in reduced flower and grain formation, resulting in smaller and fewer grains. Reduced assimilate partitioning and activity of sucrose and starch production enzymes cause a decrease in grain filling.

KEYWORDS

- **climate changes**
- **plant growth**
- **temperature**
- **productivity**
- **atmosphere**

REFERENCES

Abdallah, M. B.; Trupiano, D.; Polzella, A.; De Zio, E.; Sassi, M.; Scaloni, A. et al. Unraveling Physiological, Biochemical and Molecular Mechanisms Involved in Olive (*Olea europaea* L. cv. Chétoui) Tolerance to Drought and Salt Stresses. *J. Plant Physiol.* **2018,** *220*, 83–95. https://doi.org/10.1016/j.jplph.2017.10.009

AbdElgawad, H.; Farfan-Vignolo, E. R.; De Vos, D.; Asard, H. Elevated CO2 Mitigates Drought and Temperature-Induced Oxidative Stress Differently in Grasses and Legumes. *Plant Sci.* **2015,** *231*, 1–10. https://doi.org/10.1016/j.plantsci.2014.11.001

Abebe, A.; Pathak, H.; Singh, S. D.; Bhatia, A.; Harit, R. C.; Kumar, V. Growth, Yield and Quality of Maize with Elevated Atmospheric Carbon Dioxide and Temperature in North–West India. *Agric. Ecosyst. Environ.* **2016,** *218*, 66–72. https://doi.org/10.1016/j.agee.2015.11.014

Abid, G.; M'hamdi, M.; Mingeot, D.; Aouida, M.; Aroua, I.; Muhovski, Y. et al. Effect of Drought Stress on Chlorophyll Fluorescence, Antioxidant Enzyme Activities and Gene Expression Patterns in Faba Bean (*Vicia faba* L.). *Arch. Agron. Soil Sci.* **2017,** *63* (4), 536–552. https://doi.org/10.1080/03650340.2016.1224857

Ahamed, K. U.; Nahar, K.; Fujita, M.; Hasanuzzaman, M. Variation in Plant Growth, Tiller Dynamics and Yield Components of Wheat (*Triticum aestivum* L.) Due to High Temperature Stress. *Adv. Agric. Botan.* **2010,** *2* (3), 213–224.

Ahanger, M. A.; Tomar, N. S.; Tittal, M.; Argal, S.; Agarwal, R. Plant Growth Under Water/Salt Stress: ROS Production; Antioxidants and Significance of Added Potassium Under Such Conditions. *Physiol. Mol. Biol. Plants* **2017,** *23* (4), 731–744. https://doi.org/10.1007/s12298-017-0462-7

Ahmadi, A.; Baker, D. A. The Effect of Water Stress on the Activities of Key Regulatory Enzymes of the Sucrose to Starch Pathway in Wheat. *Plant Growth Regul.* **2001,** *35* (1), 81–91. https://doi.org/10.1023/A:1013827600528

Ahuja, I.; de Vos, R. C.; Bones, A. M.; Hall, R. D. Plant Molecular Stress Responses Face Climate Change. *Trends Plant Sci.* **2010,** *15* (12), 664–674. https://doi.org/10.1016/j.tplants.2010.08.002

Akhtar, I.; Nazir, N. Effect of Waterlogging and Drought Stress in Plants. *Int. J. Water Res. Environ. Eng.* **2013,** *2*, 34–40.

Allakhverdiev, S. I.; Kreslavski, V. D.; Klimov, V. V.; Los, D. A.; Carpentier, R.; Mohanty, P. Heat Stress: An Overview of Molecular Responses in Photosynthesis. *Photosynth. Res.* **2008,** *98* (1), 541–550. https://doi.org/10.1007/s11120–008–9331–0

Alonso, A.; Pérez, P.; Martínez-Carrasco, R. Growth in Elevated CO2 Enhances Temperature Response of Photosynthesis in Wheat. *Physiol. Plantarum* **2009,** *135* (2), 109–120. https://doi.org/10.1111/j.1399–3054.2008.01177.x

Alsajri, F. A.; Singh, B.; Wijewardana, C.; Irby, J. T.; Gao, W.; Reddy, K. R. Evaluating soybean Cultivars for Low-and High-Temperature Tolerance During the Seedling Growth Stage. *Agronomy* **2019,** *9* (1), 13. https://doi.org/10.3390/agronomy9010013

Aphalo, P. J.; Jarvis, P. G. Do Stomata Respond to Relative Humidity? *Plant Cell Environ.* **1991,** *14* (1), 127–132. https://doi.org/10.1111/j.1365–3040.1991.tb01379.x

Aranjuelo, I.; Arrese-Igor, C.; Molero, G. Nodule Performance Within a Changing Environmental Context. *J. Plant Physiol.* **2014,** *171* (12), 1076–1090. https://doi.org/10.1016/j.jplph.2014.04.002

Asada, K. Production and Scavenging of Reactive Oxygen Species in Chloroplasts and Their Functions. *Plant Physiol.* **2006,** *141* (2), 391–396. https://doi.org/10.1104/pp.106.082040

Asadi, S.; Lebaschy, M. H.; Khourgami, A.; Rad, A. H. S. Effect of Drought Stress on the Morphology of Three Salvia Sclarea Populations. *Ann. Biol. Res.* **2012,** *3* (9), 4503–4507.

Ashraf, M.; Hafeez, M. Thermotolerance of Pearl Millet and Maize at Early Growth Stages: Growth and Nutrient Relations. *Biol. Plantarum* **2004,** *48* (1), 81–86. https://doi.org/10.1023/B:BIOP.0000024279.44013.61

Ashraf, M.; Harris, P., Eds. *Abiotic Stresses: Plant Resistance Through Breeding and Molecular Approaches*; CRC Press, 2005.

Atteya, A. M. Alteration of Water Relations and Yield of Corn Genotypes in Response to Drought Stress. *Bulg. J. Plant Physiol.* **2003,** *29* (1–2), 63–76.

Ayenan, M. A. T.; Danquah, A.; Hanson, P.; Ampomah-Dwamena, C.; Sodedji, F. A. K.; Asante, I. K.; Danquah, E. Y. Accelerating Breeding for Heat Tolerance in Tomato (*Solanum lycopersicum* L.): An Integrated Approach. *Agronomy* **2019,** *9* (11), 720. https://doi.org/10.3390/agronomy9110720

Baena-González, E.; Rolland, F.; Thevelein, J. M.; Sheen, J. A Central Integrator of Transcription Networks in Plant Stress and Energy Signalling. *Nature* **2007,** *448* (7156), 938–942. https://doi.org/10.1038/nature06069

Bahreininejad, B.; Razmjou, J.; Mirza, M. Influence of Water Stress on Morpho-Physiological and Phytochemical Traits in *Thymus daenensis*, 2013.

Bal, S. K.; Minhas, P. S. Atmospheric Stressors: Challenges and Coping Strategies. In *Abiotic Stress Management for Resilient Agriculture*; Springer: Singapore, 2017; pp 9–50. DOI: 10.1007/978–981–10–5744–1_2

Ballhorn, D. J.; Kautz, S.; Jensen, M.; Schmitt, I.; Heil, M.; Hegeman, A. D. Genetic and Environmental Interactions Determine Plant Defences Against Herbivores. *J. Ecol.* **2011,** *99* (1), 313–326. https://doi.org/10.1111/j.1365–2745.2010.01747.x

Bardgett, R. D.; Mommer, L.; De Vries, F. T. Going Underground: Root Traits as Drivers of Ecosystem Processes. *Trends Ecol. Evol.* **2014,** *29* (12), 692–699. https://doi.org/10.1016/j.tree.2014.10.006

Baslam, M.; Garmendia, I.; Goicoechea, N. Elevated CO2 May Impair the Beneficial Effect of Arbuscular Mycorrhizal Fungi on the Mineral and Phytochemical Quality of Lettuce. *Ann. Appl. Biol.* **2012,** *161* (2), 180–191. https://doi.org/10.1111/j.1744–7348.2012.00563.x

Basnayake, J.; Fukai, S.; Ouk, M. Contribution of Potential Yield, Drought Tolerance and Escape to Adaptation of 15 Rice Varieties in Rainfed Lowlands in Cambodia. In *Proceedings of the Australian Agronomy Conference, Australian Society of Agronomy, Birsbane, Australia, Sept.*, 2006; pp 10–14.

Bathellier, C.; Yu, L. J.; Farquhar, G. D.; Coote, M. L.; Lorimer, G. H.; Tcherkez, G. Ribulose 1, 5-Bisphosphate Carboxylase/Oxygenase Activates O2 by Electron Transfer. *Proc. Natl. Acad. Sci.* **2020,** *117* (39), 24234–24242. https://doi.org/10.1073/pnas.2008824117

Battaglia, M.; Lee, C.; Thomason, W.; Van Mullekom, J. Effects of Corn Row Width and Defoliation Timing and Intensity on Canopy Light Interception. *Crop Sci.* **2019,** *59* (4), 1718–1731. https://doi.org/10.2135/cropsci2018.05.0337

Battaglia, M.; Lee, C.; Thomason, W.; Fike, J.; Sadeghpour, A. Hail Damage Impacts on Corn Productivity: A Review. *Crop Sci.* **2019,** *59* (1), 1–14. https://doi.org/10.2135/cropsci2018.04.0285

Bauer, H.; Ache, P.; Wohlfart, F.; Al-Rasheid, K. A.; Sonnewald, S.; Sonnewald, U. et al. How Do Stomata Sense Reductions in Atmospheric Relative Humidity? *Mol. Plant* **2013,** *6* (5), 1703–1706. DOI: 10.1093/mp/sst055

Bavita, A.; Shashi, B.; Navtej, S. B. Nitric Oxide Alleviates Oxidative Damage Induced by High Temperature Stress in Wheat, 2012. http://nopr.niscpr.res.in/handle/123456789/14132

Becker, C.; Kläring, H. P. CO2 Enrichment Can Produce High Red Leaf Lettuce Yield While Increasing Most Flavonoid Glycoside and Some Caffeic Acid Derivative Concentrations. *Food Chem.* **2016,** *199*, 736–745. https://doi.org/10.1016/j.foodchem.2015.12.059

Begcy, K.; Nosenko, T.; Zhou, L. Z.; Fragner, L.; Weckwerth, W.; Dresselhaus, T. Male Sterility in Maize After Transient Heat Stress During the Tetrad Stage of Pollen Development. *Plant Physiol.* **2019,** *181* (2), 683–700. https://doi.org/10.1104/pp.19.00707

Behboudian, M. H.; Tod, C. Postharvest Attributes of Virosa' Tomato Fruit Produced in an Enriched Carbon Dioxide Environment. *HortScience* **1995,** *30* (3), 490–491. https://doi.org/10.21273/HORTSCI.30.3.490

Beidler, K. V.; Taylor, B. N.; Strand, A. E.; Cooper, E. R.; Schönholz, M.; Pritchard, S. G. Changes in Root Architecture Under Elevated Concentrations of CO2 and Nitrogen Reflect Alternate Soil Exploration Strategies. *New Phytol.* **2015,** *205* (3), 1153–1163. https://doi.org/10.1111/nph.13123

Bell, K. L.; Bliss, L. C. Root Growth in a Polar Semidesert Environment. *Can. J. Bot.* **1978,** *56* (20), 2470–2490. https://doi.org/10.1139/b78–299

Bernacchi, C. J.; Singsaas, E. L.; Pimentel, C.; Portis Jr, A. R.; Long, S. P. Improved Temperature Response Functions for Models of Rubisco-Limited Photosynthesis. *Plant Cell Environ.* **2001,** *24* (2), 253–259. https://doi.org/10.1111/j.1365–3040.2001.00668.x

Bhargava, S.; Sawant, K. Drought Stress Adaptation: Metabolic Adjustment and Regulation of Gene Expression. *Plant Breed.* **2013,** *132* (1), 21–32. https://doi.org/10.1111/pbr.12004

Bhatt, R. M.; Rao, N. S. Influence of Pod Load on Response of Okra to Water Stress. *Indian J. Plant Physiol.* **2005,** *10* (1), 54.

Birami, B.; Gattmann, M.; Heyer, A. G.; Grote, R.; Arneth, A.; Ruehr, N. K. Heat Waves Alter Carbon Allocation and Increase Mortality of Aleppo Pine Under Dry Conditions. *Front. Forests Glob. Change* **2018,** *1*, 8. https://doi.org/10.3389/ffgc.2018.00008

Bishop, K. A.; Betzelberger, A. M.; Long, S. P.; Ainsworth, E. A. Is There Potential to Adapt Soybean (Glycine max Merr.) to Future [CO2]? An Analysis of the Yield Response of 18 Genotypes in Free-Air CO2 Enrichment. *Plant Cell Environ.* **2015,** *38* (9), 1765–1774. https://doi.org/10.1111/pce.12443

Bota, J.; Medrano, H.; Flexas, J. Is Photosynthesis Limited by Decreased Rubisco Activity and RuBP Content Under Progressive Water Stress? *New Phytol.* **2004,** *162* (3), 671–681. https://doi.org/10.1111/j.1469–8137.2004.01056.x

Boyd, R. A.; Gandin, A.; Cousins, A. B. Temperature Responses of C4 Photosynthesis: Biochemical Analysis of Rubisco, Phospho Enolpyruvate Carboxylase, and Carbonic Anhydrase in *Setaria viridis*. *Plant Physiol.* **2015,** *169* (3), 1850–1861. https://doi.org/10.1104/pp.15.00586

Brito, C.; Dinis, L. T.; Moutinho-Pereira, J.; Correia, C. M. Drought Stress Effects and Olive Tree Acclimation Under a Changing Climate. *Plants* **2019,** *8* (7), 232. https://doi.org/10.3390/plants8070232

Brodribb, T. J.; McAdam, S. A. Passive Origins of Stomatal Control in Vascular Plants. *Science* **2011,** *331* (6017), 582–585. https://doi.org/10.1126/science.1197985

Budak, H.; Kantar, M.; Yucebilgili Kurtoglu, K. Drought Tolerance in Modern and Wild Wheat. *Sci. World J.* **2013,** *2013*. https://doi.org/10.1155/2013/548246

Bueckert, R. A.; Wagenhoffer, S.; Hnatowich, G.; Warkentin, T. D. Effect of Heat and Precipitation on Pea Yield and Reproductive Performance in the Field. *Can. J. Plant Sci.* **2015,** *95* (4), 629–639. https://doi.org/10.4141/cjps-2014-342

Bukhari, S. A. H.; Peerzada, A. M.; Javed, M. H.; Dawood, M.; Hussain, N.; Ahmad, S. Growth and Development Dynamics in Agronomic Crops Under Environmental Stress. In *Agronomic Crops*; Springer: Singapore, 2019; pp 83–114. DOI: 10.1007/978-981-32-9151-5_6

Butterly, C. R.; Armstrong, R.; Chen, D.; Tang, C. Carbon and Nitrogen Partitioning of Wheat and Field Pea Grown with Two Nitrogen Levels Under Elevated CO2. *Plant Soil* **2015,** *391* (1), 367–382. https://doi.org/10.1007/s11104-015-2441-5

Calleja-Cabrera, J.; Boter, M.; Oñate-Sánchez, L.; Pernas, M. Root Growth Adaptation to Climate Change in Crops. *Front. Plant Sci.* **2020,** *11*, 544. https://doi.org/10.3389/fpls.2020.00544

Camejo, D.; Jiménez, A.; Alarcón, J. J.; Torres, W.; Gómez, J. M.; Sevilla, F. Changes in Photosynthetic Parameters and Antioxidant Activities Following Heat-Shock Treatment in Tomato Plants. *Funct. Plant Biol.* **2006,** *33* (2), 177–187. https://doi.org/10.1071/FP05067

Cao, Y.; Luo, Q.; Tian, Y.; Meng, F. Physiological and Proteomic Analyses of the Drought Stress Response in *Amygdalus Mira* (Koehne) Yü et Lu Roots. *BMC Plant Biol.* **2017,** *17* (1), 1–16. https://doi.org/10.1186/s12870–017–1000-z

Caverzan, A.; Casassola, A.; Brammer, S. P. Reactive Oxygen Species and Antioxidant Enzymes Involved in Plant Tolerance to Stress. *Abiotic Biotic Stress Plants-Recent Adv. Fut. Persp.* **2016,** *17*, 463–480.

Chapman, S. C.; Edmeades, G. O. Selection Improves Drought Tolerance in Tropical Maize Populations: II—Direct and Correlated Responses Among Secondary Traits. *Crop Sci.* **1999,** *39* (5), 1315–1324. https://doi.org/10.2135/cropsci1999.3951315x

Chater, C.; Kamisugi, Y.; Movahedi, M.; Fleming, A.; Cuming, A. C.; Gray, J. E.; Beerling, D. J. Regulatory Mechanism Controlling Stomatal Behavior Conserved Across 400 Million Years of Land Plant Evolution. *Curr. Biol.* **2011,** *21* (12), 1025–1029. https://doi.org/10.1016/j.cub.2011.04.032

Chavan, S. G.; Duursma, R. A.; Tausz, M.; Ghannoum, O. Elevated CO2 Alleviates the Negative Impact of Heat Stress on Wheat Physiology But Not on Grain Yield. *J. Exp. Bot.* **2019,** *70* (21), 6447–6459. https://doi.org/10.1093/jxb/erz386

Chaves, M. M.; Flexas, J.; Pinheiro, C. Photosynthesis Under Drought and Salt Stress: Regulation Mechanisms from Whole Plant to Cell. *Ann. Bot.* **2009,** *103* (4), 551–560. https://doi.org/10.1093/aob/mcn125

Chaves, M. M.; Maroco, J. P.; Pereira, J. S. Understanding Plant Responses to Drought—From Genes to the Whole Plant. *Funct. Plant Biol.* **2003,** *30* (3), 239–264. https://doi.org/10.1071/FP02076

Cheng, L.; Zou, Y.; Ding, S.; Zhang, J.; Yu, X.; Cao, J.; Lu, G. Polyamine Accumulation in Transgenic Tomato Enhances the Tolerance to High Temperature Stress. *J. Integr. Plant Biol.* **2009,** *51* (5), 489–499. https://doi.org/10.1111/j.1744-7909.2009.00816.x

Chowdhury, J. A.; Karim, M. A.; Khaliq, Q. A.; Ahmed, A. U. Effect of Drought Stress on Bio-Chemical Change and Cell Membrane Stability of Soybean Genotypes. *Bangladesh J. Agric. Res.* **2017,** *42* (3), 475–485. https://doi.org/10.3329/bjar.v42i3.34506

Cicchino, M.; Edreira, J. R.; Uribelarrea, M.; Otegui, M. E. Heat Stress in Field-Grown Maize: Response of Physiological Determinants of Grain Yield. *Crop Sci.* **2010,** *50* (4), 1438–1448. https://doi.org/10.2135/cropsci2009.10.0574

Cochard, H.; Martin, R.; Gross, P.; Bogeat-Triboulot, M. B. Temperature Effects on Hydraulic Conductance and Water Relations of Quercus robur L. *J. Exp. Bot.* **2000,** *51* (348), 1255–1259. https://doi.org/10.1093/jexbot/51.348.1255

Corso, D.; Delzon, S.; Lamarque, L. J.; Cochard, H.; Torres-Ruiz, J. M.; King, A.; Brodribb, T. Neither Xylem Collapse, Cavitation, or Changing Leaf Conductance Drive Stomatal Closure in Wheat. *Plant Cell Environ.* **2020,** *43* (4), 854–865. https://doi.org/10.1111/pce.13722

Crafts-Brandner, S. J.; Salvucci, M. E. Sensitivity of Photosynthesis in a C4 Plant, Maize, to Heat Stress. *Plant Physiol.* **2002,** *129* (4), 1773–1780. https://doi.org/10.1104/pp.002170

Crafts-Brandner, S. J.; Salvucci, M. E. Sensitivity of Photosynthesis in a C4 Plant, Maize, to Heat Stress. *Plant Physiol.* **2002,** *129* (4), 1773–1780. https://doi.org/10.1104/pp.002170

Crookshanks, M.; Taylor, G.; Dolan, L. A Model System to Study the Effects of Elevated CO2 on the Developmental Physiology of Roots: The Use of *Arabidopsis thaliana*. *J. Exp. Bot.* **1998,** *49* (320), 593–597. https://doi.org/10.1093/jxb/49.320.593

Cuéllar, T.; Pascaud, F.; Verdeil, J. L.; Torregrosa, L.; Adam-Blondon, A. F.; Thibaud, J. B. et al. A Grapevine Shaker Inward K+ Channel Activated by the Calcineurin B-Like Calcium Sensor 1–Protein Kinase CIPK23 Network Is Expressed in Grape Berries Under Drought Stress Conditions. *Plant J.* **2010,** *61* (1), 58–69. https://doi.org/10.1111/j.1365-313X.2009.04029.x

Cuellar-Ortiz, S. M.; De La Paz Arrieta-Montiel, M. A. R. I. A.; Acosta-Gallegos, J. O. R. G. E.; Covarrubias, A. A. Relationship Between Carbohydrate Partitioning and Drought Resistance in Common Bean. *Plant Cell Environ.* **2008,** *31* (10), 1399–1409. https://doi.org/10.1111/j.1365-3040.2008.01853.x

Da Silva, E. C.; Nogueira, R. J. M. C.; da Silva, M. A.; de Albuquerque, M. B. Drought Stress and Plant Nutrition. *Plant Stress* **2011,** *5*, 32–41.

David, R. H. A.; Ceasar, S. A.; Thirugnanasambantham, K.; Ignacimuthu, S. Genetic Engineering of Crop Plants for Drought Tolerance: Role of Transcription Factors. *Ind. J. Biol. Sci.* **2016,** *2*, 272–286.

de Abreu, I. N.; Mazzafera, P. Effect of Water and Temperature Stress on the Content of Active Constituents of *Hypericum brasiliense* Choisy. *Plant Physiol. Biochem.* **2005,** *43* (3), 241–248. https://doi.org/10.1016/j.plaphy.2005.01.020

Deepak, S. B.; Thakur, A.; Singh, S.; Bakshi, M.; Bansal, S. Changes in Crop Physiology Under Drought Stress: A Review. *J. Pharmacogn. Phytochem.* **2019,** *8*, 1251–1253.

Demirevska-Kepova, K.; Holzer, R.; Simova-Stoilova, L.; Feller, U. Heat Stress Effects on Ribulose-1, 5-Bisphosphate Carboxylase/Oxygenase, Rubisco Binding Protein and Rubisco Activase in Wheat Leaves. *Biol. Plantarum* **2005,** *49* (4), 521–525. https://doi.org/10.1007/s10535-005-0045-2

Dermody, O.; Long, S. P.; DeLucia, E. H. How Does Elevated CO2 or Ozone Affect the Leaf-Area Index of Soybean When Applied Independently? *New Phytol.* **2006,** *169* (1), 145–155. https://doi.org/10.1111/j.1469–8137.2005.01565.x

Devasirvatham, V.; Gaur, P. M.; Mallikarjuna, N.; Tokachichu, R. N.; Trethowan, R. M.; Tan, D. K. Effect of High Temperature on the Reproductive Development of Chickpea Genotypes Under Controlled Environments. *Funct. Plant Biol.* **2012,** *39* (12), 1009–1018. https://doi.org/10.1071/FP12033

Dey, S. K.; Chakrabarti, B.; Prasanna, R.; Pratap, D.; Singh, S. D.; Purakayastha, T. J.; Pathak, H. Elevated Carbon Dioxide Level Along with Phosphorus Application and Cyanobacterial Inoculation Enhances Nitrogen Fixation and Uptake in Cowpea Crop. *Arch. Agron. Soil Sci.* **2017,** *63* (13), 1927–1937. https://doi.org/10.1080/03650340.2017.1315105

Dietterich, L. H.; Zanobetti, A.; Kloog, I.; Huybers, P.; Leakey, A. D.; Bloom, A. J. et al. Impacts of Elevated Atmospheric CO2 on Nutrient Content of Important Food Crops. *Sci. Data* **2015,** *2* (1), 1–8. https://doi.org/10.1038/sdata.2015.36

Djanaguiraman, M.; Prasad, P. V. V.; Al-Khatib, K. Ethylene Perception Inhibitor 1-MCP Decreases Oxidative Damage of Leaves Through Enhanced Antioxidant Defense Mechanisms in Soybean Plants Grown Under High Temperature Stress. *Environ. Exp. Bot.* **2011,** *71* (2), 215–223. https://doi.org/10.1016/j.envexpbot.2010.12.006

Djanaguiraman, M.; Prasad, P. V.; Schapaugh, W. T. High Day-or Nighttime Temperature Alters Leaf Assimilation, Reproductive Success, and Phosphatidic Acid of Pollen Grain in Soybean [Glycine max (L.) Merr.]. *Crop Sci.* **2013,** *53* (4), 1594–1604. https://doi.org/10.2135/cropsci2012.07.0441

Djanaguiraman, M.; Prasad, P. V.; Seppanen, M. Selenium Protects Sorghum Leaves from Oxidative Damage Under High Temperature Stress by Enhancing Antioxidant Defense System. *Plant Physiol. Biochem.* **2010,** *48* (12), 999–1007. https://doi.org/10.1016/j.plaphy.2010.09.009

Dolferus, R.; Ji, X.; Richards, R. A. Abiotic Stress and Control of Grain Number in Cereals. *Plant Sci.* **2011,** *181* (4), 331–341. https://doi.org/10.1016/j.plantsci.2011.05.015

Draeger, T.; C Martin, A.; Alabdullah, A. K.; Pendle, A.; Rey, M. D.; Shaw, P.; Moore, G. Dmc1 Is a Candidate for Temperature Tolerance During Wheat Meiosis. *Theor. Appl. Genet.* **2020,** *133* (3), 809–828. https://doi.org/10.1007/s00122–019–03508–9

Edreira, J. I. R.; Otegui, M. E. Heat Stress in Temperate and Tropical Maize Hybrids: Differences in Crop Growth, Biomass Partitioning and Reserves Use. *Field Crops Res.* **2012,** *130*, 87–98. https://doi.org/10.1016/j.fcr.2012.02.009

Elemike, E. E.; Uzoh, I. M.; Onwudiwe, D. C.; Babalola, O. O. The Role of Nanotechnology in the Fortification of Plant Nutrients and Improvement of Crop Production. *Appl. Sci.* **2019,** *9* (3), 499. https://doi.org/10.3390/app9030499

Engineer, C. B.; Ghassemian, M.; Anderson, J. C.; Peck, S. C.; Hu, H.; Schroeder, J. I. Carbonic Anhydrases, EPF2 and a Novel Protease Mediate CO2 Control of Stomatal Development. *Nature* **2014,** *513* (7517), 246–250. https://doi.org/10.1038/nature13452

Essemine, J.; Ammar, S.; Bouzid, S. Impact of Heat Stress on Germination and Growth in Higher Plants: Physiological, Biochemical and Molecular Repercussions and Mechanisms of Defence. *J. Biol. Sci.* **2010,** *10* (6), 565–572.

Estrada-Campuzano, G.; Miralles, D. J.; Slafer, G. A. Genotypic Variability and Response to Water Stress of Pre-and Post-Anthesis Phases in Triticale. *Eur. J. Agron.* **2008,** *28* (3), 171–177. https://doi.org/10.1016/j.eja.2007.07.005

Farooq, M.; Irfan, M.; Aziz, T.; Ahmad, I.; Cheema, S. A. Seed Priming with Ascorbic Acid Improves Drought Resistance of Wheat. *J. Agron. Crop Sci.* **2013,** *199* (1), 12–22. https://doi.org/10.1111/j.1439–037X.2012.00521.x

Farquhar, G. D.; von Caemmerer, S. V.; Berry, J. A. A Biochemical Model of Photosynthetic CO2 Assimilation in Leaves of C3 Species. *Planta* **1980,** *149* (1), 78–90. https://doi.org/10.1007/BF00386231

Ferguson, J. N.; McAusland, L.; Smith, K. E.; Price, A. H.; Wilson, Z. A.; Murchie, E. H. Rapid Temperature Responses of Photosystem II Efficiency Forecast Genotypic Variation in Rice Vegetative Heat Tolerance. *Plant J.* **2020,** *104* (3), 839–855. https://doi.org/10.1111/tpj.14956

Ferris, R.; Long, L.; Bunn, S. M.; Robinson, K. M.; Bradshaw, H. D.; Rae, A. M.; Taylor, G. Leaf Stomatal and Epidermal Cell Development: Identification of Putative Quantitative Trait Loci in Relation to Elevated Carbon Dioxide Concentration in Poplar. *Tree Physiol.* **2002,** *22* (9), 633–640. https://doi.org/10.1093/treephys/22.9.633

Ferris, R.; Sabatti, M.; Miglietta, F.; Mills, R. F.; Taylor, G. Leaf Area Is Stimulated in Populus by Free Air CO2 Enrichment (POPFACE), Through Increased Cell Expansion and Production. *Plant Cell Environ.* **2001,** *24* (3), 305–315. https://doi.org/10.1046/j.1365–3040.2001.00684.x

Flamholz, A. I.; Prywes, N.; Moran, U.; Davidi, D.; Bar-On, Y. M.; Oltrogge, L. M. et al. Revisiting Trade-Offs Between Rubisco Kinetic Parameters. *Biochemistry* **2019,** *58* (31), 3365–3376. https://doi.org/10.1021/acs.biochem.9b00237

Flexas, J.; Bota, J.; Galmés, J.; Medrano, H.; Ribas-Carbó, M. Keeping a Positive Carbon Balance Under Adverse Conditions: Responses of Photosynthesis and Respiration to Water Stress. *Physiol. Plantarum* **2006,** *127* (3), 343–352. https://doi.org/10.1111/j.1399–3054.2006.00621.x

Flexas, J.; Gallé, A.; Galmés, J.; Ribas-Carbo, M.; Medrano, H. The Response of Photosynthesis to Soil Water Stress. In *Plant Responses to Drought Stress*; Springer: Berlin, Heidelberg, 2012; pp 129–144.

Florian, A.; Nikoloski, Z.; Sulpice, R.; Timm, S.; Araújo, W. L.; Tohge, T. et al. Analysis of Short-Term Metabolic Alterations in Arabidopsis Following Changes in the Prevailing Environmental Conditions. *Mol. Plant* **2014,** *7* (5), 893–911. https://doi.org/10.1093/mp/ssu008

Foolad, M. R. *Breeding for Abiotic Stress Tolerance in Tomato* (No. BOOK); Food Products Press, 2005.

Franks, P. J. Passive and Active Stomatal Control: Either or Both? *New Phytol.* **2013,** *198* (2), 325–327. https://www.jstor.org/stable/newphytologist.198.2.325

Frederick, J. R.; Camp, C. R.; Bauer, P. J. Drought-Stress Effects on Branch and Mainstem Seed Yield and Yield Components of Determinate Soybean. *Crop Sci.* **2001,** *41* (3), 759–763. https://doi.org/10.2135/cropsci2001.413759x

Fukayama, H.; Sugino, M.; Fukuda, T.; Masumoto, C.; Taniguchi, Y.; Okada, M. et al. Gene Expression Profiling of Rice Grown in Free Air CO2 Enrichment (FACE) and Elevated Soil Temperature. *Field Crops Res.* **2011,** *121* (1), 195–199. https://doi.org/10.1016/j.fcr.2010.11.018

Gao, S.; Wang, Y.; Yu, S.; Huang, Y.; Liu, H.; Chen, W.; He, X. Effects of Drought Stress on Growth, Physiology and Secondary Metabolites of Two Adonis Species in Northeast China. *Sci. Horticult.* **2020,** *259*, 108795. https://doi.org/10.1016/j.scienta.2019.108795

García-Caparrós, P.; Romero, M. J.; Llanderal, A.; Cermeño, P.; Lao, M. T.; Segura, M. L. Effects of Drought Stress on Biomass, Essential Oil Content, Nutritional Parameters, and Costs of Production in Six Lamiaceae Species. *Water* **2019,** *11* (3), 573. https://doi.org/10.3390/w11030573

Garrett, K. A.; Dendy, S. P.; Frank, E. E.; Rouse, M. N.; Travers, S. E. Climate Change Effects on Plant Disease: Genomes to Ecosystems, 2006.doi:10.1146/ annurev.phyto.44.070505.143420

Gautam, H. R.; Bhardwaj, M. L.; Kumar, R. Climate Change and Its Impact on Plant Diseases. *Curr. Sci.* **2013,** 1685–1691. https://www.jstor.org/stable/24099750

Ge, T.; Sui, F.; Bai, L.; Tong, C.; Sun, N. Effects of Water Stress on Growth, Biomass Partitioning, and Water-Use Efficiency in Summer Maize (Zea Mays L.) Throughout the Growth Cycle. *Acta Physiol. Plantarum* **2012,** *34* (3), 1043–1053. https://doi.org/10.1007/s11738–011–0901-y

Gharibi, S.; Tabatabaei, B. E. S.; Saeidi, G.; Goli, S. A. H. Effect of Drought Stress on Total Phenolic, Lipid Peroxidation, and Antioxidant Activity of Achillea Species. *Appl. Biochem. Biotechnol.* **2016,** *178* (4), 796–809. https://doi.org/10.1007/s12010–015–1909–3

Gharibi, S.; Tabatabaei, B. E. S.; Saeidi, G.; Talebi, M.; Matkowski, A. The Effect of Drought Stress on Polyphenolic Compounds and Expression of Flavonoid Biosynthesis Related Genes in *Achillea pachycephala* Rech. f. *Phytochemistry* **2019,** *162*, 90–98. https://doi.org/10.1016/j.phytochem.2019.03.004

Ghatak, A.; Chaturvedi, P.; Weckwerth, W. Cereal Crop Proteomics: Systemic Analysis of Crop Drought Stress Responses Towards Marker-Assisted Selection Breeding. *Front. Plant Sci.* **2017,** *8*, 757. https://doi.org/10.3389/fpls.2017.00757

Gheidary, S.; Akhzari, D.; Pessarakli, M. Effects of Salinity, Drought, and Priming Treatments on Seed Germination and Growth Parameters of *Lathyrus sativus* L. *J. Plant Nutr.* **2017,** *40* (10), 1507–1514. https://doi.org/10.1080/01904167.2016.1269349

Gladish, D. K.; Rost, T. L. The Effects of Temperature on Primary Root Growth Dynamics and Lateral Root Distribution in Garden Pea (*Pisum sativum* L.; cv. "Alaska"). *Environ. Exp. Bot.* **1993,** *33* (2), 243–258. https://doi.org/10.1016/0098–8472 (93)90070-V

Goche, T.; Shargie, N. G.; Cummins, I.; Brown, A. P.; Chivasa, S.; Ngara, R. Comparative Physiological and Root Proteome Analyses of Two Sorghum Varieties Responding to Water Limitation. *Sci. Rep.* **2020,** *10* (1), 1–18. https://doi.org/10.1038/s41598–020–68735–3

Gou, W. E. I.; Tian, L. I.; Ruan, Z. H. I.; Zheng, P. E. N. G.; Chen, F. U. C. A. I.; Zhang, L. et al. Accumulation of Choline and Glycinebetaine and Drought Stress Tolerance Induced in

Maize (*Zea mays*) by Three Plant Growth Promoting Rhizobacteria (PGPR) Strains. *Pak. J. Bot.* **2015,** *47* (2), 581–586.

Gray, J. E.; Holroyd, G. H.; Van Der Lee, F. M.; Bahrami, A. R.; Sijmons, P. C.; Woodward, F. I. et al. The HIC Signalling Pathway Links CO2 Perception to Stomatal Development. *Nature* **2000,** *408* (6813), 713–716. https://doi.org/10.1038/35047071

Gray, S. B.; Brady, S. M. Plant Developmental Responses to Climate Change. *Dev. Biol.* **2016,** *419* (1), 64–77. https://doi.org/10.1016/j.ydbio.2016.07.023

Gray, S. B.; Brady, S. M. Plant Developmental Responses to Climate Change. *Dev. Biol.* **2016,** *419* (1), 64–77. https://doi.org/10.1016/j.ydbio.2016.07.023

Gray, S. B.; Dermody, O.; Klein, S. P.; Locke, A. M.; Mcgrath, J. M.; Paul, R. E. et al. Intensifying Drought Eliminates the Expected Benefits of Elevated Carbon Dioxide for Soybean. *Nat. Plants* **2016,** *2* (9), 1–8. https://doi.org/10.1038/nplants.2016.132

Gray, S. B.; Strellner, R. S.; Puthuval, K. K.; Ng, C.; Shulman, R. E.; Siebers, M. H. et al. Minirhizotron Imaging Reveals That Nodulation of Field-Grown Soybean Is Enhanced by Free-Air CO2 Enrichment Only When Combined with Drought Stress. *Funct. Plant Biol.* **2012,** *40* (2), 137–147. https://doi.org/10.1071/FP12044

Greer, D. H.; Weedon, M. M. Modelling Photosynthetic Responses to Temperature of Grapevine (*Vitis vinifera* cv. Semillon) Leaves on Vines Grown in a Hot Climate. *Plant Cell Environ.* **2012,** *35* (6), 1050–1064. https://doi.org/10.1111/j.1365–3040.2011.02471.x

Grossiord, C.; Buckley, T. N.; Cernusak, L. A.; Novick, K. A.; Poulter, B.; Siegwolf, R. T. et al. Plant Responses to Rising Vapor Pressure Deficit. *New Phytol.* **2020,** *226* (6), 1550–1566. https://doi.org/10.1111/nph.16485

Guilioni, L.; Wéry, J.; Lecoeur, J. High Temperature and Water Deficit May Reduce Seed Number in Field Pea Purely by Decreasing Plant Growth Rate. *Funct. Plant Biol.* **2003,** *30* (11), 1151–1164. https://doi.org/10.1071/FP03105

Guilioni, L.; Wery, J.; Tardieu, F. Heat Stress-Induced Abortion of Buds and Flowers in Pea: Is Sensitivity Linked to Organ Age or to Relations Between Reproductive Organs? *Ann. Bot.* **1997,** *80* (2), 159–168. https://doi.org/10.1006/anbo.1997.0425

Gunawardhana, M. D. M.; De Silva, C. S. Impact of Temperature and Water Stress on Growth Yield and Related Biochemical Parameters of Okra, 2011.

Guo, Y. P.; Zhou, H. F.; Zhang, L. C. Photosynthetic Characteristics and Protective Mechanisms Against Photooxidation During High Temperature Stress in Two Citrus Species. *Sci. Horticult.* **2006,** *108* (3), 260–267. https://doi.org/10.1016/j.scienta.2006.01.029

Gurumurthy, S.; Sarkar, B.; Vanaja, M.; Lakshmi, J.; Yadav, S. K.; Maheswari, M. Morpho-Physiological and Biochemical Changes in Black Gram (*Vigna mungo* L. Hepper) Genotypes Under Drought Stress at Flowering Stage. *Acta Physiol. Plantarum* **2019,** *41* (3), 1–14. https://doi.org/10.1007/s11738–019–2833-x

Hafez, E. M.; Seleiman, M. F. Response of Barley Quality Traits, Yield and Antioxidant Enzymes to Water-Stress and Chemical Inducers. *Int. J. Plant Product.* **2017,** *11* (4), 477–490. DOI: 10.22069/IJPP.2017.3712

Hafez, E. M.; Abou El Hassan, W. H.; Gaafar, I. A.; Seleiman, M. F. Effect of Gypsum Application and Irrigation Intervals on Clay Saline-Sodic Soil Characterization, Rice Water Use Efficiency, Growth, and Yield. *J. Agric. Sci.* **2015,** *7* (12), 208. http://dx.doi.org/10.5539/jas.v7n12p208

Halliwell, B. Oxidative Stress and Neurodegeneration: Where Are We Now? *J. Neurochem.* **2006,** *97* (6), 1634–1658.https://doi.org/10.1111/j.1471–4159.2006.03907.x

Hasanuzzaman, M.; Hakim, K. R.; Nahar, K.; Alharby, H. F. *Plant Abiotic Stress Tolerance*; Springer: Cham, 2019.

Hasanuzzaman, M.; Nahar, K.; Fujita, M. Extreme Temperature Responses, Oxidative Stress and Antioxidant Defense in Plants. *Abiotic Stress-Plant Responses App. Agric.* **2013,** *13*, 169–205.

Hasanuzzaman, M.; Nahar, K.; Fujita, M. Plant Response to Salt Stress and Role of Exogenous Protectants to Mitigate Salt-Induced Damages. In *Ecophysiology and Responses of Plants Under Salt Stress*; Springer: New York, 2013; pp 25–87. DOI: 10.1007/978-1-4614-4747-4_2

Hasanuzzaman, M.; Nahar, K.; Alam, M. M.; Roychowdhury, R.; Fujita, M. Physiological, Biochemical, and Molecular Mechanisms of Heat Stress Tolerance in Plants. *Int. J. Mol. Sci.* **2013,** *14* (5), 9643–9684. https://doi.org/10.3390/ijms14059643

Hatfield, J. L.; Boote, K. J.; Kimball, B. A.; Ziska, L. H.; Izaurralde, R. C.; Ort, D. et al. Climate Impacts on Agriculture: Implications for Crop Production. *Agron. J.* **2011,** *103* (2), 351–370. https://doi.org/10.2134/agronj2010.0303

Hay, R.; Porter, J. *The Physiology of Crop Yield*, 2nd ed.; Blackwell Publishing: Oxford, 2006.

Helyes, L.; Lugasi, A.; Neményi, A.; Pék, Z. The Simultaneous Effect of Elevated CO2-Level and Nitrogen-Supply on the Fruit Components of Tomato. *Acta Alimentaria* **2012,** *41* (2), 265–271. https://doi.org/10.1556/aalim.41.2012.2.13

Hoegh-Guldberg, O.; Jacob, D.; Bindi, M.; Brown, S.; Camilloni, I.; Diedhiou, A. et al. Impacts of 1.5 C Global Warming on Natural and Human Systems. *Glob. Warm. 1.5°C*, 2018. http://hdl.handle.net/10138/311749

Högy, P.; Poll, C.; Marhan, S.; Kandeler, E.; Fangmeier, A. Impacts of Temperature Increase and Change in Precipitation Pattern on Crop Yield and Yield Quality of Barley. *Food Chem.* **2013,** *136* (3–4), 1470–1477. https://doi.org/10.1016/j.foodchem.2012.09.056

Hoogenboom, G.; Peterson, C. M.; Huck, M. G. Shoot Growth Rate of Soybean as Affected by Drought Stress 1. *Agron. J.* **1987,** *79* (4), 598–607. https://doi.org/10.2134/agronj1987.00021962007900040003x

Hossain, Z.; Nouri, M. Z.; Komatsu, S. Plant Cell Organelle Proteomics in Response to Abiotic Stress. *J. Proteome Res.* **2012,** *11* (1), 37–48. https://doi.org/10.1021/pr200863r

Hosseini, M. S.; Samsampour, D.; Ebrahimi, M.; Abadía, J.; Khanahmadi, M. Effect of Drought Stress on Growth Parameters, Osmolyte Contents, Antioxidant Enzymes and Glycyrrhizin Synthesis in Licorice (*Glycyrrhiza glabra* L.) Grown in the Field. *Phytochemistry* **2018,** *156*, 124–134. https://doi.org/10.1016/j.phytochem.2018.08.018

Hovenden, M.; Newton, P. Plant Responses to CO2 Are a Question of Time. *Science* **2018,** *360* (6386), 263–264. https://doi.org/10.1126/science.aat2481

Hu, L.; Wang, Z.; Huang, B. Effects of Cytokinin and Potassium on Stomatal and Photosynthetic Recovery of Kentucky Bluegrass from Drought Stress. *Crop Sci.* **2013,** *53* (1), 221–231. https://doi.org/10.2135/cropsci2012.05.0284

Hu, S.; Tong, K.; Chen, W.; Wang, Y.; Wang, Y.; Yang, L. Response of Rice Grain Quality to Elevated Atmospheric CO2 Concentration: A Meta-Analysis of 20-Year FACE Studies. *Field Crops Res.* **2002,** *284*, 108562.

Hu, Y.; Schmidhalter, U. Drought and Salinity: A Comparison of Their Effects on Mineral Nutrition of Plants. *J. Plant Nutr. Soil Sci.* **2005,** *168* (4), 541–549. https://doi.org/10.1002/jpln.200420516

Huang, B.; Xu, C. Identification and Characterization of Proteins Associated with Plant Tolerance to Heat Stress. *J. Integr. Plant Biol.* **2008,** *50* (10), 1230–1237.https://doi.org/10.1111/j.1744–7909.2008.00735.x

Humplík, J. F.; Bergougnoux, V.; Van Volkenburgh, E. To Stimulate or Inhibit? That Is the Question for the Function of Abscisic Acid. *Trends Plant Sci.* **2017,** *22* (10), 830–841. https://doi.org/10.1016/j.tplants.2017.07.009

Hund, A.; Fracheboud, Y.; Soldati, A.; Stamp, P. Cold Tolerance of Maize Seedlings as Determined by Root Morphology and Photosynthetic Traits. *Eur. J. Agron.* **2008,** *28* (3), 178–185. https://doi.org/10.1016/j.eja.2007.07.003

Hurkman, W. J.; Vensel, W. H.; Tanaka, C. K.; Whitehand, L.; Altenbach, S. B. Effect of High Temperature on Albumin and Globulin Accumulation in the Endosperm Proteome of the Developing Wheat Grain. *J. Cereal Sci.* **2009,** *49* (1), 12–23. https://doi.org/10.1016/j.jcs.2008.06.014

Huseynova, I. M.; Rustamova, S. M.; Suleymanov, S. Y.; Aliyeva, D. R.; Mammadov, A. C.; Aliyev, J. A. Drought-Induced Changes in Photosynthetic Apparatus and Antioxidant Components of Wheat (*Triticum durum* Desf.) Varieties. *Photosynth. Res.* **2016,** *130* (1), 215–223. https://doi.org/10.1007/s11120–016–0244-z

Hussain, S.; Rao, M. J.; Anjum, M. A.; Ejaz, S.; Zakir, I.; Ali, M. A. et al. Oxidative Stress and Antioxidant Defense in Plants Under Drought Conditions. In *Plant Abiotic Stress Tolerance*; Springer: Cham, 2019; pp 207–219.Impa, S. M.; Nadaradjan, S.; Jagadish, S. V. K. Drought Stress Induced Reactive Oxygen Species and Anti-Oxidants in Plants. In *Abiotic Stress Responses in Plants*; Springer: New York, 2012; pp 131–147.

Islam, M. S.; Matsui, T.; Yoshida, Y. Effect of Carbon Dioxide Enrichment on Physico-Chemical and Enzymatic Changes in Tomato Fruits at Various Stages of Maturity. *Sci. Horticult.* **1996,** *65* (2–3), 137–149. https://doi.org/10.1016/0304-4238 (95)00867-5

Jaafar, H. Z.; Ibrahim, M. H.; Fakri, N. F. M. Impact of Soil Field Water Capacity on Secondary Metabolites, Phenylalanine Ammonia-Lyase (PAL), Maliondialdehyde (MDA) and Photosynthetic Responses of Malaysian Kacip Fatimah (Labisia pumila Benth). *Molecules* **2012,** *17* (6), 7305–7322. https://doi.org/10.3390/molecules17067305

Jaafar, H. Z.; Ibrahim, M. H.; Karimi, E. Phenolics and Flavonoids Compounds, Phenylanine Ammonia Lyase and Antioxidant Activity Responses to Elevated CO2 in *Labisia pumila* (Myrisinaceae). *Molecules* **2012,** *17* (6), 6331–6347. https://doi.org/10.3390/molecules17066331

Jablonski, L. M.; Wang, X.; Curtis, P. S. Plant Reproduction Under Elevated CO2 Conditions: A Meta-Analysis of Reports on 79 Crop and Wild Species. *New Phytol.* **2002,** *156* (1), 9–26. https://doi.org/10.1046/j.1469-8137.2002.00494.x

Jagadish, S. K.; Bahuguna, R. N.; Djanaguiraman, M.; Gamuyao, R.; Prasad, P. V.; Craufurd, P. Q. Implications of High Temperature and Elevated CO2 on Flowering Time in Plants. *Front. Plant Sci.* **2016,** *7*, 913. https://doi.org/10.3389/fpls.2016.00913

Jarvi, M. P.; Burton, A. J. Root Respiration and Biomass Responses to Experimental Soil Warming Vary with Root Diameter and Soil Depth. *Plant Soil* **2020,** *451* (1), 435–446. https://doi.org/10.1007/s11104-020-04540-1

Jeuffroy, M. H.; Duthion, C.; Meynard, J. M.; Pigeaire, A. Effect of a Short Period of High Day Temperatures During Flowering on the Seed Number Per Pod of Pea (*Pisum sativum* L). *Agronomie* **1990,** *10* (2), 139–145. https://hal.archives-ouvertes.fr/hal-00885276

Jiang, Y. *Effect of Heat Stress on Pollen Development and Seed Set in Field Pea (Pisum sativum L.)* (Doctoral dissertation, University of Saskatchewan), 2016.

Jiang, Y.; Bueckert, R. A.; Warkentin, T. D.; Davis, A. R. High Temperature Effects on In Vitro Pollen Germination and Seed Set in Field Pea. *Can. J. Plant Sci.* **2017,** *98* (1), 71–80. https://doi.org/10.1139/cjps-2017–0073

Jiang, Y.; Lahlali, R.; Karunakaran, C.; Kumar, S.; Davis, A. R.; Bueckert, R. A. Seed Set, Pollen Morphology and Pollen Surface Composition Response to Heat Stress in Field Pea. *Plant Cell Environ.* **2015,** *38* (11), 2387–2397. https://doi.org/10.1111/pce.12589

Jiang, Y.; Lahlali, R.; Karunakaran, C.; Warkentin, T. D.; Davis, A. R.; Bueckert, R. A. Pollen, Ovules, and Pollination in Pea: Success, Failure, and Resilience in Heat. *Plant Cell Environ.* **2019,** *42* (1), 354–372. https://doi.org/10.1111/pce.13427

Jin, C.; Du, S.; Wang, Y.; Condon, J.; Lin, X.; Zhang, Y. Carbon Dioxide Enrichment by Composting in Greenhouses and Its Effect on Vegetable Production. *J. Plant Nutr. Soil Sci.* **2009,** *172* (3), 418–424. https://doi.org/10.1002/jpln.200700220

Johkan, M.; Oda, M.; Maruo, T.; Shinohara, Y. Crop Production and Global Warming. In *Global Warming Impacts—Case Studies on the Economy, Human Health, and on Urban and NaturalEnvironments*; Casalegno, S.; Ed.; InTech: Rijeka, Croatia, 2011; pp 139–152.

Johnson, S. N.; Ryalls, J. M.; Gherlenda, A. N.; Frew, A.; Hartley, S. E. Benefits from Below: Silicon Supplementation Maintains Legume Productivity Under Predicted Climate Change Scenarios. *Front. Plant Sci.* **2018,** *9*, 202. https://doi.org/10.3389/fpls.2018.00202

Joshi, M.; Fogelman, E.; Belausov, E.; Ginzberg, I. Potato Root System Development and Factors that Determine Its Architecture. *J. Plant Physiol.* **2016,** *205*, 113–123. https://doi.org/10.1016/j.jplph.2016.08.014

Jungqvist, G.; Oni, S. K.; Teutschbein, C.; Futter, M. N. Effect of Climate Change on Soil Temperature in Swedish Boreal Forests. *PloS One* **2014,** *9* (4), e93957. https://doi.org/10.1371/journal.pone.0093957

Kalra, N.; Chakraborty, D.; Sharma, A.; Rai, H. K.; Jolly, M.; Chander, S. et al. Effect of Increasing Temperature on Yield of Some Winter Crops in Northwest India. *Curr. Sci.* **2008,** 82–88. https://www.jstor.org/stable/24102032

Kamara, A. Y.; Menkir, A.; Badu-Apraku, B.; Ibikunle, O. The Influence of Drought Stress on Growth, Yield and Yield Components of Selected Maize Genotypes. *J. Agric. Sci.* **2003,** *141* (1), 43–50. https://doi.org/10.1017/S0021859603003423

Kang, G. Z.; Li, G. Z.; Liu, G. Q.; Xu, W.; Peng, X. Q.; Wang, C. Y. et al. Exogenous Salicylic Acid Enhances Wheat Drought Tolerance by Influence on the Expression of Genes Related to Ascorbate-Glutathione Cycle. *Biol. Plantarum* **2013,** *57* (4), 718–724. https://doi.org/10.1007/s10535-013-0335-z

Karimpour, M. Effect of Drought Stress on RWC and Chlorophyll Content on Wheat (*Triticum durum* L.) Genotypes. *World. Ess. J.* **2019,** *7*, 52–56.

Karuppanapandian, T.; Moon, J. C.; Kim, C.; Manoharan, K.; Kim, W. Reactive Oxygen Species in Plants: Their Generation, Signal Transduction, and Scavenging Mechanisms. *Aust. J. Crop Sci.* **2011,** *5* (6), 709–725.

Karuppanapandian, T.; Wang, H. W.; Prabakaran, N.; Jeyalakshmi, K.; Kwon, M.; Manoharan, K.; Kim, W. 2, 4-Dichlorophenoxyacetic Acid-Induced Leaf Senescence in Mung Bean (*Vigna radiata* L. Wilczek) and Senescence Inhibition by Co-Treatment with Silver Nanoparticles. *Plant Physiol. Biochem.* **2011,** *49* (2), 168–177. https://doi.org/10.1016/j.plaphy.2010.11.007

Kattge, J.; Knorr, W. Temperature Acclimation in a Biochemical Model of Photosynthesis: A Reanalysis of Data from 36 Species. *Plant Cell Environ.* **2007,** *30* (9), 1176–1190. https://doi.org/10.1111/j.1365–3040.2007.01690.x

Kawakami, J.; Iwama, K.; Jitsuyama, Y. Soil Water Stress and the Growth and Yield of Potato Plants Grown from Microtubers and Conventional Seed Tubers. *Field Crops Res.* **2006,** *95* (1), 89–96.

Keyvan, S. The Effects of Drought Stress on Yield, Relative Water Content, Proline, Soluble Carbohydrates and Chlorophyll of Bread Wheat Cultivars. *J. Anim. Plant Sci.* **2010,** *8* (3), 1051–1060.

Khalid, K. A. Influence of Water Stress on Growth, Essential Oil, and Chemical Composition of Herbs [Ocimum sp.]. *Int. Agrophys.* **2006,** *20* (4).

Khan, A.; Pan, X.; Najeeb, U.; Tan, D. K. Y.; Fahad, S.; Zahoor, R.; Luo, H. Coping with Drought: Stress and Adaptive Mechanisms, and Management Through Cultural and Molecular Alternatives in Cotton as Vital Constituents for Plant Stress Resilience and Fitness. *Biol. Res.* **2018,** 51. http://dx.doi.org/10.1186/s40659-018-0198-z

Kheradmand, M. A.; Fahraji, S. S.; Fatahi, E.; Raoofi, M. M. Effect of Water Stress on Oil Yield and Some Characteristics of *Brassica napus*. *Int. Res. J. Appl. Basic Sci.* **2014,** *8* (9), 1447–1453.

Kimball, B. A. Crop Responses to Elevated CO2 and Interactions with H2O, N, and Temperature. *Curr. Opin. Plant Biol.* **2016,** *31*, 36–43. https://doi.org/10.1016/j.pbi.2016.03.006

Kleinwächter, M.; Selmar, D. New Insights Explain That Drought Stress Enhances the Quality of Spice and Medicinal Plants: Potential Applications. *Agron. Sustain. Dev.* **2015,** *35* (1), 121–131. https://doi.org/10.1007/s13593-014-0260-3

Koini, M. A.; Alvey, L.; Allen, T.; Tilley, C. A.; Harberd, N. P.; Whitelam, G. C.; Franklin, K. A. High Temperature-Mediated Adaptations in Plant Architecture Require the bHLH Transcription Factor PIF4. *Curr. Biol.* **2009,** *19* (5), 408–413. https://doi.org/10.1016/j.cub.2009.01.046

Kumar, A.; Omae, H.; Egawa, Y.; Kashiwaba, K.; Shono, M. Some Physiological Responses of Snap Bean (*Phseolus vulgaris* L.) to Water Stress During Reproductive Period. In *Proceedings of the International Conference on Sustainable Crop Production in Stress Environment: Management and Genetic Option*; JNKVV: Jabarpur, India, 2005; pp 226–227.

Kumar, S.; Kaur, R.; Kaur, N.; Bhandhari, K.; Kaushal, N.; Gupta, K. et al. Heat-Stress Induced Inhibition in Growth and Chlorosis in Mungbean (*Phaseolus aureus* Roxb.) Is Partly Mitigated by Ascorbic Acid Application and Is Related to Reduction in Oxidative Stress. *Acta Physiologiae Plantarum* **2011,** *33* (6), 2091–2101. https://doi.org/10.1007/s11738–011–0748–2

Kumarathunge, D. P.; Medlyn, B. E.; Drake, J. E.; Tjoelker, M. G.; Aspinwall, M. J.; Battaglia, M. et al. Acclimation and Adaptation Components of the Temperature Dependence of Plant Photosynthesis at the Global Scale. *New Phytol.* **2019,** *222* (2), 768–784. https://doi.org/10.1111/nph.15668

Kumarathunge, D. P.; Medlyn, B. E.; Drake, J. E.; Tjoelker, M. G.; Aspinwall, M. J.; Battaglia, M. et al. Acclimation and Adaptation Components of the Temperature Dependence of Plant Photosynthesis at the Global Scale. *New Phytol.* **2019,** *222* (2), 768–784. https://doi.org/10.1111/nph.15668

Kumari, S.; Agrawal, M. Growth, Yield and Quality Attributes of a Tropical Potato Variety (*Solanum tuberosum* L. cv *Kufri chandramukhi*) Under Ambient and Elevated Carbon Dioxide and Ozone and Their Interactions. *Ecotoxicol. Environ. Saf.* **2014,** *101*, 146–156. https://doi.org/10.1016/j.ecoenv.2013.12.021

Kutcher, H. R.; Warland, J. S.; Brandt, S. A. Temperature and Precipitation Effects on Canola Yields in Saskatchewan, Canada. *Agric. Forest Meteorol.* **2010,** *150* (2), 161–165. https://doi.org/10.1016/j.agrformet.2009.09.011

Lafitte, H. R.; Yongsheng, G.; Yan, S.; Li, Z. K. Whole Plant Responses, Key Processes, and Adaptation to Drought Stress: The Case of Rice. *J. Exp. Bot.* **2007,** *58* (2), 169–175. https://doi.org/10.1093/jxb/erl101

Lai, S. K.; Zhuang, S. T.; Wu, Y. Z. et al. Impact of Elevated Atmospheric CO2 Concentration and Temperature on Growth and Development of Super Rice. *Chinese J. Ecol.* **2015,** *34,* 1253–1262.

Lam, V. P.; Kim, S. J.; Bok, G. J.; Lee, J. W.; Park, J. S. The Effects of Root Temperature on Growth, Physiology, and Accumulation of Bioactive Compounds of *Agastache rugosa. Agriculture* **2020,** *10* (5), 162.doi:10.3390/agriculture10050162

Lata, C.; Prasad, M. Role of DREBs in Regulation of Abiotic Stress Responses in Plants. *J. Exp. Bot.* **2011,** *62* (14), 4731–4748. https://doi.org/10.1093/jxb/err210

Leakey, A. D.; Xu, F.; Gillespie, K. M.; McGrath, J. M.; Ainsworth, E. A.; Ort, D. R. Genomic Basis for Stimulated Respiration by Plants Growing Under Elevated Carbon Dioxide. *Proc. Natl. Acad. Sci.* **2009,** *106* (9), 3597–3602. https://doi.org/10.1073/pnas.0810955106

Li, X.; Kang, S.; Li, F.; Zhang, X.; Huo, Z.; Ding, R. et al. Light Supplement and Carbon Dioxide Enrichment Affect Yield and Quality of Off-Season Pepper. *Agron. J.* **2017,** *109* (5), 2107–2118. https://doi.org/10.2134/agronj2017.01.0044

Li, Y.; Sun, C.; Huang, Z.; Pan, J.; Wang, L.; Fan, X. Mechanisms of Progressive Water Deficit Tolerance and Growth Recovery of Chinese Maize Foundation Genotypes Huangzao 4 and Chang 7–2, Which Are Proposed on the Basis of Comparison of Physiological and Transcriptomic Responses. *Plant Cell Physiol.* **2009,** *50* (12), 2092–2111. https://doi.org/10.1093/pcp/pcp145

Li, Y.; Yu, Z.; Liu, X.; Mathesius, U.; Wang, G.; Tang, C. et al. Elevated CO2 Increases Nitrogen Fixation at the Reproductive Phase Contributing to Various Yield Responses of Soybean Cultivars. *Front. Plant Sci.* **2017,** *8,* 1546. https://doi.org/10.3389/fpls.2017.01546

Lobell, D. B.; Schlenker, W.; Costa-Roberts, J. Climate Trends and Global Crop Production Since 1980. *Science* **2011,** *333* (6042), 616–620. https://doi.org/10.1126/science.1204531

Loladze, I. Hidden Shift of the Ionome of Plants Exposed to Elevated CO2 Depletes Minerals at the Base of Human Nutrition. *elife* **2014,** *3,* e02245. https://doi.org/10.7554/eLife.02245

Long, S. P.; Ainsworth, E. A.; Rogers, A.; Ort, D. R. Rising atmospheric carbon dioxide: plants FACE the future. *Annu. Rev. Plant Biol.* **2004,** *55,* 591.doi: 10.1146/annurev.arplant.55.031903.141610

Loutfy, N.; El-Tayeb, M. A.; Hassanen, A. M.; Moustafa, M. F.; Sakuma, Y.; Inouhe, M. Changes in the Water Status and Osmotic Solute Contents in Response to Drought and Salicylic Acid Treatments in Four Different Cultivars of Wheat (*Triticum aestivum*). *J. Plant Res.* **2012,** *125* (1), 173–184. https://doi.org/10.1007/s10265–011–0419–9

Lu, Z.; Quiñones, M. A.; Zeiger, E. Temperature Dependence of Guard Cell Respiration and Stomatal Conductance Co-Segregate in an F2 Population of Pima Cotton. *Funct. Plant Biol.* **2000,** *27* (5), 457–462. https://doi.org/10.1071/PP98128

Lundholm, J. T. Plant Species Diversity and Environmental Heterogeneity: Spatial Scale and Competing Hypotheses. *J. Veg. Sci.* **2009,** *20* (3), 377–391. https://doi.org/10.1111/j.1654–1103.2009.05577.x

Luo, H.; Xu, H.; Chu, C.; He, F.; Fang, S. High Temperature Can Change Root System Architecture and Intensify Root Interactions of Plant Seedlings. *Front. Plant Sci.* **2020,** *11,* 160. https://doi.org/10.3389/fpls.2020.00160

Luo, M.; Liang, X. Q.; Dang, P.; Holbrook, C. C.; Bausher, M. G.; Lee, R. D.; Guo, B. Z. Microarray-Based Screening of Differentially Expressed Genes in Peanut in Response to

Aspergillus parasiticus Infection and Drought Stress. *Plant Sci.* **2005,** *169* (4), 695–703. https://doi.org/10.1016/j.plantsci.2005.05.020

Ma, X.; Sukiran, N. L.; Ma, H.; Su, Z. Moderate Drought Causes Dramatic Floral Transcriptomic Reprogramming to Ensure Successful Reproductive Development in Arabidopsis. *BMC Plant Biol.* **2014,** *14* (1), 1–16. https://doi.org/10.1186/1471–2229–14–164

Madhu, M.; Hatfield, J. L. Dynamics of Plant Root Growth Under Increased Atmospheric Carbon Dioxide. *Agron. J.* **2013,** *105* (3), 657–669.https://doi.org/10.2134/agronj2013.0018

Mafakheri, A.; Siosemardeh, A. F.; Bahramnejad, B.; Struik, P. C.; Sohrabi, Y. Effect of Drought Stress on Yield, Proline and Chlorophyll Contents in Three Chickpea Cultivars. *Aust. J. Crop Sci.* **2010,** *4* (8), 580–585.

Mahdid, M.; Kameli, A.; Ehlert, C.; Simonneau, T. Rapid Changes in Leaf Elongation, ABA and Water Status During the Recovery Phase Following Application of Water Stress in Two Durum Wheat Varieties Differing in Drought Tolerance. *Plant Physiol. Biochem.* **2011,** *49* (10), 1077–1083. https://doi.org/10.1016/j.plaphy.2011.08.002

Maheswari, M.; Yadav, S. K.; Shanker, A. K.; Kumar, M. A.; Venkateswarlu, B. Overview of Plant Stresses: Mechanisms, Adaptations and Research Pursuit. In *Crop Stress and Its Management: Perspectives and Strategies*; Springer: Dordrecht, 2012; pp 1–18. DOI: 10.1007/978-94-007-2220-0_1

Maleki, M.; Ebrahimzade, H.; Gholami, M.; Niknam, V. The Effect of Drought Stress and Exogenous Abscisic Acid on Growth, Protein Content and Antioxidative Enzyme Activity in Saffron (*Crocus sativus* L.). *Afr. J. Biotechnol.* **2011,** *10* (45), 9068–9075. DOI: 10.5897/AJB10.676

Manderscheid, R.; Bender, J.; Jäger, H. J.; Weigel, H. J. Effects of Season Long CO2 Enrichment on Cereals: II—Nutrient Concentrations and Grain Quality. *Agric. Ecosyst. Environ.* **1995,** *54* (3), 175–185. https://doi.org/10.1016/0167–8809 (95)00602-O

Marchand, F. L.; Mertens, S.; Kockelbergh, F.; Beyens, L.; Nijs, I. Performance of High Arctic Tundra Plants Improved During But Deteriorated After Exposure to a Simulated Extreme Temperature Event. *Glob. Change Biol.* **2005,** *11* (12), 2078–2089. https://doi.org/10.1111/j.1365–2486.2005.01046.x

Marcińska, I.; Czyczyło-Mysza, I.; Skrzypek, E.; Filek, M.; Grzesiak, S.; Grzesiak, M. T. et al. Impact of Osmotic Stress on Physiological and Biochemical Characteristics in Drought-Susceptible and Drought-Resistant Wheat Genotypes. *Acta Physiol. Plantarum, 35* (2), 451–461. https://doi.org/10.1007/s11738-012-1088-6

Martínez, J. P.; Silva, H. F. L. J.; Ledent, J. F.; Pinto, M. Effect of Drought Stress on the Osmotic Adjustment, Cell Wall Elasticity and Cell Volume of Six Cultivars of Common Beans (*Phaseolus vulgaris* L.). *Eur. J. Agron.* **2007,** *26* (1), 30–38. https://doi.org/10.1016/j.eja.2006.08.003

Martínez-Vilalta, J.; Garcia-Forner, N. Water Potential Regulation, Stomatal Behaviour and Hydraulic Transport Under Drought: Deconstructing the Iso/Anisohydric Concept. *Plant Cell Environ.* **2017,** *40* (6), 962–976. https://doi.org/10.1111/pce.12846

Masle, J. The Effects of Elevated CO2 Concentrations on Cell Division Rates, Growth Patterns, and Blade Anatomy in Young Wheat Plants Are Modulated by Factors Related to Leaf Position, Vernalization, and Genotype. *Plant Physiol.* **2000,** *122* (4), 1399–1416. https://doi.org/10.1104/pp.122.4.1399

Masoud, S. M. The Effects of Water Deficit During Growth Stages of Canola (*Brassica napus* L.), 2007.

Matthews, J. S.; Lawson, T. Climate Change and Stomatal Physiology. *Annu. Plant Rev. Online* **2018,** 713–752. https://doi.org/10.1002/9781119312994.apr0667

Mazaheri, L. H.; Nouri, F.; Zare, A. H. Effects of the Reduction of drought stress using Supplementary Irrigation for Sunflower (*Helianthus annuus*) in Dry Farming Conditions, 2003.

McAdam, S. A.; Brodribb, T. J. Separating Active and Passive Influences on Stomatal Control of Transpiration. *Plant Physiol.* **2014,** *164* (4), 1578–1586. https://doi.org/10.1104/pp.113.231944

McClung, C. R.; Davis, S. J. Ambient Thermometers in Plants: From Physiological Outputs Towards Mechanisms of Thermal Sensing. *Curr. Biol.* **2010,** *20* (24), R1086-R1092. https://doi.org/10.1016/j.cub.2010.10.035

McMichael, B. L.; Quisenberry, J. E. The Impact of the Soil Environment on the Growth of Root Systems. *Environ. Exp. Bot.* **1993,** *33* (1), 53–61. https://doi.org/10.1016/0098–8472(93)90055-K

Merilo, E.; Yarmolinsky, D.; Jalakas, P.; Parik, H.; Tulva, I.; Rasulov, B. et al. Stomatal VPD Response: There Is More to the Story Than ABA. *Plant Physiol.* **2018,** *176* (1), 851–864. https://doi.org/10.1104/pp.17.00912

Miller, G.; Schlauch, K.; Tam, R.; Cortes, D.; Torres, M. A.; Shulaev, V. et al. The Plant NADPH Oxidase RBOHD Mediates Rapid Systemic Signaling in Response to Diverse Stimuli. *Sci. Signal.* **2009,** *2* (84), ra45–ra45. https://doi.org/10.1126/scisignal.2000448

Miranda, M. T.; Da Silva, S. F.; Silveira, N. M.; Pereira, L.; Machado, E. C.; Ribeiro, R. V. Root Osmotic Adjustment and Stomatal Control of Leaf Gas Exchange Are Dependent on Citrus Rootstocks Under Water Deficit. *J. Plant Growth Regul.* **2021,** *40* (1), 11–19. https://doi.org/10.1007/s00344-020-10069-5

Miranda-Apodaca, J.; Pérez-López, U.; Lacuesta, M.; Mena-Petite, A.; Muñoz-Rueda, A. The Interaction Between Drought and Elevated CO2 in Water Relations in Two Grassland Species Is Species-Specific. *J. Plant Physiol.* **2018,** *220*, 193–202. https://doi.org/10.1016/j.jplph.2017.11.006

Mirza, H.; Hossain, M. A.; Fujita, M. Physiological and Biochemical Mechanisms of Nitric Oxide Induced Abiotic Stress Tolerance in Plants. *Am. J. Plant Physiol.* **2010,** *5* (6), 295–324.DOI : 10.3923/ajpp.2010.295.324

Mitra, R.; Bhatia, C. R. Bioenergetic Cost of Heat Tolerance in Wheat Crop. *Curr. Sci.* **2008,** 1049–1053. https://www.jstor.org/stable/24100801

Mittler, R.; Blumwald, E. Genetic Engineering for Modern Agriculture: Challenges and Perspectives. *Annu. Rev. Plant Biol.* **2010,** *61* (1), 443–462. DOI: 10.1146/annurev-arplant-042809–112116

Mohammed, A. R.; Tarpley, L. Effects of High Night Temperature and Spikelet Position on Yield-Related Parameters of Rice (*Oryza sativa* L.) Plants. *Eur. J. Agron.* **2010,** *33* (2), 117–123. https://doi.org/10.1016/j.eja.2009.11.006

Moller, I. M.; Jensen, P. E.; Hansson, A. Oxidative Modifications to Cellular Components in Plants. *Annu. Rev. Plant Biol.* **2007,** *58* (1), 459–481.s doi: 10.1146/annurev.arplant.58.032806.103946

Monneveux, P.; Sanchez, C.; Beck, D.; Edmeades, G. O. Drought Tolerance Improvement in Tropical Maize Source Populations: Evidence of Progress. *Crop Sci.* **2006,** *46* (1), 180–191. https://doi.org/10.2135/cropsci2005.04–0034

Morales, D.; Rodríguez, P.; Dell'Amico, J.; Nicolas, E.; Torrecillas, A.; Sánchez-Blanco, M. J. High-Temperature Preconditioning and Thermal Shock Imposition Affects Water

Relations, Gas Exchange and Root Hydraulic Conductivity in Tomato. *Biologia Plantarum* **2003,** *47* (2), 203–208. https://doi.org/10.1023/B:BIOP.0000022252.70836.fc

Morgan, P. B.; Bollero, G. A.; Nelson, R. L.; Dohleman, F. G.; Long, S. P. Smaller Than Predicted Increase in Aboveground Net Primary Production and Yield of Field-Grown Soybean Under Fully Open-Air [CO2] Elevation. *Glob. Change Biol.* **2005,** *11* (10), 1856–1865.https://doi.org/10.1111/j.1365–2486.2005.001017.x

Morshedloo, M. R.; Craker, L. E.; Salami, A.; Nazeri, V.; Sang, H.; Maggi, F. Effect of Prolonged Water Stress on Essential Oil Content, Compositions and Gene Expression Patterns of Mono-and Sesquiterpene Synthesis in Two Oregano (*Origanum vulgare* L.) Subspecies. *Plant Physiol. Biochem.* **2017,** *111*, 119–128. https://doi.org/10.1016/j.plaphy.2016.11.023

Mostofa, M. G.; Ghosh, A.; Li, Z. G.; Siddiqui, M. N.; Fujita, M.; Tran, L. S. P. Methylglyoxal–a Signaling Molecule in Plant Abiotic Stress Responses. *Free Radic. Biol. Med.* **2018,** *122*, 96–109. https://doi.org/10.1016/j.freeradbiomed.2018.03.009

Mott, K. A.; Peak, D. Testing a Vapour-Phase Model of Stomatal Responses to Humidity. *Plant Cell Environ.* **2013,** *36* (5), 936–944. https://doi.org/10.1111/pce.12026

Moussa, H. R.; Abdel-Aziz, S. M. Comparative Response of Drought Tolerant and Drought Sensitive Maize Genotypes to Water Stress. *Australian J. Crop Sci.* **2008,** *1* (1), 31–36.

Myers, S. S.; Zanobetti, A.; Kloog, I.; Huybers, P.; Leakey, A. D.; Bloom, A. J. et al. Increasing CO2 Threatens Human Nutrition. *Nature* **2014,** *510* (7503), 139–142. https://doi.org/10.1038/nature13179

Nagel, K. A.; Kastenholz, B.; Jahnke, S.; Van Dusschoten, D.; Aach, T.; Mühlich, M. et al. Temperature Responses of Roots: Impact on Growth, Root System Architecture and Implications for Phenotyping. *Funct. Plant Biol.* **2009,** *36* (11), 947–959. https://doi.org/10.1071/FP09184

Nahar, K.; Ahamed, K. U.; Fujita, M. Phenological Variation and Its Relation with Yield in Several Wheat (*Triticum aestivum* L.) Cultivars Under Normal and Late Sowing Mediated Heat Stress Condition. *Notulae Sci. Biol.* **2010,** *2* (3), 51–56. https://doi.org/10.15835/nsb234723

Nakashima, K.; Yamaguchi-Shinozaki, K.; Shinozaki, K. The Transcriptional Regulatory Network in the Drought Response and Its Crosstalk in Abiotic Stress Responses Including Drought, Cold, and Heat. *Front. Plant Science* **2014,** *5*, 170. https://doi.org/10.3389/fpls.2014.00170

Nam, N. H.; Chauhan, Y. S.; Johansen, C. Effect of Timing of Drought Stress on Growth and Grain Yield of Extra-Short-Duration Pigeonpea Lines. *J. Agric. Sci.* **2001,** *136* (2), 179–189. https://doi.org/10.1017/S0021859601008607

Nasrollahi, V.; Mirzaie-Asl, A.; Piri, K.; Nazeri, S.; Mehrabi, R. The Effect of Drought Stress on the Expression of Key Genes Involved in the Biosynthesis of Triterpenoid Saponins in Liquorice (*Glycyrrhiza glabra*). *Phytochemistry* **2014,** *103*, 32–37. https://doi.org/10.1016/j.phytochem.2014.03.004

Nayyar, H.; Kaur, S.; Singh, S.; Upadhyaya, H. D. Differential Sensitivity of Desi (Small-Seeded) and Kabuli (Large-Seeded) Chickpea Genotypes to Water Stress During Seed Filling: Effects on Accumulation of Seed Reserves and Yield. *J. Sci. Food Agric.* **2006,** *86* (13), 2076–2082. https://doi.org/10.1002/jsfa.2574

Nezhadahmadi, A.; Prodhan, Z. H.; Faruq, G. Drought Tolerance in Wheat. *Sci. World J.* **2013,** *2013*. https://doi.org/10.1155/2013/610721

Nezhadahmadi, A.; Prodhan, Z. H.; Faruq, G. Drought Tolerance in Wheat. Sci World J 2013: 610721.

Nikolaeva, M. K.; Maevskaya, S. N.; Shugaev, A. G.; Bukhov, N. G. Effect of Drought on Chlorophyll Content and Antioxidant Enzyme Activities in Leaves of Three Wheat Cultivars Varying in Productivity. *Russian J. Plant Physiol.* **2010,** *57* (1), 87–95. https://doi.org/10.1134/S1021443710010127

Novick, K. A.; Ficklin, D. L.; Stoy, P. C.; Williams, C. A.; Bohrer, G.; Oishi, A. C. et al. The Increasing Importance of Atmospheric Demand for Ecosystem Water and Carbon Fluxes. *Nat. Clim. Change* **2016,** *6* (11), 1023–1027. https://doi.org/10.1038/nclimate3114

Ober, E. S.; Setter, T. L.; Madison, J. T.; Thompson, J. F.; Shapiro, P. S. Influence of Water Deficit on Maize Endosperm Development: Enzyme Activities and RNA Transcripts of Starch and Zein Synthesis, Abscisic Acid, and Cell Division. *Plant Physiol.* **1991,** *97* (1), 154–164. https://doi.org/10.1104/pp.97.1.154

Ogren, W. L.; Bowes, G. Ribulose Diphosphate Carboxylase Regulates Soybean Photorespiration. *Nat. New Biol.* **1971,** *230* (13), 159–160. https://doi.org/10.1038/newbio230159a0

Okay, S.; Derelli, E.; Unver, T. Transcriptome-Wide Identification of Bread Wheat WRKY Transcription Factors in Response to Drought Stress. *Mol. Genet. Genom.* **2014,** *289* (5), 765–781. https://doi.org/10.1007/s00438–014–0849-x

Omae, H.; Kumar, A.; Shono, M. Adaptation to High Temperature and Water Deficit in the Common Bean (*Phaseolus vulgaris* L.) During the Reproductive Period. *J. Bot.* **2012,** *2012*. doi:10.1155/2012/803413

Onwuka, B.; Mang, B. Effects of Soil Temperature on Some Soil Properties and Plant Growth. *Adv. Plants Agric. Res.* **2018,** 8 (1), 34–37.

Orr, D. J.; Alcântara, A.; Kapralov, M. V.; Andralojc, P. J.; Carmo-Silva, E.; Parry, M. A. Surveying Rubisco Diversity and Temperature Response to Improve Crop Photosynthetic Efficiency. *Plant Physiol.* **2016,** *172* (2), 707–717. https://doi.org/10.1104/pp.16.00750

Osuagwu, G. G. E.; Edeoga, H. O.; Osuagwu, A. N. The Influence of Water Stress (Drought) on the Mineral and Vitamin Potential of the Leaves of *Ocimum gratissimum* L. *Recent Res. Sci. Technol.* **2010,** *2* (2), 27–33.

Pachauri, R. K.; Allen, M. R.; Barros, V. R.; Broome, J.; Cramer, W.; Christ, R. et al. *Climate Change 2014: Synthesis Report: Contribution of Working Groups I, II and III to the Fifth Assessment Report of the Intergovernmental Panel on Climate Change*, 2014; p 151. Ipcc. hdl:10013/epic.45156

Pagamas, P.; Nawata, E. Sensitive Stages of Fruit and Seed Development of Chili Pepper (*Capsicum annuum* L. var. Shishito) Exposed to High-Temperature Stress. *Sci. Horticul.* **2008,** *117* (1), 21–25. https://doi.org/10.1016/j.scienta.2008.03.017

Pallas Jr, J. E.; Samish, Y. B. Photosynthetic Response of Peanut 1. *Crop Sci.* **1974,** *14* (3), 478–482. https://doi.org/10.2135/cropsci1974.0011183X001400030042x

Pande, S.; Sharma, M. Climate Change: Potential Impact on Chickpea and Pigeonpea Diseases in the Rainfed Semi-Arid Tropics (SAT). In *Proceedings of the 5th International Food Legumes Research Conference (IFLRC V) and 7th European Conference on Grain Legumes (AEP VII), Antalya, Turkey*, 2010.

Pardales Jr, J. R.; Kono, Y.; Yamauchi, A. Epidermal Cell Elongation in Sorghum Seminal Roots Exposed to High Root-Zone Temperature. *Plant Sci.* **1992,** *81* (2), 143–146. https://doi.org/10.1016/0168–9452 (92)90035-K

Parvaneh, R.; Seyed, M. H.; Shahrokh, T. The Studying Effect of Drought Stress on Germination, Proline, Sugar, Lipid, Protein and Chlorophyll Content in Purslane (*Portulaca oleracea* L.) Leaves. *J. Med. Plants Res.* **2012,** *6* (9), 1539–1547.DOI: 10.5897/JMPR11.698

Parvin, S.; Uddin, S.; Bourgault, M.; Roessner, U.; Tausz-Posch, S.; Armstrong, R. et al. Water Availability Moderates N2 Fixation Benefit from Elevated [CO2]: A 2-Year Free-Air CO2 Enrichment Study on Lentil (*Lens culinaris* MEDIK.) in a Water Limited Agroecosystem. *Plant Cell Environ.* **2018,** *41* (10), 2418–2434. https://doi.org/10.1111/pce.13360

Parvin, S.; Uddin, S.; Tausz-Posch, S.; Fitzgerald, G.; Armstrong, R.; Tausz, M. Elevated CO2 Improves Yield and N2 Fixation But Not Grain *N* Concentration of Faba Bean (*Vicia faba* L.) Subjected to Terminal Drought. *Environ. Exp. Bot.* **2019,** *165*, 161–173. https://doi.org/10.1016/j.envexpbot.2019.06.003

Patel, D.; Franklin, K. A. Temperature-Regulation of Plant Architecture. *Plant Signal. Behav.* **2009,** *4* (7), 577–579. https://doi.org/10.4161/psb.4.7.8849

Perdomo, J. A.; Capó-Bauçà, S.; Carmo-Silva, E.; Galmés, J. Rubisco and Rubisco Activase Play an Important Role in the Biochemical Limitations of Photosynthesis in Rice, Wheat, and Maize Under High Temperature and Water Deficit. *Front. Plant Science* **2017,** *8*, 490. https://doi.org/10.3389/fpls.2017.00490

Pérez-López, U.; Miranda-Apodaca, J.; Lacuesta, M.; Mena-Petite, A.; Muñoz-Rueda, A. Growth and Nutritional Quality Improvement in Two Differently Pigmented Lettuce Cultivars Grown Under Elevated CO2 and/or salinity. *Sci. Horticult.* **2015,** *195*, 56–66. https://doi.org/10.1016/j.scienta.2015.08.034

Perez-Lopez, U.; Miranda-Apodaca, J.; Munoz-Rueda, A.; Mena-Petite, A. Interacting Effects of High Light and Elevated CO2 on the Nutraceutical Quality of Two Differently Pigmented *Lactuca sativa* Cultivars (Blonde of Paris Batavia and Oak Leaf). *Sci. Horticult.* **2015,** *191*, 38–48. https://doi.org/10.1016/j.scienta.2015.04.030

Pérez-López, U.; Miranda-Apodaca, J.; Muñoz-Rueda, A.; Mena-Petite, A. Lettuce Production and Antioxidant Capacity are Differentially Modified by Salt Stress and Light Intensity Under Ambient and Elevated CO2. *J. Plant Physiol.* **2013,** *170* (17), 1517–1525. https://doi.org/10.1016/j.jplph.2013.06.004

Pérez-López, U.; Sgherri, C.; Miranda-Apodaca, J.; Micaelli, F.; Lacuesta, M.; Mena-Petite, A. et al. Concentration of Phenolic Compounds Is Increased in Lettuce Grown Under High Light Intensity and Elevated CO2. *Plant physiol. Biochem.* **2018,** *123*, 233–241. https://doi.org/10.1016/j.plaphy.2017.12.010

Pettigrew, W. T. Physiological Consequences of Moisture Deficit Stress in Cotton. *Crop Sci.* **2004,** *44* (4), 1265–1272. https://doi.org/10.2135/cropsci2004.1265

Piramila, B. H. M.; Prabha, A. L.; Nandagopalan, V.; Stanley, A. L. Effect of Heat Treatment on Germination, Seedling Growth and Some Biochemical Parameters of Dry Seeds of Black Gram. *Int. J. Pharm. Phytopharmacol. Res.* **2012,** *1*, 194–202.

Plaut, Z. Plant Exposure to Water Stress During Specific Growth Stages. *Encyclopedia Water Sci.* **2003,** 673–675.

Plett, J. M.; Kohler, A.; Khachane, A.; Keniry, K.; Plett, K. L.; Martin, F.; Anderson, I. C. The Effect of Elevated Carbon Dioxide on the Interaction Between *Eucalyptus grandis* and Diverse Isolates of Pisolithus sp. Is Associated with a Complex Shift in the Root Transcriptome. *New Phytol.* **2015,** *206* (4), 1423–1436. https://doi.org/10.1111/nph.13103

Pokharel, M.; Chiluwal, A.; Stamm, M.; Min, D.; Rhodes, D.; Jagadish, S. K. High Night-Time Temperature During Flowering and Pod Filling Affects Flower Opening, Yield and

Seed Fatty Acid Composition in Canola. *J. Agron. Crop Sci.* **2020,** *206* (5), 579–596. https://doi.org/10.1111/jac.12408
Poorter, H. Plant Growth and Carbon Economy. *Encyclopedia Life Sci.* **2002,** 1–6.
Prasad, P. V. V.; Djanaguiraman, M.; Perumal, R.; Ciampitti, I. A. Impact of High Temperature Stress on Floret Fertility and Individual Grain Weight of Grain Sorghum: Sensitive Stages and Thresholds for Temperature and Duration. *Front. Plant Sci.* **2015,** *6*, 820. https://doi.org/10.3389/fpls.2015.00820
Prasad, P. V. V.; Staggenborg, S. A.; Ristic, Z. Impacts of Drought and/or Heat Stress on Physiological, Developmental, Growth, and Yield Processes of Crop Plants. *Response Crops Limited Water* **2008,** *1*, 301–355. https://doi.org/10.2134/advagricsystmodel1.c11
Prasad, P. V.; Djanaguiraman, M. High Night Temperature Decreases Leaf Photosynthesis and Pollen Function in Grain Sorghum. *Funct. Plant Biol.* **2011,** *38* (12), 993–1003. https://doi.org/10.1071/FP11035
Prasad, P. V.; Boote, K. J.; Vu, J. C.; Allen Jr, L. H. The Carbohydrate Metabolism Enzymes Sucrose-P Synthase and ADG-Pyrophosphorylase in Phaseolus Bean Leaves Are Up-Regulated at Elevated Growth Carbon Dioxide and Temperature. *Plant Sci.* **2004,** *166* (6), 1565–1573. https://doi.org/10.1016/j.plantsci.2004.02.009
Prasad, P. V.; Pisipati, S. R.; Ristic, Z.; Bukovnik, U. R. S. K. A.; Fritz, A. K. Impact of Nighttime Temperature on Physiology and Growth of Spring Wheat. *Crop science*, *48* (6), 2372–2380. https://doi.org/10.2135/cropsci2007.12.0717
Prathyusha, I. V. S. N.; Chaitanya, K. V. Effect of Water Stress on the Physiological and Biochemical Responses of Two Different Coleus (Plectranthus) Species. *Biol. Futura* **2019,** *70* (4), 312–322. https://doi.org/10.1556/019.70.2019.35
Pritchard, S. G.; Rogers, H. H. Spatial and Temporal Deployment of Crop Roots in CO2-Enriched Environments. *New Phytol.* **2000,** *147* (1), 55–71. https://doi.org/10.1046/j.1469–8137.2000.00678.x
Proietti, S.; Bertini, L.; Van der Ent, S.; Leon-Reyes, A.; Pieterse, C. M. J.; Tucci, M. et al. Cross Activity of Orthologous WRKY Transcription Factors in Wheat and Arabidopsis. *J. Exp. Bot.* **2011,** *62* (6), 1975–1990. https://doi.org/10.1093/jxb/erq396
Qayyum, A.; Razzaq, A.; Ahmad, M.; Jenks, M. A. Water Stress Causes Differential Effects on Germination Indices, Total Soluble Sugar and Proline Content in Wheat (*Triticum aestivum* L.) Genotypes. *Afr. J. Biotechnol.* **2011,** *10* (64), 14038–14045. DOI: 10.5897/AJB11.2220
Qi, J.; Sun, S.; Yang, L.; Li, M.; Ma, F.; Zou, Y. Potassium Uptake and Transport in Apple Roots Under Drought Stress. *Horticult. Plant J.* **2019,** *5* (1), 10–16. https://doi.org/10.1016/j.hpj.2018.10.001
Qi, Y.; Wang, H.; Zou, Y.; Liu, C.; Liu, Y.; Wang, Y.; Zhang, W. Over-Expression of Mitochondrial Heat Shock Protein 70 Suppresses Programmed Cell Death in Rice. *Febs Lett.* **2011,** *585* (1), 231–239. https://doi.org/10.1016/j.febslet.2010.11.051
Queiroz, M. S.; Oliveira, C. E.; Steiner, F.; Zuffo, A. M.; Zoz, T.; Vendruscolo, E. P. et al. Drought Stresses on Seed Germination and Early Growth of Maize and Sorghum. *J. Agric. Sci.* **2019,** *11* (2), 310–318. https://doi.org/10.5539/jas.v11n2p310
Rae, A. M.; Tricker, P. J.; Bunn, S. M.; Taylor, G. Adaptation of Tree Growth to Elevated CO2: Quantitative Trait Loci for Biomass in Populus. *New Phytol.* **2007,** *175* (1), 59–69. https://doi.org/10.1111/j.1469–8137.2007.02091.x
Rahaie, M.; Xue, G. P.; Schenk, P. M. The Role of Transcription Factors in Wheat Under Different Abiotic Stresses. *Abiotic Stress-Plant Responses App. Agric.* **2013,** *2*, 367–385.

Rahman, M. A.; Chikushi, J.; Yoshida, S.; Karim, A. J. M. S. Growth and Yield Components of Wheat Genotypes Exposed to High Temperature Stress Under Control Environment. *Bangladesh J. Agric. Res.* **2009,** *34* (3), 360–372. https://doi.org/10.3329/bjar.v34i3.3961

Rahmati, M.; Mirás-Avalos, J. M.; Valsesia, P.; Lescourret, F.; Génard, M.; Davarynejad, G. H. et al. Disentangling the Effects of Water Stress on Carbon Acquisition, Vegetative Growth, and Fruit Quality of Peach Trees by Means of the QualiTree Model. *Front. Plant Sci.* **2018,** *9*, 3. https://doi.org/10.3389/fpls.2018.00003

Ranasinghe, S.; Taylor, G. Mechanism for Increased Leaf Growth in Elevated CO2. *J. Exp. Bot.* **1996,** *47* (3), 349–358. https://doi.org/10.1093/jxb/47.3.349

Raven, P. H.; Evert, R. F.; Eichhorn, S. E. *Biology of Plants*; W.H. Freeman and Co. Ltd.: New York, 2005.

Razmjoo, K. H. O. R. S. H. I. D.; Heydarizadeh, P. A. R. I. S. A.; Sabzalian, M. R. Effect of Salinity and Drought Stresses on Growth Parameters and Essential Oil Content of *Matricaria chamomile*. *Int. J. Agric. Biol.* **2008,** *10* (4), 451–454.

Reddy, S. K.; Liu, S.; Rudd, J. C.; Xue, Q.; Payton, P.; Finlayson, S. A. et al. Physiology and Transcriptomics of Water-Deficit Stress Responses in Wheat Cultivars TAM 111 and TAM 112. *J. Plant Physiol.* **2014,** *171* (14), 1289–1298. https://doi.org/10.1016/j.jplph.2014.05.005

Rehman, A.; Farooq, M.; Asif, M.; Ozturk, L. Supra-Optimal Growth Temperature Exacerbates Adverse Effects of Low Zn Supply in Wheat. *J. Plant Nutr. Soil Sci.* **2019,** *182* (4), 656–666. https://doi.org/10.1002/jpln.201800654

Reich, P. B.; Hobbie, S. E.; Lee, T. D.; Pastore, M. A. Unexpected Reversal of C3 Versus C4 Grass Response to Elevated CO2 During a 20-Year Field Experiment. *Science* **2018,** *360* (6386), 317–320. https://doi.org/10.1126/science.aas9313

Rellán-Álvarez, R.; Lobet, G.; Lindner, H.; Pradier, P. L.; Sebastian, J.; Yee, M. C. et al. GLO-Roots: An Imaging Platform Enabling Multidimensional Characterization of Soil-Grown Root Systems. *elife* **2015,** *4*, e07597. https://doi.org/10.7554/eLife.07597.002

Reynolds, M. P.; Pierre, C. S.; Saad, A. S.; Vargas, M.; Condon, A. G. Evaluating Potential Genetic Gains in Wheat Associated with Stress-Adaptive Trait Expression in Elite Genetic Resources Under Drought and Heat Stress. *Crop Sci.* **2007,** *47*, S–172. https://doi.org/10.2135/cropsci2007.10.0022IPBS

Ribeiro, P. R.; Fernandez, L. G.; de Castro, R. D.; Ligterink, W.; Hilhorst, H. W. Physiological and Biochemical Responses of *Ricinus communis* Seedlings to Different Temperatures: A Metabolomics Approach. *BMC Plant Biol.* **2014,** *14* (1), 1–14. https://doi.org/10.1186/s12870-014-0223-5

Ristvey, A. G.; Belayneh, B. E.; Lea-Cox, J. D. A Comparison of Irrigation-Water Containment Methods and Management Strategies Between Two Ornamental Production Systems to Minimize Water Security Threats. *Water* **2019,** *11* (12), 2558. https://doi.org/10.3390/w11122558

Rodríguez, M.; Canales, E.; Borrás-Hidalgo, O. Molecular Aspects of Abiotic Stress in Plants. *Biotecnol. Aplicada* **2005,** *22* (1), 1–10.

Rogers, A.; Ainsworth, E. A.; Leakey, A. D. Will Elevated Carbon Dioxide Concentration Amplify the Benefits of Nitrogen Fixation in Legumes? *Plant Physiol.* **2009,** *151* (3), 1009–1016. https://doi.org/10.1104/pp.109.144113

Rogers, E. D.; Benfey, P. N. Regulation of Plant Root System Architecture: Implications for Crop Advancement. *Curr. Opin. Biotechnol.* **2015,** *32*, 93–98. https://doi.org/10.1016/j.copbio.2014.11.015

Rogers, H. H.; Peterson, C. M.; McCrimmon, J. N.; Cure, J. D. Response of Plant Roots to Elevated Atmospheric Carbon Dioxide. *Plant Cell Environ.* **1992,** *15* (6), 749–752. https://doi.org/10.1111/j.1365–3040.1992.tb01018.x

Román-Avilés, B.; Snapp, S. S.; Kelly, J. D. Assessing Root Traits Associated with Root Rot Resistance in Common Bean. *Field Crops Res.* **2004,** *86* (2–3), 147–156. https://doi.org/10.1016/j.fcr.2003.08.001

Ruehr, N. K.; Grote, R.; Mayr, S.; Arneth, A. Beyond the Extreme: Recovery of Carbon and Water Relations in Woody Plants Following Heat and Drought Stress. *Tree Physiol.* **2019,** *39* (8), 1285–1299. https://doi.org/10.1093/treephys/tpz032

Ruelland, E.; Zachowski, A. How Plants Sense Temperature. *Environ. Exp. Bot.* **2010,** *69* (3), 225–232. https://doi.org/10.1016/j.envexpbot.2010.05.011

Ruiz-Vera, U. M.; Siebers, M. H.; Drag, D. W.; Ort, D. R.; Bernacchi, C. J. Canopy Warming Caused Photosynthetic Acclimation and Reduced Seed Yield in Maize Grown at Ambient and Elevated [CO2]. *Glob. Change Biol.* **2015,** *21* (11), 4237–4249. https://doi.org/10.1111/gcb.13013

Ruiz-Vera, U. M.; Siebers, M. H.; Jaiswal, D.; Ort, D. R.; Bernacchi, C. J. Canopy Warming Accelerates Development in Soybean and Maize, Offsetting the Delay in Soybean Reproductive Development by Elevated CO2 Concentrations. *Plant Cell Environ.* **2018,** *41* (12), 2806–2820. https://doi.org/10.1111/pce.13410

Ruiz-Vera, U. M.; Siebers, M.; Gray, S. B.; Drag, D. W.; Rosenthal, D. M.; Kimball, B. A. et al. Global Warming Can Negate the Expected CO2 Stimulation in Photosynthesis and Productivity for Soybean Grown in the Midwestern United States. *Plant Physiol.* **2013,** *162* (1), 410–423. https://doi.org/10.1104/pp.112.211938

Sadras, V. O.; Lake, L.; Chenu, K.; McMurray, L. S.; Leonforte, A. Water and Thermal Regimes for Field Pea in Australia and Their Implications for Breeding. *Crop Pasture Sci.* **2012,** *63* (1), 33–44. https://doi.org/10.1071/CP11321

Sage, R. F.; Kubien, D. S. The Temperature Response of C3 and C4 Photosynthesis. *Plant Cell Environ.* **2007,** *30* (9), 1086–1106. https://doi.org/10.1111/j.1365–3040.2007.01682.x

Sage, R. F.; Sharkey, T. D. The Effect of Temperature on the Occurrence of O2 and CO2 Insensitive Photosynthesis in Field Grown Plants. *Plant Physiol.* **1987,** *84* (3), 658–664. https://doi.org/10.1104/pp.84.3.658

Saha, S. A. U. R. A. V.; Sehgal, V. K.; Chakraborty, D. E. B. A. S. I. S. H.; Pal, M. A. D. A. N. Growth Behavior of Kabuli Chickpea Under Elevated Atmospheric CO2. *J. Agric. Phys.* **2013,** *13* (1), 55–61.

Saha, U.; Hancock, D.; Stewart, L.; Kissel, D.; Sonon, L. The Effect of Drought on Lignin Content and Digestibility of Tifton-85 and Coastal Bermudagrass (*Cynodon dactylon* L.) Hays Produced in Georgia. *Int. J. Appl. Agric. Sci.* **2016,** *2*, 69.

Saini, H. S.; Westgate, M. E. Reproductive Development in Grain Crops During Drought. *Adv. Agron.* **1999,** *68*, 59–96. https://doi.org/10.1016/S0065-2113 (08)60843-3

Saitoh, H. *Ecological and Physiology of Vegetable*; Nousangyoson Bunka Kyoukai: Tokyo, Japan, 2008.

Sakata, T.; Takahashi, H.; Nishiyama, I.; Higashitani, A. Effects of High Temperature on the Development of Pollen Mother Cells and Microspores in Barley *Hordeum vulgare* L. *J. Plant Res.* **2000,** *113* (4), 395.DOI:10.1007/PL00013947

Salazar, C.; Hernández, C.; Pino, M. T. Plant Water Stress: Associations Between Ethylene and Abscisic Acid Response. *Chilean J. Agric. Res.* **2015,** *75*, 71–79.http://dx.doi.org/10.4067/S0718-58392015000300008

Salvucci, M. E.; Osteryoung, K. W.; Crafts-Brandner, S. J.; Vierling, E. Exceptional Sensitivity of Rubisco Activase to Thermal Denaturation In Vitro and In Vivo. *Plant Physiol.* **2001,** *127* (3), 1053–1064. https://doi.org/10.1104/pp.010357

Samarah, N. H. Effects of Drought Stress on Growth and Yield of Barley. *Agron. Sustain. Dev.* **2005,** *25* (1), 145–149.HAL Id : hal-00886257

Samarah, N. H. Effects of Drought Stress on Growth and Yield of Barley. *Agron. Sustain. Dev.* **2020,** *25* (1), 145–149.

Samarah, N. H.; Mullen, R. E.; Cianzio, S. R.; Scott, P. Dehydrin-Like Proteins in Soybean Seeds in Response to Drought Stress During Seed Filling. *Crop Sci.* **2006,** *46* (5), 2141–2150. https://doi.org/10.2135/cropsci2006.02.0066

Samarah, N.; Mullen, R.; Cianzio, S. Size Distribution and Mineral Nutrients of Soybean Seeds in Response to Drought Stress. *J. Plant Nutr.* **2004,** *27* (5), 815–835. https://doi.org/10.1081/PLN-120030673

Santrucek, J.; Sage, R. F. Acclimation of Stomatal Conductance to a CO2-Enriched Atmosphere and Elevated Temperature in Chenopodium Album. *Funct. Plant Biol.* **1996,** *23* (4), 467–478. https://doi.org/10.1071/PP9960467

Sarani, M.; Namrudi, M.; Hashemi, S. M.; Raoofi, M. M. The Effect of Drought Stress on Chlorophyll Content, Root Growth, Glucosinolate and Proline in Crop Plants. *Int. J. Farm. Allied Sci.* **2014,** *3* (9), 994–997.

Sato, S.; Kamiyama, M.; Iwata, T.; Makita, N.; Furukawa, H.; Ikeda, H. Moderate Increase of Mean Daily Temperature Adversely Affects Fruit Set of *Lycopersicon esculentum* by Disrupting Specific Physiological Processes in Male Reproductive Development. *Ann. Bot.* **2006,** *97* (5), 731–738. https://doi.org/10.1093/aob/mcl037

Sattelmacher, B.; Marschner, H.; Kühne, R. Effects of the Temperature of the Rooting Zone on the Growth and Development of Roots of Potato (*Solanum tuberosum*). *Ann. Bot.* **1990,** *65* (1), 27–36. https://doi.org/10.1093/oxfordjournals.aob.a087974

Savicka, M.; Škute, N. Effects of High Temperature on Malondialdehyde Content, Superoxide Production and Growth Changes in Wheat Seedlings (*Triticum aestivum* L.). *Ekologija* **2010,** *56* (1), 26–33. DOI: 10.2478/v10055-010-0004-x

Savin, R.; Stone, P. J.; Nicolas, M. E.; Wardlaw, I. F. Effects of Heat Stress and Moderately High Temperature on Grain Growth and Malting Quality of Barley. *Aust. J. Agric. Res.* **1997,** *48*, 615–624.

Schrader, S. M.; Wise, R. R.; Wacholtz, W. F.; Ort, D. R.; Sharkey, T. D. Thylakoid Membrane Responses to Moderately High Leaf Temperature in Pima Cotton. *Plant Cell Environ.* **2004,** *27* (6), 725–735. https://doi.org/10.1111/j.1365-3040.2004.01172.x

Schulze, E. D.; Lange, O. L.; Kappen, L.; Evenari, M.; Buschbom, U. The Role of Air Humidity and Leaf Temperature in Controlling Stomatal Resistance of *Prunus armeniaca* L. Under Desert Conditions. *Oecologia* **1975,** *18* (3), 219–233. https://doi.org/10.1007/BF00345424

Seiler, G. J. Influence of Temperature on Primary and Lateral Root Growth of Sunflower Seedlings. *Environ. Exp. Bot.* **1998,** *40* (2), 135–146. https://doi.org/10.1016/S0098–8472(98)00027–6

Seleiman, M. F.; Kheir, A. M. Saline Soil Properties, Quality and Productivity of Wheat Grown with Bagasse Ash and Thiourea in Different Climatic Zones. *Chemosphere* **2018,** *193*, 538–546. https://doi.org/10.1016/j.chemosphere.2017.11.053

Seleiman, M. F.; Refay, Y.; Al-Suhaibani, N.; Al-Ashkar, I.; El-Hendawy, S.; Hafez, E. M. Integrative Effects of Rice-Straw Biochar and Silicon on Oil and Seed Quality, Yield and

Physiological Traits of *Helianthus annuus* L. Grown Under Water Deficit Stress. *Agronomy* **2019,** *9* (10), 637. https://doi.org/10.3390/agronomy9100637

Selmar, D.; Kleinwächter, M. Stress Enhances the Synthesis of Secondary Plant Products: The Impact of Stress-Related Over-Reduction on the Accumulation of Natural Products. *Plant Cell Physiol.* **2013,** *54* (6), 817–826. https://doi.org/10.1093/pcp/pct054

Shakeel, A. A.; Xiao-yu, X.; Long-chang, W.; Muhammad, F. S.; Chen, M.; Wang, L. Morphological, Physiological and Biochemical Responses of Plants to Drought Stress. *Afr. J. Agric. Res.* **2011,** *6* (9), 2026–2032.DOI: 10.5897/AJAR10.027

Sharma, A.; Zheng, B. Melatonin Mediated Regulation of Drought Stress: Physiological and Molecular Aspects. *Plants* **2019,** *8* (7), 190. https://doi.org/10.3390/plants8070190

Sharma, A.; Shahzad, B.; Rehman, A.; Bhardwaj, R.; Landi, M.; Zheng, B. Response of Phenylpropanoid Pathway and the Role of Polyphenols in Plants Under Abiotic Stress. *Molecules* **2019,** *24* (13), 2452. https://doi.org/10.3390/molecules24132452

Sharma, A.; Wang, J.; Xu, D.; Tao, S.; Chong, S.; Yan, D. et al. Melatonin Regulates the Functional Components of Photosynthesis, Antioxidant System, Gene Expression, and Metabolic Pathways to Induce Drought Resistance in Grafted *Carya cathayensis* Plants. *Sci. Total Environ.* **2020,** *713*, 136675. https://doi.org/10.1016/j.scitotenv.2020.136675

Sharma, H. C.; Pathania, M.; War, A. R.; Pavani, T.; Vashisth, S. Climate Change Effects on Pest Spectrum and Incidence in Grain Legumes, 2015 .http://oar.icrisat.org/id/eprint/9406

Sheen, J. Master Regulators in Plant Glucose Signaling Networks. *J. Plant Biol.* **2014,** *57* (2), 67–79. https://doi.org/10.1007/s12374-014-0902-7

Shi, S.; Fan, M.; Iwama, K.; Li, F.; Zhang, Z.; Jia, L. Physiological Basis of Drought Tolerance in Potato Grown Under Long-Term Water Deficiency. *Int. J. Plant Product.* **2015,** *9* (2), 305–320.

Sicher, R. C.; Barnaby, J. Y. Impact of Carbon Dioxide Enrichment on the Responses of Maize Leaf Transcripts and Metabolites to Water Stress. *Physiol. Plantarum* **2012,** *144* (3), 238–253. https://doi.org/10.1111/j.1399-3054.2011.01555.x

Siddique, K. H. M.; Loss, S. P.; Regan, K. L.; Jettner, R. L. Adaptation and Seed Yield of Cool Season Grain Legumes in Mediterranean Environments of South-Western Australia. *Aust. J. Agric. Res.* **1999,** *50* (3), 375–388. https://doi.org/10.1071/A98096

Siddiqui, M. H.; Al-Khaishany, M. Y.; Al-Qutami, M. A.; Al-Whaibi, M. H.; Grover, A.; Ali, H. M. et al. Response of Different Genotypes of Faba Bean Plant to Drought Stress. *Int. J. Mol. Sci.* **2015,** *16* (5), 10214–10227. https://doi.org/10.3390/ijms160510214

Siebers, M. H.; Yendrek, C. R.; Drag, D.; Locke, A. M.; Rios Acosta, L.; Leakey, A. D. et al. Heat Waves Imposed During Early Pod Development in Soybean (*Glycine max*) Cause Significant Yield Loss Despite a Rapid Recovery from Oxidative Stress. *Glob. Change Biol.* **2015,** *21* (8), 3114–3125. https://doi.org/10.1111/gcb.12935

Singh, J.; Thakur, J. K. Photosynthesis and Abiotic Stress in Plants. In *Biotic and Abiotic Stress Tolerance in Plants*; Springer: Singapore, 2018; pp 27–46.

Sinsawat, V.; Leipner, J.; Stamp, P.; Fracheboud, Y. Effect of Heat Stress on the Photosynthetic Apparatus in Maize (*Zea mays* L.) Grown at Control or High Temperature. *Environ. Exp. Bot.* **2004,** *52* (2), 123–129. https://doi.org/10.1016/j.envexpbot.2004.01.010

Smith, N. G.; Dukes, J. S. Short-Term Acclimation to Warmer Temperatures Accelerates Leaf Carbon Exchange Processes Across Plant Types. *Glob. Change Biol.* **2017,** *23* (11), 4840–4853. https://doi.org/10.1111/gcb.13735

Smith, S.; De Smet, I. Root System Architecture: Insights from Arabidopsis and Cereal Crops. *Phil. Trans. R. Soc. B* **2012,** *367* (1595), 1441–1452. https://doi.org/10.1098/rstb.2011.0234

Sobhanian, H.; Pahlavan, S.; Meyfour, A. How Does Proteomics Target Plant Environmental Stresses in a Semi-Arid Area? *Mol. Biol. Rep.* **2020,** *47* (4), 3181–3194. https://doi.org/10.1007/s11033–020–05406–6

Soliman, W. S.; Fujimori, M.; Tase, K.; Sugiyama, S. I. Oxidative Stress and Physiological Damage Under Prolonged Heat Stress in C3 Grass *Lolium perenne*. *Grassland Sci.* **2011,** *57* (2), 101–106. https://doi.org/10.1111/j.1744–697X.2011.00214.x

Sonawane, B. V.; Sharwood, R. E.; Von Caemmerer, S.; Whitney, S. M.; Ghannoum, O. Short-Term Thermal Photosynthetic Responses of C4 Grasses Are Independent of the Biochemical Subtype. *J. Exp. Bot.* **2017,** *68* (20), 5583–5597. https://doi.org/10.1093/jxb/erx350

Sourour, A.; Afef, O.; Mounir, R.; Mongi, B. Y. A Review: Morphological, Physiological, Biochemical and Molecular Plant Responses to Water Deficit Stress. *Int J Eng Sci.* **2017,** *6* (1), 1–4.DOI : 10.9790/1813–0601010104

Spollen, W. G. Regulation of Cell Expansion in Roots and Shoots at Low Water Potentials. *Water Deficits: Plant Responses from Cell to Community*, 1993; pp 37–52.

Srivastava, S.; Pathak, A. D.; Gupta, P. S.; Shrivastava, A. K.; Srivastava, A. K. Hydrogen Peroxide-Scavenging Enzymes Impart Tolerance to High Temperature Induced Oxidative Stress in Sugarcane. *J. Environm. Biol.* **2012,** *33* (3), 657.

Stitt, M.; Krapp, A. The Interaction Between Elevated Carbon Dioxide and Nitrogen Nutrition: The Physiological and Molecular Background. *Plant Cell Environ.* **1999,** *22* (6), 583–621. https://doi.org/10.1046/j.1365–3040.1999.00386.x

Su, Z.; Ma, X.; Guo, H.; Sukiran, N. L.; Guo, B.; Assmann, S. M.; Ma, H. Flower Development Under Drought Stress: Morphological and Transcriptomic Analyses Reveal Acute Responses and Long-Term Acclimation in Arabidopsis. *Plant Cell* **2013,** *25* (10), 3785–3807. https://doi.org/10.1105/tpc.113.115428

Suwa, R.; Hakata, H.; Hara, H.; El-Shemy, H. A.; Adu-Gyamfi, J. J.; Nguyen, N. T. et al. High Temperature Effects on Photosynthate Partitioning and Sugar Metabolism During Ear Expansion in Maize (*Zea mays* L.) Genotypes. *Plant Physiol. Biochem.* **2010,** *48* (2–3), 124–130. https://doi.org/10.1016/j.plaphy.2009.12.010

Suzuki, N.; Koussevitzky, S. H. A. I.; Mittler, R. O. N.; Miller, G. A. D. ROS and Redox Signalling in the Response of Plants to Abiotic Stress. *Plant Cell Environ.* **2012,** *35* (2), 259–270. https://doi.org/10.1111/j.1365-3040.2011.02336.x

Suzuki, N.; Miller, G.; Morales, J.; Shulaev, V.; Torres, M. A.; Mittler, R. Respiratory Burst Oxidases: The Engines of ROS Signaling. *Curr. Opin. Plant Biol.* **2011,** *14* (6), 691–699. https://doi.org/10.1016/j.pbi.2011.07.014

Taiz, L.; Zeiger, E. *Plant Physiology*; Sinauer Associates Inc.: Sunderland, MA, 2006.

Takatani, N.; Ito, T.; Kiba, T.; Mori, M.; Miyamoto, T.; Maeda, S. I.; Omata, T. Effects of High CO2 on Growth and Metabolism of Arabidopsis Seedlings During Growth with a Constantly Limited Supply of Nitrogen. *Plant Cell Physiol.* **2014,** *55* (2), 281–292. https://doi.org/10.1093/pcp/pct186

Tan, W.; wei Meng, Q.; Brestic, M.; Olsovska, K.; Yang, X. Photosynthesis Is Improved by Exogenous Calcium in Heat-Stressed Tobacco Plants. *J. Plant Physiol.* **2011,** *168* (17), 2063–2071. https://doi.org/10.1016/j.jplph.2011.06.009

Taub, D. R.; Miller, B.; Allen, H. Effects of Elevated CO2 on the Protein Concentration of Food Crops: A Meta-Analysis. *Glob. Change Biol.* **2008,** *14* (3), 565–575. https://doi.org/10.1111/j.1365-2486.2007.01511.x

Taylor, G.; Tricker, P. J.; Zhang, F. Z.; Alston, V. J.; Miglietta, F.; Kuzminsky, E. Spatial and Temporal Effects of Free-Air CO2Enrichment (POPFACE) on Leaf Growth, Cell

Expansion, and Cell Production in a Closed Canopy of Poplar. *Plant Physiol.* **2003,** *131* (1), 177–185. https://doi.org/10.1104/pp.011296

Tcherkez, G. The Mechanism of Rubisco-Catalysed Oxygenation. *Plant Cell Environ.* **2016,** *39* (5), 983–997. https://doi.org/10.1111/pce.12629

Teskey, R.; Wertin, T.; Bauweraerts, I.; Ameye, M.; McGuire, M. A.; Steppe, K. Responses of Tree Species to Heat Waves and Extreme Heat Events. *Plant Cell Environ.* **2015,** *38* (9), 1699–1712. https://doi.org/10.1111/pce.12417

Thomason, W.; Battaglia, M. Early Defoliation Effects on Corn Plant Stands and Grain Yield. *Agron. J.* **2020,** *112* (6), 5024–5032. https://doi.org/10.1002/agj2.20402

Thomey, M. L.; Slattery, R. A.; Köhler, I. H.; Bernacchi, C. J.; Ort, D. R. Yield Response of Field-Grown Soybean Exposed to Heat Waves Under Current and Elevated [CO2]. *Glob. Change Biol.* **2019,** *25* (12), 4352–4368. https://doi.org/10.1111/gcb.14796

Timm, S.; Woitschach, F.; Heise, C.; Hagemann, M.; Bauwe, H. Faster removal of 2-Phosphoglycolate Through Photorespiration Improves Abiotic Stress Tolerance of Arabidopsis. *Plants* **2019,** *8* (12), 563. https://doi.org/10.3390/plants8120563

Toh, S.; Imamura, A.; Watanabe, A.; Nakabayashi, K.; Okamoto, M.; Jikumaru, Y. et al. High Temperature-Induced Abscisic Acid Biosynthesis and Its Role in the Inhibition of Gibberellin Action in Arabidopsis Seeds. *Plant Physiol.* **2008,** *146* (3), 1368–1385. https://doi.org/10.1104/pp.107.113738

Toscano, S.; Ferrante, A.; Romano, D. Response of Mediterranean Ornamental Plants to Drought Stress. *Horticulturae* **2019,** *5* (1), 6. https://doi.org/10.3390/horticulturae5010006

Tricker, P. J.; ElHabti, A.; Schmidt, J.; Fleury, D. The Physiological and Genetic Basis of Combined Drought and Heat Tolerance in Wheat. *J. Exp. Bot.* **2018,** *69* (13), 3195–3210. https://doi.org/10.1093/jxb/ery081

Tschaplinski, T. J.; Stewart, D. B.; Norby, R. J. Interactions Between Drought and Elevated CO2 on Osmotic Adjustment and Solute Concentrations of Tree Seedlings. *New Phytol.* **1995,** *131* (2), 169–177.https://doi.org/10.1111/j.1469–8137.1995.tb05718.x

Tubiello, F. N.; Soussana, J. F.; Howden, S. M. Crop and Pasture Response to Climate Change. *Proc. Natl. Acad. Sci.* **2007,** *104* (50), 19686–19690. https://doi.org/10.1073/pnas.0701728104

Uga, Y.; Sugimoto, K.; Ogawa, S.; Rane, J.; Ishitani, M.; Hara, N. et al. Control of Root System Architecture by Deeper Rooting 1 Increases Rice Yield Under Drought Conditions. *Nat. Genet.* **2013,** *45* (9), 1097–1102. https://doi.org/10.1038/ng.2725

Ullah, A.; Sun, H.; Yang, X.; Zhang, X. Drought Coping Strategies in Cotton: Increased Crop Per Drop. *Plant Biotechnol. J.* **2017,** *15* (3), 271–284. https://doi.org/10.1111/pbi.12688

Ulrich, W.; Soliveres, S.; Maestre, F. T.; Gotelli, N. J.; Quero, J. L.; Delgado-Baquerizo, M. et al. Climate and Soil Attributes Determine Plant Species Turnover in Global Drylands. *J. Biogeogr.* **2014,** *41* (12), 2307–2319. https://doi.org/10.1111/jbi.12377

Urban, J.; Ingwers, M. W.; McGuire, M. A.; Teskey, R. O. Increase in Leaf Temperature Opens Stomata and Decouples Net Photosynthesis from Stomatal Conductance in *Pinus taeda* and *Populus deltoides* × Nigra. *J. Exp. Bot.* **2017,** *68* (7), 1757–1767. https://doi.org/10.1093/jxb/erx052

Vahdati, K.; Leslie, C., Eds. *Abiotic Stress: Plant Responses and Applications in Agriculture*; BoD–Books on Demand, 2013.

Vara Prasad, P. V.; Boote, K. J.; Hartwell Allen Jr, L.; Thomas, J. M. Super-Optimal Temperatures Are Detrimental to Peanut (*Arachis hypogaea* L.) Reproductive Processes and Yield at Both Ambient and Elevated Carbon Dioxide. *Glob. Change Biol.* **2003,** *9* (12), 1775–1787.https://doi.org/10.1046/j.1365-2486.2003.00708.x

Vasseur, F.; Pantin, F.; Vile, D. Changes in Light Intensity Reveal a Major Role for Carbon Balance in Arabidopsis Responses to High Temperature. *Plant Cell Environ.* **2011,** *34* (9), 1563–1576.https://doi.org/10.1111/j.1365–3040.2011.02353.x

Venuprasad, R.; Lafitte, H. R.; Atlin, G. N. Response to Direct Selection for Grain Yield Under Drought Stress in Rice. *Crop Sci.* **2007,** *47* (1), 285–293. https://doi.org/10.2135/cropsci2006.03.0181

Vico, G.; Way, D. A.; Hurry, V.; Manzoni, S. Can Leaf Net Photosynthesis Acclimate to Rising and More Variable Temperatures? *Plant Cell Environ.* **2019,** *42* (6), 1913–1928. https://doi.org/10.1111/pce.13525

von Caemmerer, S. Rubisco Carboxylase/Oxygenase: From the Enzyme to the Globe: A Gas Exchange Perspective. *J. Plant Physiol.* **2020,** *252*, 153240. https://doi.org/10.1016/j.jplph.2020.153240

von Caemmerer, S.; Evans, J. R. Temperature Responses of Mesophyll Conductance Differ Greatly Between Species. *Plant Cell Environ.* **2015,** *38* (4), 629–637. https://doi.org/10.1111/pce.12449

Vurayai, R.; Emongor, V.; Moseki, B. Effect of Water Stress Imposed at Different Growth and Development Stages on Morphological Traits and Yield of Bambara Groundnuts (*Vigna subterranea* L. Verdc). *Am. J. Plant Physiol.* **2011,** *6* (1), 17–27.DOI : 10.3923/ajpp.2011.17.27

Wahid, A. Physiological Implications of Metabolite Biosynthesis for Net Assimilation and Heat-Stress Tolerance of Sugarcane (*Saccharum officinarum*) Sprouts. *J. Plant Res.* **2007,** *120* (2), 219–228. https://doi.org/10.1007/s10265–006–0040–5

Wahid, A.; Gelani, S.; Ashraf, M.; Foolad, M. R. Heat Tolerance in Plants: An Overview. *Environ. Exp. Bot.* **2007,** *61* (3), 199–223. https://doi.org/10.1016/j.envexpbot.2007.05.011

Walker, B. J.; VanLoocke, A.; Bernacchi, C. J.; Ort, D. R. The Costs of Photorespiration to Food Production Now and in the Future. *Annu. Rev. Plant Biol.* **2016,** *67* (1), 107–29.doi: 10.1146/annurev-arplant-043015–111709

Walker, B.; Ariza, L. S.; Kaines, S.; Badger, M. R.; Cousins, A. B. Temperature Response of In Vivo Rubisco Kinetics and Mesophyll Conductance in *Arabidopsis thaliana*: Comparisons to *Nicotiana tabacum*. *Plant Cell Environ.* **2013,** *36* (12), 2108–2119. https://doi.org/10.1111/pce.12166

Walter, A.; Silk, W. K.; Schurr, U. Environmental Effects on Spatial and Temporal Patterns of Leaf and Root Growth. *Annu. Rev. Plant Biol.* **2009,** *60* (1), 279–304.

Wang, J. Z.; Cui, L. J.; Wang, Y.; Li, J. L. Growth, Lipid Peroxidation and Photosynthesis in Two Tall Fescue Cultivars Differing in Heat Tolerance. *Biol. Plantarum* **2009,** *53* (2), 237–242. https://doi.org/10.1007/s10535-009-0045-8

Wang, Q.; Zhao, R.; Chen, Q.; da Silva, J. A. T.; Chen, L.; Yu, X. Physiological and Biochemical Responses of Two Herbaceous Peony Cultivars to Drought Stress. *HortScience* **2019,** *54* (3), 492–498. https://doi.org/10.21273/HORTSCI13704-18

Wang, S. Y.; Bunce, J. A.; Maas, J. L. Elevated Carbon Dioxide Increases Contents of Antioxidant Compounds in Field-Grown Strawberries. *J. Agric. Food Chem.* **2003,** *51* (15), 4315–4320. https://doi.org/10.1021/jf021172d

Wang, X.; Cai, J.; Liu, F.; Jin, M.; Yu, H.; Jiang, D. et al. Pre-Anthesis High Temperature Acclimation Alleviates the Negative Effects of Post-Anthesis Heat Stress on Stem Stored

Carbohydrates Remobilization and Grain Starch Accumulation in Wheat. *J. Cereal Sci.* **2012,** *55* (3), 331–336. https://doi.org/10.1016/j.jcs.2012.01.004

Wang, X.; Liu, H.; Yu, F.; Hu, B.; Jia, Y.; Sha, H.; Zhao, H. Differential Activity of the Antioxidant Defence System and Alterations in the Accumulation of Osmolyte and Reactive Oxygen Species Under Drought Stress and Recovery in Rice (*Oryza sativa* L.) Tillering. *Sci. Rep.* **2019,** *9* (1), 1–11. https://doi.org/10.1038/s41598–019–44958-x

Wardlaw, I. F.; Willenbrink, J. Mobilization of Fructan Reserves and Changes in Enzyme Activities in Wheat Stems Correlate with Water Stress During Kernel Filling. *New Phytol.* **2000,** *148* (3), 413–422. https://doi.org/10.1046/j.1469–8137.2000.00777.x

Warland, J. S.; McDonald, M. R.; McKeown, A. M. Annual Yields of Five Crops in the family Brassicacae in Southern Ontario in Relation to Weather and Climate. *Can. J. Plant Sci.* **2006,** *86*, 1209–1215.

Way, D. A. The Bigger They Are, the Harder They Fall: CO2 Concentration and Tree Size Affect Drought Tolerance. *Tree Physiol.* **2011,** *31* (2), 115–116. https://doi.org/10.1093/treephys/tpr009

Way, D. A.; Yamori, W. Thermal Acclimation of Photosynthesis: On the Importance of Adjusting Our Definitions and Accounting for Thermal Acclimation of Respiration. *Photosynth. Res.* **2014,** *119* (1), 89–100. https://doi.org/10.1007/s11120–013–9873–7

Woodward, F. I.; Kelly, C. K. The Influence of CO2 Concentration on Stomatal Density. *New Phytol.* **1995,** *131* (3), 311–327. https://doi.org/10.1111/j.1469–8137.1995.tb03067.x

Xiang, D. B.; Peng, L. X.; Zhao, J. L.; Zou, L.; Zhao, G.; Song, C. Effect of Drought Stress on Yield, Chlorophyll Contents and Photosynthesis in Tartary Buckwheat (*Fagopyrum tataricum*). *J. Food Agric. Environ.* **2013,** *11*, 1358–1363.

Xiao, W.; Sheen, J.; Jang, J. C. The Role of Hexokinase in Plant Sugar Signal Transduction and Growth and Development. *Plant Mol. Biol.* **2000,** *44* (4), 451–461. https://doi.org/10.1023/A:1026501430422

Xie, X.; Wang, Y.; Williamson, L.; Holroyd, G. H.; Tagliavia, C.; Murchie, E. et al. The Identification of Genes Involved in the Stomatal Response to Reduced Atmospheric Relative Humidity. *Curr. Biol.* **2006,** *16* (9), 882–887. https://doi.org/10.1016/j.cub.2006.03.028

Xu, Z.; Shimizu, H.; Yagasaki, Y.; Ito, S.; Zheng, Y.; Zhou, G. Interactive Effects of Elevated CO2, Drought, and Warming on Plants. *J. Plant Growth Regul.* **2013,** *32* (4), 692–707. https://doi.org/10.1007/s00344-013-9337-5

Yadav, R. S.; Hash, C. T.; Bidinger, F. R.; Devos, K. M.; Howarth, C. J. Genomic Regions Associated with Grain Yield and Aspects of Post-Flowering Drought Tolerance in Pearl Millet Across Stress Environments and Tester Background. *Euphytica* **2004,** *136* (3), 265–277. https://doi.org/10.1023/B:EUPH.0000032711.34599.3a

Yamori, W.; Noguchi, K.; Hanba, Y. T.; Terashima, I. Effects of Internal Conductance on the Temperature Dependence of the Photosynthetic Rate in Spinach Leaves from Contrasting Growth Temperatures. *Plant Cell Physiol.* **2006,** *47* (8), 1069–1080. https://doi.org/10.1093/pcp/pcj077

Yang, J.; Zhang, J.; Wang, Z.; Zhu, Q.; Wang, W. Remobilization of Carbon Reserves in Response to Water Deficit During Grain Filling of Rice. *Field Crops Res.* **2001,** *71* (1), 47–55. https://doi.org/10.1016/S0378–4290 (01)00147–2

Yang, S.; Vanderbeld, B.; Wan, J.; Huang, Y. Narrowing Down the Targets: Towards Successful Genetic Engineering of Drought-Tolerant Crops. *Mol. Plant* **2010,** *3* (3), 469–490. https://doi.org/10.1093/mp/ssq016

Yang, X.; Chen, X.; Ge, Q.; Li, B.; Tong, Y.; Zhang, A. et al. Tolerance of Photosynthesis to Photoinhibition, High Temperature and Drought Stress in Flag Leaves of Wheat: A Comparison Between a Hybridization Line and Its Parents Grown Under Field Conditions. *Plant Sci.* **2006,** *171* (3), 389–397. https://doi.org/10.1016/j.plantsci.2006.04.010

Yigit, N.; Sevik, H.; Cetin, M.; Kaya, N. Determination of the Effect of Drought Stress on the Seed Germination in Some Plant Species. *Water Stress Plants* **2016,** *43*, 62.

Yin, J.; Liang, T.; Wang, S.; Zhang, M.; Xiao, J.; Zhan, Y.; Li, C. Effect of Drought and Nitrogen on Betulin and Oleanolic Acid Accumulation and OSC Gene Expression in White Birch Saplings. *Plant Mol. Biol. Rep.* **2015,** *33* (3), 705–715. https://doi.org/10.1007/s11105–014–0778–1

Yin, Y.; Li, S.; Liao, W.; Lu, Q.; Wen, X.; Lu, C. Photosystem II Photochemistry, Photoinhibition, and the Xanthophyll Cycle in Heat-Stressed Rice Leaves. *J. Plant Physiol.* **2010,** *167* (12), 959–966. https://doi.org/10.1016/j.jplph.2009.12.021

Young, L. W.; Wilen, R. W.; Bonham-Smith, P. C. High Temperature Stress of *Brassica napus* During Flowering Reduces Micro-and Megagametophyte Fertility, Induces Fruit Abortion, and Disrupts Seed Production. *J. Exp. Bot.* **2004,** *55* (396), 485–495. https://doi.org/10.1093/jxb/erh038

Yun-Ying, C. A. O.; Hua, D. U. A. N.; Li-Nian, Y. A. N. G.; Zhi-Qing, W. A. N. G.; Shao-Chuan, Z. H. O. U.; Jian-Chang, Y. A. N. G. Effect of Heat Stress During Meiosis on Grain Yield of Rice Cultivars Differing in Heat Tolerance and Its Physiological Mechanism. *Acta Agronomica Sinica* **2008,** *34* (12), 2134–2142. https://doi.org/10.1016/S1875-2780(09)60022-5

Yun-Ying, C.; Hua, D. U. A. N.; Li-Nian, Y. A. N. G.; Zhi-Qing, W. A. N. G.; Li-Jun, L. I. U.; Jian-Chang, Y. A. N. G. Effect of High Temperature During Heading and Early Filling on Grain Yield and Physiological Characteristics in Indica Rice. *Acta Agronomica Sinica* **2009,** *35* (3), 512–521. https://doi.org/10.1016/S1875-2780 (08)60071-1

Yuriko, O.; Keishi, O.; Kazuo, S.; Tran, L. Response of Plants to Water Stress. *Front. Plant Sci.* **2014,** *5*, 86.

Zahoor, I.; Hasan, H.; Gul, A.; Khursheed, A.; Ali, M.; Amir, R. et al. Molecular Mechanism of Drought Tolerance in Wheat. In *Climate Change and Food Security with Emphasis on Wheat*; Academic Press, 2020; pp 129–154. https://doi.org/10.1016/B978–0-12–819527–7.00008-X

Zanetti, S.; Hartwig, U. A.; Van Kessel, C.; Lüscher, A.; Hebeisen, T.; Frehner, M. et al. Does Nitrogen Nutrition Restrict the CO2 Response of Fertile Grassland Lacking Legumes? *Oecologia* **1997,** *112* (1), 17–25. https://doi.org/10.1007/s004420050278

Zare, M.; Azizi, M. H.; Bazrafshan, F. Effect of Drought Stress on Some Agronomic Traits in Ten Barley (*Hordeum vulgare*) Cultivars. *Tech. J. Eng. Appl. Sci.* **2011,** *1* (3), 57–62.

Zargar, S. M.; Gupta, N.; Nazir, M.; Mahajan, R.; Malik, F. A.; Sofi, N. R. et al. Impact of Drought on Photosynthesis: Molecular Perspective. *Plant Gene* **2017,** *11*, 154–159. https://doi.org/10.1016/j.plgene.2017.04.003

Zhang, X.; Cai, J.; Wollenweber, B.; Liu, F.; Dai, T.; Cao, W.; Jiang, D. Multiple Heat and Drought Events Affect Grain Yield and Accumulations of High Molecular Weight Glutenin Subunits and Glutenin Macropolymers in Wheat. *J. Cereal Sci.* **2013,** *57* (1), 134–140. https://doi.org/10.1016/j.jcs.2012.10.010

Zhang, Y.; Mian, M. A. R.; Bouton, J. H. Recent Molecular and Genomic Studies on Stress Tolerance of Forage and Turf Grasses. *Crop Sci.* **2006,** *46* (2), 497–511.https://doi.org/10.2135/cropsci2004.0572

Zhang, Z.; Liu, L.; Zhang, M.; Zhang, Y.; Wang, Q. Effect of Carbon Dioxide Enrichment on Health-Promoting Compounds and Organoleptic Properties of Tomato Fruits Grown in Greenhouse. *Food Chem.* **2014,** *153,* 157–163. https://doi.org/10.1016/j.foodchem.2013.12.052

Zheng, D.; Hunt Jr, E. R.; Running, S. W. A Daily Soil Temperature Model Based on Air Temperature and Precipitation for Continental Applications. *Clim. Res.* **1993,** *2* (3), 183–191.http://dx.doi.org/10.3354/cr002183

Zheng, Y.; Xu, M.; Hou, R.; Shen, R.; Qiu, S.; Ouyang, Z. Effects of Experimental Warming on Stomatal Traits in Leaves of Maize (*Zea may* L.). *Ecol. Evol.* **2013,** *3* (9), 3095–3111. https://doi.org/10.1002/ece3.674

Zhou, R.; Kjær, K. H.; Rosenqvist, E.; Yu, X.; Wu, Z.; Ottosen, C. O. Physiological Response to Heat Stress During Seedling and Anthesis Stage in Tomato Genotypes Differing in Heat Tolerance. *J. Agron. Crop Sci.* **2017,** *203* (1), 68–80. https://doi.org/10.1111/jac.12166

Zhu, J.; Ingram, P. A.; Benfey, P. N.; Elich, T. From Lab to Field, New Approaches to Phenotyping Root System Architecture. *Curr. Opin. Plant Biol.* **2011,** *14* (3), 310–317. https://doi.org/10.1016/j.pbi.2011.03.020

Zhu, P.; Jin, Z.; Zhuang, Q.; Ciais, P.; Bernacchi, C.; Wang, X. et al. The Important But Weakening Maize Yield Benefit of Grain Filling Prolongation in the US Midwest. *Glob. Change Biol.* **2018,** *24* (10), 4718–4730. https://doi.org/10.1111/gcb.14356

Zlatev, Z.; Lidon, F. C. An Overview on Drought Induced Changes in Plant Growth, Water Relations and Photosynthesis. *Emirates J. Food Agric.* **2012,** 57–72. https://doi.org/10.9755/ejfa.v24i1.10599

Zoghi, Z.; Hosseini, S. M.; Kouchaksaraei, M. T.; Kooch, Y.; Guidi, L. The Effect of Biochar Amendment on the Growth, Morphology and Physiology of *Quercus castaneifolia* Seedlings Under Water-Deficit Stress. *Eur. J. Forest Res.* **2019,** *138* (6), 967–979. https://doi.org/10.1007/s10342–019–01217-y

Zsófi, Z. S.; Tóth, E.; Rusjan, D.; Bálo, B. Terroir Aspects of Grape Quality in a Cool Climate Wine Region: Relationship Between Water Deficit, Vegetative Growth and Berry Sugar Concentration. *Sci. Horticult.* **2011,** *127* (4), 494–499. https://doi.org/10.1016/j.scienta.2010.11.014

Zulfiqar, F.; Younis, A.; Riaz, A.; Mansoor, F.; Hameed, M.; Akram, N. A.; Abideen, Z. Morpho-Anatomical Adaptations of Two *Tagetes erecta* L. Cultivars with Contrasting Response to Drought Stress. *Pak. J. Bot.* **2020,** *52* (3), 801–810.http://dx.doi.org/10.30848/PJB2020–3 (35)

CHAPTER 3

Impact of Climate Change on Disease Incidence of Crop Plants

JYOTSANA TILGAM[1,2,3], BHAVYASREE R. K.[4], MANJESH SAAKRE[1,2], and M. ASHAJYOTHI[2,5]

[1]*ICAR - National Institute for Plant Biotechnology, New Delhi, India*

[2]*ICAR - Indian Agricultural Research Institute, New Delhi, India*

[3]*ICAR - National Bureau of Agriculturally Important Microorganisms, Maunath Bhanjan, UP, India*

[4]*Panjab Agricultural University, Regional Research Station, Gurdaspur, Punjab, India*

[5]*ICAR- Central Agroforestry Research Institute, Jhansi, UP, India*

ABSTRACT

Earth's environment is expected to ensue to alter throughout this century and beyond. The extent of climate change in the coming years mainly depends on the anthropogenic emissions and sensitiveness of climate toward those emissions. During the last few decades, our knowledge about climate change has expanded considerably in terms of report documentation of climate change and development of models predicting the future alteration in climate. Meanwhile, modeling studies have become better to predict the influence of climate change on plant diseases. Nonetheless, various uncertainties could affect these estimations. Climate change can have a positive, negative, or neutral impact on plant disease based on its spatial and temporal distribution. Moreover, climate change affects the incidence, prevalence, and severity of plant pathogen/disease. It is

Climate-Resilient Agriculture: A Molecular Perspective. Ashutosh Singh, Saurabh Pandey (Eds.)

imperative to investigate for such models that could envisage the incidence and/or severity of plant pathogens in the actual field condition. Schemes for phytopathogen management must be restructured in changing context with the combination of innovative plans for sustainable food production. This chapter highlights different facets allied with the influences of climate change on pathogen incidence and their mitigation strategies for sustainable crop production.

3.1 INTRODUCTION

Scientific experts are noticing the global climate change in every region and across the climate system. The intergovernmental panel on climate change projects that with the current emission scenario, Earth's mean temperature would hike by up to 4°C above the pre-industrial level by 2100, with related shifts in precipitation patterns. Alterations in climate cause a rapid increase in atmospheric carbon dioxide (CO_2, 320–412 ppm) and Ozone (O_3, 10–40 ppb) concentration during the last few years (Brauer et al., 2016; Ziganshin, 2020).All life forms on Earth are substantially influenced by these changes. Global agricultural communities are facing a challenge because agriculture is reliant on specific conditions. Among the climate change drivers, elevated atmospheric CO_2, temperature, shifts in precipitation patterns, and frequency of extreme weather events are major hurdles. Additionally, climate change affects plant health by increasing disease susceptibility and these changes bring modifications in the agricultural practices to manage the effect of these changes and to avoid the reduction in productivity. Plant pathologists have long acknowledged the role of environment in governing disease development. According to classic disease, a plant disease results of interaction among three factors, namely, a susceptible host, viable pathogen and environment (Fig. 3.1). Plant–pathogen interactions are altered by changing environmental conditions including elevated temperature and atmospheric CO_2, SO_2, ozone levels, flood and drought. Change in environment variables reshapes the plant–pathogen interactions and impacts disease development by altering host physiology and pathogen virulence. Due to the rise in Earth's temperature, the Agroclimatic zone will move near the poles, thus shifting the geographic distribution of existing diseases by introduction of pathogen into the new region. The effects of these climatic changes will differ by pathosystem and geographical area.

These changes may affect not only the optimal conditions for infection but also affect the host specificity and mechanisms of plant infection. Pathogen development and survival rates are influenced with change in host susceptibility, altering the disease condition. Climate change can have a positive, negative, or neutral impact on plant health based on its spatial and temporal distribution. The effect of high temperature on plant is highly reliant on optimum growth and reproduction temperature of the plants. However, yield may get decreased and disease infection may occur if the temperature overruns a plant's the optimum temperature. Under the controlled conditions, few researchers have proved that elevated levels of CO_2 could positively influence plant growth (Zayan, 2019). Though the presence of other variables (extreme temperature, insufficiency of water, nutrient availability, etc.) may decline the crop growth and yield. Many studies have appeared to predict the effect of changing climate on plant health at different Spatiotemporal levels. The basis of predictions includes (1) Previously experienced climate change influences on plant diseases, (2) Presumptions using scientific expertise and carrying out trials, and (3) Computer models (Pautasso et al., 2012). Forecasting and management of environmental shift implications on plant health are intricated by indirect influences and the interplay with global change factors. The linkage between climate change and other factors of global change in affecting plant health must be studied as climate change feedbacks (by alteration in surface albedo patterns and carbon sequestration) play role in deteriorating plant health (O'Halloran et al., 2012).

The progress in molecular breeding and genetic engineering provided many opportunities to develop climate-resilient and disease resistant. Moreover, genetic tools like RNAi and genome editing have become routine practices to play with the genome of pathogens or plants. For instance, RNAi gene silencing in filamentous fungi offers a reference for essential biological processes, along with pathogenicity. In disease management, genome editing tools are acclaimed to completely silence the negative regulators and susceptibility genes in pathogen. Several reports describe that the Clustered Regularly Interspaced Short Palindromic Repeat (CRISPR)/Cas9 system is being used to edit the genomes of plants indicating that this method has a high success rate. Additionally, using CRISPR/Cas9 system, novel and efficient biocontrol strains could be created and released to control phytopathogens, hence preventing the introduction of a foreign gene in the environment.

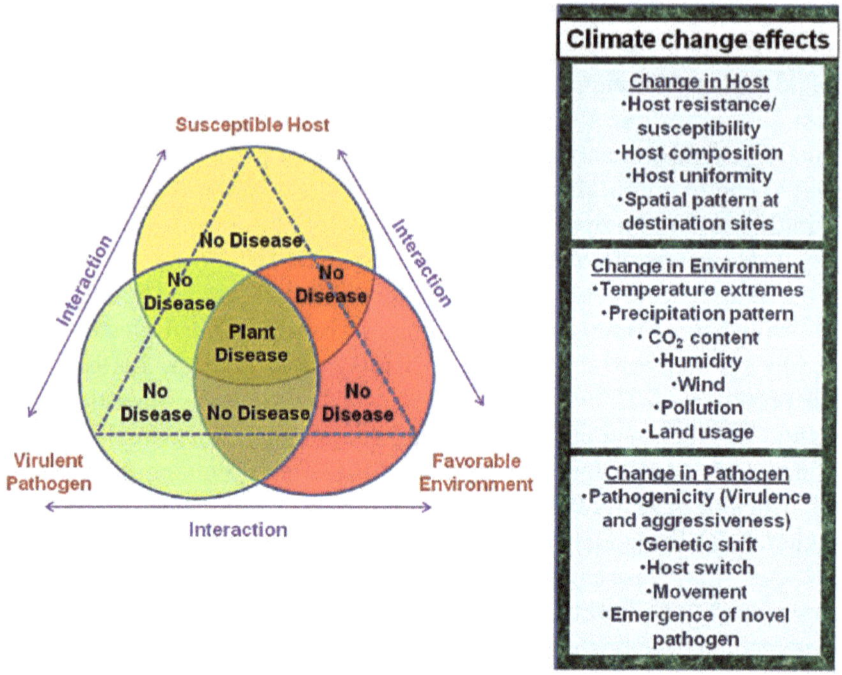

FIGURE 3.1 A schematic representation of the disease triangle where interaction among the components decides the fate of disease development. Effects of climate change in different components.

3.2 IMPACT OF CLIMATE CHANGE ON PLANT PATHOGENIC ORGANISMS

Climate change is the most challenging phenomenon for agriculture all over the world. The change in climate and the climatic aberrations are affecting the living system directly and indirectly. The atmospheric changes are not only affecting the crop plants but also the pathogenic organisms that affect the crop plants. Plant disease is the combined outcome of the susceptible host virulent pathogen and favorable environmental conditions. Any change in these factors will affect the intensity and spread of the diseases.

Climate can change the disease infection by changing the host phenology as well as the effect on the aggressiveness of the pathogens. Also, climatic variations affect the reproductive rate, evolution, and distribution of pathogens in different regions. Some of the pathogens which were minor in some areas become a major problem in that area due to climate change. Meanwhile, some new pathotypes or races of pathogens will get emerged and cause epidemics. The impact of different climatic variations on the pathogen will depend upon biological cycle, epidemiology, survival, and dispersal modes of the pathogen (Hunjan and Lore, 2020).

3.2.1 IMPACT OF CLIMATIC VARIATIONS ON FUNGAL PATHOGENS

The most dominant group of plant pathogen is fungi which is having diverse types of interactive effects with their host plant. They are highly affected by the weather parameters such as temperature and humidity. Temperature and humidity highly influence the different phases of fungal life, including their reproduction, infection, dispersal, and off-season survival. The increase in temperature will break the dormancy of the fungal spores. The high temperature coupled with high humidity is very contingent on the fungal attack (Hunjan and Lore, 2020). Temperature is a key factor which regulates the reproductive rates of many plant pathogenic fungi by increasing sporulation and spore germination. Such effects are also complemented by an increase in CO_2 concentration. This type of variation was observed in the leaf blast and leaf blight pathogens which increased the disease severity as well as the ability of a pathogen to cause an epidemic (Kobayashi et al., 2006; Rajput et al., 2017). Several new races have invaded some areas due to an increase in temperature as a result of global warming. The early infection of yellow rust and the incidence of powdery mildew and foliar diseases are observed in wheat field. The high temperature and moisture not only increase the disease initiation and development but also have favorable effects on the germination and proliferation of fungal spores (Yáñez-López et al., 2012). The elevated CO_2 levels cause so many phenological changes in host plants and also affect the susceptibility of the host to various pathogenic fungi. The infection rate of rice pathogen *Xanthomonas oryzae, Magnaporthe grisea* have increased with the elevated concentration of CO_2 (Kobayashi et al., 2006). Greater fecundity of fungal pathogens has been observed due to increased CO_2 levels and temperature, for example, *Blumeria graminis* f. sp. *Tritici* in wheat (Matić et al., 2018). Also, the life cycles of the pathogen are shortened which increases the number of generations in polycyclic pathogens and also accelerates the evolution with respect to climate change (Elad and Pertot, 2014). Global warming increases the risk of epidemic outbreaks of different pathogens. The regional spread also increases along with the severity of the epidemics. And several models are also being formulated to access the risks of such epidemics (Yáñez-López et al., 2012).The increase in temperature, CO_2, ozone, and drought spells affect plant growth and accelerate tissue damage/tissue death, and this condition increases the infection of necrotrophic pathogens which derive nutrition from dead tissues. These conditions will favor the biotrophs by altering the colonization (Elad and Pertot, 2014).

3.2.2 IMPACT OF CLIMATIC VARIATIONS ON BACTERIAL PATHOGENS

The rise in the atmospheric temperature levels due to global warming also increases the intensity of the bacterial pathogens, which lead a minor pathogen to cause an epidemic. *Burkholderia glumae* causing bacterial panicle blight in rice can cause an epidemic leading to yield losses up to 75% if prolonged hot and humid condition occurs (Ham et al., 2011). The increase in temperature sometimes breakdown the resistance of a host gene by efficiently evolving the pathogen. Such a beneficial effect for the pathogen was observed in the case of bacterial blight of rice (Webb et al., 2010). Elevated temperature favors the infection of heat-loving bacterial pathogens. Some of the bacteria like *Ralstonia, Acidovorax,* and *Burkholderia* which were previously classified as nonpathogenic and non-epidemic emerge as serious problems due to global warming (Priyanka et al., 2020).

3.2.3 IMPACT OF CLIMATIC VARIATIONS ON VIRAL PATHOGENS

As viral diseases have an extra dimension of the vector other than host, pathogen, and environment, the interactions are far more complicated in the case of viral diseases. Any change in pathogen, vector, or host will affect the disease intensity of the viral diseases. Some of the viruses will shift to the newer region due to warmer climates while there is an increase in severity of the symptoms in the viruses like barley yellow dwarf virus and yellow dwarf disease in wheat due to increased temperature and CO_2 concentration (Hunjan and Lore, 2020).

3.3 EFFECT OF CLIMATE CHANGE ON PLANT–PATHOGEN INTERACTION

The interaction between plants and pathogens is highly influenced by the environment, in fact, it is the climate which decides the compatible and incompatible interactions in plant disease development. Some of the climatic factors, such as temperature, humidity, rainfall pattern, and light CO_2. play a key role in plant growth and development, and because of this reason, crops

TABLE 3.1 Effect of Climate Change on Pathogen/Disease.

Sl. No.	Crop	Pathogen	Climate change	Influence on pathogen	References
1	Rice	*Magnaporthe grisea*	Increase in temperature and humidity	Increased potential of pathogen	Hunjan and Lore (2020)
2	Rice	*Xanthomonas oryzae*, *Magnaporthe grisea*	Elevated CO_2	Higher infection rates	Kobayashi et al. (2006)
3	Rice	*Burkholderia glumae*, *B. gladioli*	Increased temperature and humidity	Increased intensity of infection	Ham et al. (2011); Nandakumar et al. (2009)
4	Rice	*Xanthomonas oryzae*	Increased temperature	Overcome disease resistance gene	Webb et al. (2010)
5	Wheat	*Blumeria graminis* f.sp. *tritici*	Increased temperature in winters	Increased disease epidemics	Tang et al. (2017)
6	Wheat	*Puccinia striiformis*	Increased temperature	Increased intensity	Nnadi and Id (2021)
7	Wheat	*Fusarium graminearum* complex	Increased temperature	Increased intensity	Magan et al. (2011)
8	Wheat	*Puccinia graminis* f. sp. *tritici*	Increased temperature and decreased humidity	Increased spread potential	Prank et al. (2019)
9	Wheat	*Magnaporthe oryzae Triticum*	Increased temperature and humidity	Emerged as a new disease	Islam et al. (2019)
10	Wheat	*Puccinia triticina*	Increased temperature	Spread to new regions	Kolmer (2019)
11	Wheat	*Fusarium poae*	Increased temperature	Increased fungal toxin	Nazari et al. (2018)
12	Wheat	*Blumeria graminis* f. sp. *tritici*	Increased temperature and CO_2	Increased pathogenicity	Matić et al. (2018)
13	Wheat	barley yellow dwarf virus and yellow dwarf	increased temperature and CO_2 concentration	Spread to new regions	Hunjan and Lore (2020)

and tree species distribution vary differently in different climatic zones. Similarly, the same environmental factors act and influence on fitness and survival of plant pathogens. Hence, the variability and distribution of plants, plant resistance mechanisms, including PAMP-triggered immunity (PTI) to effector-triggered immunity (ETI), RNA interference, and defense hormone networks, and variability in pathogens, their virulence mechanisms, and toxin production are by and large decided by their tropical, subtropical, arid, temperate, and Mediterranean climate (Cheng et al., 2019). Also, it is known that the disease resistance, tolerance, escape, and susceptibility of host plants are manifested and favored by the environment at certain of their growth stages. However, the rapidly changing climatic scenario globally in the last few decades poses a significant change in plant–pathogen interactions, their survival, fitness, and distribution. Few of the recent studies highlighting the effect of climate change on plant pathogens and their ability to induce disease development are discussed here.

3.3.1 EFFECT OF TEMPERATURE ON PLANT–PATHOGEN INTERACTION

Temperature is a single major factor that influences the plant–pathogen interaction either positively or negatively. It is essential for both plants as well as pathogens' growth and reproduction. The vegetative mycelial growth, conidia formation, germination, and spore dispersal in fungal pathogens are mainly temperature-dependent. The rising global temperature are known to inhibit the ETI in plants which is a major factor for disease resistance in crop plants and also it can cause fitness loss in pathogens. In general, the ETI will be initiated at 10–23°C and the PTI at 23–32°C. ETI-responsive *WRKY46 is* known to dampen at elevated temperatures in *Arabidopsis thaliana* (Chen et al., 2013). Similarly, *Fusarium pseudograminearum*, wheat crown rot causing important necrotrophic plant pathogen has shown reduced fitness at increasing diurnal temperatures (15/15°C, 20/15°C, 25/15°C, and 28/15°C). In addition, the performance of resistant wheat lines was also coming down under higher temperatures, hence there was no reduction in overall disease with reduced pathogen fitness (Sabburg et al., 2015). Also, the disease susceptibility seems to be enhanced at elevated temperatures as seen in the case of *Arabidopsis-Pseudomonas syringae* pv. *Tomato* DC3000 (PstDC3000) interaction with the absence of ETI activation (Gangappa et al., 2017; Huot et al., 2017).

3.3.2 EFFECT OF DROUGHT ON PLANT–PATHOGEN INTERACTION

Moisture plays a crucial role in plant–pathogen interactions, especially initial germination and recognition process during disease development is highly influenced by the presence of moisture at the point of infection. The natural lifecycles of plant pathogens, influence on host predisposition to infection or disease expression, a shift in the natural range of the pathogens, and alteration of rate of genetic change in pathogen populations are the most expected events in plant–pathogen interactions. Climate change-induced prolonged drought became a common global event. Studies in New Zealand on pea root rot (caused by *Aphanomyces euteiches*), wheat take-all (*Gaeumannomyces graminis* var. *tritici*), kiwifruit sclerotinia rot *(Sclerotinia sclerotiorum*), wheat crown rot (*Fusarium* spp.), brassica black leg (*Leptosphaeria maculans*), onion white rot (*Sclerotium cepivorum*), grapevine black foot (*Ilyonectria/Dactylonectria* spp.), and radiata pine red needle cast (*Phytophthora pluvialis*) have shown increased disease expression with increased drought condition (Wakelin et al., 2018). The enhanced disease is a reason for two events: firstly, the effect of multiple stresses including both drought and pathogenic infection on host at a particular time and second, modifying host susceptibility to infection and disease development.

3.3.3 EFFECT OF RISING CO2 AND O3 ON PLANT–PATHOGEN INTERACTION

Climate change induced a rapid increase in atmospheric CO_2 (320–412 ppm) and O_3 (10–40 ppb) concentration in the recent past (Brauer et al., 2016; Ziganshin, 2020). The elevated levels of CO_2 can be related to increased productivity for a few crops but the atmospheric rise in O_3 is detrimental to the plants. Hence, this condition will definitely influence the disease development by damaging plant first-line physical defense structures, such as cuticles (Berner et al., 2015). A study on the incidence of chickpea *Fusarium* wilt caused by *Fusarium oxysporum,* sp. *ciceris* revealed a significant reduction of disease severity under elevated as compared with ambient $O_{3,}$ CO_2 levels (EO+ECO_2). The activities of pathogenesis-related proteins chitinase and β-1,3- glucanase, involved in plant defense mechanisms were enhanced under elevated O_3. The activity of superoxide dismutase increased significantly ($p < 0.5$, 40%) as compared with catalase (12.5%) and peroxidase (17.5%) without any significant increase under EO + ECO_2 (Bhatia et al., 2021). The deleterious effects on plant growth by elevated O_3 may be subsided by the elevated CO_2 level. However, in-depth studies need to be conducted to better understand the extreme weather events affect plant–pathogen interactions.

3.4 PREDICTION OF EFFECT OF CLIMATE CHANGE ON PATHOGEN

Plant disease is an interactive effect of host, pathogen and environment. The effect of the unpredictable and varying environment due to climate change will definitely affect the pathogen and pathogenicity. So, for studying and predicting the patterns of plant diseases, it is inevitable to study the pattern of climate changes in a specific area. As the climate is a complex phenomenon and most of the studies regarding environmental variations are performed in controlled conditions over a short period, it is a very challenging task to build up a model for the prediction of climate influenced changes in living systems. Also the complexity increase with the interactive effects of various climatic factors (Elad and Pertot, 2014).

3.4.1 NEED FOR PLANT DISEASE PREDICTION MODELS

Plant disease forecasting is to predict the occurrence or spread of disease so that the farmers can make economic decisions and also can adopt preventive measures. There are several factors including climate change as well as the complex interaction patterns which challenge the predictability of the plant diseases. But the quantifying the predictability factor is the only lead for designing solutions to many novel plant diseases which are likely to arise or to check the severity of old diseases (Pautasso et al., 2012). The general approach is to predict the behavior of a pathogen or the pattern of epidemics under projected climatic changes based on the model created on the basis of behavior of pathogen or the epidemic under known conditions (Elad and Pertot, 2014).

3.4.2 COMPLEXITY AND LIMITATIONS IN DESIGNING PLANT DISEASE PREDICTION MODELS

Although there is an expanding need of creating reliable models for predicting plant disease behavior in the prevailing condition of variable climate, there are several challenges faced by scientists all over the world in this aspect. There are several models for predicting the changes in climate change, especially for temperature change and rainfall, which are comparatively predictable (Chakraborty and Newton, 2011). But the prediction or forecasting of plant diseases is very complex due to the high-level complexities in ecological interactions. Both the crops and agricultural practices should be considered while constructing a model. So, for the prediction of disease incidence and pattern of spread, the weather predictions should be downscaled into micro-climate prediction at the level of the plant stand. This calls for the need of models which also consider complex ecological interactions (Chakraborty et al., 2000; Chakraborty, 2005; Elad and Pertot, 2014).

The disease incidence or spread depends directly on pathogen, host, and environment. Therefore, due to erratic changes in the climate, there are effects on both pathogen and the host. The climatic variations may contribute to the emergence of new pathogens, which are not subjected to studies to create a model. The host plants also undergo several phenological changes due to climate change. These changes in the host plant affect all the biological processes of the pathogen and thus make the process more complicated. This situation is furthermore complicated for the diseases (e.g., viral diseases) in which the vectors are involved.

The climate prediction models are not very reliable as there are sudden and unpredictable changes in the atmosphere and there are changes in between the adjacent areas of a region. So, all the models related to climate should be regionalized. As the pathogen growth and behavior are also dependent on climatic variations, there is a need for formulating regional models for each area. So the integration of multiscale epidemic simulations also becoming a challenging task in this aspect (Holdenrieder et al., 2004).

3.4.3 GENERALLY USED PREDICTION MODELS FOR PLANT DISEASES AND THEIR EXAMPLES

Mathematical modeling for crop diseases was initiated by Van der Plank to predict the temporal development of plant epidemics (Van Maanen and Xu, 2003). The crop prediction models have been suggested to elucidate the effects of climate change on plant diseases even back from 1966 (Chakraborty et al., 1998). Earlier, there were two important models for predicting plant diseases with respect to climate change viz., Climate matching algorithms and linked disease-crop models. The choice of the model depends on objective and the availability of data for formulating a model. The climate matching algorithm was a comparatively easy model used when detailed data were not available to build a linked disease-crop model. CLIMEX, BIOCLIM, and DOMAIN were the computer packages that were used for linking the meteorological data with pathogen distribution to determine the climatic preferences in crop prediction model. While the linked disease-crop model explains the effect of disease on crop growth and development, which are linked to pathogens (Chakraborty et al., 1998).

There are three steps in modeling. The initial process is the model development in which the biological characters of the pathogen are explained in a mathematical equation. The next step is the model analysis which follows hypothesis testing. In model analysis, the growth parameters or the model are tested against the external parameters or the environment. In hypothesis testing, the results from the analysis are compared against the actual results. There are several mathematical representations of the epidemic development based on modeling starting from Van der Plank's Disease progress curve, then more efficient, area under progress curve and then advanced to linked differential equations. Then the linked differential equations are extended to computer simulation models which have been developed in the past decades (Van Maanen and Xu, 2003).

Later, several other models were developed for analyzing epidemic dynamics and patterns of plant diseases. One of such models was Kermack and McKendrick epidemic model (K&M) which was modified to approximate the epidemic curves. The special cases of K&M such as the Vanderplank differential-delay equation and the compartmental healthy, latent, infectious and post-infectious (H-L-S-R epidemic model) were used for the efficient approximation of epidemic curve (Segarra et al., 2001).

Modern computational and simulation techniques including artificial intelligence have been incorporated into the disease prediction models too. These methods increased the efficiency and reliabilities of the models and they also incorporated multiple factors for the prediction. Some of such models are support vector machine (SVM), support vector regression (SVR), artificial neural network (ANN), extreme learning machine (ELM) which are coupled with linear and regression models (regularized regression models, logistic regression, etc.). These models are found to be efficient in forecasting plant diseases in climate change scenarios. Another model long short-term memory network (LSTM) is also used and found to have an accuracy of up to 78% in various regions (Fenu and Malloci, 2021).

3.5 DEVELOPMENT OF CLIMATE-RESILIENT AND DISEASE-RESISTANT PLANT

Climate change is expected to lead to an upsurge in plant diseases that threaten food security. The development of climate-resilient crops is vital to sustaining food security for the increasing global population. In this regard, scientists and breeders are striving to increase the yield of major food crops to meet the growing global food demand. In this situation, agricultural biotechnology and molecular breeding play an immense role in mitigating the most potentially devastating challenge of all time. In addition, the existing germplasms also provide the genes that can assist plants to tolerate various abiotic and biotic stresses. Through genetic engineering, beneficial genes can be transferred directly or existing genes can be manipulated to create the desired phenotype(s) in the crops of interest, whereas the breeding approach involves the introduction of novel alleles into the target crop through the use of available germplasms.

3.5.1 MOLECULAR BREEDING APPROACHES TO DEVELOP CLIMATE-RESILIENT CROPS

Traditional plant breeding methods rely on phenotypic selection; however, it is difficult, expensive, and unreliable to determine the nature of many traits. In addition, due to the risk of linkage drag, it is highly difficult to introgress the desirable genes from related species in an efficient manner (Singh and Singh, 2015). Furthermore, conventional breeding involves time-consuming methods and takes multiple generations to transfer the desired gene(s) (Mafakheri and Kordrostami, 2020). Certainly, to combat climate change, considering the time constraints, it is nearly impossible to enhance the production of field crops with conventional methods of plant breeding. Auspiciously, the advanced breeding techniques and genomics approaches offer new opportunities to increase food production, quality even under various biotic and abiotic stress conditions. Furthermore, Marker-assisted selection (MAS) and marker-assisted backcrossing (MABC) has been successfully applied for crop improvement to select superior genotypes and to introgress the desirable genes using DNA markers. In the case of disease resistance, multidisciplinary research is being conducted to determine how plants respond to a pathogen attack at the cellular and molecular levels. Germplasm resources from wild crop species have been successfully characterized and contribute as gene banks leading to breeding of disease-resistant cultivars. Furthermore, using genome mapping and quantitative trait loci (QTL) analysis, via biparental or association mapping (AM) populations, it has been possible to uncover genetic control in various biotic stresses, allowing us to carry out MAS, MABC, gene pyramiding or to determine genotypes of high value, and to make direct selections for breeding purposes. QTLs for sheath blight of rice (SBR) (caused by *Rhizoctonia solani*) have been identified in several rice accessions. So far, 50 QTLs conferring resistance to SBR has been dissected and showed resistance in field inoculation tests (Ramalingam et al., 2020). Among these QTLs, several loci were successfully characterized and the functions of many QTLs are decoded. Molecular characterization and sequence analysis of QTL *qSBR11-1* revealed that there is a high degree of homology in 11 tandem repeats with the chitinase defense response genes. The study also interpreted that the promoters of the QTL qSBR11-1 contain several ABA and JA-related elements which indicate the significance of ABA and JA in plant defense against *R. soloni* (Richa et al., 2016). In another study, the functional annotation of QTL *qshb9.2* using bioinformatics tools revealed that a defense-responsive gene encoding β 1–3 glucanase-like protein is responsible for SBR resistance in rice (Yadav et al., 2015). Another major devastating disease, bacterial leaf blight (BLB), is caused by the proteobacteria *Xanthomonas oryzae* pv. *oryzae* that causes up to 80% yield loss in susceptible varieties, and in addition, chemical control is ineffective. From the different rice sources, 46 resistance genes have been identified till date. Among these, genes such as the *Xa4, xa5, Xa7, xa13, Xa21, Xa33*, and *Xa38* are most commonly utilized in hybridization programs for developing BLB-resistant cultivars (Hsu et al., 2020). The gene *Xa21* is noted as a major resistant gene in dominant condition conferred resistance to several *Xanthomonas* isolates. The gene encodes both leucine-rich repeats (LRR) and serine–threonine kinases. These are present on the surface of the cell, act as a receptor for pathogen ligands, and as a result of signal perception, R-proteins are activated subsequently during the intracellular defense response (Song et al., 1995). High-efficiency PCR-based codominant markers were developed for the identification of *Xa21* genes and successfully employed for various MAS and MABC programs. Rice blast is another major disease of rice caused by *Magnaporthe oryzae* causes up to 30% yield loss annually (Asibi et al., 2019). Approximately, 100 resistance genes have been identified so far, and 37 have been characterized (Zhang et al., 2019). However, a high level of resistance to *M. oryzae* is conferred by the *Pi54* gene and it is the most widely used gene for breeding programs against *M. oryzae* fungus. Pi54 is an R gene that encodes NBS-LRR proteins attached with a distinctive zinc finger domain. Pi54 genes induce β-1,3-glucan (callose) synthesis during host–pathogen interactions and the callose acts as a physical barrier by inhibiting hyphal penetration of fungi (Ramalingam et al., 2020). The gene-based markers are developed for the identification of Pi54 and are being used in MAS (Ramkumar et al., 2011). A number of exceptional examples of QTLs involved in various disease resistances have been identified in other crops by the combined efforts of breeders and biotechnologists. A major approach for disease control is to use race-specific resistance genes; however, recent studies indicated that the resistant genes are vulnerable to counter pathogens' evolution. Therefore, most resistant cultivars turn susceptible because of novel pathogen races. To overcome these obstacles, studies have demonstrated that there are both major and minor genes which are involved in complex regulatory pathways that influence the genetic control of disease resistance. These genes are other than R genes and perhaps contribute to durable and broad-spectrum disease resistance. For instance, 3 major and 13 minor QTLs were identified in two advanced backcross rice populations CR071 and QingGuAi3. The findings illustrated that the genetic basis and durable blast resistance in rice populations could be due to the interaction between the major and minor genes. Although, there are several breeding approaches to combat disease resistance in crops, carries certain drawbacks including introduction of unfavorable tightly linked genes into the favorable ones, highly complex, laborious and time-consuming breeding methodologies. In order to overcome these drawbacks, high-throughput genotyping and phenotyping are the most promising and cost-effective technique to identify superior alleles of specific characteristics. Moreover, genome-wide association studies (GWAS) could determine the genomic regions responsible for disease resistance using a natural gene pool. In addition, genotyping-by-sequencing (GBS) is another NGS-based approach utilized to discover and genotype SNPs within crop genomes and populations.

3.5.2 PLANT DISEASE RESISTANCE THROUGH CRISPR/CAS9 TECHNOLOGY

Crop improvement has been revolutionized by advances in functional genomics, especially after the discovery of genome editing. The CRISPR/Cas9 system is a powerful tool for targeted gene editing. CRISPR/Cas9 system is a naturally occurring defense mechanism in bacteria against phages. Biotechnological advancement uses CRISPR to its full potential as a genetic editing tool in plants to develop genome-edited plants for disease resistance. CRISPR/Cas genome editing is considerably more efficient and convenient than TALENs or ZFNs, and does not require protein engineering. In addition,

CRISPR/Cas9 accomplishes cleavage of targets efficiently and a relatively small protein is utilized (Paul et al., 2021). CRISPR editing systems include a single-guide RNA molecule with a defined spacer sequence that guides the Cas protein with nuclease activity to cleave the target DNA. The chopped DNA re-joins using non-homologous recombination with insertion–deletion that cause changes in the reading frame and therefore achieve target gene inactivation. Genome editing requires molecular information about the target gene and the host genome sequence. Fortunately, whole genome sequence information, intensive, and genetic and molecular studies expanded the possibilities for disease control. Genome editing would be highly effective on genes that negatively regulate plant disease resistance, especially those that might be regarded as host susceptibility (S) genes (Yin and Qiu, 2019). S genes play a crucial role in elucidating the host-bacterial pathogen interactions and therefore become common targets to develop genetically resistant crops to bacterial diseases. The gene OsSWEET14 codes for a sucrose-efflux transporter that transports sugars such as sucrose and glucose in rice plants and necessary plant growth and development. However, the TALE protein from *Xanthomonas oryzaepv. oryzae* PXO86 hijacks OsSWEET14 for transporting sugars from host cells to endure pathogen growth and virulence. The specific binding sites of OsSWEET14 for PXO86 targeted and mutated CRISPR/Cas9 system without affecting the function of gene (Jiang et al., 2013). *Xanthomonas citri* spp. *citri (Xcc).* causes the devastating disease of citrus canker. CsLOB1 is identified as an S gene for citrus canker. CsLOB1 codes for plant transcription factor family genes containing the lateral organ boundaries domain (LBD). The bacterial effector PthA4 recognizes an effector-binding element in the CsLOB1 promoter, thereby causing CsLOB1 expression to be activated to facilitate canker development (Jia et al., 2017). Using CRISPR/Cas9 system, null mutants were developed by knocking out CsLOB1 and the edited plants exhibited resistance to *Xcc* (Peng et al., 2017). Surprisingly, there was no phenotypic difference between edited and wild plants signifying that CsLOB1 is an ideal candidate for engineering citrus varieties that could resist canker. A similar strategy of editing S genes has been achieved in most economically important crops including tomato against *Xanthomonas* spp., *Pseudomonas syringae* and *Phytophthora* spp. (Schwartz et al., 2015).

Plant diseases are predominantly caused by fungi and oomycetes; which have a huge impact on agriculture. Although they are taxonomically separated, they have similar filamentous growth and host infection structures (Paul et al., 2021). Through genome editing, fungal disease resistance can be achieved by knockout the negative regulators and susceptibility genes. Several reports describe that the CRISPR/Cas9 system being used to edit the genomes of plants indicating that this method has a high success rate. Powdery mildew is a fungal disease that affects a wide range of crop plants all over the world. Developing a broad-spectrum resistance to powdery mildew was achieved with the discovery of the *mlo (mildew resistance locus o)* mutant locus for barley mildew. Barley *Mlo* gene was cloned and the sequence analysis showed that the *Mlo* gene is evolutionarily conserved in monocots and dicots. The *Mlo* genes were edited in various crops using CRISPR and the edited plants showed resistance to powdery mildew. *SlMlo1* was targeted with two single-guide RNAs in tomato to generate *Mlo* knockout mutants. Consequently, the resulting tomato mutant is completely resistant to the fungus *Oidium neolycopersici* (Nekrasov et al., 2017). In one study, a rice gene *OsERF922* belonging to the gene family of APETELA2/ethylene responsive factors (AP2/ERF) was targeted for rice blast resistance. In rice, the gene *OsERF922* plays a crucial role in adapting to biotic and abiotic stresses. In addition to abscisic acid and salt stress, *OsERF922* is induced by *Magnaporthe oryzae* indicating that rice blast resistance is negatively regulated by *OsERF922* gene expression. Oserf922 knockout mutants were generated by CRISPR-mediated targeted modification of OsERF922 and blast resistance was enhanced without compromising agronomic traits (Wang et al., 2016).

Plant viruses often evolve rapidly and are usually involved in their transmission as insect vectors. Plant viral diseases are difficult to manage due to the virus uses host translational machinery to replicate its genome. Several investigations have shown that during viral infections, eIF4G and its regulatory cellular proteins are modified (López et al., 2022). eIF4G and its isoforms are translation initiation factors that are involved in recruitment of ribosomes and initiate translation process. Genetic screening for rice mutants that have enhanced resistance to rice tungro, spherical virus found that a loss-of-function mutation in the eIF4G gene led to rice tungro disease (RTD) resistance. Furthermore, knockout of rice eIF4G with CRISPR/Cas9 resulted in resistance to RTD and increased yield under glasshouse conditions (Macovei et al., 2018). CRISPR/Cas9-generated mutant cucumbers with mutations at two sites in the eIF4E gene were shown resistant to cucumber vein yellowing virus (Chandrasekaran et al., 2016). The modified versions of Cas9 such as FnCas9 and LshCas13a have been described to specifically target RNA in vivo (Price et al., 2015). By harnessing the CRISPR/FnCas9 system researchers established virus-resistant *Nicotiana benthamiana* and *Arabidopsis* plants against tobacco mosaic virus (TMV) and cucumber mosaic virus (CMV), respectively. When plants expressed FnCas9 and sgRNA specific to TMV or CMV, infection symptoms were attenuated and viral RNA accumulated was reduced in these plants (Zhang et al., 2018). CRISPR/Cas13a is another effective tool which directly modifies single-stranded RNA. CRISPR/Cas13a targeting the Coat protein coding sequence in the potato virus Y (PVY) viral genome led to resistance against PVY in transgenic potato lines and successfully demonstrated the ability of CRISPR/Cas13a to interfere with and attenuate PVY infection (Zhan et al., 2019).

Genome editing has emerged as a great promise to overcome the drawbacks of traditional resistance breeding. It is possible to manipulate the genome of plants by introducing restriction enzymes to knock out or replace the gene of interest without introgressing unrelated genes. CRISPR/Cas system directly modifies target genes in elite varieties, bypassing the breeding process and therefore requires less time to develop a resistant variety without disturbing the other agronomic characters.

3.5.3 RNA INTERFERENCE (RNAI) FOR DISEASE RESISTANCE

Gene silencing usually refers to the mode of "switching off" of a gene by interfering with its expression at the transcriptional or post-transcriptional level. RNAi-mediated gene silencing is accomplished through small RNAs mainly by siRNAs and miRNAs. The process of gene silencing starts with the cleavage of long dsRNAs into 21–25 nt fragments of small RNAs in cytoplasm. The targeted dsRNA may originate from convergent transcription, inverted repeat genes, intermediates of viral genome replication or through the action of RNA-dependent RNA polymerases (RDRP) on single-stranded RNAs (Pattanayak et al., 2013). The enzyme called dicer catalyzes the cleavage of long dsRNA into small RNAs. The processed small RNAs are introduced into a multiprotein silencing complex called RNA-induced silencing complex (RISC). The subsequent unwinding of siRNA duplex, in turn, leads to active confirmation of RISC complex (RISC*). Then, the target mRNA is recognized by antisense RNA, which signals RISC complex for the endonucleolytic degradation of the target mRNA (Jain et al., 2018). For the last two decades, RNAi is regarded as one of the central reverse genetics tools and has been widely used in crop improvement as a reverse genetics tool to decode gene functions, to defense against different biotic stresses, such as viruses, fungi, nematodes, and insect pests (Sen et al., 2021). The research on the RNAi for the control of various fungal and viral diseases has made phenomenal growth in recent years. The host-delivered RNAi (HD-RNAi) or host-induced gene silencing (HIGS) is a transgenic-based strategy in which the transgenic plant expresses a gene coding for double-stranded RNA (dsRNA) which is processed into siRNA or miRNA that specifically silences the target gene of infesting fungi or virus. Here, the processed dsRNA could be transferred from the host plant cell to target cell through some channel proteins, hence the mechanism is known as host-delivered gene silencing or HIGS. Due to the rapid evolution of virulent strains and associated toxins, resistance (R) genes are not always useful for every fungus. The advantage of using RNAi technology over the other approaches is that the RNAi is being used to target both host genes and/or target fungal/viral genes to combat disease resistance. For instance, the dsRNA targeting the avirulence gene Avra10 significantly reduced the development of fungal pathogen *Blumeria graminis* in transgenic barley and wheat plants (Nowara et al., 2010). *Magnaporthe oryzae* resistance was conferred in rice by HIGS of fungal transcription factor *MoAP1*, which plays an important role production of conidia, hyphal growth, and pathogenicity (Guo et al., 2019). Additionally, RNAi constructs of MAP kinase and the cyclophilin of leaf rust were expressed stably in the transgenic wheat, resulting in a drastic reduction in fungal biomass and restricting fungal development (Panwar et al., 2018). In a transgenic study to target the wheat streak mosaic virus (WSMV) nuclear inclusion protein “a” (NIa) gene, an RNAi construct was designed. Transgenic wheat plants expressing dsRNA to target viral NIa gene confers immunity to WSMV infection (Fahim et al., 2010). Furthermore, a variety of conserved sequence elements in viruses can be targeted by using artificial microRNAs (amiRNAs). These are endogenous pre-miRNAs but contain the customized sequences in place of original miRNA:miRNA* duplex region. Potential amiRNA sequences are being used to minimize the off-target effects (Qi et al., 2019). Gene constructs harboring the rice miR528, along with amiRNAs were expressed in transgenic tobacco (*Nicotiana benthamiana*) plants to target the RNA-dependent RNA polymerase (RdRp) gene of *Cymbidium* mosaic virus (CymMV) and *Odontoglossum* ringspot virus (ORSV). Transgenic tobacco lines with artificial miR528 showed resistance to CymMV and ORSV (Petchthai et al., 2018). RNAi approach has been highly effective in gaining disease-tolerant transgenic plants and in determining the function of vital genes. HIGS is an extremely useful and powerful approach to achieve disease and insect resistant plants and for advancing genetic research. The development of RNAi-derived strategies to develop durable disease and insect resistant plants will be enhanced by the development of RNAi gene targets, transformation methods, highly efficient vector constructs, stable transgenic systems, and other novel technologies such as artificial microRNAs. HIGS has proven to be an innovative method for combating disease resistance and improving yield quality in an eco-friendly and sustainable manner.

3.6 CONCLUSION AND FUTURE PROSPECTS

The past two centuries have witnessed a prodigious change in Earth's climate caused by anthropogenic activity. Climate change ultimately threatens food security and sustainability. Meanwhile, with the help of different approaches of breeding, genetic engineering, scientific communities are trying to overcome the losses caused by devastating diseases. Furthermore, crop disease modeling has become the need of the hour in this drastic climate changing scenario. This helps the growers to help with the decision making both economical and management decisions. Various studies suggest that to produce an accurate and reliable model, there is a need for more and better empirical data. This documented impact on plant pathosystems should be processed with expert knowledge along with computer simulations to develop a model (Pautasso et al., 2012). Several user-friendly computational models are also available now which can be utilized even by non-specialists and even the users can set the parameters according to the prevailing conditions (Cunniffe et al., 2015). Such models can overcome the challenge of regionalizing the models. In this era of computational modeling and simulation, it is possible to develop more accurate models with enhanced accuracy and reliability. This is possible by coupling several models and by expanding the data sets. Also, till now, only a few pathogens and crops are being focused on. So, studies are required to reach more pathogens, as "the minor disease can become a major problem" in this climate changing scenario.

KEYWORDS

- **climate change**
- **environment**
- **anthropogenic**
- **crop**
- **disease**

REFERENCES

Asibi, A. E.; Chai, Q.; Coulter, J. A. Rice Blast: A Disease with Implications for Global Food Security. *Agronomy* **2019,** *9* (8), 451.

Berner, J. M.; Maliba, B.; Inbaraj, P. Impact of Elevated Levels of CO2 and O3 on the Yield and Photosynthetic Capabilities of *Brassica napus*. *Procedia Environ. Sci.* **2015,** *29*, 255.

Bhatia, A.; Mina, U.; Kumar, V.; Tomer, R.; Kumar, A.; Chakrabarti, B. et al. Effect of Elevated Ozone and Carbon Dioxide Interaction on Growth, Yield, Nutrient Content and Wilt Disease Severity in Chickpea Grown in Northern India. *Heliyon* **2021,** *7* (1), e06049.

Brauer, M.; Freedman, G.; Frostad, J.; Van Donkelaar, A.; Martin, R. V.; Dentener, F.; Cohen, A. Ambient Air Pollution Exposure Estimation for the Global Burden of Disease 2013. *Environ. Sci. Technol.* **2016,** *50* (1), 79–88.

Chakraborty, S. Potential Impact of Climate Change on Plant – Pathogen Interactions Presented as a Keynote Address at the 15th Biennial Conference of the Australasian Plant Pathology Society. *Australasian Plant Pathol.* **2005,** *34*, 443–448.

Chakraborty, S.; Tiedemann, V.; Teng, S. Environmental Pollution Keynote Review Climate Change: Potential Impact on Plant Diseases. *Environ. Pollut.* **2000,** *108* (3), 317–326.

Chakraborty, S.; Newton, A. C. Climate Change, Plant Diseases and Food Security: An Overview. *Plant Pathol.* **2011,** 60 (1), 2–14. https://doi.org/10.1111/j.1365-3059.2010.02411.x

Chakraborty, S.; Murray, G. M.; Magarey, P. A.; Yonow, T.; O'Brien, R. G.; Croft, B. J.; Barbetti, M. J.; Sivasithamparam, K.; Old, K. M.; Dudzinski, M. J.; Sutherst, R. W.; Penrose, L. J.; Archer, C.; Emmett, R. W. Potential Impact of Climate Change on Plant Diseases of Economic Significance to Australia. *Australasian Plant Pathol.* **1998,** *27* (1), 15–35. https://doi.org/10.1071/AP98001

Chandrasekaran, J.; Brumin, M.; Wolf, D.; Leibman, D.; Klap, C.; Pearlsman, M.; Sherman, A.; Arazi, T.; Gal-On, A. Development of Broad Virus Resistance in Non-Transgenic Cucumber Using CRISPR/Cas9 Technology. *Mol. Plant Pathol.* **2016,** *17*, 1140–1153.

Cheng, C.; Gao, X.; Feng, B.; Sheen, J.; Shan, L.; He, P. Plant Immune Response to Pathogens Differs with Changing Temperatures. *Nat. Commun.* **2013,** *4* (1), 1–9.

Cheng, Y. T.; Zhang, L.; He, S. Y. Plant-Microbe Interactions Facing Environmental Challenge. *Cell Host Microbe* **2019,** *26* (2), 183–192.

Cunniffe, N. J.; Stutt, R. O. J. H.; DeSimone, R. E.; Gottwald, T. R.; Gilligan, C. A. Optimising and Communicating Options for the Control of Invasive Plant Disease When There Is Epidemiological Uncertainty. *PLoS Comput. Biol.* **2015,** *11* (4), 1–24. https://doi.org/10.1371/journal.pcbi.1004211

Elad, Y.; Pertot, I. Climate Change Impacts on Plant Pathogens and Plant Diseases. *J. Crop Improvement* **2014,** *28* (1), 99–139. https://doi.org/10.1080/15427528.2014.865412

Fahim, M.; Navarrete, A. L.; Millar, A. A.; Larkin, P. J. Hairpin RNA Derived from Viral *NIa* Gene Confers Immunity to Wheat Streak Mosaic Virus Infection in Transgenic Wheat Plants. *Plant Biotechnol. J.* **2019,** *8* (7), 821–834.

Fenu, G.; Malloci, F. M. Forecasting Plant and Crop Disease: An Explorative Study on Current Algorithms. *Big Data Cogn. Comput.* **2021,** *5* (1) 2.

Gangappa, S. N.; Berriri, S.; Kumar, S. V. PIF4 Coordinates Thermosensory Growth and Immunity in *Arabidopsis*. *Curr. Biol.* **2017,** *27* (2), 243–249.

Guo, X. Y.; Li, Y.; Fan, J.; Xiong, H.; Xu, F. X.; Shi, J.; Shi, Y.; Zhao, J. Q.; Wang, Y. F.; Cao, X. L.; Wang, W. M. Host-Induced Gene Silencing of MoAP1 Confers Broad-Spectrum Resistance to *Magnaporthe oryzae*. *Front. Plant Sci.* **2019,** *10*, 433.

Ham, J. H.; Melanson, R. A.; Rush, M. C. *Burkholderiaglumae*: Next Major Pathogen of Rice? *Mol. Plant Pathol.* **2011,** *12* (4), 329–339.

Holdenrieder, O.; Pautasso, M.; Weisberg, P. J.; Lonsdale, D. Tree Diseases and Landscape Processes: The Challenge of Landscape Pathology. *Trends Ecol. Evol.* **2004,** *19* (8), 446–452. https://doi.org/10.1016/j.tree.2004.06.003

Hsu, Y. C.; Chiu, C. H.; Yap, R.; Tseng, Y. C.; Wu, Y. P. Pyramiding Bacterial Blight Resistance Genes in Tainung82 for Broad-Spectrum Resistance Using Marker-Assisted Selection. *Int. J. Mol. Sci.* **2020,** *21*, 1281.

Hunjan, M. S.; Lore, J. S. Climate Change: Impact on Plant Pathogens, Diseases, and Their Management. In: *Crop Protection Under Changing Climate*; Springer, 2020; pp 85–100.

Huot, B.; Castroverde, C. D. M.; Velásquez, A. C.; Hubbard, E.; Pulman, J. A.; Yao, J.; He, S. Y. Dual Impact of Elevated Temperature on Plant Defence and Bacterial Virulence in *Arabidopsis*. *Nat. Commun.* **2017,** *8* (1), 1–12.

Islam, M. T.; Kim, K.-H.; Choi, J. 2019. Wheat blast in Bangladesh: the current situation and future impacts. The Plant Pathology Journal, 35 (1), 1.

Jain, P. K.; Bhattacharya, R.; Kohli, D.; Aminedi, R.; Agrawal, P. K. RNAi for Resistance Against Biotic Stresses in Crop Plants. In: *Biotechnologies of Crop Improvement*; Springer: Cham, 2018; 2:67–112.

Jia, H.; Zhang, Y.; Orbović, V.; Xu, J.; White, F. F.; Jones, J. B.; Wang, N. Genome Editing of the Disease Susceptibility Gene Cs LOB 1 in Citrus Confers Resistance to Citrus Canker. *Plant Biotechnol. J.* **2017,** *15* (7), 817–823.

Jiang, W.; Zhou, H.; Bi, H.; Fromm, M.; Yang, B. and Weeks, D. P. Demonstration of CRISPR/Cas9/sgRNA-Mediated Targeted Gene Modification in Arabidopsis, Tobacco, Sorghum and Rice. *Nucleic Acids Res.* **2013,** *41* (20), 188.

Kobayashi, T.; Ishiguro, K.; Nakajima, T.; Kim, H. Y.; Okada, M.; Kobayashi, K. Effects of Elevated Atmospheric CO2 Concentration on the Infection of Rice Blast and Sheath Blight. *Phytopathology* **2006,** *96* (4), 425–431.

Kolmer, J. A. Virulence of *Puccinia triticina*, the Wheat Leaf Rust Fungus, in the United States in 2017. *Plant Dis.* **2019,** *103* (8), 2113–2120.

López-Ulloa, B.; Fuentes, Y.; Pizarro-Ortega, M. S. and López-Lastra, M. RNA-Binding Proteins as Regulators of Internal Initiation of Viral mRNA Translation. *Viruses* **2022,** *14* (2), 188.

Macovei, A.; Sevilla, N. R.; Cantos, C.; Jonson, G. B.; Slamet-Loedin, I.; Čermák, T.; Voytas, D. F.; Choi, I. R.; Chadha-Mohanty, P. Novel Alleles of Rice eIF4G Generated by CRISPR/Cas9-Targeted Mutagenesis Confer Resistance to Rice Tungro Spherical Virus. *Plant Biotechnol. J.* **2018,** *16* (11), 1918–1927.

Mafakheri, M.; Kordrostami, M. In Plant Ecophysiology and Adaptation under Climate Change: Mechanisms and Perspectives II. In: *Role of Molecular Tools and Biotechnology in Climate-Resilient Agriculture*; Springer: Singapore, 2020; pp 491–529.

Magan, N.; Medina, A.; Aldred, D. Possible Climate-Change Effects on Mycotoxin Contamination of Food Crops Pre-and Postharvest. *Plant Pathol.* **2011,** *60* (1), 150–163.

Matić, S.; Cucu, M. A.; Garibaldi, A.; Gullino, M. L. Combined Effect of CO2 and Temperature on Wheat Powdery Mildew Development. *Plant Pathol. J.* **2018,** *34* (4), 316.

Nandakumar, R.; Shahjahan, A. K. M.; Yuan, X. L.; Dickstein, E. R.; Groth, D. E.; Clark, C. A.; Cartwright, R. D.; Rush, M. C. *Burkholderia glumae* and *B. gladioli* Cause Bacterial Panicle Blight in Rice in the Southern United States. *Plant Dis.* **2009,** *93* (9), 896–905.

Nazari, L.; Pattori, E.; Manstretta, V.; Terzi, V.; Morcia, C.; Somma, S.; Moretti, A.; Ritieni, A.; Rossi, V. Effect of Temperature on Growth, Wheat Head Infection, and Nivalenol Production by *Fusarium poae*. *Food Microbiol.* **2018,** *76*, 83–90.

Nekrasov, V.; Wang, C.; Win, J.; Lanz, C.; Weigel, D.; Kamoun, S. Rapid Generation of a Transgenefree Powdery Mildew Resistant Tomato by Genome Deletion. *Sci. Rep.* **2017,** *7*, 482.

Nnadi, N. E.; Id, D. A. C. Climate Change and the Emergence of Fungal Pathogens. *PLOS Pathologens* **2021,** *17* (4), 1–6. https://doi.org/10.1371/journal.ppat.1009503

Nowara, D.; Gay, A.; Lacomme, C.; Shaw, J.; Ridout, C.; Douchkov, D.; Hensel, G.; Kumlehn, J.; Schweizer, P. HIGS: Host-Induced Gene Silencing in the Obligate Biotrophic Fungal Pathogen *Blumeria graminis*. *Plant Cell* **2010,** *22* (9), 3130–3141.

O'Halloran, T. L.; Law, B. E.; Goulden, M. L.; Wang, Z.; Barr, J. G.; Schaaf, C.; Brown, M.; Fuentes, J. D.; Göckede, M.; Black, A. and Engel, V. Radiative Forcing of Natural Forest Disturbances. *Glob. Change Biol.* **2012,** *18* (2), 555–565.

Panwar, V.; Jordan, M.; McCallum, B.; Bakkeren, G. Host-Induced Silencing of Essential Genes in *Puccinia triticina* Through Transgenic Expression of RNAi Sequences Reduces Severity of Leaf Rust Infection in Wheat. *Plant Biotechnol. J.* **2018,** *16*, 1013–1023.

Pattanayak, D.; Solanke, A. U.; Kumar, P. A. Plant RNA Interference Pathways: Diversity in Function, Similarity in Action. *Plant Mol. Biol. Rep.* **2013,** *31* (3), 493–506.

Paul, N. C.; Park, S. W.; Liu, H.; Choi, S.; Ma, J.; MacCready, J. S.; Chilvers, M. I.; Sang, H. Plant and Fungal Genome Editing to Enhance Plant Disease Resistance Using the CRISPR/Cas9 System. *Front. Plant Sci.* **2021,** 1534.

Pautasso, M.; Döring, T. F.; Garbelotto, M.; Pellis, L.; Jeger, M. J. Impacts of Climate Change on Plant Diseases—Opinions and Trends. *Eur. J. Plant Pathol.* **2012,** *133* (1), 295–313. https://doi.org/10.1007/s10658–012–9936–1

Peng, A.; Chen, S.; Lei, T.; Xu, L.; He, Y.; Wu, L.; Yao, L.; Zou, X. Engineering Canker-Resistant Plants Through CRISPR/Cas9-Targeted Editing of the Susceptibility Gene CsLOB1 Promoter in Citrus. *Plant Biotechnol. J.* **2017,** *15*, 1509–1519.

Petchthai, U.; Yee, C. S. L.; Wong, S. M. Resistance to CymMV and ORSV in Artificial MicroRNA Transgenic *Nicotiana benthamiana* Plants. *Sci. Rep.* **2018,** *8* (1), 1–8.

Prank, M.; Kenaley, S. C.; Bergstrom, G. C.; Acevedo, M.; Mahowald, N. M. Climate Change Impacts the Spread Potential of Wheat Stem Rust, a Significant Crop Disease. *Environ. Res. Lett.* **2019,** *14* (12), 124053.

Price, A. A.; Sampson, T. R.; Ratner, H. K.; Grakoui, A.; Weiss, D. S. Cas9-Mediated Targeting of Viral RNA in Eukaryotic Cells. *Proc. Natl. Acad. Sci.* **2015,** *112* (19), 6164–6169.

Priyanka, A. K. M.; Varma, S.; Kumar, V.; Sharma, R. S. Impact of Climate Change on Plant Diseases and Management Strategies: A Review. *IJCS* **2020,** *8* (2), 2968–2973.

Qi, T.; Guo, J.; Peng, H.; Liu, P.; Kang, Z.; Guo, J. Host-Induced Gene Silencing: A Powerful Strategy to Control Diseases of Wheat and Barley. *Int. J. Mol. Sci.* **2019,** *20* (1), 206.

Rajput, L. S.; Sharma, T.; Madhusudhan, P.; Sinha, P. Effect of Temperature on Growth and Sporulation of Rice Leaf Blast Pathogen *Magnaporthe oryzae*. *Int. J. Curr. Microbiol. Appl. Sci.* **2017,** *6* (3), 394–401.

Ramalingam, J.; Raveendra, C.; Savitha, P.; Vidya, V.; Chaithra, T. L.; Velprabakaran, S.; Saraswathi, R.; Ramanathan, A.; Arumugam Pillai, M. P.; Arumugachamy, S.; Vanniarajan, C. Gene Pyramiding for Achieving Enhanced Resistance to Bacterial Blight, Blast, and Sheath Blight Diseases in Rice. *Front. Plant Sci.* **2020,** *11*, 1662.

Ramkumar, G.; Srinivasarao, K.; Mohan, K. M.; Sudarshan, I.; Sivaranjani, A. K. P.; Gopalakrishna, K. Development and Validation of Functional Marker Targeting an InDel in the Major Rice Blast Disease Resistance Gene Pi54 (Pikh). *Mol. Breed.* **2011,** *27*, 129–135.

Richa, K.; Tiwari, I. M.; Kumari, M.; Devanna, B. N.; Sonah, H.; Kumari, A.; Nagar, R.; Sharma, V.; Botella, J. R.; Sharma, T. R. Functional Characterization of Novel Chitinase Genes Present in the Sheath Blight Resistance QTL: qSBR11–1 in Rice Line Tetep. *Front. Plant Sci.* **2016,** *7*, 244.

Sabburg, R.; Obanor, F.; Aitken, E.; Chakraborty, S. Changing Fitness of a Necrotrophic Plant Pathogen Under Increasing Temperature. *Glob. Change Biol.* **2015,** *21* (8), 3126–3137.

Schwartz, A. R.; Potnis, N.; Timilsina, S.; Wilson, M.; Patané, J.; Martins Jr, J.; Minsavage, G. V.; Dahlbeck, D.; Akhunova, A.; Almeida, N.; Vallad, G. E. Phylogenomics of *Xanthomonas* Field Strains Infecting Pepper and Tomato Reveals Diversity in Effector Repertoires and Identifies Determinants of Host Specificity. *Front. Microbial.* **2015,** *6*, 535.

Segarra, J.; Jeger, M. J.; Van den Bosch, F. Epidemic Dynamics and Patterns of Plant Diseases. *Phytopathology* **2001,** *91* (10), 1001–1010. https://doi.org/10.1094/PHYTO.2001.91.10.1001

Sen, P.; Lata, C.; Kiran, K.; Mondal, T. K. RNA Interference (RNAi) in Functional Genomics of Wheat. In: *Genome Engineering for Crop Improvement*; Springer: Cham, 2021; pp. 239–264.

Singh, B. D.; Singh, A. K. *Marker-Assisted Plant Breeding: Principles and Practices*; Springer: New Delhi, 2015.

Song, W. Y.; Wang, G. L.; Chen, L. L.; Kim, H. S.; Pi, L. Y.; Holsten, T.; Gardner, J.; Wang, B.; Zhai, W. X.; Zhu, L. H.; Fauquet, C. A Receptor Kinase-Like Protein Encoded by the Rice Disease Resistance Gene, Xa21. *Science* **1995,** *270* (5243), 1804–1806.

Tang, X.; Cao, X.; Xu, X.; Jiang, Y.; Luo, Y.; Ma, Z.; Fan, J.; Zhou, Y. Effects of Climate Change on Epidemics of Powdery Mildew in Winter Wheat in China. *Plant Dis.* **2017,** *101* (10), 1753–1760.

Van Maanen, A.; Xu, X. M. Modelling Plant Disease Epidemics. *Eur. J. Plant Pathol.* **2003,** *109*, 669–682. https://doi.org/10.1023/A

Wakelin, S. A.; Gomez-Gallego, M.; Jones, E.; Smaill, S.; Lear, G.; Lambie, S. Climate Change Induced Drought Impacts on Plant Diseases in New Zealand. *Australasian Plant Pathol.* **2018,** *47* (1), 101–114.

Wang, F.; Wang, C.; Liu, P.; Lei, C.; Hao, W.; Gao, Y.; Liu, Y. G.; Zhao, K. Enhanced Rice Blast Resistance by CRISPR/Cas9-Targeted Mutagenesis of the ERF Transcription Factor Gene OsERF922. *PLoS One* **2016,** *11*, 0154027.

Webb, K. M.; Ona, I.; Bai, J.; Garrett, K. A.; Mew, T.; Vera Cruz, C. M.; Leach, J. E. A Benefit of High Temperature: Increased Effectiveness of a Rice Bacterial Blight Disease Resistance Gene. *New Phytologist* **2010,** *185* (2), 568–576.

Yadav, S.; Anuradha, G.; Kumar, R. R.; Vemireddy, L. R.; Sudhakar, R.; Donempudi, K.; Venkata, D.; Jabeen, F.; Narasimhan, Y. K.; Marathi, B.; Siddiq, E. A. Identification of QTLs and Possible Candidate Genes Conferring Sheath Blight Resistance in Rice (*Oryza sativa* L.). *SpringerPlus* **2015,** *4* (1), 1–12.

Yáñez-López, R.; Torres-Pacheco, I.; Guevara-González, R. G.; Hernández-Zul, M. I.; Quijano-Carranza, J. A.; Rico-García, E. The Effect of Climate Change on Plant Diseases. *Afr. J. Biotechnol.* **2012,** *11* (10), 2417–2428. https://doi.org/10.5897/ajb10.2442

Yin, K.; Qiu, J. L. Genome Editing for Plant Disease Resistance: Applications and Perspectives. *Phil. Trans. R. Soc. B* **2019,** *374* (1767), 20180322.

Zayan, S. A. Impact of Climate Change on Plant Diseases and IPM Strategies. In *Plant Diseases-Current Threats and Management Trends*; IntechOpen, 2019.

Zhan, X.; Zhang, F.; Zhong, Z.; Chen, R.; Wang, Y.; Chang, L.; Bock, R.; Nie, B.; Zhang, J. Generation of Virus-Resistant Potato Plants by RNA Genome Targeting. *Plant Biotechnol. J.* **2019,** *17* (9), 1814–1822.

Zhang, M.; Wang, S.; Yuan, M. An Update on Molecular Mechanism of Disease Resistance Genes and Their Application for Genetic Improvement of Rice. *Mol. Breed.* **2019,** *39* (10), 1–3.

Zhang, T.; Zheng, Q.; Yi, X.; An, H.; Zhao, Y.; Ma, S.; Zhou, G. Establishing RNA Virus Resistance in Plants by Harnessing CRISPR Immune System. *Plant Biotechnol. J.* **2018,** *16* (8), 1415–1423.

Ziganshin, M. G. Method of Assessment Generation Efficiency at Thermal Power Plants Considering Emissions of Pollutants. *Power Eng.: Res. Equip. Technol.* **2020,** *21* (6), 29–38.

CHAPTER 4

Impact of Biotic and Abiotic Interaction on Crop Plants

AMRESH KUMAR,[1,4] CHIRAG MAHESHWARI,[2] SACHIN PHOGAT,[3] ANKUR POUDEL,[3,5] ADITI ARYA,[4] and SUBODH KUMAR SINHA[1]

[1]*ICAR-National Institute for Plant Biotechnology, Pusa Campus, New Delhi, India*

[2]*Division of Biochemistry, ICAR-Indian Agricultural Research Institute, New Delhi, India*

[3]*ICAR-Indian Agricultural Research Institute, New Delhi, India*

[4]*Department of Biotechnology, Deenbandhu Chottu Ram University of Science and Technology, Murthal, Haryana, India*

[5]*Lamjung Campus, Institute of Agriculture & Animal Sciences, Tribhuvan University, Nepal*

ABSTRACT

Plants are continually subjected to abiotic and biotic stressors, which decreases crop output. Plants respond to these challenges in a variety of ways at the molecular, cellular, and physiological levels. This demonstrates that simultaneously exposing crop plants to abiotic and biotic stress may improve plant performance by increasing tolerance to biotic stress. Hormone signaling pathways that have synergistic and/or antagonistic effects coordinate such an interaction between biotic and abiotic stressors. This interaction of multiple stress responses is further controlled by transcription factors (TF), kinase cascades, and reactive oxygen species (ROS). In some cases, stress interaction might result in cross-tolerance and increased plant resistance

Climate-Resilient Agriculture: A Molecular Perspective. Ashutosh Singh, Saurabh Pandey (Eds.)

to pathogens. This chapter focuses on the interactions of abiotic and biotic stresses at the molecular level and covers regulatory elements of crop plants, with particular emphasis on the simultaneous effects of abiotic and biotic stresses. Recognize critical regulators that are associated with different types of stress response pathways, allowing physiologists and molecular biologists to devise agronomically suitable approaches for creating stress-resistant crop plants.

4.1 INTRODUCTION

Due to global warming, plants are prone to be exposed to different types of environmental stresses, that is, biotic and abiotic. Crop plants have developed mechanisms to manage these stress conditions. Stresses in plants adversely affect growth, development, and altimetry economic losses by reducing the yield and quality of crop plants. Heat, drought, cold, freezing, chilling, salinity, nutrient, ozone (O3), high light intensity, and anaerobic stress are examples of abiotic stress. Biotic stress refers to living organisms, that is, bacteria, viruses, nematodes, and insects, and these agents cause a variety of infections, diseases, and damage to crop plants. A survey indicated that plant losses yield from 17.2 (in potatoes) to 30.0% (in rice) due to pathogens, weeds, and insects (Savary et al., 2019). Similarly, abiotic stresses, that is, drought and heat or high temperature have been shown 51–82% annual loss of crop production globally (Oshunsanya et al., 2019). Simultaneous effect of biotic and abiotic stresses is more harmful than when they occur separately at different stages of plant growth (Prasad et al., 2011). Likewise, combinations of stresses, including salinity and heat, salinity, and drought impact crop productivity (Suzuki, 2014). Pathogens, insects, weeds, and herbivore pest attacks are affected by drought, low and high temperatures, and salinity (Atkinson and Urwin, 2012; Peters et al., 2014). In addition, drought increases weed competition with crops as some weeds are more efficient at utilizing water than crop plants (Valerio et al., 2013). Furthermore, many abiotic stresses suppress the defense system of crops and make them more vulnerable to pathogens (Atkinson and Urwin, 2012). There is a higher risk of abiotic and biotic stresses for major crop plants to be exposed to, as well as a combination of them. It is decisive to understand how stress factors link with each other in order to determine how they affect plants and how they work together in combination. The combined effect of different abiotic and biotic stresses on a plant is not every time additive, because results are determined by the nature of the interaction of stress (Pandey et al., 2015; Ramu et al., 2016). A matrix of stress interactions evolved by Mittler and colleagues demonstrates that stress combinations can either have negative or positive effects on plants based on how different stresses interact and affect plant growth, development, and productivity (Mittler, 2006; Suzuki et al., 2014). The action of identifying the stress in time, and responding in a timely and efficient manner, is a critical component of plant defense. When plants encounter different stresses, they activate complex signaling cascades of defense (Miao et al., 2009), which result in increased plant resistance to stress (Rivero et al., 2014).

Upon exposure of the plant to biotic and/or abiotic stresses, particular ion channels, kinase cascades, and phytohormone levels are increased. These compounds include salicylic acid, abscisic acid, ethylene, and jasmonic acid, reactive oxygen species (ROS) (Rizhsky et al., 2002, 2004), adjusted the genetic machinery of crop plants resulting in suitable defense mechanism and plant resistance increases to overcome the damage caused by the stresses (Choudhury et al., 2017). However, most of the mechanisms responsible for the resistance of crop plants to combinations of stress are still not known, and more research is needed to give them a better understanding.

Few studies have been shown to understand the simultaneous tolerance of crops to biotic and abiotic stresses. Our aim in the current chapter is to provide a better understanding of recent studies about the plant growth, development, physiological, and molecular aspect of combinations of abiotic and biotic stresses. In addition, we are looking at how different types of stress responses interact and affect the crop plant growth, development, and physiological traits.

4.2 PLANTS RESPONSE TO SIMULTANEOUS EFFECT OF ABIOTIC AND BIOTIC STRESSES

Different types of stress interactions have distinct impacts on crop plants depending on the nature, degree, and time period of the stress. Increase the possibility of the incident of abiotic stresses in the plant due to climatic change in the future. Climate change can broaden the host range of infections, increasing the likelihood of the emergence of virulent strains (Garrett et al., 2006). Although the combination of abiotic and biotic factors in crop plants has been addressed for several years, responses to physiological and molecular traits after crop plants encounter the combined effect of biotic and abiotic stresses remain unclear. The majority of abiotic–biotic stress combinations, as well as some abiotic–abiotic interactions, occur not only between the plant and the stressors, but also among the stress factors at or outside the plant interface. Depending on the stress type and plant pathogen, the simultaneous appearance of biotic and abiotic stresses can show a negative or positive effect on plants (Tippmann et al., 2006). In addition, the complexity of their interactions determines the degree of impact on plant response. Hot-wave and drought, for instance, may cause higher water evaporation from the soil. Apart from that, heat and drought stress shows synergistic effects on the physiology of plants, arising in a high rate of negative impact, and simultaneous weed and drought stress reduce the high rate of water availability to crop plants, increasing the competitiveness of crop plants and weeds, and finally negatively impacting yields (Mittler, 2006). The combined effect of the plant pathogen and high temperature increases the chance of disease in crop plants. Plants are affected by high temperatures because they decrease tolerance to plant viruses like Tobacco mosaic virus (TMV) and Tomato spotted wilt virus (TSWV) (Kiraly et al., 2008). The combination or interaction of two stresses does not all the time lead to a negative effect on crop plants. Combinations of two stresses can neutralize one another effect, resulting in a net positive or neutral effect on crop plants. One stress may tolerate against another stress to plants and net no effect on growth, development, and yield. For example, alone drought and ozone stresses are a negative impact on the growth of plants, but the combination of both stresses (drought and ozone) enhanced the resistance of plants (Puckette et al., 2007). These findings suggest that with the simultaneous effect of abiotic and biotic stresses, crop plants show a very complex and distinctive response showing a positive or negative effect on crop plants.

4.3 EFFECTS OF A COMBINATION OF STRESS IN CROP PLANTS

While abiotic stress-related studies have increased significantly over the past decade, the vast majority of them focus on individual plant responses under controlled conditions (Obata et al., 2015). However, under certain field conditions, plants respond differently to a variety of stresses simultaneously or in combination, and the specific stress treatment will affect the plant metabolism (Rizhsky et al., 2004; Tambussi et al., 2020). Thus, to gain a more comprehensive understanding of how crop plants survive abiotic stresses, it is necessary to investigate individual, sequential, and combined stresses under a field environmental condition (Sewelam et al., 2020). A plant's basal level defense response is activated when it encounters mild drought stress in terms of individual stress is a concern but in the case of severe drought stress causes nutrients leakage from a cell into the apoplast which promotes pathogen infection (Ramegowda and Senthil-Kumar, 2014). In case of sequential and combined abiotic stresses rate of transpiration increases by stomatal opening in the course of plant encounter with heat stress, while simultaneous effect of heat and drought stress, closer of stomata to lower the loss of water from crop plants (Zandalinas et al., 2021) we know less about the molecular aspect of the additive response to sequential and combined stresses (Obata et al., 2015). In addition to numerous combined effects of biotic and abiotic stresses, plant interactions with a pathogen and drought stress are well studied (Ramegowda et al., 2013; Wang et al., 2009). In the natural environment, plants may be infected by pathogens during drought stress development, resulting in the combined effects of the plant pathogen and drought and the resulting interaction depends on different stress intensities (Achuo et al., 2006; Xu et al., 2008). Stress combination of a plant is very complex and involves different types of pathways during multiple stress crosstalk. In a consequence of crosstalk, the results of combined stresses may impact negative or positive on the growth, physiology, and yield of a plant.

4.3.1 THE NEGATIVE EFFECT OF MULTIPLE STRESSES

When various stresses give a negative effect on plant growth and reproduction, the potential of the crop to identify and counter a particular combination of stress (Table 4.1) is very crucial. Droughts and heat waves will grow more regular and extreme in the future, lowering agricultural productivity reported by the climate change model (IPCC, 2008). Plants are restricted in growth and yield due to high temperatures and droughts globally (Vile et al., 2012). It will be necessary to the development of crop that is resistant to heat stress, drought, and their combination in order to meet the requirement of future food of an increasing world population. There have been numerous studies exploring the effects of different stress and their combinations on plant development and physiological traits. Plant growth decreased under different stresses but in the combined effect of stresses resulted in more adverse effects, that is, the combined stress effect was generally not additive in nature. In fact, some growth traits of plants were responding differently to one stress factor than others, that is, in response to drought, root biomass increases, while in response to heat stress, in reproductive tissues of plant leaf area increase. Drought decreases the plant height, plant biomass, and spike numbers, but there is no effect on these traits by heat stress but on the contrary, heat stress decreased kernel weight and aborted spikes, while drought did not have effects on these traits. Alone heat stress and drought and their combination of stress decrease in crop yield, number of grains, spikelet number, and harvest index and chlorophyll contents in bread wheat (Prasad et al., 2011). Heat stress caused reduced grain yield, relative water content, pollen fertility, rate of photosynthesis, and chlorophyll content while increasing malondialdehyde and proline content in maize and wheat.

The drought and heat stress act on a plant in a combination; it also exhibits antagonistic responses in crop plants (Mittler and Blumwald, 2010), that is, plants increase their opening of stomata in order to maintain the optimum temperature of a plant by transpiration. Stressed plants may be less vulnerable to proline toxicity, so sucrose (here acts as an osmoprotectant) might accumulate in the plant instead of proline, possibly to protect hyperactive mitochondria from toxicpyrroline-5 carboxylate (P5C) which is synthesis from proline (Rizhsky et al., 2004). When salinity stress is added to heat stress, the salt absorption capacity might be increased since increased transpiration would result in increased transpiration (Wen et al., 2005). Root system architecture (RSA) is a prime aspect of the crop and abiotic and biotic stress factors. The nutritional status of the crop plant induces changes in the overall root morphology, RSA is associated with nitrate concentrations in soil (Sinha et al., 2018; Sinha et al., 2020). RSA is a potential trait used for screening of plant variety for resistance to the combined effect of drought stress and plant pathogen infection.

Drought and cold stress could be adversely affected crops by high light intensity. Low temperatures or lack of carbon dioxide retard the dark reactions, and high light intensities increase the reduction of oxygen, as a result, the production of reactive oxygen species (ROS) (Mittler et al., 2004). The deficiency of nutrients complicates the acclimation response in plants when combined with other stresses; as a result, activation of ROS scavenging enzymes is dependent on energy and resources and micronutrient (Mittler and Blumwald, 2010). Both salinity and drought, individually and in combination, reduced plant growth, photosynthetic rate, chlorophyll content, water potential, osmotic potential, and maximal photochemical efficiency of PSII and more effect by stress combination (Ahmed et al., 2013). Wild barleys

TABLE 4.1 Negative Interactions of Different Stress Combinations on Crop Plants and Their Outcome.

Combination of stress	Crop plants	Outcomes of combination of stress	References
Drought and high temperature	Wheat	Combined stress affects the fertility of spikelet, dry weight, leaf chlorophyll content, and harvest index (HI)	Prasad et al. (2011)
	Maize	Combined stress affected the decreasing dry weight, plant length, surface area, seminal lateral roots, and root mass ratio (RMR)	Vescio et al. (2021)
Drought and salinity	Wheat	Depletion in biomass, plant development, and the photosynthetic rate are related to root length	Dugasa et al. (2018)
	Maize	Decreased primary root length under combined stress	Li et al. (2021)
Drought and chilling	Sugarcane	The outcome of abiotic stress is variety dependent, combined stress have more effects on sensitive genotypes than tolerant genotypes	Sales et al. (2013)
Drought and low N	Wheat	Low nitrogen leads to an increase in ABA concentration in wheat seedlings, but the combination of stress interaction regulates genes for uptake, signaling, and assimilation of nitrate	Mahmoud et al. (2020)
Drought and nutrient	Mungbean	Under a combination of stress, decreases in gas exchange by photosynthesis, stomatal opening and closing, transpiration, water use efficiency, and Phosphorus uptake	Meena et al. (2021)
Salinity and high/low temperature	Wheat	High/low temperatures possess a collective effect on growth retardation under salt stress. High alpha-tocopherol amount in the combination of drought and salt stress but low under high temperature	Kale et al. (2002) Tewari et al. (2018)
Salinity and pathogen	Rice	Low expression of OsMAPK5 increased expression of pathogenesis-related genes and as results increase tolerance to fungal and bacterial attacks but lowered resistance to drought, cold, and salt stress. In contrast, increased OsMAPK5 kinase activity in overexpression lines and resulting in higher resistance to drought, cold, and salt stresses	Li et al. (2011)
High temperature and UV-C	Strawberry	Reduced fungal attack, delayed in vitro germination, and decreases sugar content when the occurrence of stress combination	Pan et al. (2004)

TABLE 4.1 *(Continued)*

Combination of stress	Crop plants	Outcomes of combination of stress	References
High temperature and high light	Sunflower	It showed that 89, 113, and 186 genes were differentially expressed in leaves and seeds in response to high light, high temperature, and simultaneously high light and high temperature, respectively	Hewezi et al. (2008)
High temperature and CO_2	Soybean and maize	Increasing the concentrations of CO_2 can reduce the effects of quite raised temperatures on crop yield	Sicher et al. (2015)
UV-B and heavy metals	Pea	UV-B and 0.01 mM Nickel showed negative effects when it is given simultaneously	Srivastava et al. (2015)

show high resistance and ROS scavenging activities such as catalase (CAT), superoxide dismutase (SOD), glutathione reductase (GR), and guaiacol peroxidase (POD) under a simultaneous occurrence of salt and drought stresses than that of cultivable variety of barley, but these stress combinations show negative interaction in bread wheat (*T. aestivum*) (Yousfi et al., 2010). Total plant biomass is directly dependent on available nitrogen in a plant (Kumar et al., 2022) as well as ratio of potassium, and sodium ion under drought, salt stress, and its combination. The correlations between plant growth and physiological traits weakened when drought and salt stress were combined, suggesting that growth and physiological characteristics have a complex relationship. Drought-tolerant vs drought-sensitive cultivars of sugarcane were compared under drought, cold stress, and their combination of stresses. Drought-sensitive cultivars had reduced biomass under drought conditions than drought-tolerant variety (Sales et al., 2013). The combined effect of drought and cold stress increases the negative effects on drought-sensitive variety but in the case of drought-resistant variety, shows a positive effect on lipid peroxidation. This outcome supports the role of antioxidant mechanisms in plant defense during the different combinations of stresses.

Heavy metal and combination with other abiotic stresses causes adverse effects on plant development. Nickel (Ni) and UV-B alone inhibited shoot and root growth in pea seedlings, but the combination of stress emerging more serious damage (Srivastava et al., 2012). Plant growth retardation was observed in a stress combination of nickel (Ni) and Ultraviolet-B (UV-B), followed by decreases in the Chlorophyll, photosynthetic activity, and carotenoid content. Plant growth is also hampered by heavy metals during droughts (de Silva et al., 2012). By altering xylem structure and hydraulic conductivity, drought and nickel (Ni), cobalt (Co), copper (Cu), and chromium (Cr) all reduced xylem structure and plant growth. Combining zinc (Zn) and cadmium (Cd) increased oxidative stress, to a greater extent than an alone of high cadmium or zinc concentration (Cherif et al., 2011). This suggests that high concentrations of zinc (Zn) and cadmium (Cd) contribute to oxidative stress in crop plants. In contrast, a low concentration of zinc (Zn) decreases oxidative stress and the outcome is a restoration of chlorophyll content. Compaction of soil is generally considered to negatively affect plant development and yield, as it does with other environmental stresses, such as drought, salt, and extreme temperatures (Alameda et al., 2012). Plants were found to be less resistant to biotic stresses under both cold and heat stresses. Gene silencing is a powerful tool against viral pathogens; its activities are demised by cold stress (Szittya et al., 2003). Reduced plant biomass under heat stress, by combining heat stress and virus stress, and finally an extreme reduction in plant biomass was observed from combining all three stresses. Furthermore, the combined effect of virus and drought, virus and heat, and triple stress as a result, stomata are close, while heat stress or virus infection acts on crop plant alone occurred in stomata opening.

4.3.2 THE POSITIVE EFFECT OF MULTIPLE STRESSES

The combination of different stresses may benefit crop plants compared to the stress applied individually, for example, drought stress would lead to a lower stomatal conductance, thereby improving crop plants' resistance to ozone stress (Low et al., 2006). Chromosis and necrotic lesions occurred under short-term ozone (O_3) stress, while drought caused entire leaves to wilt and collapse, as well as chlorosis. Interestingly, when ozone (O3) and drought stress act simultaneously, it canceled out the effects caused by either stress alone. To control the combined effects of ozone and drought, stomatal conductance may be decreased. Furthermore, the ROS concentrations under the stress combination were also lower than they were under drought or O3 alone (Table 4.2). When heat and salt stress alone on tomato plant results in a negative effect but a combined effect of heat and salt stress protects from this stress and is opposed to the combined effects of drought and heat stress (Rivero et al., 2014).

Crop plants may respond differently to stress combinations by accumulating glycine betaine and trehalose in plants cell and a combination of stress conditions; the accumulation of these compounds lowers down Na+/K+ ratio, which enhances cellular water status and photosynthesis. Among its numerous effects, glycine betaine and trehalose has involved in protecting photosystem-II (PSII) from heat and photo-induced inactivation (Allakhverdiev et al., 2003) and inhibiting salt-induced potassium ion efflux (Cuin and Shabala, 2005). Moreover, these compounds protect cells against oxidative stress. In fact, the combination of salt and heat stress inhibited H_2O_2 accumulation and protein oxidation in tomato plants, suggesting that the plants' resistance to stress combination may partly be due to the prevention of oxidative stress. It is also possible to consider elevated carbon dioxide concentrations beneficial when combined with various stresses. Carbon dioxide concentrations increase, the conductance of stomata decreases and diffusion of ozone (O_3) into plant leaves is inhibited, transpiration rate is reduced, and water-use efficiency is improved (Ainsworth et al., 2008).

Moreover, the combination of high CO_2 with salt or high light also increased the biomass of lettuce (Perez-Lopez et al., 2013). An increase in the plant biomass is caused by better water use efficiency of the plant and antioxidation capacity under combined stress conditions.

TABLE 4.2 Positive Interactions of Different Stress Combinations on Crop Plants and Their Outcome.

Combination of stress	Crop plants	Outcomes of combination of stress	References
Salinity and high temperature	Tomato	Tobacco plants are protected from salinity and the accumulation of glycine betaine and trehalose when heat and salinity are combined	Rivero et al. (2014)
Salinity and boron	Maize	Under salt stress, certain membrane components can be affected by boron, which can regulate the uptake and transport of water by controlling the functions of certain aquaporin isoforms	Martinez-Ballesta et al. (2008)
Ozone and high CO_2	Rice	Elevated CO_2 (627 ppm) increases yields, but the application of O_3 decreases the rice yields	Ainsworth et al. (2008)

4.3.3 ABA PLAYS A POSITIVE ROLE IN DIFFERENT STRESSES

In addition to the role of resistance to drought and low-temperature stress in a plant, abscisic acid (ABA) plays an important role in the plant defense mechanism (Melotto et al., 2006). High concentrations of ABA in plants are known to contribute to less disease resistance, whether this occurs naturally, through drought-induced accumulation, or exogenously, during pathogen infection (Mohr et al., 2003). ABA is responsible for both positive and negative effects in a plant from infection of a plant pathogen. Mutants lacking ABA biosynthesis and perception have also been shown to have negative effects on disease resistance. Tomato mutants infected by B. cinerea inhibit pathogen growth by deficient of ABA biosynthesis (Audenaert et al., 2002). Elicitors derived from plant pathogens such as flagellin 22 (flg22) and lipopolysaccharide (LPS) demonstrated that ABA exerts a positive effect predominantly in preinvasive defense against plant pathogens through rapid closure of stomata (Melotto et al., 2006). The enhanced resistance of plants against stress could be due to ABA-mediated callose biosynthesis or low degradation of it in the plant cell (Jacobs et al., 2003).

4.4 MOLECULAR ASPECT OF SIMULTANEOUS EFFECT OF DIFFERENT BIOTIC AND ABIOTIC STRESSES ON THE PLANT

Plants must deal with a wide range of complicated interactions involving several environmental elements. They have evolved specialized systems that allow them to adapt to and withstand stressful experiences throughout evolution. Plants subjected to both biotic and abiotic stresses have metabolic disruptions, implying physiological consequences (Bolton et al., 2009; Massad et al., 2012), resulting in growth and productivity loss (Shao et al., 2008). Abiotic stress is one of the greatest setbacks to plant development and has a considerable impact on growth and as a result significant decreases in the yield. In most plant species, the consequent growth decreases can exceed 50% (Francini and Sebastiani, 2019). Furthermore, biotic stress is an additional obstacle, exerting significant strain on plants and compounding the loss caused by plant pathogen or herbivore attacks (Zandalinas et al., 2021). The prompt recognition of stress to respond in a timely and efficient manner is a critical stage in plant defense. Following detection, the plants constitutive basal defense systems activate complicated defense signaling cascades that differ based on the type of stress (Lin et al., 2021). Abiotic and biotic stresses activate specific kinase cascades and ion channels for the flow of signals. Reactive oxygen species (ROS), phytohormones accumulate, and reprogramming of genetic material are also part of the defense machinery (Devireddy et al., 2021).

Presently, the researcher has primarily focused on understanding the plant responses against specific abiotic and/or biotic challenges, even though responses to concurrent stresses are guaranteed to result in a far more complex scenario. Plants use a variety of signaling routes depending on the challenge, from stimulus perception to cell reaction (s). Plants appear to behave differently when subjected to several stresses at the same time, and this response cannot be predicted simply to separate single or multiple stresses. Multiple stress study has attempted to imitate natural settings, but control of field conditions is not our hand, that is, it is not controlled and single stress can substantially alter the defense reaction of principal stress of the plants. Furthermore, depending on the field conditions and the plant developmental stage, plants can exhibit varying degrees of sensitivity. The severity of the stress and the plant type are two more aspects that can influence an interaction. They are affected by the exact combination of stresses as well as the degree of coexistence of stress. The interaction of concurrent stresses is unknown in terms of whether they are antagonistic, synergistic, or additive, resulting in increased or decreased vulnerability to a certain form of stress. Combining two different types of stresses results in a detrimental and cumulative effect on crop plants, with the second stress causing the most harmful effect. A combination of stressors, on the other hand, may result in an antagonistic effect in plants (Yasuda et al., 2008). Common beans treated with drought exhibit greater symptoms when infected with *Macrophomina phaseolina* (Suleman et al., 2001), while treating tomato leaves with abscisic acid (ABA) exogenously makes wild-type plants more susceptible to *Botrytis cinerea* (Audenaert et al., 2002).

Interestingly, plants that can protect themselves from one stress may grow more tolerant to other stresses as a result of cumulative stress exposure (Bowler et al., 2000); this is called cross-tolerance. It demonstrates that plants have a strong regulatory mechanism that permits them to adjust swiftly to changing environments. Wounding, for example, improves tomato plant to salt tolerance. Furthermore, *Pseudomonas syringae* infection confers resistance from the herbivore insect *Helicoverpa zea* in tomato plants (Stout et al., 1999). After drought stress treatment, tomato infection by Botrytis cinerea and Oidium neolycopersici decreased, giving evidence of an association between abiotic and biotic stresses (Achuo et al., 2006). In grapevines, ozone (O_3) exposure at low and nontoxic doses can induce the expression of resistance-related genes in the vine leaves (Modesti et al., 2022). In turn, biotic stress can help to develop tolerance to abiotic stress and this effect is seen when plants are attacked by pathogens. Infection of a plant may trigger the closing of stomata to prevent pathogen entrance, resulting in reduced water loss and better plant resilience to abiotic stress (Goel et al., 2008). It is also demonstrated that viral infection of plants protects from drought stress (Xu et al., 2008). In Arabidopsis, Verticillium infection increased expression of Vascular-Related no Apical Meristem (NAM) and Cup-Shaped Cotyledon (CUC) domain transcription factor VND7. VND7 increased water storage capacity and plant drought resistance by stimulating de novo xylem formation (Reusche et al., 2012). Stress produces distinct signaling pathways, some of which share components and outcomes. This could help plants save energy while also creating a more adaptable signaling network. Defense priming has been found to improve resistance to both biotic and abiotic stresses in a variety of crops. Genetic modification of genes and proteins, BABA (beta-aminobutyric acid) or BTH (benzothiadiazole) use, or exposure to a pathogen previously can produce induced resistance. We are focusing on cross-talk between the abiotic and biotic stresses to induce plant resistance to these different stresses.

4.4.1 ABIOTIC AND BIOTIC STRESS CROSS-TOLERANCE

Plants can deal with different abiotic and biotic stresses simultaneously, and there is a lot of proof of relation between these stressful situations. Plant resistance is typically induced by environmental pressures such as abiotic and biotic stresses. However, certain plants exposed to each stress individually have been observed to be more vulnerable than plants exposed to different stresses concurrently. Furthermore, some environmental pressures have the potential to predispose the plant to respond rapidly and more resiliently to future difficulties. As a result, cross-tolerance between environmental effects and biotic stress may benefit plant resilience and have substantial agricultural ramifications. Abiotic stress, it turns out, alters defense systems both at the pathogen infection site and in systemic areas, resulting in an improvement in the plants innate immunity system.

Similarly, osmotic stress increases tolerance to powdery mildew in barley. A callose-containing papilla that is able to block fungal growth is needed to induce this resistance (Weise et al., 2004). Drought increased the ABA concentration of tomato leaves while lowering sensitivity to the necrotrophic fungus Botrytis cinerea and increasing resistance to fungus *Odium neolycopersici* but does not affect on *Botrytis cinerea*, according to Achuo et al. (2006). The discovery that drought and salt stress cause unique gene expression patterns supports this distinction. Furthermore, exposing *Nicotiana benthamiana* to extreme drought suppressed *P. syringae* pv. *tabaci* development (Ramegowda et al., 2013). Recently, investigating the interplay of abiotic and biotic stresses, as a result, many detoxifying enzymes, regulatory hormones, signaling pathways, as well as expression of a gene must be triggered to prepare the plant for battle (Atkinson and Urwin, 2013). The defense mechanism of plants to many stresses is anticipated to be very complex, with the integration of diverse signaling pathways that regulate several metabolic pathways.

4.4.2 MULTIPLE STRESS RESPONSES TRIGGERED SIGNALING PATHWAYS

The combined effect of different abiotic and biotic stresses shows complicated reactions to the many stresses. Accumulation of particular metabolites after the encounter of stress improves a plant's response to different stresses and hence defends it from dual aggressors. The first responses triggered to battle to stress are callose buildup, change in ion fluxes, phytohormones and reactive oxygen species (ROS), and the ensuing signal transduction initiates metabolic reprogramming toward defense.

4.4.2.1 REACTIVE OXYGEN SPECIES (ROS)

As a result of stress sensing, there is a fast production of ROS. One of the most important tasks of ROS in cells is to act as signaling molecules. The plant fine-tunes ROS generation to avoid tissue damage. For a long time, ROS have been recognized as toxic and damaging chemicals in stressed organisms. However, whereas high amounts of ROS cause cell death, lesser levels are managed to regulate the plants stress. ROS are primarily engaged in signaling during biotic stress and may reduce the oxidative damage generated by abiotic stress (Kissoudis et al., 2014). By enhancing de novo xylem synthesis and the subsequent enhancement of water flow, ROS is involved in improving drought resistance in Arabidopsis thaliana after infection with the vascular pathogen Verticillium spp. (Xia et al., 2012). Furthermore, the synthesis of ROS can facilitate communication between cells by increasing the signal via Respiratory Burst Oxidase Homologue D (RBOHD) (Miller et al., 2010) and in addition, it is a secondary messenger, altering the structure of proteins and activating defensive genes (Spoel et al., 2010). It shows that ZAT12 plays a role in biological stress response in both abiotic and biotic as well as in ROS scavenging and ROS may facilitate cross-tolerance in between abiotic and biotic environments (Davletova et al., 2005). ROS production in the Arabidopsis plant can be identified by ROS responsive to transcription factors (TF), resulting in stress-related genes being activated. ROS induces resistance through activation of stress-related factors such as mitogen-activated protein kinases (MAPKs), antioxidant enzymes, transcription factors, antioxidant enzymes, dehydrins and low temperature induced-, heat shock-, and pathogenesis-related proteins (Gechev et al., 2006). Stress tolerance priming, which happens after the application of specific drugs, causes certain modifications in ROS signaling. Brassinosteroids were applied to cucumber plants, which boosted hydrogen peroxide (H_2O_2) levels; as a result plants were resistant from biotic and abiotic stresses (Xia et al., 2012). Hydrogen peroxide (H_2O_2) priming used for the salt stress resistance in citrus plant consequences increased the amount of oxidized S-nitrosylated proteins somewhat after stress application, but untreated plants were more susceptible to the stress (Tanou et al., 2009).

4.4.2.2 MITOGEN-ACTIVATED PROTEIN KINASE (MAPK) CASCADES

Mitogen-Triggered Protein Kinase (MAPK) cascades are activated in response to the perception and acknowledgment of stress stimuli and they also control the stress response pathways. A wide variety of cellular processes are signaling by MAPKs in response to different abiotic and biotic stresses, with certain kinases necessary in response to different types of stresses (Brader et al., 2007). Because MAPKs are involved in a variety of stresses, they may play a role in the combined effect of abiotic and biotic stresses. GhMPK6a, for example, controls the biotic and abiotic stresses in cotton (Li et al., 2013). SA mediates pathogen-activated MAPK pathways, and the consequent activation of PR genes causes defense responses (Xiong et al., 2006). After phosphorylation by MPK3, the Arabidopsis VIP1 protein is translocated into the nucleus and indirectly induces PR1 (Pitzschke et al., 2009). Furthermore, MAPKs such as MPK3, MPK4, and MPK6 reacted to a wide range of abiotic stimuli (Gudesblat et al., 2007). MAPK cascades have the main role in cross-talk of stress response because MPK3 and MPK6 are necessary for fully primed defensive reactions (Beckers et al., 2009), and these two MPKs play a key role in modifying resistance from to future stress. Overexpression of rice MPK5 (OsMPK5) gene and ABA-induced OsMPK5 activity are responsible for the improvement in resistance to abiotic and biotic stresses. OsMPK5 is thought to have two roles in its response to stress, the first one is a positive regulator from the Cochliobolus miyabeanus (necrotrophic brown spot pathogen) resistant plant, and the second one is a moderator of abiotic stress resistance (Sharma et al., 2013). To strengthen their defense responses, activation of MPK1 and MPK2 in tomato plants occurs in reply to UV-B stress, wounds and infections (Holley et al., 2003). The MAPK signaling pathway is also involved in ROS and ABA signaling, as a result, better plant defense and cross-adaptation to abiotic and biotic stresses (Miura et al., 2014; Zhou et al., 2014).

4.4.2.3 ROLE OF HORMONE IN SIGNALING UNDER THE INTERACTION OF STRESS

Regulation of all types of stress by certain hormones enables the defense system to determine environmental situations. The major hormone implicated in the perception of numerous abiotic stressors is ABA. Abiotic stress regulation network affected by ABA concentration increases, while biotic stress is mostly mediated by antagonistic interactions between other hormones stresses such as salicylic acid and jasmonic acid/ethylene (Liu et al., 2009). ABA has been shown to build after infection in some circumstances. For example, after Pst DC 3000 infection, increased levels of ABA were detected (Truman et al., 2006), which resulted in the inhibition of other defense responses (De Torres et al., 2007). Recent research, however, indicates that ABA has a favorable effect on biotic stress tolerance. Because of its dual impact, ABA is a contentious chemical that can turn from "good to negative" depending on the surroundings (type of stress and timing). Furthermore, when combined with abiotic and biotic stresses, ABA mainly functions antagonistically oppositely with salicylic acid/jasmonic acid/ethylene, making the plant more sensitive to disease caused by pathogen and herbivore assault (Robert et al., 2007). However, because a rise in ABA caused by abiotic stress causes stomatal closure as a "secondary impact," biotic invaders are prevented from entering the plant through these passive ports. As a result, in such conditions, the plant is shielded from both abiotic and biotic stresses.

There are three stages to ABA influence on pathogen infection. Consequently, ABA's first effect on the combination of both abiotic and biotic stresses is only allied to abiotic stress since an infection takes longer to determine itself in establishment, so plants respond to it subsequently. At this point, ABA is responsible for the closing of stomata, which reduces water loss and, as a result, maintains a favorable water potential. Salicylic acid/jasmonic acid/ethylene may not yet be activated during this early phase, and ABA inhibits their induction. In this case, future stress responses to prospective infections are altered and the second stage is concerned with the post infection effects. Callose induces the defense system that can help to keep pathogens at bay. Callose accumulation in plants requires an intact ABA signaling pathway following infection. ABA, depending on environmental conditions, may either induce or suppress further callose growth. As a result, a change in ABA concentration produced by earlier stress may affect the eventual outcome after biotic stress, such as boosting the resistance of plants via callose accumulation or activating additional defensive pathways. Finally at the beginning of the third phase, in order to modify the defense response, PAMPs increase the accumulation of specific hormones such as salicylic acid, jasmonic acid, and ethylene (Nishimura et al., 2003). It is highly dependent on the insight of the stress timing whether ABA acts as a regulator of abiotic and biotic stresses and salicylic acid involvement in bridging the gap between plants and pathogens has gotten a lot of attention.

ABA and salicylic acid act antagonistically in plant stress resistance (Yasuda et al., 2008). Miura and Tada (2014), on the other hand, demonstrated that, in addition to ABA, salicylic acid appears to have a role in a drought of the plant.

4.4.2.4 ROLE OF TRANSCRIPTION FACTORS AND MOLECULAR RESPONSES

Transcriptional factors regulate the reprogramming of the molecular machinery and gene expression following the identification of specific stresses. Some genes changed expression is crucial in enabling plants to create an efficient defensive mechanism, and there is persuasive evidence that many genes have more than one function and are able to establish plants' resistance to many stresses. Phytohormones such as ABA, salicylic acid, jasmonic acid, and Ethylene are involved in the activation of such defense-related genes. The BOTRYTIS SUSCEPTIBLE1 (BOS1) gene is a transcription factor protein, for example, it is controlled by both ABA and jasmonic acid and confers tolerance to osmotic stress and necrotrophic diseases (Mengiste et al., 2003), bos1 mutant plants are more sensitive to different stresses. A transcription factor called MYB96 in Arabidopsis reduces the molecular connection between drought stress-induced ABA and pathogen-induced salicylic acid in plants defending against infection by pathogens (Seo et al., 2010). Tobacco's SlAIM1 is influenced positively by abiotic stress and Botrytis cinerea infection, while rice's OsMAPK5 is positively regulated by abiotic stress, salt, and cold withstand, and disease resistance (Abou et al., 2009).

Surprisingly, when a plant is an encounter to abiotic stress, numerous pathogen-related genes are activated to secure disease resistance. Plant resistance to pathogens requires pathogen-related proteins, and their expression is high when plants are attacked. Transgenic Arabidopsis that overproduces the NAC transcription factor NTL6 increases its defense response to pathogen attack by increasing the expression of pathogen-related genes (Kim et al., 2007). Transcription factor (TF) like DREB controls the response to abiotic and biotic stresses, enabling Arabidopsis plant resistance to cold and infections (Tsutsui et al. 2009). The WHIRLY1 (DNA-binding protein) protein and REDOX-RESPONSIVE TRANSCRIPTION FACTOR1 (RRTF1) (involved in regulation in redox homeostasis) have recently been proposed to have a role in link between plastids and the nucleus of a cell (Foyer et al.

2014). WHIRLY1 is responsible for a change in redox reaction in plastids and sends them to the nucleus of a cell in an independent NPR1 (Nonexpresser of pathogen related gene1). This protein is important for retrograde signaling, which leads to acclimatization and adaptation to new stimuli. Similarly, RRTF1, which is triggered by both biotic and abiotic stimulus responses, may help distant leaves prepare to protect themselves from future threats.

4.5 CONCLUSION

The majority of abiotic and biotic stresses that act on crop plants in the natural environmental condition are being research conducted in the laboratory condition. Contrarily to this, several groups have only recently undertaken the study of various combinations of stress that mimic the natural environmental condition much more realistically. In some cases, plants were reported to be less vulnerable to biotic stress as a consequence of specific abiotic stress. When plants are repeatedly exposed to abiotic stresses, namely drought, salinity, extremely low and high temperatures, and different nutrient stresses, etc., these generally weaken their defenses and they become more susceptible to biotic stresses. Stress responses are prioritized by plants based on their individual severity when faced with abiotic and biotic stress combinations. The interactions between the different stresses do not adversely affect the yield; they have actually a positive effect on each other by the interaction of ozone and UV, drought and ozone, heat and salt, and high carbon dioxide combined with drought, ozone, salt or high intensity of light. To date, most of the studies on the combination of stress have a negative effect on crop yield in natural environmental conditions; it is suggested that crops plants with increased resistance to combined stress are urgently needed. In addition, reproductive tissues are more sensitive to the combined stress of heat and drought stress, as compared with vegetative tissues, highlighting the negative effect of this stress combination in the field for major crop plants. It is critical to study the reaction of reproductive tissues to stress combinations in order to produce crops that will perform better in present and future field conditions. Different pathways and signals coordinate to regulate the plants' response to different combined stresses. Plant response to stress combinations is still not fully understood, and more research is needed to address the molecular and biochemical mechanisms involved.

KEYWORDS

- **abiotic stress**
- **biotic stress**
- **multiple stress**
- **stress interaction**
- **stress combinations**
- **crop production**
- **cross-tolerance**
- **plant hormones signaling**
- **plant resistance**

REFERENCES

Abou Qamar, S.; Luo, H.; Laluk, K.; Mickelbart, V. M.; Mengiste, T. Crosstalk Between Biotic and Abiotic Stress Responses in Tomato is Mediated by AIM1 Transcription Factor. *Plant J.* **2009,** *58*, 1–12.

Achuo, E. A.; Prinsen, E.; Hofte, M. Influence of Drought; Salt Stress and Abscisic Acid on the Resistance of Tomato to Botrytis cinerea and Oidium neolycopersici. *Plant Pathol.* **2006,** *55* (2), 178–186. https://doi.org/10.1111/j.1365-3059.2006.01340.x

Ahmed, I. M.; Dai, H.; Zheng, W.; Cao, F.; Zhang, G.; Sun, D.; Wu, F. Genotypic Differences in Physiological Characteristics in the Tolerance to Drought and Salinity Combined Stress Between Tibetan Wild and Cultivated Barley. *Plant Physiol. Biochem.* **2013,** *63*, 49–60. https://doi.org/10.1016/j.plaphy.2012.11.004

Ainsworth, E. A.; Rogers, A.; Leakey, A. D. Targets for Crop Biotechnology in a Future High-CO2 and High-O3 World. *Plant Physiol.* **2008,** *147* (1), 13–19. https://doi.org/10.1104/pp.108.117101

Alameda, D.; Anten, N. P. R.; Villar, R. Soil Compaction Effects on Growth and Root Traits of Tobacco Depend on Light, Water Regime and Mechanical Stress. *Soil Tillage Res.* **2012,** *120*, 121–129. https://doi.org/10.1016/j.still.2011.11.013

Allakhverdiev, S. I.; Hayashi, H.; Nishiyama, Y.; Ivanov, A. G.; Aliev, J. A.; Klimov, V. V.; Murata, N.; Carpentier, R. Glycinebetaine Protects the D1/D2/Cytb559 Complex of Photosystem II Against Photo-Induced and Heat-Induced Inactivation. *J. Plant Physiol.* **2003,** *160* (1), 41–49. https://doi.org/10.1078/0176-1617-00845

Atkinson, N. J.; Urwin, P. E. The Interaction of Plant Biotic and Abiotic Stresses: From Genes to the Field. *J. Exp. Bot.* **2012,** *63* (10), 3523–3543. https://doi.org/10.1093/jxb/ers100

Audenaert, K.; De Meyer, G. B.; Höfte, M. M. Abscisic Acid Determines Basal Susceptibility of Tomato to Botrytis Cinerea and Suppresses Salicylic Acid-Dependent Signaling Mechanisms. *Plant Physiol.* **2002,** *128* (2), 491–501. https://doi.org/10.1104/pp.010605

Bahuguna, R. N.; Gupta, P.; Bagri, J.; Singh, D.; Dewi, A. K.; Tao, L.; Islam, M.; Sarsu, F.; Singla-Pareek, S. L.; Pareek, A. Forward and Reverse Genetics Approaches for Combined Stress Tolerance in Rice. *Indian J. Plant Physiol.* **2018,** *23* (4), 630–646. https://doi.org/10.1007/s40502-018-0418-0

Beckers, G. J. M.; Jaskiewicz, M.; Liu, Y.; Underwood, W. R.; He, S. Y.; Zhang, S.; Conrath, U. Mitogen-Activates Protein Kinases 3 and 6 are Required for Full Priming of Stress Responses in Arabidopsis Thaliana. *Plant Cell* **2009,** *21* (3), 944–953. https://doi.org/10.1105/tpc.108.062158

Besser, K.; Jarosch, B.; Langen, G.; Kogel, K. H. Expression Analysis of Genes Induced in Barley after Chemical Activation Reveals Distinct Disease Resistance Pathways. *Mol. Plant Pathol.* **2000,** *5*, 277–286.

Bolton, M. D. Primary Metabolism and Plant Defense-Fuel for the Fire. *Mol. Plant Microbe Interact.* **2009,** *22* (5), 487–497. https://doi.org/10.1094/MPMI-22-5-0487

Bowler, C.; Fluhr, R. The Role of Calcium and Activated Oxygens as Signals for Controlling Cross-Tolerance. *Trends Plant Sci.* **2000,** *5* (6), 241–246. https://doi.org/10.1016/s1360-1385(00)01628-9

Brader, G.; Djamei, A.; Teige, M.; Palva, E. T.; Hirt, H. The MAP Kinase Kinase MKK2 Affects Disease Resistance in Arabidopsis. *Mol. Plant Microbe Interact.* **2007,** *20* (5), 589–596. https://doi.org/10.1094/MPMI-20-5-0589

Cherif, J.; Mediouni, C.; Ben Ammar, W.; Jemal, F. Interactions of Zinc and Cadmium Toxicity in their Effects on Growth and in Antioxidative Systems in Tomato Plants (Solanum Lycopersicum). *J. Environ. Sci.* **2011,** *23* (5), 837–844. https://doi.org/10.1016/s1001-0742(10)60415-9

Choudhury, F. K.; Rivero, R. M.; Blumwald, E.; Mittler, R. Reactive Oxygen Species, Abiotic Stress and Stress Combination. *Plant J. Cell Mol. Biol.* **2017,** *90* (5), 856–867. https://doi.org/10.1111/tpj.13299

Coakley, S. M.; Scherm, H.; Chakraborty, S. Climate Change and Plant Disease Management. *Ann. Rev. Phytopathol.* **1999,** *37*, 399–426. https://doi.org/10.1146/annurev.phyto.37.1.399

Cuin, T. A.; Shabala, S. Exogenously Supplied Compatible Solutes Rapidly Ameliorate NaCl-Induced Potassium Efflux from Barley Roots. *Plant Cell Physiol.* **2005,** *46* (12), 1924–1933. https://doi.org/10.1093/pcp/pci205

Davletova, S.; Schlauch, K.; Coutu, J.; Mittler, R. The Zinc-Finger Protein Zat12 Plays a Central Role in Reactive Oxygen and Abiotic Stress Signaling in Arabidopsis. *Plant Physiol.* **2005,** *139* (2), 847–856. https://doi.org/10.1104/pp.105.068254

de Silva, N. D.; Cholewa, E.; Ryser, P. Effects of Combined Drought and Heavy Metal Stresses on Xylem Structure and Hydraulic Conductivity in Red Maple (Acer rubrum L.). *J. Exp. Bot.* **2012,** *63* (16), 5957–5966. https://doi.org/10.1093/jxb/ers241

De Torres-Zabala, M.; Truman, W.; Bennett, M. H.; Lafforgue, G.; Mansfield, J. W.; Rodriguez Egea, P. R.; Bögre, L.; Grant, M. Pseudomonas syringae pv. Tomato hijicks the Arabidopsis Abscisic Acid Signaling Pathway to Cause Disease. *EMBO J.* **2007,** *26* (5), 1434–1443. https://doi.org/10.1038/sj.emboj.7601575

Devireddy, A. R.; Zandalinas, S. I.; Fichman, Y.; Mittler, R. Integration of Reactive Oxygen Species and Hormone Signaling During Abiotic Stress. *Plant J. Cell Mol. Biol.* **2021,** *105* (2), 459–476. https://doi.org/10.1111/tpj.15010

Dugasa, M. T.; Cao, F.; Ibrahim, W.; Wu, F. Differences in Physiological and Biochemical Characteristics in Response to Single and Combined Drought and Salinity Stresses between Wheat Genotypes Differing in Salt Tolerance. *Physiol. Plant.* **2018,** *165*, 134–143.

Foyer, C. H.; Karpinska, B.; Krupinska, K. The Functions of WHIRLY1 and REDOX-RESPONSIVE Transcription Factor 1 in Crosstolerance Responses in Plants: A Hypothesis. *Philos. Trans. R. Soc. B* **2014,** *369* (1640). https://doi.org/10.1098/rstb.2013.0226

Francini, A.; Sebastiani, L. Abiotic Stress Effects on Performance of Horticultural Crops. *Horticulturae* **2019,** *5* (4), 67. https://doi.org/10.3390/horticulturae5040067

Garrett, K. A.; Dendy, S. P.; Frank, E. E.; Rouse, M. N.; Travers, S. E. Climate Change Effects on Plant Disease: Genomes to Ecosystems. *Ann. Rev. Phytopathol.* **2006,** *44*, 489–509. https://doi.org/10.1146/annurev.phyto.44.070505.143420

Gechev, T. S.; van Breusegem, F.; Stone, J. M.; Denev, I.; Laloi, C. Reactive Oxygen Species as Signals that Modulate Plant Stress Responses and Programmed Cell Death. *BioEssays* **2006,** *28* (11), 1091–1101. https://doi.org/10.1002/bies.20493

Goel, A. K.; Lundberg, D.; Torres, M. A.; Matthews, R.; Akimoto-Tomiyama, C.; Farmer, L.; Dangl, J. L.; Grant, S. R. The Pseudomonas Syringae Type III Effector HopAM1 Enhances Virulence on Water-Stressed Plants. *Mol. Plant Microbe Interact.* **2008,** *21* (3), 361–370. https://doi.org/10.1094/MPMI-21-3-0361

Gudesblat, G. E.; Iusem, N. D.; Morris, P. C. Guard Cell-Specific Inhibition of Arabidopsis MPK3 Expression Causes Abnormal Stomatal Responses to Abscisic Acid and Hydrogen Peroxide. *New Phytol.* **2007,** *173* (4), 713–721. https://doi.org/10.1111/j.1469-8137.2006.01953.x

Hewezi, T.; Léger, M.; Gentzbittel, L. A Comprehensive Analysis of the Combined Effects of High Light and High Temperature Stresses on Gene Expression in Sunflower. *Ann. Bot.* **2008,** *102* (1), 127–140. https://doi.org/10.1093/aob/mcn071

Holley, S. R.; Yalamanchili, R. D.; Moura, D. S.; Ryan, C. A.; Stratmann, J. W. Convergence of Signaling Pathways Induced by Systemin; Oligosaccharide Elicitors; and Ultraviolet-B Radiation at the Level of Mitogen-Activated Protein Kinases in Lycopersicon Peruvianum Suspension-Cultured Cells. *Plant Physiol.* **2003,** *132* (4), 1728–1738. https://doi.org/10.1104/pp.103.024414

IPCC. Kundzewicz, Z. W., Palutikof, J., Wu, S., Eds.; Climate Change and Water. Technical Paper of the Intergovernmental Panel on Climate Change, Cambridge University Press, 2008.

Jacobs, A. K.; Lipka, V.; Burton, R. A.; Panstruga, R.; Strizhov, N.; Schulze-Lefert, P.; Fincher, G. B. AnArabidopsis Callose Synthase, GSL5, is Required for Wound and Papillary Calloseformation. *Plant Cell* **2003,** *15* (11), 2503–2513. https://doi.org/10.1105/tpc.016097

Keleş¸ Y.; Öncel, I. Response of Antioxidative Defence System to Temperature and Water Stress Combinations in Wheat Seedlings. *Plant Sci.* **2002,** *163* (4), 783–790. https://doi.org/10.1016/S0168-9452(02)00213-3

Kim, S. Y.; Kim, S. G.; Kim, Y. S.; Seo, P. J.; Bae, M.; Yoon, H. K.; Park, C. M. Exploring Membrane-Associated NAC Transcription Factors in Arabidopsis: Implications for Membrane Biology in Genome Regulation. *Nucleic Acids Res.* **2007,** *35* (1), 203–213. https://doi.org/10.1093/nar/gkl1068

Király, L.; Hafez, Y. M.; Fodor, J.; Király, Z. Suppression of Tobacco Mosaic Virus Induced Hypersensitive-Type Necrotization in Tobacco at High Temperature is Associated with Down Regulation of NADPH Oxidase and Superoxide and Stimulation of Dehydroascorbate Reductase. *J. General Virol.* **2008,** *89* (3), 799–808. https://doi.org/10.1099/vir.0.83328-0

Kissoudis, C.; van de Wiel, C.; Visser, R. G. F.; van der Linden, G. Enhancing Crop Resilience to Combined Abiotic and Biotic Stress Through the Dissection of Physiological and Molecular Crosstalk. *Front. Plant Sci.* **2014,** *5*, 207. https://doi.org/10.3389/fpls.2014.00207

Kumar, A.; Kumar, S.; Venkatesh, K.; Singh, N. K.; Mandal, P. K.; Sinha, S. K. Physio-Molecular Traits of Contrasting Bread Wheat Genotypes Associated with 15-N Influx Exhibiting Homeolog Expression Bias in Nitrate Transporter Genes under Different External Nitrate Concentrations. *Planta* **2022,** *255* (5), 104. https://doi.org/10.1007/s00425-022-03890-7

Li, P.; Yang, X.; Wang, H.; Pan, T.; Yang, J.; Wang, Y.; Xu, Y.; Yang, Z.; Xu, C. Metabolic Responses to Combined Water Deficit and Salt Stress in Maize Primary Roots. *J. Integr. Agric.* **2021,** *20* (1), 109–119. https://doi.org/10.1016/S2095-3119(20)63242-7

Li, W.; Zhang, C.; Lu, Q.; Wen, X.; Lu, C. The Combined Effect of Salt Stress and Heat Shock on Proteome Profiling in Suaeda Salsa. *J. Plant Physiol.* **2011,** *168* (15), 1743–1752. https://doi.org/10.1016/j.jplph.2011.03.018

Li, Y.; Zhang, L.; Wang, X.; Hao, L.; Chu, X.; Guo, X. Cotton GhMPK6a Negatively Regulates Osmotis Tolerance and Bacterial Infection in Transgenic Nicotiana Benthaminana and Plays a Pivotal Role in Development. *FEBS J.* **2013,** *280*, 5128–5144.

Lin, L.; Wu, J.; Jiang, M.; Wang, Y. Plant Mitogen-Activated Protein Kinase Cascades in Environmental Stresses. *Int. J. Mol. Sci.* **2021,** *22* (4), 1543. https://doi.org/10.3390/ijms22041543

Liu, C.; Ruan, Y.; Lin, Z.; Wei, R.; Peng, Q.; Guan, C.; Ishii, H. Antagonism Between Acibenzolar-s-Methyl-Induced Systemic Acquired Resistance and Jasmonic Acid-Induced Systemic Acquired Susceptibility to Colletotrichum Orbiculare Infection in Cucumber. *Physiol. Mol. Plant Pathol.* **2008,** *72* (4–6), 141–145. https://doi.org/10.1016/j.pmpp.2008.08.001

Low, M.; Herbinger, K.; Nunn, A. J., H€aberle, K. H.; Leuchner, M.; Heerdt, C.; Werner, H.; Wipfler, P.; Pretzsch, H.; Tausz, M.; et al. Extraordinary Drought of 2003 Overrules Ozone Impact on Adult Beech Trees (Fagus sylvatica). *Trees* **2006,** *20*, 539–548.

Mahmoud, D.; Pandey, R.; Sathee, L.; Dalal, M.; Singh, M. P.; Chinnusamy, V. Regulation of Expression of Genes Associated with Nitrate Response by Osmotic Stress and Combined Osmotic and Nitrogen Deficiency Stress in Bread Wheat (Triticum aestivum L.). *Plant Physiol. Rep.* **2020,** *25* (2), 200–215. https://doi.org/10.1007/s40502-020-00503-x

Martinez-Ballesta, Mdel C.; Bastías, E.; Zhu, C.; Schäffner, A. R.; González-Moro, B.; González-Murua, C.; Carvajal, M. Boric Acid and Salinity Effects on Maize Roots. Response of Aquaporins ZmPIP1 and ZmPIP2, and Plasma Membrane H+-ATPase, In Relation to Water and Nutrient Uptake. *Physiol. Plant.* **2008,** *132* (4), 479–490. https://doi.org/10.1111/j.1399-3054.2007.01045.x

Massad, T. J.; Dyer, L. A.; Vega C., G. Cost of Defense and a Test of the Carbon-Nutrient Balance and Growth-Differentiation Balance Hypotheses for Two Co-occurring Classes of Plant Defense. *PLoS One* **2012,** *7* (10), e7554. https://doi.org/10.1371/journal.pone.0047554

McDonald, A.; Riha, S.; DiTommaso, A.; DeGaetano, A. Climate Change and the Geography of Weed Damage: Analysis of U.S. Maize Systems Suggests the Potential for Significant Range Transformations. *Agric. Ecosyst. Environ.* **2009,** *130* (3–4), 131–140. https://doi.org/10.1016/j.agee.2008.12.007

Meena, S. K.; Pandey, R.; Sharma, S.; Gayacharan; Kumar, T.; Singh, M. P.; Dikshit, H. K. Physiological Basis of Combined Stress Tolerance to Low Phosphorus and Drought in a Diverse Set of Mungbean Germplasm. *Agronomy* **2021,** *11* (1), 99. https://doi.org/10.3390/agronomy11010099

Melotto, M.; Underwood, W.; Koczan, J.; Nomura, K.; He, S. Y. Plant Stomata Function Ininnate Immunity Against Bacterial Invasion. *Cell* **2006,** *126* (5), 969–980. https://doi.org/10.1016/j.cell.2006.06.054

Mengiste, T.; Chen, X.; Salmeron, J.; Dietrich, R. The BORTYTIS SUSCEPTIBLE1 Gene Encodes an R2R3MYB Transcription Factor Protrin that is Required for Biotic and Abiotic Stress Responses in Arabidopsis. *Plant Cell* **2003,** *15* (11), 2551–2565. https://doi.org/10.1105/tpc.014167

Miao, S. L.; Zou, C. B.; Breshears, D. D. Vegetation Responses to Extreme Hydrological Events: Sequence Matters. *Am. Nat.* **2009,** *173* (1), 113–118. https://doi.org/10.1086/593307

Miller, G.; Suzuki, N.; Ciftci-Yilmaz, S.; Mittler, R. Reactive Oxygen Species Homeostasis and Signalling During Drought and Salinity Stresses. *Plant Cell Environ.* **2010,** *33* (4), 453–467. https://doi.org/10.1111/j.1365-3040.2009.02041.x

Mittler, R. Abiotic Stress, The Field Environment and Stress Combination. *Trends Plant Sci.* **2006,** *11* (1), 15–19. https://doi.org/10.1016/j.tplants.2005.11.002

Mittler, R.; Blumwald, E. Genetic Engineering for Modern Agriculture: Challenges and Perspectives. *Ann. Rev. Plant Biol.* **2010,** *61*, 443–462. https://doi.org/10.1146/annurev-arplant-042809-112116

Miura, K.; Tada, Y. Regulation of Water; Salinity and Cold Stress Responses by Salicylic Acid. *Front. Plant Sci.* **2014,** *5*, 4. https://doi.org/10.3389/fpls.2014.00004

Modesti, M.; Forniti, R.; Brunori, E.; Mencarelli, F.; Bellincontro, A.; Tonutti, P. Ozone Treatments to Induce Systemic-Acquired Resistance in Leaves of Potted Vines: Molecular Responses and NIR Evaluation for Identifying Effective Dose and Exposition Duration. *OENO One* **2022,** *56* (1), 175–187. https://doi.org/10.20870/oeno-one.2022.56.1.5373

Mohr, P. G.; Cahill, D. M. Abscisic Acid Influences the Susceptibility of Arabidopsis Thalianato Pseudomonas syringae pv. tomato and Peronospora parasitica. *Funct. Plant Biol.* **2003,** *30* (4), 461–469. https://doi.org/10.1071/FP02231

Nishimura, M. T.; Stein, M.; Hou, B. H.; Vogel, J. P.; Edwards, H.; Somerville, S. C. Loss of a Callose Synthase Results in Salicylic Acid-Dependent Disease Resistance. *Science* **2003,** *301* (5635), 969–972. https://doi.org/10.1126/science.1086716

Obata, T.; Witt, S.; Lisec, J.; Palacios-Rojas, N.; Florez-Sarasa, I.; Yousfi, S.; Araus, J. L.; Cairns, J. E.; Fernie, A. R. Metabolite Profiles of Maize Leaves in Drought, Heat, and Combined Stress Field Trials Reveal the Relationship Between Metabolism and Grain Yield. *Plant Physiol.* **2015,** *169* (4), 2665–2683. https://doi.org/10.1104/pp.15.01164

Oshunsanya, S. O.; Nwosu, N. J.; Li, Y. Abiotic Stress in Agricultural Crops under Climatic Conditions. In Sustainable Agriculture, Forest and Environmental Management; Jhariya, M. K., Banerjee, A., Meena, R. S., Eds.; Springer, 2019; pp 71–100.

Pan, J.; Vicente, A. R.; Martínez, G. A.; Chaves, A. R.; Civello, P. M. Combined use of UV-C Irradiation and Heat Treatment to Improve Postharvest Life of Strawberry Fruit. *J. Sci. Food Agric.* **2004,** *84* (14), 1831–1838. https://doi.org/10.1002/jsfa.1894

Pandey, P.; Ramegowda, V.; Senthil-Kumar, M. Shared and Unique Responses of Plants to Multiple Individual Stresses and Stress Combinations: Physiological and Molecular Mechanisms. *Front. Plant Sci.* **2015a,** *6*, 723. https://doi.org/10.3389/fpls.2015.00723

Pérez-López, U.; Miranda-Apodaca, J.; Muñoz-Rueda, A.; Mena-Petite, A. Lettuce Production and Antioxidant Capacity are Differentially Modified by Salt Stress and Light Intensity under Ambient and Elevated CO2. *J. Plant Physiol.* **2013,** *170* (17), 1517–1525. https://doi.org/10.1016/j.jplph.2013.06.004

Peters, K.; Breitsameter, L.; Gerowitt, B. Impact of Climate Change on Weeds in Agriculture: A Review. *Agron. Sustain. Dev.* **2014,** *34* (4), 707–721. DOI: 10.1007/s13593-014-0245-2

Pitzschke, A.; Djamei, A.; Teige, M.; Hirt, H. VIP1 Response Elements Mediate Mitogen-Activated Protein Kinase 3-Induced Stress Gene Expression. *Proc. Natl. Acad. Sci. U. S. A.* **2009,** *106* (43), 18414–18419. https://doi.org/10.1073/pnas.0905599106

Prasad, P. V. V.; Pisipati, S. R.; Momčilović, I.; Ristic, Z. Independent and Combined Effects of High Temperature and Drought Stress During Grain Filling on Plant Yield and Chloroplast EF-Tu Expression in Spring Wheat. *J. Agron. Crop Sci.* **2011,** *197* (6), 430–441. https://doi.org/10.1111/j.1439-037X.2011.00477.x

Puckette, M. C.; Weng, H.; Mahalingam, R. Physiological and Biochemical Responses to Acute Ozone-Induced Oxidative Stress in Medicago Truncatula. *Plant Physiol. Biochem.* **2007,** *45* (1), 70–79. https://doi.org/10.1016/j.plaphy.2006.12.004

Ramegowda, V.; Senthil-Kumar, M.; Ishiga, Y.; Kaundal, A.; Udayakumar, M.; Mysore, K. S. Drought Stress Acclimation Impacts Tolerance to Sclerotina Sclerotiorum and Pseudomonas Syringae in Nicotiana Benthamiana. *Int. J. Mol. Sci.* **2013,** *14* (5), 9497–9513. https://doi.org/10.3390/ijms14059497

Ramegowda, V.; Senthil-Kumar, M.; Udayakumar, M.; Mysore, K. S. A High-Throughput Virus-Induced Gene Silencing Protocol Identifies Genes Involved Inmulti-Stress Tolerance. *BMC Plant Biol.* **2013b,** *13*, 193. https://doi.org/10.1186/1471-2229-13-193

Ramu, V. S.; Paramanantham, A.; Ramegowda, V.; Mohan-Raju, B.; Udayakumar, M.; Senthil-Kumar, M. Transcriptome Analysis of Sunflower Genotypes with Contrasting Oxidative Stress Tolerance Reveals Individual- and Combined- Biotic and Abiotic Stress Tolerance Mechanisms.

PLoS One **2016,** *11* (6), e0157522. https://doi.org/10.1371/journal.pone.0157522

Reusche, M.; Thole, K.; Janz, D.; Truskina, J.; Rindfleisch, S.; Drübert, C.; Polle, A.; Lipka, V.; Teichmann, T. Verticillium Infection Triggers VASCULAR-RELATED NAC DOMAIN7-dependent de novo Xylem Formation and Enhances Drought Tolerance in Arabidopsis. *Plant Cell* **2012,** *24* (9), 3823–3837. https://doi.org/10.1105/tpc.112.103374

Rivero, R. M.; Mestre, T. C.; Mittler, R. O.; Rubio, F.; Garcia-Sanchez, F. R.; Martinez, V. The Combined Effect of Salinity and Heat Reveals a Specific Physiological, Biochemical and Molecular Response in Tomato Plants. *Plant Cell Environ.* **2014,** *37* (5), 1059–1073. https://doi.org/10.1111/pce.12199

Rizhsky, L.; Liang, H.; Mittler, R. The Combined Effect of Drought Stress and Heat Shock on Gene Expression in Tobacco. *Plant Physiol.* **2002,** *130* (3), 1143–1151. https://doi.org/10.1104/pp.006858

Rizhsky, L.; Liang, H.; Shuman, J.; Shulaev, V.; Davletova, S.; Mittler, R. When Defense Pathways Collide: The Response of Arabidopsis to a Combination of Drought and Heat Stress. *Plant Physiol.* **2004,** *134* (4), 1683–1696. https://doi.org/10.1104/pp.103.033431.

Robert-Seilaniantz, A.; Navarro, L.; Bari, R.; Jones, J. D. G. Pathological Hormone Imbalances. *Curr. Opin. Plant Biol.* **2007,** *10* (4), 372–379. https://doi.org/10.1016/j.pbi.2007.06.003

Sales, C. R.; Ribeiro, R. V.; Silveira, J. A.; Machado, E. C.; Martins, M. O.; Lagôa, A. M. Superoxide Dismutase and Ascorbate Peroxidase Improve the Recovery of Photosynthesis in Sugarcane Plants Subjected to Water Deficit and Low Substrate Temperature. *Plant Physiol. Biochem.* **2013,** *73*, 326–336. https://doi.org/10.1016/j.plaphy.2013.10.012

Savary, S.; Willocquet, L.; Pethybridge, S. J.; Esker, P.; McRoberts, N.; Nelson, A. The Global Burden of Pathogens and Pests on Major Food Crops. *Nat. Ecol. Evol.* **2019,** *3* (3), 430–439. https://doi.org/10.1038/s41559-018-0793-y

Seo, P. J.; Kim, M. J.; Park, J. Y.; Kim, S. Y.; Jeon, J.; Lee, Y. H.; Kim, J.; Park, C. M. Cold Activation of a Plasma Membrane-Tethered NAC Transcription Factor Induces a Pathogen Resistance Response in Arabidopsis. *Plant J. Cell Mol. Biol.* **2010,** *61* (4), 661–671. https://doi.org/10.1111/j.1365-313X.2009.04091.x

Shao, H. B.; Chu, L. Y.; Jaleel, C. A.; Zhao, C. X. Water-Deficit Stress—Induced Anatomical Changes in Higher Plants. *C. R. Biol.* **2008,** *331* (3), 215–225. https://doi.org/10.1016/j.crvi.2008.01.002

Sharma, R.; de Vleesschauwer, D.; Sharma, M. K.; Ronald, P. C. Recent Advances in Dissecting Stress-Regulatory Crosstalk in Rice. *Mol. Plant* **2013,** *6* (2), 250–260. https://doi.org/10.1093/mp/sss147

Sicher, R. C.; Bunce, J. A. The Impact of Enhanced Atmospheric CO2 Concentrations on the Responses of Maize and Soybean to Elevated Growth Temperatures. In *Combined Stresses in Plants*; Sicher, R. C., Bunce, J. A., Eds.; Springer, 2015; pp 27–48.

Sinha, S. K.; Kumar, A.; Tyagi, A.; Venkatesh, K.; Paul, D.; Singh, N. K.; Mandal, P. K. Root Architecture Traits Variation and Nitrate-Influx Responses in Diverse Wheat Genotypes under Different External Nitrogen Concentrations. *Plant Physiol. Biochem.* **2020,** *148*, 246–259. https://doi.org/10.1016/j.plaphy.2020.01.018

Sinha, S. K.; Rani, M.; Kumar, A.; Kumar, S.; Venkatesh, K.; Mandal, P. K. Natural Variation in Root System Architecture in Diverse Wheat Genotypes Grown under Different Nitrate Conditions and Root Growth Media. *Theor. Exp. Plant Physiol.* **2018,** *30* (3), 223–234. https://doi.org/10.1007/s40626-018-0117-2

Spoel, S. H;, Tada, Y.; Loake, G. J. Post-Translational Protein Modification as a Tool for Transcription Reprogramming. *New Phytol.* **2010,** *186* (2), 333–339. https://doi.org/10.1111/j.1469-8137.2009.03125.x

Srivastava, G.; Kumar, S.; Dubey, G.; Mishra, V.; Prasad, S. M. Nickel and Ultraviolet-B Stresses Induce Differential Growth and Photosynthetic Responses in Pisum sativum L. Seedlings. *Biol. Trace Elem. Res.* **2012,** *149* (1), 86–96. https://doi.org/10.1007/s12011-012-9406-9

Stout, M. J.; Fidantsef, A. L.; Duffey, S. S.; Bostock, R. M. Signal Interactions in Pathogen an Insect Attack: Systemic Plant-Mediated Interactions Between Pathogens and Herbivores of the Tomato. Lycopersicum esculentum. *Physiol. Mol. Plant Pathol.* **1999,** *54* (3–4), 115–130. https://doi.org/10.1006/pmpp.1998.0193

Suleman, P.; Al-Musallam, A.; Menezes, C. A. The Effect of Solute Potential and Water Stress on Black Scorch Caused by Chalara Paradoxa and Chalara Radicicola on Date Palms. *Plant Dis.* **2001,** *85* (1), 80–83. https://doi.org/10.1094/PDIS.2001.85.1.80

Suzuki, N.; Rivero, R. M.; Shulaev, V.; Blumwald, E.; Mittler, R. Abiotic and Biotic Stress Combinations. *New Phytol.* **2014,** *203* (1), 32–43. https://doi.org/10.1111/nph.12797

Szittya, G.; Silhavy, D.; Molnar, A.; Havelda, Z.; Lovas, A.; Lakatos, L.; Banfalvi, Z.; Burgyan, J. Low Temperature Inhibits RNA Silencing-Mediated Defence by the Control of siRNA Generation. *EMBO J.* **2003,** *22* (3), 633–640. https://doi.org/10.1093/emboj/cdg74

Tanou, G.; Molassiotis, A.; Diamantidis, G. Hydrogen Peroxide and Nitric Oxide-Induced Systemic Antioxidant Prime-Like Activity Under NaCl-Stress and Stress-Free Conditions in Citrus Plants. *J. Plant Physiol.* **2009,** *166* (17), 1904–1913. https://doi.org/10.1016/j.jplph.2009.06.012

Tewari, K.; Kumar, V.; Kumar, A.; Bansal, N.; Vinutha, T.; Ali, K.; Sachdev, A.; Kumari, S.; Dahuja, A. Molecular Cloning and Functional Analysis of the Promoter of γ-Tocopherol

Methyl Transferase (γ-TMT) Gene of Soybean (Glycine max). *3 Biotech* **2018,** *8* (8), 325. https://doi.org/10.1007/s13205-018-1347-3
Tippmann, H. F.; Schlüter, U.; Collinge, D. B. In *Common Themes in Biotic and Abiotic Stresssignalling in Plants*; Global Science Publications Books, 2006.
Truman, W.; de Zabala, M. T.; Grant, M. Type III Effectors Orchestrate a Complex Interplay Between Transcriptional Networks to Modify Basal Defence Responses During Pathogenesis and Resistance. *Plant J. Cell Mol. Biol.* **2006,** *46* (1), 14–33. https://doi.org/10.1111/j.1365-313X.2006.02672.x
Tsutsui, T.; Kato, W.; Asada, Y.; Sako, K.; Sato, T.; Sonoda, Y.; Kidokoro, S.; Yamaguchi-Shinozaki, K.; Tamaoki, M.; Arakawa, K.; Ichikawa, T.; Nakazawa, M.; Seki, M.; Shinozaki, K.; Matsui, M.; Ikeda, A.; Yamaguchi, J. DEAR1; A Transcriptional Repressor of DREB Protein that Mediates Plant Defense and Freezing Stress Responses in Arabidopsis. *J. Plant Res.* **2009,** *122* (6), 633–643. https://doi.org/10.1007/s10265-009-0252-6
Valerio, M.; Lovelli, S.; Perniola, M.; Di Tommaso, T.; Ziska, L. The Role of Water Availability on Weed–Crop Interactions in Processing Tomato for Southern Italy. *Acta Agric. Scand. B Soil Plant Sci.* **2013,** *63* (1), 62–68. https://doi.org/10.1080/09064710.2012.715184
Vescio, R.; Abenavoli, M. R.; Sorgonà, A. Single and Combined Abiotic Stress in Maize Root Morphology. *Plants* **2020,** *10* (1), 5. https://doi.org/10.3390/plants10010005
Vile, D.; Pervent, M.; Belluau, M.; Vasseur, F.; Bresson, J.; Muller, B.; Granier, C.; Simonneau, T. Arabidopsis Growth Under Prolonged High Temperature and Water Deficit: Independent or Interactive Effects? *Plant Cell Environ.* **2012,** *35* (4), 702–718. https://doi.org/10.1111/j.1365-3040.2011.02445.x
Wang, Y.; Bao, Z.; Zhu, Y.; Hua, J. Analysis of Temperature Modulation of Plant Defenseagainst Biotrophic Microbes. *Mol. Plant Microbe Interact.* **2009,** *22* (5), 498–506. https://doi.org/10.1094/MPMI-22-5-0498
Wen, X.; Qiu, N.; Lu, Q.; Lu, C. Enhanced Thermotolerance of Photosystem II in Salt-Adapted Plants of the Halophyte Artemisia Anethifolia. *Planta* **2005,** *220* (3), 486–497. https://doi.org/10.1007/s00425-004-1382-7
Wiese, J.; Kranz, T.; Schubert, S. Induction of Pathogen Resistance in Barley by Abiotic Stress. *Plant Biol.* **2004,** *6* (5), 529–536. https://doi.org/10.1055/s-2004-821176
Xia, X. J.; Wang, Y. J.; Zhou, Y. H.; Tao, Y.; Mao, W. H.; Shi, K.; Asami, T.; Chen, Z.; Yu, J. Q. Reactive Oxygen Species are Involved in Brassinosteroid-Induced Stress Tolerance in Cucumber. *Plant Physiol.* **2012,** *150*, 1034–1045
Xu, P.; Chen, F.; Mannas, J. P.; Feldman, T.; Sumner, L. W.; Roossinck, M. J. Virus Infection improves Drought Tolerance. *New Phytol.* **2008,** *180* (4), 911–921. https://doi.org/10.1111/j.1469-8137.2008.02627.x
Yasuda, M.; Ishikawa, A.; Jikumaru, Y.; Seki, M.; Umezawa, T.; Asami, T.; Maruyama-Nakashita, A.; Kudo, T.; Shinozaki, K.; Yoshida, S.; Nakashita, H. Antagonistic Interaction Between Systemic Acquired Resistance and the Abscisic Acid-Mediated Abiotic Stress Response in Arabidopsis. *Plant Cell* **2008,** *20* (6), 1678–1692. https://doi.org/10.1105/tpc.107.054296
Yousfi, S.; Serret, M. D.; Voltas, J.; Araus, J. L. Effect of Salinity and Water Stress During the Reproductive Stage on Growth, Ion Concentrations, D13C, and d15N of Durum Wheat and Related Amphiploids. *J. Exp. Bot.* **2010,** *61* (13), 3529–3542. https://doi.org/10.1093/jxb/erq184

Zandalinas, S. I.; Sen Gupta, S.; Fritschi, F. B.; Azad, R. K.; Nechushtai, R.; Mittler, R. The Impact of Multifactorial Stress Combination on Plant Growth and Survival. *New Phytol.* **2021,** *230* (3), 1034–1048. https://doi.org/10.1111/nph.17232

Zhou, J.; Xia, X. J.; Zhou, Y. H.; Shi, K.; Chen, Z.; Yu, J. Q. RBOH1-Dependent H2O2 Production and Subsequent Activation of MPK1/2 Play an Important Role in Acclimation-Induced Cross-Tolerance in Tomato. *J. Exp. Bot.* **2014,** *65* (2), 595–607. https://doi.org/10.1093/jxb/ert404.

PART II

Climate-Resilient Agriculture: Breeding and Molecular Approaches

CHAPTER 5

Marker-Assisted Selection: Strategy for Biotic and Abiotic Stress Tolerance in Plants

ANJALI JOSHI,[1] AMARJEET KUMAR,[2] SNEHA ADHIKARI,[3] and RANJIT SAROJ[4]

[1]*Genetics and Tree Improvement Division, Arid Forest Research Institute, Jodhpur, Rajasthan, India*

[2]*Department of Genetics and Plant Breeding, MTTC and VTC, Selesih, CAU, Imphal, India*

[3]*ICAR-IIWBR, Regional Station Flowerdale, Shimla, India*

[4]*Division of Genetics, Indian Agricultural Research Institute, New Delhi, IndiaPortugal*

ABSTRACT

The world population is increasing day by day along with a decrease in total cultivated area. Climatic change and evolution of new pest and disease biotypes have also raised serious challenges in front of breeders to not only increase the production and productivity but at the same time enhance resistance to biotic and abiotic stresses as well. Breeding for plant resistance to stress is the most competent and inexpensive way of maintaining and enhancing the yield of economic product obtained from plants. Tremendous advances have been made to understand the physiological and molecular mechanisms controlling resistance to stress. With the advent of molecular marker technology, the dissection of complex and polygenically inherited traits has become possible, thereby helping in the detection of numerous stress resistance genes linked to molecular markers. Genetic mapping of

Climate-Resilient Agriculture: A Molecular Perspective. Ashutosh Singh, Saurabh Pandey (Eds.)

quantitative traits loci (QTLs) and certain major genes affecting economically important traits have led to the amalgamation of biotechnology and conventional breeding procedures. The information obtained in the form of map position of economically important traits and linked molecular markers can be exploited by means of selection of traits using marker-assisted selection (MAS). MAS involves the selection of the desired individuals on the basis of DNA marker patterns rather than, or in addition to, the phenotypic values for the trait. When used appropriately, MAS is a tool that could help breeders select more efficiently, thereby increasing both the reliability and the speed of selection. Along with other advantages MAS also helps in performing selection at an early stage of growth.

5.1 INTRODUCTION

World population has increased at an alarmingly high rate in the last century and will further increase by 2050 (Foresight, 2011). Increased population have led to an increased demand for basic amenities including food and water which have further increased pressure on the existing food supply system (Foresight, 2011). Apart from increasing population, increased income, change in the requirement, and preferences of consumers have further necessitated the need for increasing food production. The situation is further complicated because of climate change which is considered a prominent threat to food security in future. Climate change has led to an increase in frequency of extreme weather phenomenon such as droughts and floods (Hay et al., 2016). It may also lead to an adverse effect on plant growth as a result of increase in amount of carbon dioxide, temperature (Nelson et al., 2009) and may also lead to resurgence of new diseases and pests. As per the estimates of FAO (2015), food production increment to the tune of 60% will be required in order to meet the increased food demand arisen due to population growth and dietary changes. In order to increase food production, expansion of the agricultural activities to previously unexplored land areas is required which further exposes the plant to altered growth conditions. Although the agricultural production in last few years has increased at a rate higher than the rate of population growth, there are still serious concerns on whether the present land mass would be sufficient to cater to the increased demand of food by the year 2050. In the present scenario, the required increase in food production will have to occur in the prevailing conditions of decreasing area, increasing scarcity of water, increasing pollution, degradation of arable land, unavoidable emergence of novel biotypes and races of pests and pathogens, and the imminent antagonistic effects of changing climatic conditions. The necessity to increase food production under impending conditions has thus posed a challenge to plant breeders, and there is an emergent need to develop improved varieties and make them available to farmers as a partial solution to the never-ending problem of food scarcity.

Plant breeding refers to the application of genetic principles to change the genotypic constitution of the plants and increase their utility for humans (Allard, 2019). It involves the selection of plants having desirable characteristics followed by controlled mating of the selected individuals and then selecting superior individuals among the progeny. Plant breeding is a coevolutionary process between plants and humans. Humans changed the genotypic constitution of the plants being grown in agricultural fields and the selected plants allowed changes in human populations. Thus, agriculture in association with modern crop varieties has been a significant aspect responsible for the sustenance of civilization (Agronomy, 1992). Keeping in view the increasing population, current agricultural yield trends, and increasing environmental pressure, plant breeding should mainly focus on improvement of traits involved in maintaining stability and sustainability of crop yield including tolerance to biotic and abiotic stresses and characters pertaining to increased water and nutrient use efficiency (Trethowan *et al.*, 2005; Slafer *et al.*, 2005). Additionally, varieties suitable for the cultivation on marginal and degraded land must be developed and greater emphasis must be given for the improvement of minor crops (Naylor *et al.*, 2004).

Plant breeding involves selection of better individuals in terms of yield, quality of edible parts, ease of cultivation, harvesting or processing, resistance to biotic and abiotic stresses among a range of variants. Each of these features has their own agronomic or food value and encompasses a number of individual traits. Improving a single trait without any regard to all other traits is straightforward and relatively easy, although it would rarely lead to the development of a variety amenable for widespread cultivation by farmers. Therefore, there is a need of simultaneously improving all the traits affecting a particular agronomic feature. This task is further complicated due to existence of genetic correlations between different traits. Genetic correlation between traits may be due to the presence of pleiotropic genes, linked genes or on account of the presence of genetic structure within the population (Hart and Clark, 1997). Selection for one trait would therefore also change other correlated traits either in the desired or in an undesired direction (Falconer and Mackay, 1996). It is one of the major bottlenecks of conventional plant breeding. In conventional plant breeding several unwanted genes may linger with the desired genes for several generations, thus limiting the genetic gain.

Also, the different procedures adopted for phenotyping of traits are generally time consuming, expensive, and unreliable for certain traits, especially those involving resistance to abiotic stresses. The existence of significant genotype x environment interactions further complicates the process of selection and reduces the selection efficiency.

The efficiency of selection and hence the probability of success from conventional plant breeding can be ensured by supplementing it with biotechnological tools (Huang *et al.*, 2002). One such biotechnological tool is DNA marker technology which traces its roots to the extensive research conducted in the area of molecular genetics and genomics. Sequences of DNA that can be traced to a specific location on the genome, transmitted generation after generation and following the laws of inheritance are called DNA markers (Singh et al., 2018). The DNA markers are not considered genes as they are generally neutral genomic regions without any biological effects (Moniruzzaman et al., 2014). DNA markers are used as tags and owing to the linkage with the trait of interest, they can be used to detect allelic variation in the genomic regions controlling the trait. Incorporation of DNA markers in the process of selection during breeding programmes greatly increases the precision and efficiency of the breeding process and is popularly called marker-assisted selection (MAS). The MAS approach changes the criteria of selection from selection of phenotypes to either direct or indirect selection of genes and therefore bypasses numerous problems associated with conventional plant breeding. DNA markers, therefore, improve the efficiency of conventional plant breeding processes and thereby are expected to play an important role in enhancing global food production (Kasha, 1999; Ortiz, 1998).

5.2 MARKER-ASSISTED SELECTION (MAS): A MOLECULAR TOOL FOR PRECISION PLANT BREEDING

Selection of individuals having required gene combination in the segregating population is a crucial step determining the success of breeding programs. Breeders work with a number of segregating populations at a time which often makes the selection of desirable individuals quite laborious and time consuming (Ribaut and Betran, 1999). MAS involves the indirect selection for the characteristics of interest which may include biotic stress tolerance, abiotic stress tolerance, quality or productivity by selecting for the linked DNA marker (Prabhu et al., 2009) rather than selection on the basis of phenotypic observation for the trait. Molecular marker is helpful in the selection of traits controlled by either genes or QTLs (Choudhary et al.,

2008). The construction of dense genetic maps along with the availability of an array of molecular markers in different crops has revolutionized the application of MAS in crop plants. The total number of genes that has to be incorporated in a genotype and the distance between the target gene and flanking markers affects the success of MAS (Perumalsamy et al., 2010). Lesser the distance between flanking marker and the gene of interest, tighter is the linkage, lower are the chances of recombination between them, and more reliable is the MAS. MAS has numerous advantages over phenotypic selection. MAS helps in precise transfer of the genomic region controlling the trait of interest. It speeds up the recovery of recurrent parent genome and hence greatly enhances the effectiveness and efficiency of plant breeding programs (Wijerathna, 2015).

After the markers linked tightly with the genomic regions of interest are identified, breeders may utilize the linked marker alleles for the identification of plant carrying the desired genes even before the plants are grown under field conditions thereby helping in fast tracking of certain traits leading to quicker development of lines and release of variety. MAS technology also assists the breeder in undertaking more informed crosses. As MAS does not require phenotypic evaluation and trait development, selection can be performed in nurseries during off season, thereby making the process more time saving and cost effective. MAS reduces the total number of lines that need to be evaluated. Based on the marker profile a number of lines are discarded during early stage of breeding cycle and only desirable lines are advanced to subsequent generations. This saves precious resources and helps in more efficient utilization of the field space/polyhouse, a resource which is often limited in a breeding program. MAS also substitutes field trials which have to be undertaken at a specific time, in particular location and are technically complicated, with molecular tests which do not have any specific spatial or temporal requirements. Phenotypic evaluation of a trait is further convoluted by environmental variations. Molecular markers, however, are neutral to environmental fluctuations and can also be used for efficient selection of trait with low heritability. MAS also ensures biosafety as it helps in selecting disease-resistant plants on the basis of their molecular data and does not require inoculation of the pathogen in field or greenhouse condition. Additionally, DNA markers also help in the incorporation of resistant traits into elite cultivars even before a particular race or biotype of disease or pests have affected the crop. DNA markers linked to QTL controlling a particular trait behave as single Mendelian factor and hence help in easy selection of polygenic trait which is difficult to select using conventional breeding approaches. As MAS does not involve phenotypic screening, it is an ideal method to select for traits whose phenotypic evaluation is difficult, involves destructive bioassays or in situation where phenotypic evaluation is not feasible as in case of screening for resistance to pathogen where quarantine restriction exists. In case of conventional phenotypic screening single plant selection is not reliable due to the effect of environment, and it cannot distinguish between the heterozygous and homozygous individuals. However, single plant can be selected using DNA markers as it selects the individual on the basis of genotype and also it helps in distinguishing a homozygote from the heterozygote.

5.3 MARKERS AND DETECTION OF MARKER TRAIT ASSOCIATION

5.3.1 GENETIC MARKERS

Genetic markers act as signs or flags for the target gene and represent genetic differences between individuals. Genetic markers are linked tightly with the gene of interest and may be used as tags to follow the inheritance of gene generation after generation; however, they do not themselves affect the phenotype of the trait of interest. Genetic markers are situated at specific positions within the genome called locus. Genetic markers fall into two broad categories namely, classical markers and molecular/DNA markers. Classical marker further includes morphological, biochemical, and cytological markers. The morphological markers include easily observable phenotypic traits like flower color, pigmentation, seed shape, growth habits, etc., that can be observed visually and differentiate between individuals of a species. Morphological markers can be scored visually and do not involve the use of sophisticated techniques. However, the major disadvantage is that morphological markers are limited in number and may not be linked to an economically important trait. Some morphological marker expresses at a particular developmental stage, and their expression is affected by environmental factors (Eagles et al., 2001). Morphological markers are being used since ancient times to assess the variation in different crop plants (Karaköy et al., 2014). Cytological markers are related to the variations involving number, shape, size, banding patterns, position, and order of chromosomes. These are generally used for physical mapping and identification of linkage groups (Jiang, 2013). Biochemical markers constitute allelic variant of an enzyme known as isozymes and are detected by electrophoresis and staining. These are controlled by numerous genes and have similar functions (Bayley, 1983). They are codominant in nature and are generally used for the detection of population structure, population subdivision, gene flow, and genetic diversity (Mateu-Andres and De Paco, 2005). However, these markers are limited in number and are tissue and plant growth stage specific and their scoring is affected by extraction methodologies (Mondini et al., 2009). DNA markers are nucleotide sequences which detect polymorphism between individuals. They generally do not have any biological effect and therefore cannot be considered genes. They signify constant landmarks in the genome and are easily identifiable and transmitted as per standard laws of inheritance. DNA markers are valuable and informative if they are able to reveal variation among individuals of either the same or of different species and such markers are called polymorphic DNA markers. Monomorphic markers on the other hand are not able to discriminate between individuals. If a polymorphic marker is able to discriminate between homozygote and heterozygote, it is referred to as codominant marker. Codominant marker represents differences in length. If the marker is not able to differentiate between homozygote and heterozygote, it is called dominant marker. Dominant markers are either present or absent. An ideal molecular marker must be codominant, highly reproducible, polymorphic, and should be distributed evenly and densely throughout the species genome (Mondini et al., 2009).

Numerous molecular markers have been used in different breeding programmes involving various crops. Microsatellite markers also called as Simple sequence repeats (SSRs), short tandem repeats, or simple sequence length polymorphism markers are used most widely in plant breeding. They are tandem repeats usually 1–6 nucleotide long and are evenly distributed throughout the genome of a species (Beckmann and Weber, 1992; Sara and Barat, 2014). SSRs may also be a part of expressed sequence tags and genes coding proteins (Morgante et al., 2002). They are highly polymorphic, reliable, codominant, reproducible, and allow multistage selection which makes the selection process more cost effective (Witcombe et al., 2008). Besides, microsatellite markers differentiate between homozygotes and heterozygotes and are easily available through public databases and published linkage maps of various species (McCouch et al., 2002; Caixeta et al., 2009). The only disadvantages of microsatellites are that in case the size differences are minute, polyacrylamide gel electrophoresis is required to resolve the bands. Also, SSR markers provide information on single locus per assay which increases the cost of genotyping. However, multiplexing can be done and the amount of information generated per PCR reaction can be increased (Collard and Mackill, 2008). On account of their high reproducibility and genome abundance, SSR markers have been considered a marker of choice for plant mapping studies and are also used for the evaluation of genetic diversity in crop plants. Besides microsatellites, restriction fragment length polymorphism (RFLP), random amplified polymorphic DNA (RAPD), Amplified fragment length polymorphism (AFLP), Inter simple sequence repeat (ISSR), Sequence characterized amplified region (SCAR), Cleaved amplified polymorphic sequences (CAPS), Sequence-related amplified polymorphism (SRAP), and Single nucleotide polymorphism (SNP) can also be used. RFLP helps in the detection of variation at DNA sequence level and involves treating the DNA with restriction enzymes. Mutations involving base-pair deletions, duplications, inversions, translocations, and transpositions are the chief reasons responsible for the appearance of variation in the pattern of RFLP markers. RAPD employs a single, short (usually 10 nucleotide long), arbitrary primer for PCR amplification (Sara and Barat, 2014). Polymorphism existing either between or at the site of primer binding can be perceived by electrophoresis in the form of either presence or absence of particular band. AFLP markers are codominant markers which combine both RFLP and PCR technology (Lynch and Walsh, 1998). They are cost effective and do not require any prior information of the sequence. SRAP is a dominant marker system using a CCGG sequence containing forward primer and AATT sequence containing reverse primer. It involves low annealing temperature of 35°C for first five cycles, while the annealing temperature of 50°C is used for the remaining 35 cycles. ISSR are dominant marker system and constitute amplification of the region located between two oppositely oriented but identical microsatellite repeats regions (Zietkiewicz et al., 1994). They do not require any prior sequence information (Chatterjee et al., 2004). CAPS are codominant markers and employ PCR amplification of DNA followed by restriction enzyme digestion (Michaels and Amasino, 1998). The primer used in CAPS is fetched from either databank or cDNA sequences or cloned RAPD bands. SCAR are codominant markers produced by the purification of the PCR fragments followed by SCAR primer designing (Bhagyawant, 2016). They are more reproducible and specific when compared to RAPD markers.

Most of the above-discussed markers are random DNA markers (RDMs) which are developed from anonymous regions in the genome irrespective of the genic information. The trait association with such markers may be lost as a result of recombination (Salgotra and Stewart, 2020). Genic molecular markers (GMMs) though present within the target gene may still be derived from the noncoding region of the gene and therefore, may not be linked directly with the target trait. This may increase the chances of false selection during MAS. Thus, if the candidate gene controlling a specific trait is known, DNA markers designed from functionally characterized gene sequence, popularly called the functional markers (FMs) or perfect markers, may be synthesized for increased reliability of MAS (Andersen and Lubberstedt, 2003). FMs are increasingly being used for precision breeding of various biotic and abiotic stress-resistant traits. SNPs as FMs are more valuable and reliable in plant breeding in comparison to RDMs and genic molecular markers. SNPs are widely used DNA markers owing to a number of advantages over previous generation markers. On account of high density of SNPs within the genome there is high probability of finding SNPs within a gene of interest. SNPs as FMs are situated within the gene of interest and cannot be lost as a result of recombination and therefore can be utilized for MAS in populations having different recombination patterns. Also, detection of SNP marker may be automated, increasing the throughput of the analysis for quick utilization in plant breeding programs.

5.4 REQUIREMENTS FOR MAS

Three main factors affect the success of an MAS program. First availability of a genetic map containing an adequate number of evenly spaced polymorphic markers that can be used for locating desired genes or QTLs accurately, secondly there should be close linkage between the adjacent markers and the gene or QTL of interest, thirdly, recombination should be there between the markers and the rest of the genome. The success of MAS also depends on the relative location of the markers with respect to target gene. Three types of associations between marker and gene of interest can be seen. The most favorable association involves the location of the marker within the gene of interest commonly known as the gene-assisted selection. Such markers are generally rare. There is also a possibility when the markers are in linkage disequilibrium (LD) with target gene throughout the population. Linkage disequilibrium is the nonrandom association of alleles of two or more than two genes. It usually is observed in case of genes which are located physically close to each other. Sometimes, the marker is in linkage equilibrium (LE) with the target gene. It is one of the most challenging and difficult situations for the application of MAS. In certain cases, especially involving polygenic traits, the target gene situated within the QTL has not been characterized. Therefore, MAS for such traits involves selection of the entire chromosomal segment containing the QTL. In such cases it is often preferable to have either two polymorphic markers flanking the QTL in situations where the chromosomal segment to be selected for is more than 20 cM in length. It usually eliminates the possibility of selecting genotypes possessing double recombination between flanking markers.

5.5 DETECTION OF MARKER TRAIT ASSOCIATION

One of the preliminary requirements for the application of MAS is the assessment of marker trait association (MTA). MTA helps in the identification of marker tightly associated with the trait of interest and suitable for carrying out MAS. MTA can be identified using two broad approaches. The first approach involves using markers evenly and densely covering the entire genome for genotyping an entire mapping population followed by phenotyping of the population. The association between marker genotypes and phenotypic differences is then assessed to detect MTA. The genotyping for this approach is expensive, time-consuming, and extensive. Also, phenotyping at this scale is rather difficult, laborious, lacks precision and for most of the traits may be practically impossible. In the second approach to detect MTA only the most informative individuals possessing extreme phenotypes for the target trait are genotyped. The approach is popularly known as the bulk segregant analysis and the MTA is inferred by finding the differences in allelic frequency between the plant groups possessing contrasting phenotypes.

The first approach of QTL mapping includes family-based linkage mapping and association mapping studies (Cadic et al., 2013). Both of these methods differ in the type of mapping population used which ultimately determines the mapping power and resolution. Family-based mapping starts with the construction of mapping population which may be biparental populations derived from two parents and includes doubled haploids (DH), F2, backcrosses (BC), near-isogenic lines (NIL), and recombinant inbred line (RIL) population or multiparent population derived from the crossing of multiple parents and includes multiparent advanced generation intercross (MAGIC) and Nested Association Mapping (NAM) populations. Biparental mapping involves: (1) Identifying the parental lines that show contrasting phenotypes for the target trait, (2) selection of the type of molecular markers that express polymorphism between parental lines and can be used for mapping, (3) hybridization of the parental lines and mapping population development, (4) using molecular markers for genotyping of the mapping population followed by phenotyping for the trait of interest; and (5) detection of MTA using a suitable statistical method. Genes/QTLs can be identified relative to the linkage map using different statistical methods including single-marker analysis or interval mapping (Kearsey and Farquhar, 1998).

The QTL detection power is affected by a number of different factors such as type and size of mapping population, allele frequencies, and QTL effects.

Though biparental mapping has many advantages, it is also characterized with certain limitations. As biparental mapping population only covers few recent recombination events, occurred during the process of mapping population development, the QTL identified from the population localizes to a relatively larger genomic segment spanning an interval of 10–20 cM. The population only covers a small part of the genetic variation in the species as it is derived from only two parental lines and only two allele per locus can be mapped. On account of allelic diversity, the MTA detected in one mapping population may not be identified in another breeding population. Therefore, the MTA detected in one biparental mapping population must be validated before being utilized for MAS in an unrelated population. In certain case the MTA is lost during the validation step. In such cases the appropriate step is to detect new markers through either fine mapping or candidate gene analysis, in the genomic region around the target locus and identify MTAs that are common across diverse breeding populations. This would facilitate identification of numerous markers either within or around a particular gene, and the chances of parents of any population to be polymorphic for at least one of the markers will be very high. This will therefore help breeders in tracking the parental alleles during the entire breeding process and speeding up the process of MAS in any cross.

The limitation of biparental mapping population can be easily overcome by utilization of multiparent mapping populations. Multiparent mapping population includes multiparent advanced generation intercross (MAGIC) and nested association mapping (NAM) populations. Due to involvement of multiple parents the phenotypic diversity of such population is very large making them suitable for high resolution mapping of QTLs. The design for NAM was proposed by Yu et al. (2008) for identifying the genomic regions associated with flowering time in maize. MAGIC population was used for the first time for mapping QTL in mouse (Threadgill et al., 2002). In plants it was first used in *Arabidopsis thaliana* (Kover et al., 2009). MAGIC population is formed by intermating multiple inbred parents for several generations followed by construction of inbred lines. Both NAM and MAGIC populations increase the power and resolution of detecting QTLs. Statistical approaches for QTL mapping using MAGIC populations have now become available. Multiparent mapping population also helps in increasing the detection power of rare alleles (Yu and Buckler, 2006; Xiao et al., 2016).

Genes controlling complex traits can also be identified by association mapping or linkage disequilibrium (LD) mapping or natural population-based mapping. Linkage disequilibrium denotes nonrandom association of alleles of two or more than two genes. In contrast to the linkage mapping, association mapping deduces associations between marker genotype and phenotypic variation by investigating the genetic polymorphisms present in different genetic backgrounds that have been created as a result of historic recombination events (Nordborg and Tavare, 2002). As association mapping does not require mapping population it can be performed quickly and is relatively inexpensive. In LD mapping either markers densely covering the entire genome or markers associated with few candidate genes are used for genotyping in order to assess their association with the phenotype (Rafalski, 2002). The main steps of LD mapping include assembling the mapping population constituting of wild relatives, elite cultivars, exotic accessions, landraces, breeding lines, etc. The mapping population is then phenotype for the trait of interest and broad sense heritability is calculated. The mapping population is genotyped with either genome wide or candidate gene-based markers. Thereafter, the extent of LD is quantified and the influence of kinship and population structure is ascertained. Appropriate statistical methods are then employed to test the association between phenotype and genotype. The power of association mapping for the detection of QTLs is strongly affected by sample size, minor allele frequency (MAF), extent of LD, and population structure. Presence of subgroups within the population containing unequal allele distribution is a major impediment in LD mapping studies. The presence of subgroups leads to the detection of false association between marker and trait phenotype even when the marker is not linked with the genomic regions responsible for variation in phenotypic trait values (Buckler and Thornsberry, 2002). Therefore, the first step in LD mapping is to identify a set of subpopulations by using genetic similarity matrix produced by using marker data. This significantly reduces the detection of false positive associations. The amount of LD is determined by the frequency of recombination that has occurred in the examined segment of genome (Rafalski, 2002) with plant species subjected to population bottlenecks in the past, as soybean, sugarbeet, and sugarcane having large LD blocks extending over several cM (Jannoo et al., 1999; Zhu et al., 2003) while in species with high recombination rates as maize, LD tends to decay over few base pairs. Association mapping when compared to linkage mapping shows reduced efficiency when MTA have to be assessed in species characterized with low genetic diversity (Myles et al., 2009). Linkage mapping on the other hand has its own limitations. Therefore, a combination of linkage and association mapping must be used for the purpose of QTL detection and validation in plants (Zhang et al., 2011).

The second QTL mapping approach namely bulk segregant analysis involves genotyping of only two pools of DNA derived from individuals belonging to two extreme phenotypes and is cost effective as well as more efficient for traits controlled by major genes or major QTLs. Bulk segregant analysis can now be attempted in both biparental and natural populations (Zou et al., 2016). The power of bulked segregant analysis is significantly affected by sample size and sample selection. The sample for BSA can be generated by selecting individuals from either two tails of the distribution (bidirectional selection) or from single tail (single selection). Bidirectional selection prevents the effect of segregation distortion and therefore is more effective than single selection. However, for traits subjected to either negative or fatal selection pressure, single selection is usually preferred. With the expansion of high-throughput genotyping platforms, chip-based BSA are being successfully employed to detect QTL controlling various traits. Pooled DNA analysis has also been used in plants for mapping of major genes or two or three major effect QTLs. Certain drawbacks of BSA include insufficient density of markers; inaccurate allele frequency estimation within pools because of gel-based marker systems; low QTL detection power due to small-sized population suitable for detection of only large effect QTLs and detection of a high level of false-positive associations. The false-positive association can be effectively eliminated by validating the detected QTLs by screening the entire population with the markers that shows polymorphism between the two bulks.

5.6 SCHEMES OF MAS

MAS can be employed at different stages of a breeding scheme. Following MAS schemes can be followed for more efficient selection of desirable trait. However, there may be overlap between these schemes.

5.6.1 MARKER-ASSISTED EVALUATION OF BREEDING MATERIAL

MAS finds its utility in a number of procedures that precede hybridization and line development including identification of cultivar or purity assessment, valuation of genetic diversity, selection of parents, and confirmation of hybrid. Conventionally, these procedures were implemented on the basis of phenotypic assessment; however, the intervention of molecular marker has significantly improved their reliability and efficiency. Genetic purity is an indication of the trueness to type and shows the amount of contamination of genetic materials with different varieties or species. Genetic purity of the parents is one of the key determinants of hybrid breeding programme. It also determines seed quality in seed multiplication programmes. Conventionally, the genetic purity of any material is determined by grow out test which involves raising the plant to maturity and assessing its floral and morphological characteristics (Yashitola et al., 2002). Apart from being time consuming, labor intensive, and tedious, grow out tests are also illusive due to the presence of G × E interaction. Molecular markers, in contrast, are based on variation in DNA sequence and provide an unbiased means of classifying crop varieties. The sequence tagged microsatellite (STMS) markers are most extensively used for the assessment of genetic purity of lines (Sundaram et al., 2008). High level of genetic diversity is required to achieve progress in breeding programmes. For broadening the genetic base of the breeding materials diverse strains must be identified and hybridized with elite cultivars (Xu et al., 2004; Reif et al., 2005). DNA markers have played a vital role in assessing genetic diversity and assisting breeders by providing a detailed information about the genetic resources and helping in selection of parents. Besides molecular markers also help in classifying parental lines into different heterotic groups for exploitation of heterosis and help in identifying the genomic regions subjected to selection. The detection of shift in the allele frequencies signals the breeders to follow specific haplotypes or alleles which can help in designing suitable strategies for breeding plants (Steele et al., 2004). Besides this the genomic regions being subjected to selection can also be targeted for QTL analysis and validation and the information can also be utilized for developing new varieties having specific combination of alleles by using different MAS schemes (Steele et al., 2004).

5.6.2 MARKER-ASSISTED BACKCROSSING

A number of previously released crop varieties are tried and tested by farmers. The agricultural operations pertaining to these varieties including the sowing rates, date of sowing, rate of fertilizer application, and irrigation timing have already been standardized. Therefore, there is reluctance among farmers to let go off these varieties and adopt new ones. Also, the processing industry does not accept new varieties easily as they have already established protocols for manufacturing the processed end product. However, with widespread cultivation the older varieties become susceptible to newly evolved races of pathogen and hence breeding new varieties is prerequisite to ensure future food security. Backcrossing is a valuable plant breeding method for instances where a superior performing variety is available for cultivation but needs to be improved for few simply inherited traits. It helps in incorporating one or few genes into the background of an otherwise elite variety. Marker-assisted backcrossing (MABC) involves the transfer of a recessive or dominant allele controlled phenotypic trait from a donor parent (nonrecurrent parent) into a superior crop variety (recurrent parent) by repeated backcrossing and utilizing molecular markers for selection. It comprises selection at both the phenotypic and genotypic level. MABC had been used in the improvement of many elite varieties that needed to be improved for a single simply inherited character (Niebur et al., 2004; Crosbie et al., 2006; Dwivedi et al., 2007). MABC involves three levels of MAS (Holland, 2004), that is, foreground selection, recombinant selection, and background selection. Foreground selection constitutes the selection of plants having donor parent marker allele at the target locus. It helps to maintain the heterozygous state of the target locus until the final backcross, after which the selected plants are selfed and progenies homozygous for the target donor parent allele are identified. Markers which are tightly linked to the gene/QTL of interest are used for foreground selection which is specifically useful in case of traits whose phenotypic screening is laborious or time consuming, for seedling stage selection of traits expressed in reproductive stage and for the selection of recessive alleles. Second level of MABC is the recombinant selection in which the backcross progeny having the donor parent allele at the target locus and allele of the recipient parent at the linked flanking marker locus is selected. Recombinant selection helps in the reduction of the size of integrated segment of donor chromosome containing the target allele. This is significant as with each backcross there is slower reduction in size of the donor fragment compared to other unlinked regions; as a result many undesirable genes linked with the target locus also tend to be inherited, popularly known as linkage drag. In case of conventional backcrossing the donor segment remains considerably larger (i.e., more than 10 cM) even after many generations of backcrossing; however, by selecting for markers flanking a target gene (i.e., located at less than 5 cM from the target gene) the problem of linkage drag can be reduced significantly. As the possibility of double recombination is very rare, at least two backcross generations are subjected to recombinant selection. The third level of MABC is the background selection which involves selection of back cross progeny having the highest percentage of recurrent parent genome except for the target locus at which donor parent allele is desired. Background selection involves the use of tightly linked flanking markers for recombinant selection, whereas unlinked markers are used to select for the genome of the recurrent parent thereby accelerating the recovery of the genome of recurrent parent. Six generations of backcrossing are required to recover the genome of the recurrent parent, while several fragments of the donor chromosome may be present. The use of molecular markers significantly reduces the number of backcross generations required for the recovery of the recurrent parent genome. For the majority of crop plants, greater than 90% of the recurrent parent genome can be recovered within two backcross generations provided a suitable number of progeny and an adequate number of markers (one marker every 10 cM of genome) are used for the purpose of background selection (Tanksley et al., 1989).

5.6.3 MARKER-ASSISTED GENE PYRAMIDING

Gene pyramiding refers to the process of assembling multiple genes in a single individual. Several genes may be pyramided by using conventional methods, which have their own limitations. Phenotyping does not allow identification of plant containing more than one gene, each of them producing the same phenotypic effect. As for phenotypic selection each of the traits to be selected for needs to be tested and it becomes problematic to evaluate plants for characteristics which involve destructive bioassays. The DNA marker assays provide a solution to this problem as they are nondestructive and a single DNA sample may be used for assaying markers linked to different desirable genes. As the molecular markers have become accessible, combining desirable alleles at different locus in a short time period has become much easier (Tester and Langridge, 2010). Marker-assisted gene pyramiding has been employed successfully in several important crop species and a number of genes have been pyramided (Gupta et al., 2010a, 2010b). Some studies recommend that combining multiple resistance genes operative against different races of pathogen can offer durable resistance against diseases (Singh et al., 2001). Single-gene resistance is easily overcome by evolution of new races of pathogen. However, it is difficult to overcome the resistance contributed by multiple resistance genes. The integration of QTLs controlling quantitative resistance is a promising strategy to breed for durable disease resistance.

Quantitative resistance can help to sustain crop productivity in case of the failure of qualitative resistance (Castro et al., 2003). The most widespread application of gene pyramiding is therefore, in breeding for durable disease resistance by combining multiple resistance genes into a single genotype. Conventional gene pyramiding for multiple resistance genes governing the same phenotype was difficult as it demanded progeny tests to identify plants possessing more than one gene. The utilization of linked DNA markers had substituted the requirement of progeny testing. Different strategies for marker-assisted pyramiding of target genes have been evaluated by Servin et al. (2004), and it has been found that pyramiding over consecutive generations is better in order to minimize marker genotyping. Marker-assisted gene pyramiding may also be used for combining abiotic stress tolerance QTLs, especially those that are effective at different plant growth stages. It can also facilitate pyramiding of epistatic QTLs.

5.6.4 MARKER-ASSISTED RECURRENT SELECTION

Marker-assisted recurrent selection (MARS) is a breeding strategy that helps in the targeted recombination of different genes through recurrent selection (Bernardo and Charcosset, 2006). MARS involves one cycle of MAS which is based on both the marker scores and phenotypic data followed by two or three cycles of marker-based selection which takes into consideration only the marker scores (Prasanna et al., 2010). MARS utilizes molecular markers for selection of individuals having desirable genomic regions responsible for expression of a complex trait (Ribaut et al., 2010) followed by intermating of the selected individuals. The selection and intermating are done in the same crop season (Jiang et al., 2007). It helps in assembling best performing genotypes within or across populations and increases the frequency of desirable alleles in the population. MARS also improves the efficiency of recurrent selection by accelerating it mainly by integrating numerous desirable genes/QTLs from different sources. MARS overcomes one of the major shortcomings of MABC related to its inefficiency to transfer QTLs for complex traits governed by multiple genes or large number of QTLs. For certain complex traits including grain yield, abiotic and biotic stress resistance, etc., MARS has been projected for "forward breeding" of genes and pyramiding of multiple QTLs (Eathington, 2005; Crosbie et al., 2006; Ribaut et al., 2010). In MARS, de novo QTL mapping is done in order to identify marker trait association for each breeding population. This assists in assembling multiple QTLs including QTLs with small genetic effect. MARS reduces the number of breeding cycles required and at the same time it also increases the precision of selection. Among various genetic markers, functional markers help in the most efficient accumulation of favorable alleles for a complex trait. MARS was used in sweet corn and increased the frequency of favorable allele from 0.50 to 0.80 at 18 loci in one F2 population and at 11 loci in another F2 population (Edwards and Johnson, 1994). The drawback of MARS is that the generated QTL information may not be transferable to other populations as every MARS population is managed independently (Rahman et al., 2020).

5.6.5 EARLY GENERATION MARKER-ASSISTED SELECTION

MAS can be used at early stage of a breeding programme so as to eliminate plants with an undesirable combination of genes. This facilitates utilization of precious resources in the succeeding generations. Also, if the marker is not linked tightly to the gene/QTL, the selection must be performed in early generations as the chances of recombination between genes/QTLs and marker increase with the advancing generation. MAS in early generations, however, requires genotyping of a large number of plants. As stated by Ribaut and Betran (1999), single large-scale MAS is a strategy of early generation MAS in which single-step MAS was performed in either F2 or F3 population by using flanking markers (located < 5 cM on both sides of the target locus) and up to three stable genes/QTLs with large phenotypic effect were selected in a single MAS step. As early generation MAS reduces the population size due to high selection pressure there are high chances of occurrence of genetic drift at non-target loci and hence large-sized population must be used for initial screening (Ribaut and Betran, 1999). The problem of genetic drift can also be minimized by performing selection in F3 population instead of F2 population. Ribaut and Betran (1999) also suggested that, similar to MAB, flanking markers surrounding the target locus could be used for selection so as to minimize linkage drag. In case of self-pollinated crops, the primary aim is the early fixation of alleles in homozygous state. Therefore, in breeding method involving self-pollinated crops such as bulk method and single-seed descent method, the screening is performed in the homozygous F5 or F6 generations. The alleles can be fixed in their homozygous state in the F2 generation itself by using co-dominant DNA markers. But it requires large population sizes and practically only small number of loci can be fixed in each generation. The size of the breeding population can also be reduced by selecting for both the target locus homozygotes and the heterozygotes and thereby enriching rather than fixing the alleles (Bonnett et al., 2005).

5.6.6 COMBINED MARKER-ASSISTED SELECTION

Combined MAS (Moreau et al., 2004) involves combining phenotypic screening with MAS. It is superior to phenotypic screening or MAS alone and maximizes genetic gain. This approach is adopted either when certain QTLs controlling a specific trait remain unknown or when a trait is controlled by a large number of QTLs. It is more efficient than phenotypic screening in instances where large population sizes and lower heritability of trait are present (Hospital et al., 1997). Bohn et al. (2001) studied MAS for breeding insect tolerance in maize and stated that though phenotypic selection was more efficient than MAS, an increased selection efficiency was observed in case of combined MAS. Zhou et al. (2003) also reported the superiority of combined MAS over conventional phenotypic selection for a major QTL controlling resistance to Fusarium head blight disease in wheat. In practice, MAS schemes can be used in different breeding programme whereas phenotypic selection can be performed at different breeding stages in order to confirm the observations of MAS and also select for genes whose map location is unknown. Though molecular marker-based selection may not be 100% reliable, however, it may still be useful in decreasing the number of plants eligible for phenotypic screening. The scheme is particularly advantageous in instances where cheaper marker genotyping is available (Han et al., 1997). This scheme was referred to as "tandem selection" and "stepwise selection" by Han et al. (1997) and Langridge and Chalmers (2005), respectively.

5.6.7 GENOMIC SELECTION

A quantitative trait is usually controlled by a limited number of major genes and numerous minor QTL/genes. As only a small number of QTL/genes are selected using MAS, the benefits achieved are limited to the amount of the genotypic/phenotypic variation explained by the marker-associated major QTL. However, inclusion of even the minor QTLs associated with a particular trait would make the selection more effective. Genomic selection (GS) is a type of MAS, where marker effects across the entire genome are simultaneously estimated and used to compute genome estimated breeding values (GEBVs) (Meuwissen et al., 2001). GS uses phenotypic data, genotypic data, as well as modeling to forecast the genomic estimated breeding values (GEBVs) for every individual in a breeding population (Bhat et al., 2016; Crossa et al., 2013). Individuals are then selected on the basis of this breeding value instead of marker pattern which is used in MAS. GS requires: (1) a training population; (2) genotyping and phenotyping of the training population; (3) genotypes with high values of GEBV to be selected on the basis of their genotypic data; (4) a testing population comprising of progeny of the genotypes used as study material that are taken as input for the GS model to yield GEBVs; (5) high GEBV values are selected again in genotypes; and (6) the selected genotypes are used as parents for continuous crossing and selection (Lorenz et al., 2011). GS has been instrumental in improvement of complex traits including those involving biotic and abiotic stress tolerance.

5.7 STRESS IN PLANTS

Plant stress as defined by Larcher (1987) is "a state in which increasing demands made upon a plant lead to an initial destabilization of functions, followed by normalization and improved resistance. In general, plant exposure to unfavourable conditions, either environmental or biological, is known as 'Plant Stress.' If the limits of tolerance are exceeded and the adaptive capacity is overtaxed, permanent damage or even death may result." Stress refers to certain external conditions which adversely affect the growth, development, and productivity of plants (Verma et al., 2013). Plant stress is a condition that demands specific responses from plants. If the demand does not exceed the tolerance limits of plant, normal functions are restored; however, if the demand exceeds the plant tolerance limits, a permanent damage may result. The amount of damage incurred in response to plant exposure to stress depends on the strength and duration of stress. In case the plant response does not adequately deal with the stress, it may lead to plant death (Pahlich, 1993).

Stress in plants falls under two different categories namely biotic stress and abiotic stress. Biotic stress is triggered by living organisms namely, bacteria, fungi, virus, nematodes, insects, and weeds (Kranner et al., 2010). The biotic stress agent deprives the host plant of the nutrients and can adversely affect its growth and development and under severe condition may lead to plant death. It leads to both pre- and post-harvest losses in plants. Once a plant senses stress, a suit of cellular and molecular processes is triggered (Lamers et al., 2020) which activates a number of morphological, physiological, and biochemical mechanisms (Saijo and Loo, 2020). Plants have developed mechanism to sense biotic stresses and overcome its deleterious effects on growth, survival, and yield (Lamers et al., 2020). The molecular mechanisms leading to plant defence response against biotic stresses had been explicated to great depths (Cheng et al., 2012; Wang et al., 2019).

Abiotic stress is levied on plants by the external environment and may be chemical or physical (Verma et al., 2013). They include abnormal water status (high or low), temperature (cold, frost, and heat), wind conditions, nutrient levels, and salinity (Kranner et al., 2010). Abiotic stress significantly affects the physiology of plants causing widespread changes in the cellular processes. Some of these changes do not lead to increased adaptation and merely represent the damage inflicted by the stressor, such as the disruptions in enzyme kinetics, molecular interactions as a result of toxic ions, and changes in the fluidity of membrane and the structure of protein due to heat or cold stress. However, there are abiotic stress-induced responses which are adaptive in nature and enhance the plants resistance to stress. A phytohormone called abscisic acid (ABA) also plays a significant role in plant adaptation to environmental stress such as drought, low temperature, salinity or mechanical wounding (Sah et al., 2016). The processes involved in the adaptive response to abiotic stress include rebalancing of cellular homeostasis, repair of damage induced by stress, and adjustment of growth to levels appropriate for the specific stress condition. These adaptive responses can therefore form potential targets for crop improvement.

The changing environmental conditions and emergence of new stresses demand development of new and improved approaches in order to enhance the stress survival capability of different crop varieties and at the same time maintain their production potential. Genetic improvement is an effective solution to combat both the biotic and abiotic stresses as well as to reduce the usage of harmful chemicals which adversely affects the ecology. Recent advances in breeding involving molecular markers allow tracking and more precise transfer of genes/QTLs governing a particular trait, and this has opened new avenues in the field of stress resistance breeding. The availability of a number of molecular markers and molecular genetic maps has made the application of MAS possible for characteristics administered by major genes/QTLs (Choudhary et al., 2008). The comprehensive procedure of MAS has been depicted in Figure 5.1. MAS as applied in different crops to improve stress resistance have been explained in the subsequent section and presented in Table 5.1.

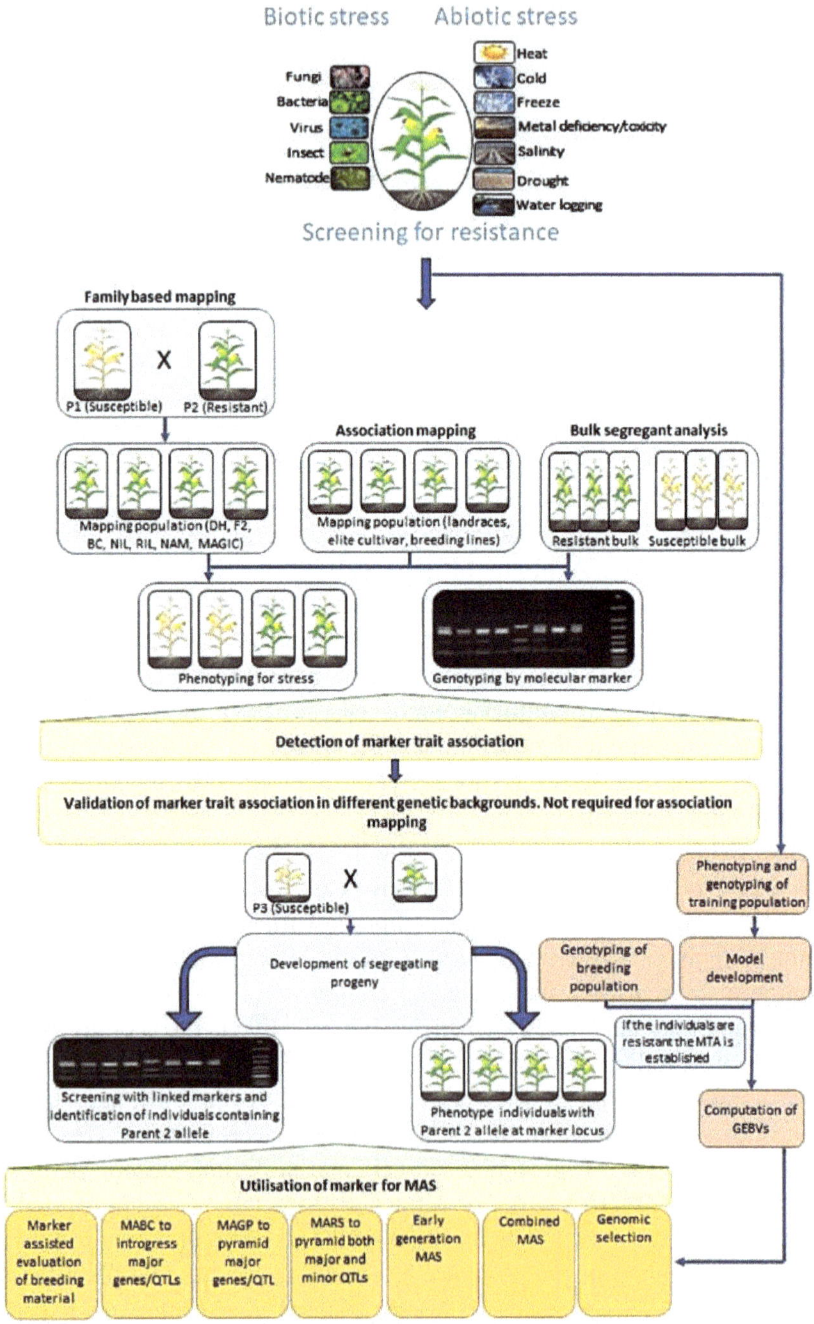

FIGURE 5.1 A comprehensive procedure for MAS.

TABLE 5.1 MAS for Biotic and Abiotic Stress Resistance in Plants.

Crop	Trait	Gene/QTLs	Marker	References
Wheat	Leaf rust	*Lr*34	SSR	Suenaga et al. (2003)
		Lr 35	STS and CAPS	Seyfarth et al. (1999)
		*Lr*47	CAPS	Helguera et al. (2000)
	Stem rust	*Sr*31	STS markers	Mago et al. (2002)
	Powdery mildew	Qpm.vt-1A, Qpm.vt-2A, Qpm.vt-2B	SSRs	Liu et al. (2001)
		Pm4a	CAPS	Ma et al. (2004)
		Pm5e	SSR	Huang et al. (2003)
	Fusarium head blight	Qfhs.ndsu-3BS	RFLPs	Anderson et al. (2001)
	Yellow rust	*Yr15*	RAPD, SSR	Chague` et al. (1999)
		YrMoro	STS	Smith et al. (2002)
	Reduces concentration of sodium in the leaf blades and retains sodium in sheaths	*TmHKT7-A2*	ESTs	Huang et al. (2006)
	Controls sodium and potassium transport from roots to shoot and maintains high potassium to sodium ratio	*Kna1*	SSR	Dubcovsky et al. (1996), Gorham et al. (1990)
	Decreases the uptake of sodium and enhances potassium loading in xylem	*Nax1*	SSR	Huang et al. (2006), Lindsay et al. (2004)
	Decreases the uptake of sodium and enhances potassium loading in xylem	*Nax2*	EST	Byrt et al. (2007)
Rice	Bacterial blight	*xa5, xa13, Xa21*	STS	Singh et al. (2001)
	Rice blast	*Pi5(t)*	CAPS	Jeon et al. (2003)
		Pi1, Piz5, Pita	RFLP, SAP	Hittalmani et al. (2000)

TABLE 5.1 *(Continued)*

Crop	Trait	Gene/QTLs	Marker	References
		Piz, Pizt	SNP	Hayashi et al. (2004)
		Pi-b, Pi-k, Pita2	SSR	Fjellstrom et al. (2004)
	Gall midge	*Gm7*	SA598 SCAR	Sardesai et al. (2002)
	Brown planthopper	*bph2*	STS	Murai et al. (2001)
	Determines root length and dry weight at seedling stage under saline conditions	qRL-7, qDWRO-9a, qDWRO-9b qBI-1a, qBI-1b	SSRs	Sabouri and Sabouri (2008)
	Affects shoot sodium and potassium concentrations, respectively	qSNC-7, qSKC-1	EST	Lin et al. (2004), Pushparajan et al. (2011)
	Regulates potassium to sodium homoeostasis	QNa, QNa:K, SKC1/OsHKT8	SSR	Ren et al. (2005)
	Enhances ST in shoots	qST1, qST3	RFLP, SSR, AFLP, isozyme	Lee et al. (2007)
	Improves sodium to potassium ratio	qNAK-2, qNAK-6	SSR	Ming-zhe et al. (2005)
	Controls shoot sodium to potassium homoeostasis	*Saltol*	SSR	Thomson et al. (2010)
	Controls shoot sodium to potassium homoeostasis	*Saltol and non-Saltol*	SSR	Alam et al. (2011
	Controls root potassium content	QKr1.2	SSR	Ahmadi and Fotokian (2011)
Barley	Barley yellow mosaic virus resistance	*rym4/rym5*	SSR	Graner et al. (1999), Williams (2003)
		rym4, rym9, rym11	SSRs	Werner et al. (2003)

TABLE 5.1 *(Continued)*

Crop	Trait	Gene/QTLs	Marker	References
	Leaf rust resistance	*Rph7*	CAPS	Graner et al. (2000)
		Rph15	CAPS	Weerasena et al. (2004)
	Improves the proline, chlorophyll, and fluorescence content and enhances the salt tolerance	QFv2H, QCh7Ha, Qch2Ha, QWSC2H	RFLP	Siahsar and Narouei (2010)
	Enhances vegetative growth under saline stress	Five QTLs for salt tolerance were identified	SSR	Zhou et al. (2012)
	Reduces sodium content of plants grown under salt stress	A locus *HvNax3* on wild barley (*Hordeum vulgare* ssp. *spontaneum*)	RFLP	Shavrukov et al. (2010)
Maize	Sugarcane mosaic virus	*Scm1, Scm2*	SCAR, CAPS	Dussle et al. (2002)
Soybean	Maintains growth under salt stress	Eight QTLs for salt tolerance were detected	SSR	Chen et al. (2008)
	Maintains growth under salt stress	A major QTL for salt tolerance was detected	SSR	Hamwieh et al. (2011)
	Common bacterial blight resistance	QTL	RAPD, SCAR, STS, SSR, RFLP	Taran et al. (2001)
Common bean	Common bacterial blight resistance	QTL	RAPD	Jung et al. (1997)
	Bean common mosaic virus resistance	*I*	SCAR	Melotto et al. (1996)
	Anthracnose resistance	*Are*	SCAR	Adam-Blondon et al. (1994)
	White mold resistance	QTL	RAPD, AFLP	Kolkman and Kelly (2003)
	Fusarium wilt resistance	*PvPR1, PvPR2*	RAPD	Schneider et al. (2001)

TABLE 5.1 *(Continued)*

Crop	Trait	Gene/QTLs	Marker	References
Cowpea	*Striga gesneriodes* resistance	*Rsg1*	SCAR	Bouker et al. (2004)
	Thrips tabaci and *Frankliniella schultzei* resistance	QTL	AFLP	Muchero et al. (2010b)
Chickpea	*Ascochyta* blight resistance	QTL	RAPD	Millan et al. (2003), Cobos et al. (2005)
		Ar19	RAPD	Rakshit et al. (2003)
		QTL	SSR	Udupa and Baum (2003)
		QTL_{AR3}	SSR	Iruela et al. (2007)
		QTL	SSR	Anbessa et al. (2009)
	Fusarium wilt resistance	$QTL_{Foc02,}$ QTL_{Foc5}	SSR	Cobos et al. (2009)
		foc1	SSR	Gowda et al. (2009)
		foc2	SSR	Gowda et al. (2009)
		foc3	SSR	Gowda et al. (2009)
	Ascochyta blight resistance	QTL	SSR	Aryamanesh et al. (2010)
Pea	Powdery mildew resistance	*er*	RFLP	Dirlewanger et al. (1994)
	Pea seed borne mosaic virus resistance	*Sbm-1*	STS	Frew et al. (2002)
	Fusarium wilt resistance	*Fw*	RFLP	Dirlewanger et al. (1994)
	Pea common mosaic virus resistance	*mo*	RFLP	Dirlewanger et al. (1994)
		Sbm-1	cDNA-AFLP	Gao et al. (2004)
	Rust resistance	*Up1*	RAPD	Barilli et al. (2010).

TABLE 5.1 *(Continued)*

Crop	Trait	Gene/QTLs	Marker	References
Mung bean	Powdery mildew resistance	QTLs	RFLP	Humphry et al. (2003)
		QTLs	RAPD, CAP, AFLP	Chen et al. (2007)
Black gram	Yellow mosaic virus resistance	Monogenic	STS-RGA	Basak et al. (2004)
Faba bean	Broomrape resistance	*Oc1, Oc2, Oc3*	RAPD	Roman et al. (2002)
	Rust resistance	*Uvf-1*	RAPD	Avila et al. (2003)
Lentil	*Ascochyta* blight resistance	*Ra/2*	RAPD, SCAR	Chowdhery et al. (2001)
	Anthracnose resistance	*LCt-2*	AFLP, RAPD	Tullu et al. (2003)
	Fusarium wilt resistance	*Fw*	RAPD	Eujayl et al. (1998)
		Fw	AFLP	Hamwieh et al. (2005)
		Fw	SSR	Hamwieh et al. (2005)
	Cold hardiness	*Frt*	RAPD	Eujayl et al. (1999)
		Frt	RAPD, SSR, AFLP	Kahraman et al. (2004)
Lathyrus	*Ascochyta* blight resistance	QTL	RAPD, STMS	Skiba et al. (2004)
Pigeon pea	*Fusarium* wilt resistance	Gene	RAPD	Kotresh et al. (2006)

5.7.1 MARKER-ASSISTED SELECTION FOR BIOTIC AND ABIOTIC STRESS RESISTANCE

5.7.1.1 RICE

Huge economic losses have been encountered in rice due to both biotic and abiotic stresses; however, molecular breeding has played a vital role in improving varieties with respect to different resistance traits. MAS have been employed for pyramiding different blast and blight-resistant genes in rice, for example, *xa5*, *xa13*, *Xa21 and xa13* and *Xa21 have been* pyramided in vulnerable rice cultivars, PR106 and Pusa Basmati (Singh et al., 2001). Jeung et al. (2006) pyramided three blight-resistant genes (*Xa4*, *xa5,* and *Xa21*) into the susceptible cultivars. Improved Samba Mahsuri was bred by MABC and introgressing three blight-resistant genes namely, *Xa21*, *xa13,* and *xa5 into Samba Mahsuri variety of rice* (Sundaram et al., 2008). Pradhan et al. (2015) introgressed *Xa21*, *xa13, and xa5 genes into Jalmagna, a deep-water rice variety, using MABC.* Hittalmani et al. (2000) introgressed three major blast-resistant genes viz., *Pi1*, *Piz-5*, *Pita* from dissimilar parents into a popular rice cultivar Co-39 using MAS. MABC was used to transfer two blast-resistant genes (*Piz-5* and *Pi54*) into the genetic background of popular rice cultivar, PRR78 (Singh et al., 2012). A co-dominant marker (pTA 248) associated with *Xa21,* a major blast-resistant gene identified in *Oryza longistaminata*, was also developed for MAS (Nguyen et al., 2018). *Pi54* gene is another blast-resistant gene that provides durable and stable resistance against diverse blast strains (Thakur et al., 2015), and a functional marker for this gene has been identified which can be used for MAS (Ramkumar et al., 2011). Quantitative resistance against rice sheath blight disease was detected in different rice landraces namely, Teqing, Tetep, and Jasmine 85 (Channamallikarjuna et al., 2010; Yadav et al., 2015). Among the detected resistant QTLs, three QTLs viz., *qSBR7-1*, *qSBR11-1*, and *qSBR11-2* were detected in Tetep (Channamallikarjuna et al., 2010) and all three were pyramided in Pusa 6B variety (Singh et al., 2015). MAS was also used to transfer sheath blight-resistant QTLs into different rice germplasm (Tan et al., 2005; Wang et al., 2012). Microsatellite markers namely, RM3754 and RM 3761 were found linked to gene controlling resistance to rice green leafhopper, *Grh5*, and can be used for insect-resistant breeding (Fujita et al., 2006). Five genes providing resistance to rice brown plant-hopper have also been utilized for MAS breeding. Two SSR markers, RM463 and RM7102, have been identified to be linked brown plant hopper-resistant gene, *bph2*, and when used for MAS have increased the efficiency of selection by 89.9% and 91.2%, respectively (Sun et al., 2006). Microsatellite markers, RM5953, RM401, RM217, RM210, RM1103, and RM242, are allied with genes encoding resistance to two different BPH biotypes namely, biotype 2 and 3 (Shabani-mofrad et al., 2015) and can be effectively used for MAS. A rice stripe disease resistance gene, *Stvb-i*, derived from, Modan, an Indica rice cultivar was fine mapped on chromosome 11 and was flanked by STS markers 7L and 21R (Hayano-Saito et al., 2000). The genes *Stv-b*^{i} and *Bph14*, *Bph15* were used to develop stripe disease and BPH-resistant rice varieties (Xu, 2013). A resistance gene against rice yellow mottle virus was identified on chromosome 4 and was flanked by microsatellite markers RM273 and RM252 (Albar et al., 2003). Rice gall midge resistance was found to be controlled by about 10 resistance genes, out of which 8 resistance genes have been mapped and tagged (Kumar et al., 2005; Himabindu et al., 2007). Flanking markers have been identified and are being used to select for lines containing the gall midge resistance genes *Gm1* and *Gm2* (Himabindu et al., 2007).

Drought stress is one of the chief constraints affecting the production of rice in rainfed areas. Many QTLs conferring resistance to drought and effective under diverse genetic backgrounds and environments have already been identified and mapped. Performance under water stress conditions may also be improved by selection of different secondary traits like root length, root thickness, root depth, and root penetration ability (Babu, 2010). Four QTLs responsible for deeper roots were introgressed from "Azucena" into "IR64" using MAS (Courtois et al., 2003). MABC was also applied in Kalinga III, an upland variety of rice, in order to improve its root morphology and resistance to drought (Roberto and Silvio, 2006). Bernier et al. (2007) detected a large effect, drought tolerance QTL (*qtl12.1*) controlling 51% of the total genetic variation. The QTL was located on chromosome 12 and SSR markers, RM28048 and RM511, linked to the QTL were also identified. A QTL located near the *sd1* gene on chromosome 1 contributed 32% to the total genetic variation for yield under the conditions of drought (Kumar et al., 2007). The *Sub1* QTL located on chromosome 9 and accounting 70% of the total phenotypic variation for survival under submerged conditions (Xu et al., 2000) was also identified. The *Sub1A* gene controlled submergence tolerance in rice (Xu et al., 2006) and a CAPS marker, GnS2, based on the SNP in the *Sub1A* gene was designed which can be used for MAS (Neeraja et al., 2007). A chromosome 1 located major QTL, *Saltol,* was identified in Pokkali, a salt-tolerant rice cultivar, and conferred salt tolerance at the vegetative stage contributing 64–80% of the total phenotypic variation for salt tolerance (Bonilla et al., 2002). This QTL has been identified in other rice varieties as well (Takehisa et al., 2004). A vegetative stage salt-tolerant gene has been detected in the cultivar Nona Bokra at the same position and has also been positionally cloned. Ren et al. (2005) also identified a salt-tolerant QTL, *SKC1,* which encoded a member of HKT-type transporters and in salt tolerant cultivar is responsible for maintaining K+ homoeostasis. Most of the salt-tolerant QTLs identified are responsible for tolerance in the vegetative phase of the plant; however, reproductive stage salt-tolerant QTLs are rare. The P deficiency, Al and Fe toxicity QTLs have also been detected in different rice cultivars (Wissuwa et al., 2002; Mackill, 2006). The gene accountable for enhanced phosphorus uptake under low phosphorus conditions, namely, *Pup1,* has been mapped on chromosome 12 and derived from Kasalath cultivar (Wissuwa et al., 1998). Cultivation of rice at a higher altitude is also affected by cold stress leading to growth retardation (Shinada et al., 2013; Shinada et al., 2014). The RAPD marker OPT8.511 and an SSR marker RM202 showed strong linkage with cold tolerance (Kim et al., 2000; Chen and Li, 2005). Another major QTL, *qCTS4*, has been fine mapped to chromosome 4 and is flanked by markers RM6770 and RM7200. This QTL explained 40% of the phenotypic variation for stunting and yellowing of seedlings under the conditions of cold stress and could be used for cold tolerance MAS in rice (Andaya and Tai, 2007).

5.7.1.2 WHEAT

Wheat (*Triticum* spp.) is a chief food crop of the world. It is affected by a number of biotic and abiotic stresses. Resistance genes providing resistance against different biotic stresses in wheat have been identified. Genes governing resistance against *Mycosphaerella* (anamorph *Septoria*) *tritici*, *B arley yellow dwarf virus*, *Wheat streak mosaic virus*, *Soilborne cereal mosaic virus*, hessian fly, cereal cyst nematode, aphids, etc., have been identified (Dubcovsky, 2011; Miedaner and Korzun, 2012). As resistance governed by single major genes is not much stable, focus has now shifted to assembling of minor genes or QTLs so as to breed for durable resistance. Miedaner et al. (2006) were able to introgress *Fusarium* head blight (FHB) resistance conferring QTLs, *Fhb1* and *Qfhs.ifa-5A* into winter and spring wheat lines resulting in enhanced resistance to disease. Marker aided transfer of rust resistance genes (*Lr19* and *Lr24*) (Singh et al., 2004) and QTL governing *Stagonospora nodorum* blotch, *Septoria tritici* blotch, *Fusarium* head blight, and tan spot resistance were achieved (Zwart et al., 2010; Miedaner et al., 2012). Marker-assisted gene pyramiding has been undertaken to breed for resistance against powdery mildew (Liu et al., 2000; Wang et al., 2001), leaf rust (Gupta et al., 2005; Nocente et al., 2007), Fusarium head blight (Badea et al., 2008; Tamburic-Ilincic et al., 2011); Hessian fly (Dweikat et al., 1997), etc. Tyagi et al. (2014) with the help of molecular markers pyramided three genes for resistance against three rusts (Lr24, Sr24, Yr36) and eight genes/QTLs controlling four grain quality traits in the background of popular wheat cultivar PBW343. Three alien genes namely Lr 9, Lr 24, Lr 28 conferring leaf rust resistance and linked to SCAR marker SCS5550, SCS1302607, and SCS421570, respectively, were pyramided in the background of an adapted but susceptible bread wheat variety HD 2329 (Charpe et al., 2012). An immensely popular and high yielding wheat variety, PBW343, became susceptible to *Yr27,* a virulence gene for stripe rust. Owing to the importance of this variety in the Indian farming system, it was subjected to marker-assisted introgression of genes governing stripe rust resistance. Modified marker-assisted backcrossing was done in order to pyramid genes present in alien segments of *Aegilops umbellulata* (*Lr76/Yr70*) and *Aegilops ventricosa* (*Lr37/Yr17/Sr38*) into PBW343. After pyramiding the resistance alien fragments, one of the lines, PBW723, was recovered possessing 81.57% of the genetic background of PBW343 and was subsequently released for cultivation as a stripe rust resistance variant of PBW343. A total of forty-two traits/genes including numerous rust resistance genes (Lr9, Lr19/Sr25, Lr24/Sr24, Lr34/Yr18, Lr46/Yr29, Lr47, Sr26, Sr32, Sr33, and Sr36) have been shortlisted for germplasm enhancement and varietal development (Cakir et al., 2008). A project, "Bringing Genomics to the Wheat Fields," was undertaken with an aim to incorporate 27 disease and pest-resistant genes and 20 genes responsible for superior pasta and breadmaking quality into a number of lines which were highly adapted to the US. This involved greater than 3000 marker-assisted backcrosses and led to development of 240 MAS derived lines of which 45 were released for cultivation (Sorrells, 2007). Gautam et al. (2020) also pyramided Lr37/Yr17/Sr38 and Yr15 rust resistance genes into "Express" variety of wheat. Genes/QTLs conferring resistance to all three rust and improved grain quality were pyramided into PBW343 by crossing two parental lines, P1 (carrying genes Yr70/Lr76 + Lr37/Yr17/Sr38) and P2 (carrying genes Gpc-B1/Yr36+QPhs.ccsu-3A.1+QGw.ccsu-1A.3+Lr24/Sr24+Glu-A1-1/Glu-A1-2) developed in the background of PBW343 by MAS. One line (CCSU-7) of the eleven selected lines showed significantly higher protein content and grain yield compared to the check PBW343. This line could be useful for breeding for both durable resistance and superior wheat grain quality. Rahman et al. (2020) used MARS to pyramid minor QTLs controlling resistance to crown rot disease. They identified a total of 23 QTLs controlling resistance to crown rot and after two cycles of recombination they were able to recombine 22 markers and a significant positive response with respect to crown rot resistance was observed for each cycle of recurrent selection due to combining of many small effect resistance QTLs. Barloy et al. (2007) used MAS involving SCAR marker to pyramid two cereal cyst nematode (CCN) resistant genes fetched from *Aegelops variabilis* namely CreX and CreY into wheat cultivars. One SSR marker, Xwmc 634, and two AFLP markers co-segregated with a green bug resistance gene, *Gb3* could be used for MAS (Weng and Lazar, 2002; Weng et al., 2005). Markers linked to the green bug-resistant genes viz., *Gbx 1*, *Gba*, *Gbb*, *Gbc*, *Gbd*, *Gbz* can also be used to pyramid these genes into adapted wheat cultivars and breed for durable insect resistance (Zhu et al., 2005). Hessian fly resistance gene, *H9, specific markers, Xbarc263, Xcfa2153, and SOPO05.909* were identified by Kong et al. (2005). Markers linked to the gene cluster including *H9*, *H10*, *H11* located on 1AS; *H13* situated on 6DS (Liu et al., 2005a; Liu et al., 2005b); *H22* situated on 1DS (Zhao et al., 2006a, 2006b); and *H32* located on 3D (Sardesai et al., 2005) have been detected. Kong et al. (2007) reported markers Xpsp2999 and Xwem6b flanking *H16 and H17* located on 1AS. Arzani et al. (2004) reported markers Xgwm106 and Xgwm337 which flank gene-governing resistance to Russian wheat aphid (RWA), *Dn4*. Marker Xgwm437 was linked to *Dn2* at a distance of 2.8 cM (Devi et al., 2017). Two microsatellite markers, Xgwm44 (14.6 cM) and Xgwm111 (3 cM), were found linked to *Dn6* located on 7DS. *Dn6* is tightly linked to other RWA-resistant genes namely, *Dn1*, *Dn2*, *and Dn5* (Liu and Gill, 2002). Kirigwi et al. (2007) detected an SSR marker, wmc89 linked to QTL located on chromosome 4A and controlling multiple traits including grain fill rate, spike density, biomass production, grain yield, and drought susceptibility index. Hence, this marker could be useful in enhancing drought tolerance. Matilda et al. (2009) used SSR marker, barc108 and wmc603, to select for lines from the segregating generation of a cross having drought-resistant variety, Izvor, as one of the parents. These markers could help in the selection of a genomic region contributed by Izvor, located on chromosome 7A and constituting of genes providing enhanced drought tolerance. Finkel (2009) used markers to select for QTL providing enhanced transpiration efficiency. Courtois et al. (2003) transferred a number of QTLs associated with deep root character from "Azucena," an upland japonica cultivar to "IR64," a lowland indica cultivar through MAS. The selected lines showed surplus root biomass and high yield under drought conditions. Sodium excluder genes, *Nax1* and *Nax2,* were also transferred from tetraploid to hexaploid wheat and were expected to enhance saline and water logging tolerance of wheat genotypes (James et al., 2011). A major gene, *Fr1,* was found to be closely linked to a vernalization gene, *Vrn1* located on chromosome 5A while *Fr2* was linked to *Vrn3* gene located on chromosome 5D. Both these genes were responsible for significantly increasing the cold tolerance ability of genotypes (Snape et al., 1997). Prostrate growth type is also associated with freeze tolerance in wheat, and a prostrate growth controlling gene was found to be associated closely with both *Fr1* and *Vrn1* genes (Roberts, 1990). Campbell et al. (2013) used mapping populations to detect interactions between two alleles of *Vrn1* and *Fr2 loci*. Both these alleles were responsible for enhancing freezing tolerance in wheat. Selection of varieties carrying *FR-A2-T* allele and recessive allele, *vrn-A1* could prove helpful for frost tolerance breeding, and molecular markers for the selection of these alleles have already been developed by Campbell et al. (2013).

5.7.1.3 MAIZE

Maize acts as a staple food crop for significant part of the world's population and is also a major source of feed for animals (Strable and Scanlon, 2009). This crop is exposed to a number of biotic and abiotic stresses worldwide. Sorghum downy mildew is an economically important maize disease and markers flanking two major resistance QTLs were used to generate QTL-NILs and CM139, a resistant line (Prasanna, 2009). Lohithaswa et al. (2015) identified QTLs conferring resistance against sorghum downy mildew on chromosome 3 and with the help of molecular markers transferred them into eight susceptible lines. Major genes governing resistance against polysora rust and turcicum leaf blight have also been pyramided into maize lines using MAS (Prasanna et al., 2010). Genes, *Htn1* and *Ht2,* conferring resistance to turcicum leaf blight and QTL (*RppQ*) providing resistance to polysora rust were pyramided by generating various breeding populations. QTLs for banded leaf and sheath blight (BLSB) (Zhao et al., 2006a; Adhikari et al., 2021) and red flour beetle (Joshi et al., 2021) were also identified. Major genes providing resistance to *Maize streak virus* (MSV) namely, *Msv1*, *Mv1*, *Wsm1*, *Wsm2*, *Wsm3*, and *Mdm1* and a viral dQTL (Wisser et al., 2006) were also detected. Qiu et al. (2021) identified two stable and novel QTLs for resistance to head smut and located within the bnlg1014 to umc2224 (qHS1) interval and in the umc1006 to umc1857 (qHS6) interval. Both these QTLs can be used for further mapping, and MAS. Zhao et al. (2012) enhanced the head smut resistance of ten inbred lines by introducing QTL *qHSR1* via MABC. The candidate gene for *qHSR1* was *ZmWAK* and it helped in improving both yield-related traits and resistance to head smut (Zuo et al., 2015; Konlasuk et al., 2015). This gene has been introgressed via MAS in local head smut susceptible Chinese germplasm "Tongsipingtou." It had also been used to produce resistant maize variety (Jidan558) and elite inbred lines (Chang7-2R and Ji853R) (Zhao et al., 2012). The major genes, *ZmCCT* and *ZmAuxRP1*, were identified and were responsible for providing stalk rot resistance in maize (Wang et al., 2017; Ye et al., 2019). MAS was also utilized to introgress nine resistant *ZmCCT* haplotypes into seven maize inbred lines. Haplotype H5 was selected from an inbred line GEMS14 and showed low photoperiodic sensitivity and increased resistance to stalk rot (Li et al., 2017). The inbred lines and hybrids into which H5 was introgressed also possessed improved agronomic traits, lower photoperiodic sensitivity, and stable resistance to stalk rot. Therefore, the H5 haplotype has been used widely to breed for stalk-rot resistance in Chinese maize lines (Lanubile et al., 2017). The stalk rot resistance QTLs viz., *Rpi1*, *RpiQI319-1/2*, and *qRfg3* have been identified in maize by Yang et al. (2005), Song et al. (2015), and Ma et al. (2017), respectively. MAS was also used to introgress *Scmv1* and *Scmv2* into susceptible maize line to produce sugarcane mosaic virus (SCMV) resistant near isogenic lines (NIL) (Xing et al., 2006). A multiple disease resistance exhibiting maize NIL possessing the *qMdr9.02* locus was produced by MAS, and the line was resistant for two important foliar diseases of maize namely, southern corn leaf blight and gray leaf spot (Yang et al., 2017). Awata et al. (2021) used two Kompetitive Allele Specific PCR (KASP) markers (*S3_146250249* and *S3_146363360*) to select for resistance to Maize lethal necrosis virus. Ren et al. (2022) identified a major QTL qRtsc8-1 conferring maize tar spot complex resistance and fine mapped it to a 721 kb region located on chromosome 8. The production markers for these resistant QTLs were validated in the breeding lines. A marker linked QTL conferring resistance to *Diatraea grandiosella*, a corn borer, contributed by maize line CML139 was identified and incorporated in MAS breeding programmes (Khairallah et al., 1998). MAS was applied to improve resistance to stem borer and different *Bacillus thuringiensis* (Bt) resistance genes were pyramided into the breeding populations with a view to producing durable pest resistance (Bergvinson and Hoisington, 2001). Samayoa et al. (2015) found significant single nucleotide polymorphisms (SNPs) associated with three traits responsible for resistance against Mediterranean corn borer. Out of these, four SNPs were associated with stem damage, ten with tunnel length, and eleven with kernel resistance. Landi et al. (2002) identified markers associated with root strength and root yield under water stress conditions. Water logging stress affects about 15% of the total maize growing areas in Asia (Rathore et al., 2000). MAS would help in the identification of water logging stress-resistant lines. Hu et al. (2016) identified six QTLs conferring seed germination at low temperature. They also detected six QTLs controlling primary root length at low temperatures. Four common QTLs controlling both low-temperature seed germination and primary root length were identified and the markers *umc1303*, *umc1*, *umc62,* and *bnl 14.28a* were found to be associated with these QTLs and can be efficiently used for MAS for cold tolerance in maize.

5.7.1.4 PIGEONPEA

Pigeonpea, a legume crop, is a source of food and fuel wood. Besides it also improves infiltration of water and conserves nutrients and water in the soil. It is also fairly adapted to low water areas and can easily thrive on degraded soils. The productivity of this crop has continued to be constant over the last few decades despite the continuous efforts to improve its genetic yield potential (Choudhary et al., 2013). *Fusarium* wilt is one of the most devastating diseases of pigeonpea and the resistance against this disease is governed by major R-genes (Singh et al., 2016). Microsatellite markers linked to different R genes have been identified and validated and can be used for marker-assisted selection for improving disease resistance of popular pigeonpea cultivar (Singh et al., 2016). The resistance against sterility mosaic disease is governed by the two nonlinked genes, *SV1* and *SV2,* and both of them display inhibitory gene interaction (Bhairappanavar et al., 2014) and though markers for their selection were identified, their application in resistance breeding has yet to be realized. Markers $OPA09_{910}$ and $SCOPA09_{942}$ were also found to be linked with the resistance to plume moth. Both screening and MAS can be undertaken with the coupling phase SCAR marker $SCOPA09_{942}$ (Mishra et al., 2015). Pigeonpea is a hardy crop and can grow under low water conditions (Magadum et al., 2013). However, if the water availability falls below a particular level yield of the crop is significantly affected. A number of candidate genes as CcHyPRP, CcCDR, DLP, CcCYP, CcMT1, LTP1, and APB were detected in stressed plants. Of these, Cyclophilin (CcCYP) gene was responsible for providing tolerance to salinity, drought, and cold stress (Pazhamala et al., 2015).

5.7.1.5 TREES

Marker-assisted selection in forest trees is highly advantageous as it can help in performing selection at an early stage of growth, thereby reducing the age of selection facilitating the wise use of limited resources. Besides MAS also increases the efficiency of selection as it significantly reduces the phenotyping cost, helps in more accurate evaluation of genotypes by combining information from both marker and phenotypic evaluation, and can effectively maintain genetic diversity and control inbreeding. Marker identification studies in the case of forest trees are limited due to a number of constraints including long generation time, longer juvenile phase which pose difficulty with the generation of mapping population and also phenotypic evaluation of traits of economic importance which generally expresses later in the life cycle. Also, most of the forest tree species are in an early domestication stage and therefore, possess low LD levels which pose a major disadvantage in MAS. MAS in forest trees is generally limited to selection for wood quality and growth traits (Grattapaglia et al., 2018). Besides genomic regions controlling resistance to various biotic and abiotic stresses have also been identified and selected for in few perennial tree species. Quesada *et al.* (2010) used 498 clonally propagated individuals of loblolly pine to detect pitch canker resistance genes and were able to identify 10 SNPs closely associated with disease resistance. Krutovsky and Neale (2005) also studied the association between markers and fifteen cold tolerance related genomic regions in Douglas fir. Newcombe et al. (1994) used *Populus trichocarpa* X *P. deltoides* progenies to identify a single gene *Mmd1* encompassing rust resistance in poplar.

5.8 CONCLUSION

Climate change has predisposed earth to drastic whether events. Extreme weather phenomena such as aberrant rains, floods, drought, and rising temperatures are some of the major outcomes of the changing climate. Moreover, the anthropogenic activities such as the use of excessive fertilizers, excessive pesticides, inappropriate irrigation practices, etc., have further complicated the situation leading to evolution of new pathogen races, evolution of insect biotypes, salt accumulation, etc. These circumstances have predisposed the plants to more frequent and many times concurrent biotic and abiotic stresses. The situation thus demands the breeders to produce stress-tolerant cultivars to ensure food security and guarantee safety to farmers. Since ancient times, plant breeders have been accumulating favorable alleles through phenotypic selection which is time consuming, less reliable, and may lead to linkage drag. Moreover, as most of the economically significant traits are controlled by polygenes, the reliability of the phenotypic selection is further compromised. MAS could serve as a powerful tool in increasing the efficiency of selection. Recent efforts in genetic analysis have allowed the identification of genomic sequences responsible for expression of target traits. The superior alleles at the target sequences are being used as DNA markers for screening for target traits. Technological developments, including automation, allele-specific diagnostics, and DNA chips, have made MAS approaches based on large-scale screening much more effective and powerful. The knowledge of DNA sequences could help in designing "perfect" or "diagnostic" markers located within the target gene, thereby eradicating the likelihood of recombination between gene and marker and increasing the selection efficiency. Technological advancement has made available the whole genome sequences of many plant species leading to the generation of KBioscience Competitive Allele-Specific PCR (KASPar) which enables genotyping of few SNPs in large samples. Incorporation of such novel markers in the breeding programs will help in achieving the breeding targets more easily. The cost of MAS has decreased significantly and with the advent of new technology the precision is expected to increase and the cost will further go down. In order to enhance the practicability of MAS, the present barriers must be analyzed and appropriate steps must be taken to overcome them. The integration of high-throughput markers in traditional breeding programs will prove an important tool in overcoming the challenges of food and nutritional deficiency in an era of climate change.

KEYWORDS

- **marker-assisted selection**
- **abiotic stress resistance**
- **biotic stress resistance**
- **genetic mapping**
- **marker-assisted backcrossing**
- **marker-assisted gene pyramiding**

REFERENCES

Adam-Blondon, A. F.; Sevigenc, M.; Bannerot, H.; Dron, M. SCAR, RAPD and RFLP Markers Linked to a Dominant Gene (*Are*) Conferring Resistance to Anthracnose in Common Bean. *Theor. Appl. Genet.* **1994,** *88*, 865–870.

Adhikari, S.; Joshi, A.; Kumar, A.; Singh, N. K.; Jaiswal, J. P.; Jeena, A. S.; Singh, R. P.; Pandey, D. Identification of QTL for Banded Leaf and Sheath Blight in Teosinte Derived Maize Population. *Agric. Res.* **2021.** https://doi.org/10.1007/s40003-021-00567-7

Agronomy. Crop Science Society: Madison, WI, 1992; p 284.

Ahmadi, J.; Fotokian, M. Identification and Mapping of Quantitative Trait Loci Associated with Salinity Tolerance in Rice (*Oryza Sativa*) Using SSR Markers. *Iranian J. Biotechnol.* **2011,** *9* (1), 21–30.

Alam, R.; Sazzadur Rahman, M.; Seraj, Z. I.; Thomson, M. J.; Ismail, A. M.; Tumimbang-Raiz, E.; Gregorio, G. B. Investigation of Seedling-Stage Salinity Tolerance QTLs Using Backcross Lines Derived from *Oryza sativa* L. Pokkali. *Plant Breed.* **2011,** *130*, 430–437. https://doi.org/10.1111/j.1439-0523.2010.01837.x

Albar, L.; Ndjiondjop, M. N.; Esshak, Z.; Berger, A.; Pinel, A.; Jones, M.; Fargette, D.; Ghesquière, A. Fine Genetic Mapping of a Gene Required for Rice Yellow Mottle Virus Cell-to-Cell Movement. Theor. Appl. Genet. **2003,** *107, 371–378.*

Allard, R. P*lant Breeding*. In *Encyclopaedia Britannica;* 2019.

Anbessa, Y.; Taran, B.; Warkentin, T. D.; Tullu, A.; Vandenberg, A. Genetic Analyses and Conservation of QTL for Ascochyta Blight Resistance in Chickpea (*Cicer arietinum* L.). *Theor. Appl. Genet.* **2009,** *119*, 757–765.

Andaya, V.; Tai, T. Fine Mapping of the qCTS12 Locus, a Major QTL for Seedling Cold Tolerance in Rice. *Theor. Appl. Genet.* **2006,** *113* (3), 467–475. https://doi.org/10.1007/s00122-006-0311-5

Andersen, J. R.; Lubberstedt, T. Functional Markers in Plants. *Trends Plant Sci.* **2003,** *8*, 554–560.

Anderson, J. A.; Stack, R. W.; Liu, S.; Waldron, B. L.; Fjeld, A. D.; Coyne, C.; Moreno-Sevilla, B.; Mitchell Fetch, J.; Song, Q. J.; Cregan, P. B.; Frohberg, R. C. DNA Markers for Fusarium Head Blight Resistance QTLs in Two Wheat Populations. *Theor. Appl. Genet.* **2001,** *102*, 1164–1168.

Aryamanesh, N.; Nelson, M. N.; Yan, G.; Clarke, H. J.; Siddique, K. H. M. Mapping a Major Gene for Growth Habit and QTLs for Ascochyta Blight Resistance and Flowering Time in a Population Between Chickpea and *Cicer reticulatum*. *Euphytica* **2010,** *173*, 307–319.

Arzani, A.; Peng, J. H.; Lapitan, N. L. V. DNA and Morphological Markers for a Russian Wheat Aphid Resistance Gene. *Euphytica* **2004,** *139*, 167–172.

Avila, C. M.; Sillero, J. C.; Rubiales, D.; Moreno, M. T.; Torres, A. M. Identification of RAPD Markers Linked to the *Uvf-1* Gene Conferring Hypersensitive Resistance Against Rust (*Uromyces viciae-fabae*) in *Vicia Faba* L. *Theor. Appl. Genet.* 2003, *107*, 353—358.

Awata, L. A. O.; Ifie, B. E.; Danquah, E.; Jumbo, M. B.; Suresh, L. M.; Gowda, M.; Marchelo-Dragga, P. W.; Olsen, M. S.; Shorinola, O.; Yao, N. K.; Boddupalli, P. M.; Tongoona, P. B. Introgression of Maize Lethal Necrosis Resistance Quantitative Trait Loci into Susceptible Maize Populations and Validation of the Resistance Under Field Conditions in Naivasha, Kenya. *Front. Plant Sci.* **2021,** *12*, 649308. DOI: 10.3389/fpls.2021.649308

Babu, R. C. Breeding for Drought Resistance in Rice: An Integrated View from Physiology to Genomics. *Electron. J. Plant Breed.* **2010,** *1*, 1133–1141.

Badea, A.; Eudes, F.; Graf, R. J.; Laroche, A.; Gaudet, D. A.; Sadasivaiah, R. S. Phenotypic and Marker-Assisted Evaluation of Spring and Winter Wheat Germplasm for Resistance to *Fusarium* Head Blight. *Euphytica* **2008,** *164*, 803–819.

Barilli, E.; Satovic, Z.; Rubiales, D.; Torres, A. M. Mapping of Quantitative Trait Loci Controlling Partial Resistance Against Rust Incited by *Uromyces pisi* (Pers.) Wint. in a *Pisum fulvum* L. Intraspecific Cross. *Euphytica* **2010,** *175*, 151–159.

Barloy, D.; Lemoine, J.; Abelard, P.; Tanguy, A. M.; Rivoal, R.; Jahier, J. Marker-Assisted Pyramiding of Two Cereal Cyst Nematode Resistance Genes from *Aegilops variabilisin* Wheat. *Mol. Breed.* **2007,** *20*, 31–40.

Basak, J.; Kundagrami, S.; Ghoose, T. K.; Pal, A. Development of Yellow Mosaic Virus (YMV) Resistance Linked DNA Marker in *Vigna mungo* from Populations Segregating for YMV-Reaction. *Mol. Breed.* **2004,** *14*, 375—382.

Bayley, D. C. Isozymic Variation and Plant Breeders' Rights. In *Isozymes in Plant Genetics and Breeding*; Tanksley, S. D., Orton, T. J., Eds Elsevier: Amsterdam, 1983; pp 425–440.

Beckmann, J. S.; Weber, J. L. Survey of Human and Rat Microsatellites. *Genomics* **1992,** *12*, 627–631.

Bergvinson, D.; Hoisington, D. Options in Developing Stem Borer-Resistant Maize: CIMMYT's Approaches and Experiences. *Insect Sci. Appl.* 2001, *21*, 409–415.

Bernardo, R.; Charcosset, A. Usefulness of Gene Information in Marker-Assisted Recurrent Selection: A Simulation Appraisal. *Crop Sci.* **2006,** *46*, 614–621.

Bernier, J.; Kumar, A.; Venuprasad, R.; Spaner, D.; Atlin, G. N. A Large-Effect QTL for Grain Yield Under Reproductive-Stage Drought Stress in Upland Rice. *Crop Sci.* **2007,** *47*, 507–516.

Bhagyawant, S. S. RAPD-SCAR Markers: An Interface Tool for Authentication of Traits. *J. Biosci. Med.* **2016,** *4*, 1–9.

Bhairappanavar, S. B.; Fakrudin, B.; Sambasiva Rao, K. R. S. Inheritance Studies of Sterility Mosaic Disease (SMD) Resistance Cross Gullyal White × BSMR736 in Pigeonpea (*Cajanus cajan* (L.) Millsp.). *Plant Gene Trait* **2014,** *5*, 27–32.

Bhat, J. A.; Ali, S.; Salgotra, R. K.; Mir, Z. A.; Dutta, S.; Jadon, V.; Tyagi, A.; Mushtaq, M.; Jain, N.; Singh, P. K.; Singh, G. P.; Prabhu, K. V. Genomic Selection in the Era of Next Generation Sequencing for Complex Traits in Plant Breeding. *Front. Genet.* **2016,** *7*, 221. https://doi.org/ 10.3389/fgene.2016.00221

Bohn, M.; Groh, S.; Khairallah, M. M.; Hoisington, D. A.; Utz, H. F.; Melchinger, A. E. Re-evaluation of the Prospects of Marker-Assisted Selection for Improving Insect Resistance Against *Diatraea* spp. in Tropical Maize by Cross Validation and Independent Validation. *Theor. Appl. Genet.* **2001,** *103*, 1059–1067. https://doi.org/10.1007/s001220100708

Bonilla, P.; Dvorak, J.; Mackill, D.; Deal, K.; Gregorio, G. RFLP and SSL Mapping of Salinity Tolerance Genes in Chromosome 1 of Rice (Oryza sativa L.) Using Recombinant Inbred Lines. *Philipp. Agric. Sci.* **2002,** *85*, 68–76.

Bouker, O.; Kong, L.; Singh, B. B.; Murdock, K.; Ohm, H. W. AFLP and AFLP-Derived SCAR Markers Associated with *Striga gesnerioides* Resistance in cowpea. *Crop Sci.* **2004,** *44*, 1259–1264.

Buckler, E. S.; Thornsberry, J. M. Plant Molecular Diversity and Applications to Genomics. *Curr. Opin. Plant Biol.* 2002, *5*, 107–111. https://doi.org/10.1016/S1369-5266(02)00238-8.

Byrt, C. S.; Zhao, M.; Kourghi, M.; Bose, J.; Henderson, S. W.; Qiu, J.; et al. Non-Selective Cation Channel Activity of Aquaporin AtPIP2; 1 Regulated by Ca2+ and pH. *Plant Cell Environ.* **2017,** *40*, 802–815. https://doi.org/10.1111/pce.12832

Cadic, E.; Coque, M.; Vear, F.; Grezes-Besset, B.; Pauquet, J.; Piquemal, J.; et al. Combined Linkage and Association Mapping of Flowering Time in Sunflower (*Helianthus annuus* L.). *Theor. Appl. Genet.* **2013,** *126*, 1337–1356. https://doi.org/10.1007/s00122-013-2056-2

Caixeta, E. T.; Oliveira, A. C. B.; Brito, G. G.; Sakiyama, N. S. Tipos de Marcadores Moleculares. In *Marcadores Moleculares*; Borem, A., Caixeta, E. T., Eds.; Editora UFV: Vicosa, Brazil, 2009; pp 11−93.

Cakir, M.; Drake-Brockman, F.; Shankar, M.; Golzar, H.; McLean, R.; Bariana, H.; Wilson, R.; Barclay, I.; Moore, C.; Jones, M.; Loughman, R. In *Molecular Mapping and Marker-Assisted Improvement of Rust Resistance in the Australian Wheat Germplasm*, Proceedings of 11th International Wheat Genet Symposium, Brisbane, Australia, August 24–29, 2008; Appels, R., Eastwood, R., Lagudah, E., Langridge, P., Mackay, M., McIntyre, L., Sharp, P., Eds.; Sydney University Press, 2008; pp 1–3. http://hdl.handle.net/2123/3317

Campbell, K. G.; Carter, A. H.; Skinner, D. Z. *Evaluation and Selection for Cold Tolerance in Wheat*; Wheat and Barley Research Annual Progress Reports and Final Reports, Washington Grain Commission, 2013; pp 46–50.

Castro, A. J.; Chen, X.; Corey, A.; Filichkina, T.; Hayes, P. M.; Mundt, C.; Richardson, K.; Sandoval-Islas, S.; Vivar, H. Pyramiding and Validation of Quantitative Trait Locus (QTL) Alleles Determining Resistance to Barley Stripe Rust. *Crop Sci.* **2003,** *43*, 2234–2239. https://doi.org/10.2135/cropsci2003.2234

Chagué, V.; Fahima, T.; Dahan, A.; Sun, G. L.; Korol, A. B.; Ronin, Y. I.; Grama, A.; Röder, M. E.; Nevo, E. Isolation of Microsatellite and RAPD Markers Flanking the *Yr15* Gene of Wheat Using NILs and Bulked Segregant Analysis. *Genome* **1999,** *42*, 1050–1056.

Channamallikarjuna, V.; Sonah, H.; Prasad, M.; Rao, G. J. N.; Chand, S.; Upreti, H. C. Identification of Major Quantitative Trait Loci *qSBR11-1* for Sheath Blight Resistance in Rice. *Mol. Breed.* **2010,** *25*, 155–166. https://doi.org/10.1007/s11032-009-9316-5.

Charpe, A.; Koul, S.; Gupta, S. K.; Singh, A.; Pallavi, J. K.; Prabhu, K. V. Marker Assisted Gene Pyramiding of Leaf Rust Resistance Genes *Lr9*, *Lr24* and *Lr28* in a Bread Wheat Cultivar HD 2329. *J. Wheat Res.* **2012,** *4*, 20–28.

Chatterjee, S. N.; Vijayan, K.; Roy, G. C.; Nair C. V. ISSR Profiling of Genetic Variability in the Ecotypes of *Antheraea mylitta* Drury, the Tropical Tasar Silkworm. *Russ. J. Genet.* **2004,** *40*, 152–159.

Chen, H.; Cui, S.; Fu, S.; Gai, J.; Yu, D. Identification of Quantitative Trait Loci Associated with Salt Tolerance During Seedling Growth in Soybean (*Glycine max* L.). *Aust. J. Agric. Res.* **2008,** *59*, 1086–1091.

Chen, W.; Li, W. Mapping of QTL Conferring Cold Tolerance at the Early Seedling Stage in Rice by Molecular Markers. *J. Wuhan Bot. Res.* **2005,** *23*, 116–120

Chen, H. M.; Liu, C. A.; Kuo, C. G.; Chein, C. M.; Sun, H. C.; Huang, C. C.; Lin, Y. C.; Ku, H. M. Development of a Molecular Marker for a Bruchid (*Callosobruchus chinensis* L.) Resistance Gene in Mungbean. *Euphytica* **2007,** *157*, 113–122.

Cheng, X.; Tian, C.; Li, A.; Qiu, J. Advances on Molecular Mechanisms of Plant-Pathogen Interactions. *Yi Chuan* **2012,** *34*, 134–144. doi: 10.3724/SP.J.1005.2012.00134

Chowdhery, M. A.; Andrahennadi, C. P.; Slinkard, A. E.; Vandenberg, A. RAPD and SCAR Markers for Resistance to Ascochyta Blight in Lentil. *Euphytica* **2001,** *118*, 331—337.

Choudhary, K.; Choudhary, O. P.; Shekhawat, N. S. Marker Assisted Selection: A Novel Approach for Crop Improvement. *Am. Eurasian J. Agric.* **2008,** *1*, 26–30.
Choudhary, A. K.; Raje, R. S.; Datta, S.; Sultana, R.; Ontagodi, T. Conventional and Molecular Approaches Towards Genetic Improvement in Pigeonpea for Insect Resistance. *Am. J. Plant Sci.* **2013,** *4*, 372–385.
Cobos, M. J.; Fernandez, M. J.; Rubio, J. K. M.; Moreno, M. T.; Gil, J.; Millan, T. A Linkage Map of Chickpea (*Cicer arietinum* L.) Based on Populations from Kabuli×Desi Crosses: Location of Genes for Resistance to Fusarium Wilt Race 0. *Theor. Appl. Genet.* **2005,** *110*, 1347–1353.
Cobos, M. J.; Winter, P.; Kharrat, M.; Cubero, J. I.; Gill, J.; Millan, T.; Rubio, J. Genetic Analysis of Agronomic Traits in a Wide Cross of Chickpea. *Field Crops Res.* **2009,** *111*, 130–136.
Collard, C. Y. B.; Mackill, D. J. Marker-Assisted Selection: An Approach for Precision Plant Breeding in the Twenty-First Century. *Philos. Trans. R. Soc. B.* **2008,** *363*, 557−572.
Courtois, B.; Shen, L.; Petalcorin, W.; Carandang, S.; Mauleon, R.; Li, Z. Locating QTLs Controlling Constitutive Root Traits in the Rice Population IAC 165 × Co39. *Euphytica* **2003,** *134*, 335–345.
Crosbie, T. M.; Eathington, S. R.; Johnson, G. R.; Edwards, M.; Reiter, R.; Stark, S.; Mohanty, R. G.; Oyervides, M.; Buehler, R. E.; Walker, A. K.; Dobert, R.; Delannay, X.; Pershing, J. C.; Hall, M. A.; Lamkey, K. R. Plant Breeding: Past, Present, and Future. In *Plant Breeding: The Arnel R. Hallauer International Symposium*; Lamkey, K. R., Lee, M., Ed.; Blackwell: Ames, IA, 2006; pp 3–50.
Crossa, J.; Beyene, Y.; Kassa, S.; Perez, P.; Hickey, J. M.; Chen, C.; de los Campos, G.; Burgueño, J.; Windhausen, V. S.; Buckler, E.; Jannink, J. L.; Cruz, M. A. L.; Babu, R. Genomic Prediction in Maize Breeding Populations with Genotyping-by-Sequencing. *G3 (Bethesda)* **2013,** *3*, 1903–1926.
Devi, E. L.; Devi, C. P.; Kumar, S.; Sharma, S. K.; Beemrote, A.; Chongtham, S. K.; Ngachan, S. V. Marker Assisted Selection (MAS) to Generate Stress Tolerant Crop Plants. *Plant Gene* **2017,** *11*, 205–218. 2:10.1016/j.plgene.2017.05.014
Dirlewanger, E.; Isaac, P. G.; Ranade, S.; Belagouza, M.; Cousin, R.; de Vienne, D. Restriction Fragment Length Polymorphism Analysis of Loci Associated with Disease Resistance Genes and Development Traits in *Pisum sativum* L. *Theor. Appl. Genet.* **1994,** *88*, 17–27.
Dubcovsky, J. Marker-Assisted Selection in Wheat. Published Online by the USDA National Institute of Food and Agriculture, 2011. http://maswheat.ucdavis.edu/.
Dubcovsky, J.; Santa Maria, G.; Epstein, E.; Luo, M. C.; Dvoøák, J. Mapping of the K+/Na+ Discrimination Locus *Kna1* in Wheat. *Theor. Appl. Genet.* **1996,** *92*, 448–454.
Dussle, C. M.; Quint, M.; Xu, M. L.; Melchinger, A. E.; Lübberstedt, T. Conversion of AFLP Fragments Tightly Linked to SCMV Resistance Genes *Scmv1* and *Scmv2* into Simple PCR-Based Markers. *Theor. Appl. Genet.* **2002,** *105*, 1190–1195.
Dweikat, I.; Ohm, H.; Patterson, F.; Cambron, S. Identification of RAPD Markers for 11 Hessian Fly Resistance Genes in Wheat. *Theor. Appl. Genet.* **1997,** *94*, 419–423.
Dwivedi, S. L.; Crouch, J. H.; Mackill, D. J.; Xu, Y.; Blair, M. W.; Ragot, M.; Upadhyaya, H. D.; Ortiz, R. The Molecularization of Public Sector Crop Breeding: Progress, Problems, and Prospects. *Adv. Agron.* **2007,** *95*, 163–318.
Eagles, H. A.; Bariana, H. S.; Ogbonnaya, F. C.; Rebetzke, G. J.; Hollamby, G. J.; Henry, R. J.; Henschke, P. H.; Carter, M. Implementation of Markers in Australian Wheat Breeding. *Aust. J. Agric. Res.* **2001,** *52*, 1349–1356.

Eathington, S. R. In *Practical Applications of Molecular Technology in the Development of Commercial Maize Hybrids*, Proceedings of the 60th Annual Corn and Sorghum Seed Research Conferences, American Seed Trade Association: Washington, D.C., 2005.

Edwards, M.; Johnson, L. In *RFLPs for Rapid Recurrent Selection*, Proceedings of the Analysis of Molecular Marker Data, Joint Plant Breeding, Symposium Series of American Society for Horticultural Science and CSSA: Madison, 1994; pp 33–40.

Eujayl, I.; Erskine, W.; Baum, M.; Pehu, E. Inheritance and Linkage Analysis of Frost Injury in Lentil. *Crop Sci.* **1999,** *39*, 639—642.

Eujayl, I.; Erskine, W.; Bayaa, B.; Baum, M.; Pehu, E. *Fusarium* Vascular Wilt in Lentil: Inheritance and Identification of DNA Markers for Resistance. *Plant Breed.* **1998,** *117*, 497—499.

Falconer, D. S.; Mackay, T. F. C. In *Introduction to Quantitative Genetics*, 4th Ed.; Pearson: London, U.K., 1996; p 464.

FAO. Climate Change and Food Security: Risks and Responses, 2015. https://www.fao.org/3/i5188e/I5188E.pdf

Finkel, E. Making Every Drop Count in the Buildup to a Blue Revolution. *Science* **2009,** *323*, 1004–1005.

Fjellstrom, R.; Conaway-Bormans, C.; McClung, A. M.; Marchetti, M. A.; Shank, A. R.; Park, W. D. Development of DNA Markers Suitable for Marker Assisted Selection of Three Pi Genes Conferring Resistance to Multiple *Pyricularia grisea* Pathotypes. *Crop Sci.* **2004,** *44*, 1790–1798.

Foresight. The Future of Food and Farming. Executive Summary, The Government Office for Science, London, 2011 [Online]. https://www.gov.uk/government/publications/future-of-foodand-farming

Frew, T. J.; Russell, A. C.; Timmerman-Vaughan, G. M. Sequence Tagged Site Markers Linked to the *smb1* Gene for Resistance to Pea Seed Borne Mosaic Virus in Pea. *Plant Breed.* **2002,** *121*, 512–516.

Fujita, D.; Doi, K.; Yoshimura, A.; Yasui, H. Molecular Mapping of a Novel Gene, *grh5*, Conferring Resistance to Green Rice Leafhopper (*Nephotettix concticeps* Uhler) in Rice, *Oryza sativa* L. *Theor. Appl. Genet.* **2006,** *113*, 567–573.

Gao, Z.; Eyers, S.; Thomas, C.; Eills, N.; Maule, A. Identification of Markers Tightly Linked to the *sbm* Recessive Genes for Resistance to Pea Seed-Borne Mosaic Virus. *Theor. Appl. Genet.* **2004,** *109*, 488–494.

Gautam, T.; Dhillon, G. S.; Saripalli, G.; Rakhi; Singh, V. P.; Prasad, P.; Kaur, S.; Chhuneja, P.; Sharma, P. K.; Balyan, H. S.; Gupta, P. K. Marker-Assisted Pyramiding of Genes/QTL for Grain Quality and Rust Resistance in Wheat (Triticum aestivum L.). *Mol. Breed.* **2020,** 40, 49. https://doi.org/10.1007/s11032-020-01125-9

Gorham, J.; Wyn Jones, R. G.; Bristol, A. Partial Characterization of the Trait for Enhanced K+-Na+ Discrimination in the D Genome of Wheat. *Planta* **1990,** *180*, 590–597.

Gowda, S. J. M.; Radhika, P. N.; Kadoo, Y.; Mhase, L. B.; Gupta, V. S. Molecular Mapping of Wilt Resistance Genes in Chickpea. *Mol. Breed.* **2009,** *24*, 177–183.

Graner, A.; Streng, S.; Drescher, A.; Jin, Y.; Borovkova, I.; Steffenson, B. Molecular Mapping of the Leaf Rust Resistance Gene *Rph7* in Barley. *Plant Breed.* **2000,** *119*, 389–392.

Graner, A.; Streng, S.; Kellermann, A.; Schiemann, A.; Bauer, E.; Waugh, R.; Pellio, B.; Ordon, F. Molecular Mapping of the rym5 Locus Encoding Resistance to Different Strains of the Barley Yellow Mosaic Virus Complex. *Theor. Appl. Genet.* **1999,** *98*, 285–290.

Grattapaglia, D.; Silva-Junior, O. B.; Resende, R. T.; Cappa, E. P.; Müller, B. S. F.; Tan, B.; Isik, F.; Ratcliffe, B.; El-Kassaby, Y. A. Quantitative Genetics and Genomics Converge to Accelerate Forest Tree Breeding. *Front. Plant Sci.* **2018**. 9.10.3389/fpls.2018.01693.

Gupta, S.; Charpe, A.; Koul, S.; Prabhu, K. V.; Haq, Q. M. Development and Validation of Molecular Markers Linked to an *Aegilops umbellulata*-Derived Leaf Rust Resistance Gene, *Lr9*, for Marker-Assisted Selection in Bread Wheat. *Genome* **2005,** *48*, 823–830.

Gupta, P. K.; Kumar, J.; Mir, R. R.; Kumar, A. Marker-Assisted Selection as a Component of Conventional Plant Breeding. *Plant Breed. Rev.* **2010a,** *33*, 145–217.

Gupta, P.K.; Langridge, P.; Mir, R. R. Marker-Assisted Wheat Breeding: Present Status and Future Possibilities. *Mol. Breed.* **2010b,** *26*, 145–161

Hamwieh, A.; Tuyen, D. D.; Cong, H.; Benitez, E. R.; Takahashi, R.; Xu, D. H. Identification and Validation of a Major QTL for Salt Tolerance in Soybean. *Euphytica* **2011,** *179*, 451–459. https://doi.org/10.1007/s10681-011-0347-8

Hamwieh, A.; Udupa, S. M.; Choumane, W.; Sarker, A.; Dreyer, F.; Jung, C.; Baum, M. A Genetic Linkage Map of *Lens* sp. Based on Microsatellite and AFLP Markers and the Localization of Fusarium Vascular Wilt Resistance. *Theor. Appl. Genet.* **2005,** *110*, 669—677.

Han, F.; Romagosa, I.; Ullrich, S. E.; Jones, B. L.; Hayes, P. M.; Wesenberg, D. M. Molecular Marker-Assisted Selection for Malting Quality Traits in Barley. *Mol. Breed.* **1997,** *3*, 427–437. https://doi.org/10.1023/A:1009608312385

Hart, D. L.; Clark, A. G. In *Principles of Population Genetics*, 3rd Ed.; Sinauer: Sunderland, M.A., 1997; p 542.

Hay, J. E.; Easterling, D.; Ebi, K. L.; Parry, M. Introduction to the Special Issue: Observed and Projected Changes in Weather and Climate Extremes. *Weather Clim. Extrem.* **2016,** *11*, 1−3.

Hayashi, K.; Hashimoto, N.; Daigen, M.; Ashikawa, I. Development of PCR-Based SNP Markers for Rice Blast Resistance Genes at the *Piz* Locus. *Theor. Appl. Genet.* **2004,** *108*, 1212–1220.

Helguera, M.; Khan, I. A.; Dubcavsky J. Development of PCR Markers for Wheat Leaf Rust Resistance Gene Lr47. *Theor. Appl. Genet.* **2000,** *101*, 625–631.

Himabindu, K.; Vijayalakshmi, P.; Sundaram, R. M.; Neeraja, C. N.; Mishra, B.; Bentur, J. S. Flanking SSR Markers for Allelism Test for the Asian Rice Gall Midge (*Orseolia oryzae*) Resistance Genes. *Euphytica* **2007,** *157*, 267–279

Hittalmani, S.; Parco, A.; Mew, T. V.; Zeigler, R. S.; Huang, N. Fine Mapping and DNA Marker-Assisted Pyramiding of the Three Major Genes for Blast Resistance in Rice. *Theor. Appl. Genet.* **2000,** *100*, 1121–1128.

Holland, J. B. Implementation of Molecular Markers for Quantitative Traits in Breeding Programs—Challenges and Opportunities. In *New Directions for a Diverse Planet*, Proceedings of the 4th International Crop Science Congress, Brisbane, QLD, Australia, Sept 26–Oct 1, 2004; Fischer, T., Ed.; The Regional Institute Ltd.: Gosford, NSW, Australia, 2004 [Online]. www. cropscience.org.au/icsc2004/ (verified Feb 4, 2008).

Hospital, F.; Moreau, L.; Lacoudre, F.; Charcosset, A.; Gallais, A. More on the Efficiency of Marker-Assisted Selection. *Theor. Appl. Genet.* 1997, *95*, 1181–1189. https://doi.org/10.1007/s001220050679

Hu, S.; Lubberstedt, T.; Zhao, G.; Lee, M. QTL Mapping of Low-Temperature Germination Ability in the Maize IBM Syn4 RIL Population. *PLoS One* **2016,** *11* (3), e0152795.

Huang, J. K.; Pray, C.; Rozelle, S. Enhancing the Crops to Feed the Poor. *Nature* **2002,** *418*, 678–684.

Huang, S.; Spielmeyer, W.; Lagudah, E. S.; James, R. A.; Platten, J. D.; Dennis, E. S.; Munns, R. A Sodium Transporter (HKT7) is a Candidate for Nax1, A Gene for Salt Tolerance in Durum Wheat. *Plant Physiol.* **2006,** *142*, 1718–1727.

Huang, X. Q.; Wang, L. X.; Xu, M. X.; Röder, M. S. Microsatellite Mapping of the Powdery Mildew Resistance Gene *Pm5e* in Common Wheat (*Triticum aestivum* L.). *Theor. Appl. Genet.* **2003,** *106*, 858–865.

Humphry, M. E.; Magner, T.; McIntyre, C. L.; Aitken, E. A. B.; Liu, C. J. Identification of a Major Locus Conferring Resistance to Powdery Mildew (*Erysiphe polygoni* DC) in Mungbean (*Vigna radiata* (L.) Wilczek) by QTL Analysis. *Genome* **2003,** *46*, 738–744.

Iruela, M.; Castro, P.; Rubio, J.; Cubero, J. I.; Jacinto, C.; Millan, T.; Gil, J. Validation of a QTL for Resistance to Ascochyta Blight Linked to Resistance to *Fusarium* Wilt Race 5 in Chickpea (*Cicer arietinum* L.). *Eur. J. Plant Pathol.* **2007,** *119*, 29–37.

James, R. A.; Blake, C.; Byrt, C. S.; Munns, R. Major Genes for Na+ Exclusion, Nax1 and Nax2 (wheat HKT1;4 and HKT1;5), Decrease Na+ Accumulation in Bread Wheat Leaves Under Saline and Waterlogged Conditions. *J. Exp. Bot.* **2011,** *62*, 2939–2947.

Jannoo, N.; Grivet, L.; Dookum, A.; D'Hont, A.; Glaszmann, J. C. Linkage Disequilibrium Among Modern Sugarcane Cultivars. *Theor. Appl. Genet.* **1999,** *99*, 1053–1060.

Jeon, J. S.; Chen, D.; Yi, G. H.; Wang, G. L.; Ronald, P. C. Genetic and Physical Mapping of Pi5(t), a Locus Associated with Broad-Spectrum Resistance to Rice Blast. *Mol. Gen. Genom.* **2003,** *269*, 280–289.

Jeung, J. U.; Heu, S. G.; Shin, M. S.; Veracruz, C. M.; Jena, K. K. Dynamics of *Xanthomonas oryzae* pv. *oryzae* Populations in Korea and Their Relations to Known Bacterial Blight Resistance Genes. *Phytopathology* **2006,** *96*, 867–875.

Jiang, G. L. Molecular Markers and Marker-Assisted Breeding in Plants. In *Plant Breeding from Laboratories to Fields*; Andersen, S. B., Ed.; InTech: Rijeka, 2013; pp 45–83.

Jiang, G. L.; Shi, J.; Ward, R. W. QTL Analysis of Resistance to Fusarium Head Blight in the Novel Wheat Germplasm CJ 9306. I. Resistance to Fungal Spread. *Theor. Appl. Genet.* **2007,** *116*, 3–13.

Joshi, A.; Adhikari, S.; Singh, N. K. Mapping Genomic Regions for Red Flour Beetle (Triboliumcastaneum (Herbst)) Resistance in Teosinte (Zea mays L. subsp. parviglumis H. H. Iltis&Doebley) Derived Maize Backcross Inbred Line Population. *Genet. Resour. Crop Evol.* **2021,** *68* (4), 1529–1544. https://doi.org/10.1007/s10722-020-01083-3.

Jung, G.; Skroch, P. W.; Coyne, D. P.; Nienhuis, J.; Arnaud-Santana, E.; Kaeppler, S. M. A.; Basset, M. J. Molecular-Marker-Based Genetic Analysis of Tepary Bean-Derived Common Bacterial Blight Resistance in Different Developmental Stages of Common Bean. *J. Am. Soc. Hort. Sci.* **1997,** *122*, 329–337.

Kahraman, A.; Kusmenoglu, I.; Aydin, N.; Aydogan, A.; Erskine, W.; Muehlbauer, F. J. QTL Mapping of Winter Hardness Genes in Lentil. *Crop Sci.* **2004,** *44*, 13—22.

Karaköy, T.; Baloch, F. S.; Toklu, F.; Ozkan, H. Variation for Selected Morphological and Quality-Related Traits Among 178 faba Bean Landraces Collected from Turkey. *Plant Genet. Resour.* **2014,** *12*, 5–13.

Kasha, K. J. Biotechnology and World Food Supply. *Genome* **1999,** *42*, 642–645.

Kearsey, M. J.; Farquhar, A. G. L. QTL Analysis in Plants, Where are we Now? *Heredity* **1998,** *80*, 137–142. https://doi.org/10.1046/j.1365-2540.1998.00500.x

Khairallah, M. M.; Bohn, M.; Jiang, C.; Deutsch, J. A.; Jewell, D. C.; Mihm, J. A.; Melchinger, A. E.; Gonzalez-de-Leon, D.; Hoisington, D. A. Molecular Mapping of QTL for Southwestern Corn Borer Resistance, Plant Height and Flowering in Tropical Maize. *Plant Breed.* **1998,** *117*, 309–318.

Kim, K. M.; Sohn, J. K.; Chung, I. K. Analysis of OPT8511 RAPD Fragments Closely Linked with Cold Sensitivity at Seedling Stage in Rice (*Oryza sativa* L.). *Mol. Cells* **2000,** *10*, 382–385.

Kirigwi, F. M.; Van Ginkel, M.; Brown-Guedira, G.; Gill, B. S.; Paulsen, G. M.; Fritz, A. K. Markers Associated with a QTL for Grain Yield in Wheat Under Drought. *Mol. Breed.* **2007,** *20*, 401–413.

Kolkman, J. M.; Kelly, J. D. QTL Conferring Resistance and Avoidance to White Mold in Common Bean. *Crop Sci.* **2003,** *43*, 539–548.

Kong, L.; Cambron, S.; Ohm, H. Hessian Fly Resistance Genes H16 and H17 are Mapped to a Resistance Gene Cluster in the Distal Region of Chromosome 1 as in Wheat. *Mol. Breed.* **2007,** *20*, 410–413.

Kong, L.; Ohm, H. W.; Cambron, S. E.; Williams, C. E. Molecular Mapping Determines that Hessian Fly Resistance Gene H9 is Located on Chromosome 1A of Wheat. *Plant Breed.* **2005,** *124*, 525–531

Konlasuk, S.; Xing, Y.; Zhang, N.; Zuo, W.; Zhang, B.; Tan, G.; Xu, M. ZmWAK, A Quantitative Resistance Gene to Head Smut in Maize, Improves Yield Performance by Reducing the Endophytic Pathogen Sporisorium reiliana. *Mol. Breed.* **2015,** *35*, 174. https://doi.org/10.1007/s11032-015-0325-2

Kotresh, H.; Fakrudin, B.; Punnuri, S. M.; Rajkumar, B. K.; Thudi, M.; Paramesh, H.; Lohithaswa, H.; Kuruvinashetti, M. S. Identification of Two RAPD Markers Genetically Linked to a Recessive Allele of a *Fusarium* Wilt Resistance Gene in Pigeon Pea (*Cajanus cajan* L. Millsp.). *Euphytica* **2006,** *149*, 113—120.

Kover, P. X.; Valdar, W.; Trakalo, J.; Scarcelli, N.; Ehrenreich, I. M.; et al. A Multiparent Advanced Generation Inter-cross Tofine-Map Quantitative Traits in *Arabidopsis thaliana. PLoS Genet.* **2009,** *5*, e1000551.

Kranner, I.; Minibayeva, F. V.; Beckett, R. P.; Seal, C. E. What is Stress? Concepts, Definitions and Applications in Seed Science. *New Phytol.* **2010,** *188*, 655–673.

Krutovsky, K. V.; Neale, D. B. Nucleotide Diversity and Linkage Disequilibrium in Cold-Hardiness- and Wood Quality-Related Candidate Genes in Douglas Fir. *Genetics* **2005,** *171* (4), 2029–2041.

Kumar, R.; Venuprasad, R.; Atlin, G. N. Genetic Analysis of Rainfed Lowland Rice Drought Tolerance Under Naturally-Occurring Stress in Eastern India: Heritability and QTL Effects. *Field Crops Res.* **2007,** *103*, 42–52.

Lamers, J.; Van Der Meer, T.; Testerink, C. How Plants Sense and Respond to Stressful Environments. *Plant Physiol.* **2020,** *182*, 1624–1635. https://doi.org/10.1104/pp.19.01464

Landi, P.; Sanguineti, M. C.; Darrah, L. L.; Giuliani, M. M.; Salvi, S.; Conti, S.; Tuberosa, R. Detection of QTLs for Vertical Root Pulling Resistance in Maize and Overlap with QTLs for Root Traits in Hydroponics and for Grain Yield Under Different Water Regimes. *Maydica* **2002,** *47*, 233–243.

Lanubile, A.; Maschietto, V.; Borrelli, V. M.; Stagnati, L.; Logrieco, A. F.; Marocco, A. Molecular Basis of Resistance to Fusarium Ear Rot in Maize. *Front. Plant Sci.* **2017,** *8*, 1774. https://doi.org/10.3389/fpls.2017.01774.

Larcher, W. Stress bei Pflanzen. *Naturwissenschaften* **1987,** *74*, 158–167. https://doi.org/10.1007/BF00372919

Lee, S. Y.; Ahn, J. H.; Cha, Y. S.; Yun, D. W.; Lee, M. C.; Ko, J. C.; Lee, K. S.; Eun, M. Y. Mapping QTLs Related to Salt Tolerance of Rice at Seedling Stage. *Plant Breed.* **2007,** *126*, 43–46.

Li, Y.; Tong, L.; Deng, L.; Liu, Q.; Xing, Y.; Wang, C.; Liu, B.; Yang, X.; Xu, M. Evaluation of ZmCCT Haplotypes for Genetic Improvement of Maize Hybrids. *Theor. Appl. Genet.* **2017,** *130*, 2587–2600.

Lin, H. X.; Zhu, M. Z.; Yano, M. J.; Gao, P.; Liang, Z. W.; et al. QTLs for Na+ and K+ Uptake of the Shoots and Roots Controlling Rice Salt Tolerance. *Theor. Appl. Genet.* **2004,** *108*, 253–260.

Lindsay, M. P.; Lagudah, E. S.; Hare, R. A.; Munns, R. A Locus for Sodium Exclusion (Nax1), A Trait for Salt Tolerance, Mapped in Durum Wheat. *Funct. Plant Biol.* **2004,** *31*, 1105–1114.

Liu, X. M.; Fritz, A. K.; Reese, J. C.; Wilde, G. E.; Gill, B. S.; Chen, M. S. H9, H10 and H11 Compose a Cluster of Hessian Fly-Resistance Genes in the Distal Gene-Rich Region of Wheat Chromosome 1AS. *Theor. Appl. Genet.* **2005a,** *110*, 1473–1480.

Liu, X. M.; Gill B. S. Identification of Microsatellite Markers Linked to Russian Wheat Aphid Resistance Genes Dn4 and Dn6. *Theor. Appl. Genet.* 2002, *104*, 1042–1048.

Liu, X. M.; Gill, B. S.; Chen, M. S. Hessian Fly Resistance Gene H13 is Mapped to a Distal Cluster of Resistance Genes in Chromosome 6DS of Wheat. *Theor. Appl. Genet.* **2005b,** *111*, 243–249.

Liu, S.; Griffey, C. A.; Shagai Maroof, M. A. Identification of Molecular Markers Associated with Adult Plant Resistance to Powdery Mildew in Common Wheat Cultivar Massey. *Crop Sci.* **2001,** *41*, 1268–1275.

Liu, J.; Liu, D.; Tao, W.; Li, W.; Wang, S.; Chen, P.; Cheng S.; Gao, D. Molecular Marker-Facilitated Pyramiding of Different Genes for Powdery Mildew Resistance in Wheat. *Plant Breed.* **2000,** *119*, 21–24.

Lorenz, A. J.; Chao, S.; Asoro, F. G.; Heffner, E. L.; Hayashi, T.; Iwata, H.; Smith, K. P.; Sorrells, M. E.; Jannink, J. L. Genomic Selection in Plant Breeding: Knowledge and Prospects. In *Advances in Agronomy*; Sparks, D. L., Ed.; Academic Press: San Diego, CA, USA, 2011; vol 110, pp 77–123.

Lynch, M.; Walsh, B. Genetics and Analysis of Quantitative Traits. Sinauer: Sunderland, MA; 1998.

Ma, C.; Ma, X.; Yao, L.; Liu, Y.; Du, F.; Yang, X.; Xu, M. qRfg3, A Novel Quantitative Resistance Locus Against Gibberella Stalk Rot in Maize. *Theor. Appl. Genet.* **2017,** *130*, 1723–1734.

Ma, Z. Q.; Wei, J. B.; Cheng, S. H. PCR-Based Markers for the Powdery Mildew Resistance Gene *Pm4a* in Wheat. *Theor. Appl. Genet.* **2004,** *109*, 140–145.

Mackill, D. J.; Collard, B. C. Y.; Neeraja, C. N.; Rodriguez, R. M.; Heuer, S.; Ismail, A. M. QTLs in Rice Breeding: Examples for Abiotic Stresses. In *Rice Genetics 5: Proceedings of the International Rice Genetics Symposium*; Brar, D. S., Mackill, D. J., Hardy, B., Eds.; Inter-national Rice Research Institute: Manila, 2006; pp 155–167.

Magadum, S. K.; Shrishail, D.; Doddabhimappa, G.; Verma, S. K. Role of Molecular Breeding in Genetic Improvement of Pigeonpea. *Int. J. Bio-resource Stress Manag.* **2013,** *4* (2), 382–388.

Mago, R.; Spielmeyer, W.; Lawrence, G. J.; Lagudah, E. S.; Ellis, J. G.; ÆPryor, A. Identification and Mapping of Molecular Markers Linked to Rust Resistance Genes

Located on Chromosome 1RS of Rye Using Wheat-Rye Translocation Lines. *Theor. Appl. Genet.* **2002,** *104*, 1317–1324.
Mateu-Andres, I.; De Paco, L. Allozymic Differentiation of the *Antirrhinum majus* and *A. siculum* Species Groups. *Ann. Bot.* **2005,** *95*, 465–473.
Matilda, C.; Elena, T.; Stanislav, K.; Roxana, N.; Ionica, G.; Nicolae, S. Marker – Assisted Selection (MAS) for Drought Tolerance in Wheat Using Markers Associated with Membrane Stability. *Genetica ùi ameliorarea plantelor* **2009,** *77*, 7–12.
McCouch, S. R.; Teytelman, L.; Xu, Y.; Lobos, K. B.; Clare, K.; Walton, M.; Fu, B.; Maghirang, R.; Li, Z.; Xing, Y. Development and Mapping of 2240 New SSR Markers for Rice (*Oryza sativa* L.). *DNA Res.* **2002,** *9*, 199–207.
Melotto, M.; Afanador, L.; Kelly, J. D. Development of a SCAR Marker Linked to the *I* Gene in Common Bean. *Genome* **1996,** *39*, 1216–1219.
Meuwissen, T. H.; Hayes, B. J.; Goddard, M. E. Prediction of Total Genetic Value Using Genome-Wide Dense Marker Maps. *Genetics* **2001,** *157*, 1819–1829.
Michaels, S. D.; Amasino, R. M. A Robust Method for Detecting Single-Nucleotide Changes as Polymorphic Markers by PCR. *Plant J.* **1998,** *14*, 381–385.
Miedaner, T.; Korzun, V. Marker-Assisted Selection for Disease Resistance in Wheat and Barley Breeding. *Phytopathology* **2012,** *102*, 560–566.
Miedaner, T.; Risser, P.; Paillard, S.; Schnurbusch, T.; Keller, B.; Hartl, L.; Holzapfel, J.; Korzu,n V.; Ebmeyer, E.; Utz, H. F. Broad Spectrum Resistance Loci for Three Quantitatively Inherited Diseases in Two Winter Wheat Populations. *Mol. Breed.* **2012,** *29*, 731–742
Miedaner, T.; Wilde, F.; Steiner, B.; Buerstmayr, H.; Korzun, V.; Ebmeyer, E. Stacking Quantitative Trait Loci (QTL) for Fusarium Head Blight Resistance from Non-adapted Sources in an European Elite Spring Wheat Background and Assessing Their Effects on Deoxynivalenol (DON) Content and Disease Severity. *Theor. Appl. Genet.* **2006,** *112*, 562–569.
Millan, T.; Rubio, J.; Iruela, M.; Daly, K.; Cubero, J. I.; Gil, J. Markers Associated with *Ascochyta* Blight Resistance in Chickpea and Their Potential in Marker Assisted Selection. *Field Crops Res.* **2003,** *84*, 373–384.
Ming-Zhe, Y.; Jian-Fei, W.; Hong-You, C.; Hu-Qu, Z.; Hong-Sheng, Z. Inheritance and QTL Mapping of Salt Tolerance in Rice. *Rice Sci.* **2005,** *12* (1), 25–32.
Mishra, R. R.; Nag, A.; Panigrahi, J. Development of Dominant Sequence Characterized Amplified Region (SCAR) Marker Linked with Plume Moth (*Exelastis atomosa* Walsingham 1886) Resistance in Pigeon-Pea. *Chilean J. Agric. Res.* **2015,** *75* (4), 497–502.
Mondini, L.; Noorani, A.; Pagnotta, M. A. Assessing Plant Genetic Diversity by Molecular Tools. *Diversity* **2009,** *1*, 19–35.
Moniruzzaman, M.; Khatun, R.; Mintoo, A. A. Application of Marker Assisted Selection for Livestock Improvement in Bangladesh. *Bangladesh Vet.* **2014,** *31*, 1–11.
Moreau, L.; Charcosset, A.; Gallais, A. Experimental Evaluation of Several Cycles of Marker-Assisted Selection in Maize. *Euphytica* 2004, *137*, 111–118. https://doi.org/10.1023/B:EUPH.0000040508.01402.21
Morgante, M.; Hanafeyk, M.; Powell, W. Microsatellites are Preferentially Associated with Nonrepetitive DNA in Plant Genomes. *Nat. Genet.* **2002,** *30*, 194–200.
Muchero, W.; Ehlers, J. D.; Roberts, P. A. QTL Analysis for Resistance to Foliar Damage Caused by *Thrips tabaci* and *Frankliniella schultzei* (Thysanoptera: Thripidae) Feeding in Cowpea [*Vigna unguiculata* (L.) Walp.]. *Mol. Breed.* **2010,** *25*, 47—56.

Murai, H.; Hashimoto, Z.; Sharma, P. N.; Shimizu, T.; Murata, K.; Takumi, S.; Mori, N.; Kawasaki, S.; Nakamura, C. Construction of a High-Resolution Linkage Map of a Rice Brown Planthopper (*Nilaparvata lugens* Stal) Resistance Gene *bph2*. *Theor. Appl. Genet.* **2001,** *103*, 526–532.

Myles, S.; Peiffer, J.; Brown, P. J.; Ersoz, E. S.; Zhang, Z.; Costich, D. E.; et al. Association Mapping: Critical Considerations Shift from Genotyping to Experimental Design. *Plant Cell* **2009,** *21*, 2194–2202. https://doi.org/10.1105/tpc.109.068437.

Naylor, R. L.; Falcon, W. P.; Goodman, R. M.; Jahn, M. M.; Sengooba, T.; Tefera, H.; Nelson, R. J. Biotechnology in the Developing World: A Case for Increased Investments in Orphan Crops. *Food Pol.* **2004,** *29*, 15–44.

Neeraja, C. N.; Maghirang-Rodriguez, R.; Pamplona, A.; Heuer, S.; Collard, B. C. Y.; Septiningsih, E. M.; Vergara, G.; Sanchez, D.; Xu, K.; Ismail, A. M.; Mackill, D. J. A Marker-Assisted Backcross Approach for Developing Submergence-Tolerant Rice Cultivars. *Theor. Appl. Genet.* **2007,** *115*, 767–776. https://doi.org/10.1007/s00122-007-0607-0.

Nelson, G. C.; Rosegrant, M. W.; Koo, J.; Robertson, R.; Sulser, T.; Zhu, T.; Ringler, C.; Msangi, S.; Palazzo, A.; Batka, M. **In** *Climate Change: Impact on Agriculture and Costs of Adaptation*; International Food Policy Research Institute (IFPRI): Washington, D.C., U.S.A., 2009.

Newcombe, G.; Bradshaw, H. D.; Chastagner, G. A.; Stettler, R. F. A Major Gene for Resistance to Melampsora Medusae fsp. Deltoidae in a Hybrid Poplar Pedigree. *Phytopathology* **1996,** *86* (1), 87–94.

Nguyen, H. T.; Vu, Q. H.; Van Mai, T.; Nguyen, T. T.; Vu, L. D.; Nguyen, T. T. Marker-Assisted Selection of *Xa21* Conferring Resistance to Bacterial Leaf Blight in Indica Rice Cultivar LT2. *Rice Sci.* **2018,** *25*, 52–56.

Niebur, W. S.; Rafalski; Smith, O. S.; Cooper, M. Applications of Genomics Technologies to Enhance Rate of Genetic Progress for Yield of Maize Within a Commercial Breeding Program. In *New Directions for a Diverse Planet: Proceedings of the 4th International Crop Scienec Congress*, Fischer, T., Ed.; Brisbane, QLD, Australia, Sept. 26Oct. 1, 2004; The Regional Institute Ltd., Gosford, NSW, Australia [Online]. www. cropscience.org.au/icsc2004/ (verified Feb. 4, 2008).

Nocente, F.; Gazza, L.; Pasquini, M. Evaluation of Leaf Rust Resistance Genes *Lr1, Lr9, Lr24, Lr47* and Their Introgression into Common Wheat Cultivars by Marker-Assisted Selection. *Euphytica* **2007,** *155*, 329–336.

Nordborg, M.; Tavaré, S. Linkage Disequilibrium: What History has to Tell Us. *Trends Genet.* **2002,** *18* (2), 83–90. https://doi.org/10.1016/s0168-9525(02)02557-x.

Ortiz, R. Critical Role of Plant Biotechnology for the Genetic Improvement of Food Crops: Perspectives for the Next Millennium. *Electron. J. Biotechnol.* [Online] **1998,** *1*. DOI: 10.2225/vol1-issue3-fulltext-7

Pahlich, E. Larcher's Definition of Plant Stress: A Valuable Principle for Metabolic Adaptability Research. *Rev. Bras. Fisiol. Veg.* **1993,** *5*, 209–216.

Pazhamala, L.; Saxena, R. K.; Singh, V. K.; Sameerkumar, C. V.; Kumar, V.; Sinha, P.; Patel. K.; Obala, J.; Kaoneka, S. R.; Tongoona, P.; Shimelis, H. A.; Gangarao, N. V. P. R.; Odeny, D.; Rathore, A.; Dharmaraj, P. S.; Yamini, K. N.; Varshney, R. K. Genomics- Assisted Breeding for Boosting Crop Improvement in Pigeonpea (*Cajanus cajan*). *Front. Plant Sci.* **2015,** *6* (50), 1–12.

Perumalsamy, S.; Bharani, M.; Sudha, M.; Nagarajan, P.; Arul, L.; Sarawathi, R.; et al. Functional Marker-Assisted Selection for Bacterial Leaf Blight Resistance

Genes in Rice (*Oryza sativa* L.). *Plant Breed.* **2010,** *129*, 400–406. https://doi.org/10.1111/j.1439-0523.2009.01705.x.

Pradhan, S. K.; Nayak, D. K.; Mohanty, S.; Behera, L.; Barik, S. R.; Pandit, E.; Lenka, S.; Anand, A. Pyramiding of Three Bacterial Blight Resistance Genes for Broad-Spectrum Resistance in Deepwater Rice Variety, Jalmagna. *Rice* **2015,** *8*, 19. https://doi.org/10.1186/s12284-015-0051-8.

Prasanna, B. M. In *Molecular Markers for Maize Improvement in Asia*, Proceedings of 10th Asian Regional Maize Workshop, Makassar, Indonesia, Oct 20–23, 2008; CIMMYT: Mexico, DF, 2009.

Prasanna, B. M.; Pixley, K.; Warburton, M. L.; Xie, C. X. Molecular Marker-Assisted Breeding Options for Maize Improvement in Asia. *Mol. Breed.* **2010,** *26*, 339–356. https://doi.org/10.1007/s11032-009-9387-3.

Pushparajan, N.; Krishnasamy, V.; Babu, R. C.; Kannanbabu, J. R. Association Mapping of Salinity Tolerance in Rice Using Molecular Markers. *Int. J. Biores. Stress. Manag.* **2011,** *2* (3), 307–312.

Qiu, H.; Tan, K.; Li, C.; Shen, T.; Zhang, Z. Identification of QTL for Resistance to Head Smut in Maize (*Zea mays*. L). *Euphytica* **2021,** *217*, 185. https://doi.org/10.1007/s10681-021-02916-7.

Quesada, T.; Gopal, V.; Cumbie, W. P.; Eckert, A. J.; Wegrzyn, J. L.; Neale, D. B.; Goldfarb, B.; Huber, D. A.; Casella, G.; Davis J. M. Association Mapping of Quantitative Disease Resistance in a Natural Population of Loblolly Pine (*Pinus taeda* L.). *Genetics* **2010,** *186*, 677–686.

Rafalski, A. Applications of Single Nucleotide Polymorphisms in Crop Genetics. *Curr. Opin. Plant Biol.* **2002,** *5* (2), 94–100. https://doi.org/10.1016/s1369-5266(02)00240-6.

Rahman, M.; Davies, P.; Bansal, U.; Pasam, R.; Hayden, M.; Trethowan, R. Marker-Assisted Recurrent Selection Improves the Crown Rot Resistance of Bread Wheat. *Mol. Breed.* **2020,** *40*, 28. https://doi.org/10.1007/s11032-020-1105-1

Rakshit, S.; Winter, P.; Tekeoglu, M.; Munoz, J. J.; Pfaff, T.; Benko-Iseppon, M.; Muehlbauer, F. J.; Kahl, G. DAF Markers Tightly Linked to a Major Locus for *Ascochyta* Blight Resistance in Chickpea (*Cicer arietinum* L.). *Euphytica* **2003,** *132*, 23–30.

Ramkumar, G.; Srinivasarao, K.; Mohan, K. M.; Sudarshan, I.; Sivaranjani, A. K. P.; Gopalakrishna, K. Development and Validation of Functional Marker Targeting an InDel in the Major Rice Blast Disease Resistance Gene *Pi54* (*Pikh*). *Mol. Breed.* **2011,** *27*, 129–135.

Rathore, T. R.; Warsi, M. Z. K.; Singh, N. N.; Vasal, S. K. In *Production of Maize Under Excess Soil Moisture (waterlogging) Conditions*, Proceedings of the Seventh Asian Regional Maize Workshop; Vasal, S. K., Gonzalez Ceniceros, F., Fan X., Eds.; PCARRD: Los Baños, Philippines, 2000; pp 232–239.

Reif, J. C.; Zhang, P.; Dreisigacker, S.; Warburton, M. L.; van Ginkel, M.; Hoisington, D.; Bohn, M.; Melchinger, A. E. 2005. Wheat Genetic Diversity Trends During Domestication and Breeding. *Theor. Appl. Genet.* **2005,** *110* (5), 859–864. https://doi.org/10.1007/s00122-004-1881-8.

Ren, Z. H.; Gao, J. P.; Li, L. G.; Cai, X. L.; Huang, W.; Chao, D. Y.; Zhu, M. Z.; Wang, Z. Y.; Luan, S.; Lin, H. X. A Rice Quantitative Trait Locus for Salt Tolerance Encodes a Sodium Transporter. *Nat. Genet.* **2005,** *37*, 1141–1146. https://doi.org/10.1038/ng1643

Ren, J.; Wu, P.; Huestis, G. M.; Zhang, A.; Qu, J.; Liu, Y.; Zheng, H.; Alakonya, A. E.; Dhliwayo, T.; Olsen, M.; Vicente, F. S.; Prasanna, B. M.; Chen, J.; Zhang, X. Identification and Fine Mapping of a Major QTL (qRtsc8-1) Conferring Resistance to Maize Tar Spot

Complex and Validation of Production Markers in Breeding Lines. *Theor. Appl. Genet.* 2022. https://doi.org/10.1007/s00122-022-04053-8.

Ribaut, J. M.; Betran, J. Single Large-Scale Marker-Assisted Selection (SLS–MAS). *Mol. Breed.* **1999,** *5*, 531–541.

Ribaut, J. M.; de Vicente, M. C.; Delannay, X. Molecular Breeding in Developing Countries: Challenges and Perspectives. *Curr. Opin. Plant Biol.* **2010,** *13*, 1–6.

Roberts, D. W. A. Identification of Loci on Chromosome 5A of Wheat Involved in Control of Cold Hardiness, Vernalization, Leaf Length, Rosette Growth Habit, and Height of Hardened Plants. *Genome* **1990,** *33*, 247–259.

Roberto, T.; Silvio, S. Genomics-Based Approaches to Improve Drought Tolerance of Crops. *Trends Plant Sci.* **2006,** *11*, 405–412.

Roman, B.; Torres, A. M.; Rubiales, D.; Cubero, J. I.; Satovic, Z. Mapping of Quantitative Trait Loci Conferring Broomrape (*Orobanche crenata* Forsk.) Resistance in Faba Bean (*Vicia faba* L.). *Genome* **2002,** *45*, 1057—1063.

Sabouri, H.; Sabouri, A. New Evidence of QTLs Attributed to Salinity Tolerance in Rice. *Afr. J. Biotechnol.* **2008,** *7*, 4376–4383.

Sah, S. K.; Reddy, K. R.; Li, J. Abscisic Acid and Abiotic Stress Tolerance in Crop Plants. *Front. Plant Sci.* **2016,** *7*, 571. https://doi.org/10.3389/fpls.2016.00571.

Saijo, Y.; Loo, E. P. I. Plant Immunity in Signal Integration Between Biotic and Abiotic Stress Responses. *New Phytol.* **2020,** *225*, 87–104. https://doi.org/10.1111/nph.15989.

Salgotra, R. K.; Stewart, C. N. Jr. Functional Markers for Precision Plant Breeding. *Int. J. Mol. Sci.* 2020, *21*, 4792. https://doi.org/10.3390/ijms21134792.

Samayoa, L. F.; Malvar, R. A.; Olukolu, B. A.; Holland, J. B.; Butron, A. Genome- Wide Association Study Reveals a Set of Genes Associated with Resistance to the Mediterranean Corn Borer (*Sesamia nonagrioides* L.) in a Maize Diversity Panel. *BMC Plant Biol.* **2015,** *15*, 35.

Sara, F.; Barat, A. F. Marker-Assisted Selection for Disease Resistance: Applications in Breeding (Review). *J. Agron. Crop Sci.* **2014,** *7*, 1392−1405

Sardesai, N.; Kumar, A.; Rajyashri, K. R.; Nair, S.; Mohan, M. Identification of an AFLP Marker Linked to *Gm7*, A Gall Midge Resistance Gene and its Conversion to a SCAR Marker for its Utility in Marker Aided Selection in Rice. *Theor. Appl. Genet.* **2002,** *105*, 691–698.

Sardesai, N.; Nemacheck, J. A.; Subramanyam, S.; Williams, C. E. Identification and Mapping of H32, A New Wheat Gene Conferring Resistance to Hessian Fly. *Theor. Appl. Genet.* **2005,** *111*, 1167–1173.

Schneider, K. A.; Grafton, K. F.; Kelly, J. D. QTL Analysis of Resistance to *Fusarium* Root Rot in Bean. *Crop Sci.* **2001,** *41*, 535–542.

Servin, B.; Martin, O. C.; Mézard, M.; Hospital, F. Toward a Theory of Marker-Assisted Gene Pyramiding. *Genetics* **2004,** *168* (1), 513–523. https://doi.org/10.1534/genetics.103.023358.

Seyfarth, R.; Feuillet, C.; Schachermayr, G.; Winzeler, M.; Keller, B. Development of Molecular Marker for the Adult Plant Leaf Rust Resistance Gene *Lr35* in Wheat. *Theor. Appl. Genet.* **1999,** *99*, 554–560.

Shabanimofrad, M.; Yusop, M. R.; Ashkani, A.; Musa, M. H.; Adam, N. A.; Haifa, I.; Harun, A. R.; Latif, M. A. Marker-Assisted Selection for Rice Brown Planthopper (*Nilaparvata lugens*) Resistance Using Linked SSR Markers. *Turk. J. Biol.* **2015,** *39*, 1–8.

Shavrukov, Y.; Bovill, J.; Afzal, I.; Hayes, J. E.; Roy, S. J.; Tester, M.; Collins, N. C. HVP10 Encoding V-PPase is a Prime Candidate for the Barley HvNax3sodium Exclusion Gene: Evidence from Fine Mapping and Expression Analysis. *Planta* **2013,** *237*, 1111–1122.

Shinada, H.; Iwata, N.; Sato, T.; Fujino, K. Genetical and Morphological Characterization of Cold Tolerance at the Fertilization Stage in Rice. *Breed. Sci.* **2013,** *63*, 197–204.

Shinada, H.; Iwata, N.; Sato, T.; Fujino, K. QTL Pyramiding for Improving of Cold Tolerance at Fertilization Stage in Rice. *Breed. Sci.* **2014,** *63*, 483–488.

Siahsar, B. A.; Narouei, M. Mapping QTLs of Physiological Traits Associated with Salt Tolerance in 'Steptoe' × 'Morex' Doubled Haploid Lines of Barley at Seedling Stage. *J. Food Agric. Environ.* **2010,** *8*, 751–759.

Singh, R.; Datta, D.; Priyamvada; Singh, S.; Tiwari, R. Marker-Assisted Selection for Leaf Rust Resistance Genes *Lr19* and *Lr24* in Wheat (*Triticum aestivum* L.). *J. Appl. Genet.* **2004,** *45*, 399–403.

Singh, S.; Sidhu, J. S.; Huang, N.; Vikal, Y.; Li, Z.; Brar, D. S.; Dhaliwal, H. S.; Khush, G. S. Pyramiding Three Bacterial Blight Resistance Genes (xa5, xa13 and Xa21) Using Marker-Assisted Selection into Indica Rice Cultivar PR106. *Theor. Appl. Genet.* 2001, *102*, 1011–1015.

Singh, D.; Sinha, B.; Rai, V. P.; Singh, M. N.; Singh, D. K.; Kumar, R.; Singh, A. K. Genetics of Fusarium Wilt Resistance in Pigeonpea (*Cajanus cajan*) and Efficacy of Associated SSR Markers. *Plant Pathol. J.* **2016,** *32* (2), 95–101.

Singh, R. P.; Singh, P. K.; Gupta, R.; Singh, R. M. Biotechnological Tools to Enhance Sustainable Production. In *Biotechnology for Sustainable Agriculture*; Singh, R. M., Mondal, S., Eds.; Woodhead Publishing, 2018; pp 19–66, ISBN: 9780128121603. DOI: 10.1016/B978-0-12-812160-3.00002-7

Singh, V. K.; Singh, A.; Singh, S. P.; Ellur, R. K.; Choudhary, V.; Sarkhel, S.; Singh, D.; Gopala Krishnan, S.; Nagarajan, M.; Vinod, K. K.; Singh, U. D.; Rathore, R.; Prasanthi, S. K.; Agrawal, P. K.; Bhatt, J. C.; Mohapatra, T.; Prabhu, K. V.; Singh, A. K. Incorporation of Blast Resistance into 'PRR78', An Elite Basmati Rice Restorer Line, Through Marker Assisted Backcross Breeding. *Field Crop Res.* **2012,** *128*, 8–16.

Singh, A. K.; Singh, V. K.; Singh, A.; Ellur, R. K.; Pandian, R. T. P.; Gopala Krishnan, S. Introgression of Multiple Disease Resistance into a Maintainer of Basmati Rice CMS Line by Marker-Assisted Backcross Breeding. *Euphytica* **2015,** *203*, 97–107. DOI: 10.1007/s10681-014-1267-1

Skiba, B.; Ford, R.; Pang, E. C. K. Construction of a Linkage Map Based on a *Lathyrus sativus* Backcross Population and Preliminary Investigation of QTLs Associated with Resistance to *Ascochyta* Blight. *Theor. Appl. Genet.* **2004,** *109*, 1726—1735.

Slafer, G. A.; Araus, J. L.; Royo, C.; Del Moral, L. F. G. Promising Eco-Physiological Traits for Genetic Improvement of Cereal Yields in Mediterranean Environments. *Ann. Appl. Biol.* 2005, *146*, 61–70.

Smith, P. H.; Koebner, R. M. D.; Boyd, L. A. The Development of a STS Marker Linked to a Yellow Rust Resistance Derived from the Wheat Cultivar Moro. *Theor. Appl. Genet.* **2002,** *104*, 1278–1282.

Snape, J. W.; Semokhoskii, A.; Fish, L.; Sarma, R. N.; Quarrie, S. A.; Galiba. G.; Sutka, J. Mapping Frost Tolerance Loci in Wheat and Comparative Mapping with Other Cereals. *Acta Agron. Hungar.* **1997,** *45*, 265–270.

Song, F. J.; Xiao, M. G.; Duan, C. X.; Li, H. J.; Zhu, Z. D.; Liu, B. T.; Sun, S. L.; Wu, X. F.; Wang, X. M. Two Genes Conferring Resistance to Pythium Stalk Rot in Maize Inbred Line Qi319. *Mol. Gen. Genom.* **2015,** *290*, 1543–1549.

Sorrells, M. E. Application of New Knowledge, Technologies, and Strategies to Wheat Improvement. *Euphytica* **2007,** *157*, 299–306.

Steele, K. A.; Edwards, G.; Zhu, J.; Witcombe, J. R. Marker-Evaluated Selection in Rice: Shifts in Allele Frequency Among Bulks Selected in Contrasting Agricultural Environments Identify Genomic Regions of Importance to Rice Adaptation and Breeding. *Theor. Appl. Genet.* **2004,** *109*, 1247–1260. https://doi.org/10.1007/s00122-004-1732-7

Strable, J.; Scanlon, M. J. Maize (*Zea mays*): A Model Organism for Basic and Applied Research in Plant Biology. *Cold Spring Harb. Protoc.* **2009,** *2009*, pdb.emo132. DOI: 10.1101/pdb.emo132

Suenaga, K.; Singh, R. P.; Huerta-Espino, J.; William, H. M. Microsatellite Markers for Genes *Lr34/Yr18* and Other Quantitative Loci for Leaf Rust and Stripe Rust Resistance in Bread Wheat. *Phytopathology* **2003,** *93*, 881–890.

Sun, L. H.; Wang, C. M.; Su, C. C.; Liu, Y. Q.; Zhai, H. Q.; Wan, J. M. Mapping and Marker-Assisted Selection of a Brown Planthopper Resistance Gene *bph2* in Rice (*Oryza sativa* L.). *Acta Genet. Sin.* **2006,** *33*, 717–723.

Sundaram, R. M.; Vishnupriya, M. R.; Biradar, S. K.; Laha, G. S.; Reddy, G. A.; Shobha Rani, N.; Sarma, N. P.; Sonti, R. V. Marker Assisted Introgression of Bacterial Blight Resistance in Samba Mahsuri, an Elite Indica Rice Variety. *Euphytica* **2008,** *160*, 411–422.

Takehisa, H.; Shimodate, T.; Fukuta, Y.; Ueda, T.; Yano, M.; Yamaya, T.; Kameya, T.; Sato, T. Identification of Quantitative Trait Loci for Plant Growth of Rice in Paddy Field Flooded with Salt Water. *Field Crops Res.* **2004,** *89*, 85–95. https://doi.org/10.1016/j.fcr.2004.01.026

Tamburic-Ilincic, L.; Falk, D.; Schaafsma, A. *Fusarium* Ratings in Ontario Winter Wheat Performance Trial (OWWPT) Using an Index that Combines Fusarium Head Blight Symptoms and Deoxynivalenol Levels. *Czech J. Genet. Plant Breed.* **2011,** *47*, S115–S122.

Tan, C. X.; Ji, X. M.; Yang, Y.; Pan, X. Y.; Zuo, S. M.; Zhang, Y. F.; Zou, J. H.; Chen, Z. X.; Zhu, L. H.; Pan, X. B. Identification and Marker-Assisted Selection of Two Major Quantitative Genes Controlling Rice Sheath Blight Resistance in Backcross Generations. *Acta Genet. Sin.* **2005,** *32*, 399–1405.

Tanksley, S. D.; Young, N. D.; Paterson, A. H.; Bonierbale, M. W. RFLP Mapping in Plant Breeding: New Tools for an Old Science. *Biotechnology* **1989,** *7*, 257–263.

Taran, B.; Michaels, T. E.; Pauls, K. P. Mapping Genetic Factors Affecting the Reaction to *Xanthomous axonopodis* pv. *phaseoli* in *Phaseolus vulgaris* L. Under Field Conditions. *Genome* **2001,** *44*, 1045—1056.

Tester, M.; Langridge, P. Breeding Technologies to Increase Crop Production in a Changing World. *Science* **2010,** *327*, 818–822.

Thakur, S.; Singh, P. K.; Das, A.; Rathour, R.; Variar, M.; Prashanthi, S. K. Extensive Sequence Variation in Rice Blast Resistance Gene *Pi54* Makes it Broad Spectrum in Nature. *Front. Plant Sci.* **2015,** *6*, s345. https://doi.org/10.3389/fpls.2015.00345

Thomson, M. J.; de Ocampo, M.; Egdane, J.; et al. Characterizing the Saltol Quantitative Trait Locus for Salinity Tolerance in Rice. *Rice* **2010,** *3*, 148–160. https://doi.org/10.1007/s12284-010-9053-8

Threadgill, D. W.; Hunter, K. W.; Williams, R. W. Genetic Dissection of Complex and Quantitative Traits: From Fantasy to Reality Via a Community Effort. *Mamm. Genome* **2002,** *13*, 175–178.

Trethowan, R. M.; Reynolds, M.; Sayre, K.; Ortiz-Monasterio, I. Adapting Wheat Cultivars to Resource Conserving Farming Practices and Human Nutritional Needs. *Ann. Appl. Biol.* **2005,** *146*, 405–413.

Tullu, A.; Buchwaldt, L.; Warkentin, T.; Taran, B.; Vandenberg, A. Genetics of Resistance to Anthracnose and Identification of AFLP and RAPD Markers Linked to the Resistance Gene in PI320937 Germplasm of Lentil (*Lens culinaris* Medikus). *Theor. Appl. Genet.* **2003,** *106*, 428—434.

Tyagi, S.; Mir, R. R.; Kaur, H.; Chhuneja, P.; Ramesh, B.; Balyan, H. S.; Gupta, P. K. Marker-Assisted Pyramiding of Eight QTLs/Genes for Seven Different Traits in Common Wheat *(Triticum aestivum* L.). *Mol. Breed.* **2014,** *34*, 167–175.

Udupa, S. M.; Baum, M. Genetic Dissection of Pathotype-Specific Resistance to Ascochyta Blight Disease in Chickpea (*Cicer arietinum* L.) Using Microsatellite Markers. *Theor. Appl. Genet.* **2003,** *106*, 1196–1202.

Verma, S.; Nizam, S.; Verma, P. K. Biotic and Abiotic Stress Signalling in Plants. In *Stress Signaling in Plants: Genomics and Proteomics Perspective*; Sarwat, M., Ahmad, A., Abdin, M., Eds.; Springer: New York, NY, 2013; vol 1, pp 25–49. https://doi.org/10.1007/978-1-4614-6372-6_2

Wang, X. Y.; Chen, P. D.; Zhang, S. Z. Pyramiding and Marker-Assisted Selection for Powdery Mildew Resistance Genes in Common Wheat. *Acta Genet. Sin.* **2001,** *28*, 640–646.

Wang, Z.; Ma, L. Y.; Cao, J.; Li, Y. L.; Ding, L. N.; Zhu, K. M.; et al. Recent Advances in Mechanisms of Plant Defense to *Sclerotinia sclerotiorum*. *Front. Plant Sci.* **2019,** *10*, 1314. https://doi.org/10.3389/fpls.2019.01314

Wang, Y.; Pinson, S. R. M.; Fjellstrom, R. G.; Tabien, R. E. Phenotypic Gain from Introgression of Two QTL, *qSB9-2* and *qSB12-1* for Rice Sheath Blight Resistance. *Mol. Breed.* **2012,** *30*, 293–303.

Wang, C.; Yang, Q.; Wang, W.; Li, Y.; Guo, Y.; Zhang, D.; Ma, X.; Song, W.; Zhao, J.; Xu, M. A Transposon-Directed Epigenetic Change in ZmCCT Underlies Quantitative Resistance to Gibberella Stalk Rot in Maize. *New Phytol.* **2017,** *215*, 1503–1515.

Weerasena, J. S.; Steffenson, B. J.; Falk, A. B. Conversion of an Amplified Fragment Lenght Polymorphism Marker into a Co-dominant Marker in the Mapping of the *Rph15* Gene Conferring Resistance to Barley Leaf Rust, *Puccinia hordei* Otth. *Theor. Appl. Genet.* **2004,** *108*, 712–719.

Weng, Y.; Lazar, M. D. Amplified Fragment Length Polymorphism- and Simple Sequence Repeat-Based Molecular Tagging and Mapping of Green Bug Resistance Gene Gb3 in Wheat. *Plant Breed.* **2002,** *121*, 218–223.

Weng, Y.; Li, W.; Devkota, R. N.; Rudd, J. C. Microsatellite Markers Associated with Two Aegilops tauschii- Derived Greenbug Resistance Loci in Wheat. *Theor. Appl. Genet.* 2005, *110*, 462–469.

Wijerathna, Y. M. A. M. Marker Assisted Selection: Biotechnology Tool for Rice Molecular Breeding. *Adv. Crop Sci. Technol.* **2015,** *3*, 187. https://doi.org/10.4172/2329-8863.1000187

Williams, K. J. The Molecular Genetics of Disease Resistance in Barley. *Australian J. Agric. Res.* **2003,** *54*, 1065–1079.

Wisser, R. J.; Balint-Kurti, P. J.; Nelson, R. J. The Genetic Architecture of Disease Resistance in Maize: A Synthesis of Published Studies. *Phytopathology* **2006,** *96*, 120–129.

Wissuwa, M.; Wegner, J.; Ae, N.; Yano, M. Substitution Mapping of Pup1: A Major QTL Increasing Phosphorus Uptake of Rice from a Phosphorus-Deficient Soil. Theor. Appl. Genet. **2002,** *105, 890–897. https://doi.org/10.1007/s00122-002-1051-9*

Wissuwa, M.; Yano, M.; Ae, N. Mapping of QTLs for Phosphorus-Deficiency Tolerance in Rice (Oryza sativa L.). Theor. Appl. Genet. **1998,** *97, 777–783.*

Witcombe, J. R.; Hollington, P. A.; Howarth, C. J.; Reader, S.; Steele, K. A. Breeding for Abiotic Stresses for Sustainable Agriculture. *Philos. Trans. R. Soc. Lond. Ser. B. Biol. Sci.* **2008,** *363*, 703−716.

Xiao, Y.; Tong, H.; Yang, X.; Xu, S.; Pan, Q.; Qiao, F.; Raihan, M. S.; Luo, Y.; Liu, H.; Zhang, X.; Yang, N.; Wang, X.; Deng, M.; Jin, M.; Zhao, L.; Luo, X.; Zhou, Y.; Li, X.; Liu, J.; Zhan, W.; Liu, N.; Wang, H.; Chen, G.; Cai, Y.; Xu, G.; Wang, W.; Zheng, D.; Yan, J. Genome-Wide Dissection of the Maize Ear Genetic Architecture Using Multiple Populations. *New Phytol.* **2016,** *210*, 1095–1106.

Xing, Y.; Ingvardsen, C.; Salomon, R.; Lübberstedt, T. Analysis of Sugarcane Mosaic Virus Resistance in Maize in an Isogenic Dihybrid Crossing Scheme and Implications for Breeding Potyvirus-Resistant Maize Hybrids. *Genome* **2006,** *49*, 1274–1282.

Xu, J. Pyramiding of Two *BPH* Resistance Genes and *Stv-bi* Gene Using Marker-Assisted Selection in *Japonica* Rice. *Crop Breed. Appl. Biotechnol.* **2013,** *13*, 99–106.

Xu, Y.; Beachell, H.; McCouch, S. R. A Marker-Based Approach to Broadening the Genetic Base of Rice in the USA. *Crop Sci.* **2004,** *44*, 1947– 1959. https://doi.org/10.2135/cropsci2004.1947

Xu, K.; Xu, X.; Fukao, T.; Canlas, P.; Maghirang-Rodriguez, R.; Heuer, S.; Ismail, A. M.; Bailey-Serres, J.; Ronald, P. C.; Mackill, D. J. Sub1A is an Ethylene-Response-Factor-Like Gene that Confers Submergence Tolerance to Rice. *Nature* **2006,** *442* (7103), 705–708. https://doi.org/10.1038/nature04920.

Xu, K.; Xu, X.; Ronald, P.; et al. A High-Resolution Linkage Map of the Vicinity of the Rice Submergence Tolerance Locus Sub1. *Mol. Gen. Genet.* **2000,** *263*, 681–689. https://doi.org/10.1007/s004380051217

Yadav, S.; Anuradha, G.; Kumar, R. R.; Vemireddy, L. R.; Sudhakar, R.; Donempudi, K. Identification of QTLs and Possible Candidate Genes Conferring Sheath Blight Resistance in Rice (*Oryza sativa* L.). *SpringerPlus* **2015,** *4*, 175. https://doi.org/10.1186/s40064-015-0954-2

Yang, Q.; Balint-Kurti, P.; Xu, M. Quantitative Disease Resistance: Dissection and Adoption in Maize. *Mol. Plant.***2017,** *10*, 402–413.

Yang, D.; Jin, D.; Wang, B.; Zhang, D.; Nguyen, H.; Zhang, C.; Chen, S. Characterization and Mapping of Rpi1, A Gene that Confers Dominant Resistance to Stalk Rot in Maize. *Mol. Gen. Genom.* 2005, *274*, 229–234.

Yashitola, J.; Thirumurugan, T.; Sundaram, R. M.; Naseerullah, M. K.; Ramesha, M. S.; Sarma, N. P.; Stone, R. V. Assessment of Purity of Rice Hybrids Using Microsatellite and STS Markers. *Crop Sci.* **2002,** *42*, 1369–1373.

Ye, J.; Zhong, T.; Zhang, D.; Ma, C.; Wang, L.; Yao, L.; Zhang, Q.; Zhu, M.; Xu, M. The Auxin-Regulated Protein ZmAuxRP1 Coordinates the Balance Between Root Growth and Stalk Rot Disease Resistance in Maize. *Mol. Plant.* 2019, *12*, 360–373.

Yu, J.; Buckler, E. S. Genetic Association Mapping and Genome Organization of Maize. *Curr. Opin. Biotechnol.* **2006,** *17*, 155–160.

Yu, J.; Holland, J. B.; McMullen, M. D.; Buckler, E. S. Genetic Design and Statistical Power of Nested Association Mapping in Maize. *Genetics* **2008,** *178* (1), 539–551. https://doi.org/10.1534/genetics.107.074245

Zhang, J. N.; Hao, C. Y.; Ren, Q.; Chang, X. P.; Liu, G. R.; Jing, R. L. Association Mapping of Dynamic Developmental Plant Height in Common Wheat. *Planta* **2011,** *234*, 891–902. https://doi.org/10.1007/s00425-011-1434-8

Zhao, H. X.; Liu, X. M.; Chen, M. S. H22 a Major Resistance Gene to the Hessian Fly (*Mayetiola destructor*) is Mapped to the Distal Region of Wheat Chromosome 1ds. *Theor. Appl. Genet.* **2006a,** *113*, 1491–1496.

Zhao, X.; Tan, G.; Xing, Y.; Wei, L.; Chao, Q.; Zuo, W.; Lübberstedt, T.; Xu, M. Marker-Assisted Introgression of qHSR1 to Improve Maize Resistance to Head Smut. *Mol. Breed.* **2012,** *30*, 1077–1088.

Zhao, M.; Zhang, Z.; Zhang, S. Quantitative Trait Loci for Resistance to Banded Leaf and Sheath Blight in Maize. *Crop Sci.* **2006b,** *46*, 1039–1045.

Zhou, W. C.; Kolb, F. L.; Bai, G. H.; Domier, L. L.; Boze, L. K.; Smith, N. J. Validation of a Major QTL for Scab Resistance with SSR Markers and Use of Marker-Assisted Selection in Wheat. *Plant Breed.* **2003,** *122*, 40–46. https://doi.org/10.1046/j.1439-0523.2003.00802.x

Zhu, J. K. Abiotic Stress Signaling and Responses in Plants. *Cell* **2016,** *167*, 313–324.

Zhu, L. C.; Smith, C. M.; Fritz, A.; Boyko, E.; Voothuluru, P.; Gill B. S. Inheritance and Molecular Mapping of New Greenbug Resistance Genes in Wheat Germplasms Derived from Aegilops tauschii. *Theor. Appl. Genet.* **2005,** *111*, 831–837.

Zhu, Y. L.; Song, Q. J.; Hyten, D. L.; Van Tassel, C. P.; MatukumalliL, K.; Grimm, D. R.; Hyatt, S. M.; Fickus, E. W.; Young, N. D.; Cregan, P. B. Single Nucleotide Polymorphisms (SNPs) in Soybean. *Genetics* **2003,** *163*, 1123–1134.

Zietkiewicz, E.; Rafalski, A.; Labuda, D. Genome Fingerprinting by Simple Sequence Repeat (SSR)-Anchored Polymerase Chain Reaction Amplification. *Genomics* **1994,** *20*, 176–183.

Zou, C.; Wang, P.; Xu, Y. Bulked Sample Analysis in Genetics, Genomics and Crop Improvement. *Plant Biotechnol. J.* **2016,** *14*, 1941–1955. https://doi.org/10.1111/pbi.12559

Zuo, W.; Chao, Q.; Zhang, N.; Ye, J.; Tan, G.; Li, B.; Xing, Y.; Zhang, B.; Liu, H.; Fengler, K. A.; Zhao, J.; Zhao, X.; Chen, Y.; Lai, J.; Yan, J.; Xu, M. A Maize Wall-Associated Kinase Confers Quantitative Resistance to Head Smut. *Nat. Genet.* **2015,** *47*, 151–157.

Zwart, R. S.; Thompson, J. P.; Milgate, A. W.; Bansal, U. K.; Williamson, P. M.; Raman, H.; Bariana, H. S. QTL Mapping of Multiple Foliar Disease and Root-Lesion Nematode Resistances in Wheat. *Mol. Breed.* **2010,** *26*, 107–124.

CHAPTER 6

Breeding and Molecular Approaches for Heat-Resilient Crops

ALKA BHARATI,[1,2] AMARJEET KUMAR,[3] AJAY KUMAR CHANDRA,[2] SACHIN PHOGAT,[2,4] ANKUR POUDEL,[2,4] SNEHA ADHIKARI,5 VISHAL DINKAR,[6] AND SHARANI CHOUDHURY[2,4]

[1]*ICAR-Central Agroforestry Research Institute, Jhansi, India*

[2]*ICAR-Indian Agricultural Research Institute, New Delhi, India*

[3]*Department of Genetics and Plant Breeding, MTTC & VTC, Selesih, CAU, Imphal, India*

[4]*ICAR-National Institute for Plant Biotechnology, New Delhi, India*

[5]*ICAR-IIWBR, Regional Station Flowerdale, Shimla, India*

[6]*ICAR-Central Institute of Temperate Horticulture, Srinagar, J&K, India*

ABSTRACT

Global warming has caused an increase in the earth's temperature by 0.8° since the pre-industrial era, and it will likely rise by 5° by the end of the 21st century if the GHG emissions rates are sustained. The temperature change will accelerate the rise in extreme temperature and erratic rainfall patterns, causing enormous loss of yield and crop failure due to heat stress, and therefore, placing global food and nutritional security at risk. Heat stress causes a cascade of morphological, physiological, biochemical, and molecular responses in plants triggering complex pathways and affecting the growth and development of plants throughout their ontogeny. The mechanisms, such as adaptation, avoidance, and acclimatization infer the different levels of heat tolerance depending upon crop and genotype. Understanding the

Climate-Resilient Agriculture: A Molecular Perspective. Ashutosh Singh, Saurabh Pandey (Eds.)

metabolic pathways, signaling mechanisms, QTLs, genes, molecular players, and interaction among them for conferring heat tolerance and integrations of this knowledge to develop climate-smart genotype is the need of the hour. Some good studies of cellular and molecular mechanisms that enable plants to thrive under heat stress include membrane stability maintenance, ROS scavenging, antioxidant production, osmoregulation, induction MAPK and CDPK cascade, Chaperone signaling, heat shock proteins production, cell detoxification, and transcriptional activation. Genetic strategies can be formulated based on this knowledge by integrating traditional breeding, molecular breeding, transgenic technologies, and genome editing.

6.1 INTRODUCTION

Global mean surface temperature has been rising at the rate of 0.2°C per decade and has risen approximately 1°C since the pre-industrial era. In addition to this, regional-scale warming has risen more than 1C exposing 20–40% of the world population to over 1.5°C warming at least in one season (Allen et al., 2018). An increase in the concentration of greenhouse gases due to anthropological activity is the major cause of global warming, which is predicted to further increase by 3°C by 2100 from the present value. Rising temperature has disrupted many natural processes leading to extreme weather conditions, such as heat waves, drought, and floods. Adaptation to the surrounding environment is a continuous process for living organisms. In the past, plants have adapted to the gradual increase in mean temperature, but the real challenge in the present era is the rapidity of temperature change and significant variability surrounding the mean temperature rise. Variability in the mean temperature is particularly dangerous because it results in sudden heat shock and heat wave, exposing the plant to extreme stress in a very short period. Field crops are extremely sensitive to heat stress, especially during the reproductive stage. High temperature induces morphological, physiological, and biochemical changes, which ultimately affect field crops' growth, development, and yields. A plant is under heat stress when the temperature rises beyond a threshold level for a time interval, adequate to cause irreversible damage to plant growth and development. Transient temperature elevation of 10–15°C above ambient temperature could cause heat shock or heat stress. In addition to temperature rise, other factors, such as high-temperature intervals, intensity, rate of increase in temperature, and the difference between day and night temperatures also influence heat stress

(Wahid et al., 2007). A threshold temperature refers to a value of the daily mean temperature at which a detectable reduction in growth begins. A lower developmental threshold or a base temperature is one below which plant growth and development stop similarly; an upper developmental threshold is a temperature above which growth and development cease. Threshold temperature varies from crop to crop and genotype to genotype; for example, the upper threshold for wheat, maize, and rice is 26, 38, 34°C, respectively. Determining the upper threshold temperature for the crop is challenging due to its dependence on other environmental factors but determining it is also essential as high-temperature stress is becoming a major limiting factor for field crop production in tropical and semitropical climates. Heat stress is particularly concerning during reproductive stag because plants divert resources to cope with heat stress as the expanse of reproductive development. Brief exposure of plants to high temperatures during seed filling can accelerate senescence, diminish seed set and weight, and reduce yield. In wheat, 2–5-day exposure to a temperature above the upper threshold during anthesis reduces floret fertility; similarly daily mean temperature of 35°C may cause total failure in the seed setting (Maestri et al., 2002; Prasad and Djanaguiraman, 2014) also, increasing duration of HT during grain filling reduces grain weight linearly due to reduction of grain filling phase and forced maturation of incompletely filled grain. Due to reduced grain filling, starch accumulation also gets affected in wheat grains (Hurkman et al., 2003)

Several physiological, biochemical, and cellular activity gests hampered due to heat stress depending on its intensity. At a very high temperature, all the cellular functions are altered within a very short period, ultimately leading to cell disruption, cell death, and catastrophic collapse of the entire cellular organization (Schöffl et al., 1999). Moderate heat stress led to the increase in membrane fluidity due to denaturation and aggregation of proteins. It may cause cell injury and cell death after long exposure. Low heat stress mainly results in the inactivation of enzymes, protein degradation, repression of protein synthesis machinery, and membrane integrity loss (Howarth, 2005). Heat stress also affects microtubule organization. All the above responses eventually produce toxic compounds and reactive oxygen species (ROS). Plants have different mechanisms to cope with heat stress which can be interpreted as avoidance, escape, or tolerance (Osmond et al., 1987). In the avoidance mechanism, plant changes its physiology to maintain the internal temperature (water accumulation, stomatal closure, transpiration cooling, shedding of older leaves, solar radiation reflection, extensive rooting, etc.) to reduce the deleterious effects of high temperature (Hasanuzzaman et al.,

2013). In escape mechanisms, plants reduce their life cycle and complete their reproductive phase before the onset of heat stress. Tolerance is the endurance of a plant to heat stress without modifying its growth habits and maintaining yield (Barnabás et al., 2008). Tolerance mechanisms are particularly of interest from an agronomic point of view. Plant tolerance to high temperature may be achieved through various mechanisms, including changes at the physiological, biochemical, cellular, molecular, and whole-plant levels. Tolerant crops and varieties tend to have higher photosynthetic activity (Allakhverdiev et al., 2008), nitrogen uptake (Xu and Huang, 2006), carbon allocation, and antioxidant enzyme activity at a higher temperature. Plant hormones such as ethylene and cytokinin also play a regulatory role in thermotolerance. At the molecular level, heat shock proteins (HSPs), known as molecular chaperones, play an important role in stabilizing protein structure and rendering protein denaturation due to high-temperature stress. Apart from chaperone-like function, HSPs also play a role in signal transduction during heat stress. Signal transduction at the molecular level starts with the sensing of heat by various sensors present in the cell (cell membrane, HSP90.2, thioredoxin, and a thioredoxin-like protein, cytoskeletons, etc.), followed by transduction of signals to the nucleus, which eventually alters the expression of genes to accumulate stress-related transcripts and proteins as a stress tolerance strategy.

6.2 PLANTS RESPONSE TO HEAT STRESS

6.2.1 MORPHOLOGICAL, ANATOMICAL, AND PHENOLOGICAL RESPONSE

Plant growth and yield are severely influenced by prolonged radiation and high temperature causing pre- and postharvest damage. Morphological modifications in plants due to heat stress depend on phenological stages at which plants are experiencing stress as well as the duration and intensity of stress. These modifications may occur due to impairment of physiological processes or due to alteration of developmental patterns. Visual morphological symptoms of heat injury include sunburn and scorching of leaves and twigs, leave abscission and senescence, growth retardation, discoloration of fruits and leaves, and reduced dry matter production (Fahad et al., 2017). Moreover, the reproductive phase is extremely sensitive to high temperatures. In wheat, every stage from flower initial to grain filling gets affected, which includes delayed flower development, impairment of reproductive organ viability, retarded pollen tube growth, reduction in fertilization efficiency, and pollen viability (Prasad and Djanaguiraman, 2014). In rice too, the number of spikes and florets per plant is reduced due to heat stress (Fahad et al., 2016). The seed set is highly affected in sorghum under similar conditions (Prasad et al., 2006). Optimal temperature requirement for proper growth and development of roots is lower than that of shoots. Any temperature changes highly affect the root system architectures (RSA), root–shoot ratio, and nutrient uptake. Effect of high temperature on root development may be promotive, inhibitive, first promotive than inhibitive, depending on the species. In adult maize plants, lateral root development is compensated by the establishment of long axial roots for water extraction from deeper soil. Similarly, in potato, lateral and adventitious root development is inhibited due to high temperature and also swelling of root cap meristem and bending of root tips have been reported (Calleja-Cabrera et al., 2020). Studies also indicated that global warming is favoring invasive species through changes in root architecture (Gavelienė et al., 2022). High temperature affects the entire developmental stage of crop plants and ultimately affects the yield. Multimethod estimate of grain yield loss from five countries due to 1° increase in global temperature showed yield loss of 8.6, 6.9, 3.6, 3.3% in maize, wheat, soybean, and rice, respectively (Zhao et al., 2017). Thousand-grain weight loss of up to 8% has been recorded in cereals after increasing 1° from the threshold temperature (Oosterhuis, 2002).

To survive under heat stress, plants exhibit many cellular subcellular and tissue level anatomical changes. Most common of which are cell size reduction, changing density orientation, and opening time of stomata to control water loss and transpiration cooling, Enlargement of xylem vessel diameter for better water transport (Banon et al., 2004). At the subcellular level, chloroplast and cell membrane are mostly affected by temperature change. Damage to the thylakoid membrane, loss of grana stacking, and grana swelling are most evident in the chloroplast, which reduces photosynthesis. In *Vitis vinifera,* swelling of stroma and lamellae of mesophyll cells and antenna and depleted photosystem II have been observed in response to heat stress (Zhang et al., 2005). Sensitivity to high temperature varies for different phenological stages, depending on species and genotypes. It will be interesting to observe changes in phenology at each stage as well as their cumulative effect on plant development (Wahid et al., 2007). During the vegetative stage, leaf gas exchange properties get highly affected due to high temperature. As mentioned earlier, reproductive stage is highly sensitive; even heat stress for a short period can cause several phenological changes like an increase in the floral bud, flower abortion, pollen and another development, fertilization, and fruit set. High temperature during grain filling in wheat also influences flour and bread quality and changes the physiochemical property of grain (Perrotta et al., 1998).

6.2.2 PHYSIOLOGICAL AND BIOCHEMICAL RESPONSES

6.2.2.1 WATER AND NUTRIENT RELATION

Plants tend to maintain leaf water potential at high temperatures when the availability of water in the soil is sufficient. However, in field conditions, heat and drought stress come in conjugation, impairing the tendency to maintain tissue water status. Few examples (sugarcane, tomato) also suggested that in spite of sufficient soil moisture and humidity, heat stress altered the components of leaf water potential, which is due to its effect on root hydraulic conductance (Wahid and Close, 2007; Morales et al., 2003). Enhanced transpiration due to high temperature results in more water loss during day time as compared with night, impairing many physiological activities due to a reduction in water potential (Tsukaguchi et al., 2003). High temperature does not have a direct relationship with the nutrition status of plants, but due to prolonged stress, altered RSA may influence the nutrient uptake (Bassirirad, 2000).

6.2.2.2 PHOTOSYNTHESIS

Maintenance of photosynthesis during stress is crucial for obtaining optimum yield. Both the stages of photosynthesis, that is, the photochemical reaction in the thylakoid lamellae and carbon metabolism in the stroma of chloroplast, get affected due to heat stress. Chlorophyll fluorescence is light re-emission by pigments of chlorophyll molecules during their return from excited to ground state. It is basically the ratio of variable fluorescence to maximum fluorescence (Fv/Fm) and is the indicator of photosynthetic conversion in plants. Chlorophyll fluorescence and base fluorescence (Fo) are shown to be correlated with heat tolerance (Yamada et al., 1996). Photosystem II (PSII) is a protein supercomplex bound to the thylakoid membrane, which is the site for initiation of the photosynthetic electron transport chain, splitting of water molecules, and evolution of oxygen. PSII is prone to irreversible damage during heat stress due to increased fluidity of the thylakoid membrane. In natural conditions, heat stress comes with excessive light stress to photosynthesis. PSII has the ability to tolerate light stress to some extent, first by dissipating excessive light capture by LHC in the form of heat and subsequently by accommodating the damage to the reaction center binding D1 protein of PSII. A specific protease recycles damaged D1 proteins up to some extent, but at an extremely high light intensity, damage to D1 protein will become irreversible (Yamamoto, 2016). Damage to the electron transport chain causes low ATP production, which further affects ribulose-1,5-bisphosphate (RuBP) regeneration, thus limiting the CO_2 fixation. Moderate heat stress also denatures Rubisco activase (Rca), a catalytic chaperone of Rubisco, and hence impairs Rubisco activation (Perdomo et al., 2017)

6.2.2.3 OSMOLYTE AND SECONDARY METABOLITE ACCUMULATION

Accumulation of low molecular mass organic compound to maintain osmotic balance is a key adaptive mechanism in many abiotic stress conditions, including drought, salinity, heavy metal, cold, and heat. These compatible osmolyte accumulations prevent cell collapse due to extensive water loss during heat stress conditions. Depending upon the plant species, the nature of osmolytes may differ, which may be sugars, sugar alcohols, amino acids (e.g., proline), tertiary and quaternary ammonium, or sulphonium compounds (e.g., glycine betaine, GB). Maize and sugarcane accumulate a high level of GB under water deficit due to high temperature (Wahid and Close, 2007; Quan et al., 2004). In contrast, Arabidopsis, rice, mustard, and tobacco do not naturally produce GB during stress. However, genetically engineered plants of these species have been developed consisting of GB biosynthesis pathway (Quan et al., 2004). GB and proline also buffer cells' redox potential, hence protecting them from stress (Wahid and Close, 2007). A four-carbon non-protein amino acid, gamma-aminobutyric acid (GABA), has its amine group at gamma carbon rather than at alpha carbon. It remains in an unbound form, it is highly soluble in water, and takes different conformation in solution, which also includes a proline-like cyclic structure. It also remains in zwitter ionic form in physiological pH, and accumulation of low concentration is nontoxic to cells (Shelp et al., 1999). LeProT1, a GABA transporter from tomato, has shown to also transport other compatible osmolytes like proline (Schwacke, 1999). GABA also showed stress-induced accumulation in plant cells also; its kinetics correlates with stress mitigation (Kinnersley and Turano, 2000). These studies indicate that GABA might have a role as compatible osmolytes.

The plant is a sessile organism that has developed secondary metabolites production, a strong defense mechanism against biotic and abiotic stresses. Secondary metabolites were synthesized through intermediates of primary carbon metabolism. Various pathways like phenylpropanoid, shikimate, and methyl erythritol phosphate (MEP) pathways are involved in the production of secondary metabolites. Principle enzyme of the phenylpropanoid pathway, phenylalanine ammonia lyase (PAL), is a marker enzyme for various biotic and abiotic stress whose activity also correlates with stress acclimatization to some extent. The xanthophyll cycle in the plant has evolved essentially for the photoprotection of cell organelles and cell membrane. Zeaxanthin present in the periphery of LHC is responsible for protecting it from ROS damage during various stresses.

6.2.2.4 PHYTOHORMONES

Phytohormones regulate almost every aspect of the growth, development, and metabolism of plants. Studies have shown that exogenous phytohormone application ameliorates heat tolerance to some extent in the plants. Extensive research in model plants has been carried out to elucidate the biosynthesis and signaling of phytohormones (Li et al., 2021). Abscisic acid (ABA) is a crucial phytohormone for abiotic stress tolerance (Suzuki et al., 2016). Rise in temperature causes oxidative damage due to disruption of several cellular activities, resulting in ROS production. ABA influences ROS production through respiratory burst oxidase homologs (RBOHs), an NADPH oxidase. *Arabidopsis* mutants impaired by the RBOHs gene displayed reduced tolerance to heat stress (Kaya et al., 2019). Similarly, the ABA biosynthesis mutant had low ROS levels during heat vis-à-vis heat tolerance (Larkindale and Knight, 2002). ABA also influences plant energy and carbon status by regulating sucrose metabolism and transport to cope with stress conditions (Rezau et al., 2019). ABA also regulates heat stress transcription factor (Hsfs); for example, *FaHSFA2c HsfA6f* in tall fescue, and wheat, respectively, are upregulated due to ABA treatment (Wang et al., 2017; Bi et al., 2020). Auxin is a key hormone for thermomorphogenesis due to temperature changes which are evident by impaired heat-induced growth in auxin mutant plants (Küpers et al., 2020). PIL (PIN-like Proteins) are the efflux carryauxin which is responsible for the intracellular distribution of auxin, and thereby, its signaling. PILS6 in *Arabidopsis* is found to be temperature-sensitive, whose concentration in the cell diminishes due to heat shock resulting in low auxin concentration in the cell (Feraru et al., 2019). In addition to it, TIR1, an auxin co-repressor F box protein, is stabilized by the HSP90 chaperone system during heat stress (Watanabe et al., 2016). Salicyclic acid is a component of systemic acquired resistance (SAR) and the hypersensitive response (HR). SA is thought to be instrumental for long-term thermotolerance by influencing Ca^{2+} ion and antioxidant activity. Experiments with SA analogues (sulfosalicylic acid, Acetyl-SA, methyl salicylate) showed increased activity of H_2O_2 scavenging enzymes, increased antioxidant production and increased thermotolerance (Shi et al., 2006; Dat et al., 1998; Llusia et al., 2005). Cytokinin and gibberellin act opposite to ABA during high-temperature stress. Cytokinin enhances stomatal conductance and leaf transpiration during an earlier stage of stress, which is the main cooling mechanism. A sudden increase in transpiration creates a water deficit which is marked by an increase in the expression of a dehydration marker gene, *ERD10B*. Water deficit further stimulates ABA level and ultimately stomatal closure. Ectopic expression of the cytokinin oxidase/dehydrogenase CKX1 gene from *Arabidopsis* in tobacco revealed enhanced heat and drought tolerance (Mackova et al., 2013). Cytokinin-responsive genes were upregulated due to heat stress (Cerny et al., 2014) and also the antioxidant system is stimulated by cytokinin upon heat stress (Xu and Huang, 2009). Exogenous cytokinin treatment resulted in the retention of photosynthetic ability by enhancing stay green phenotype in rice during heat stress (Yang et al., 2016).

6.2.2.5 MEMBRANE THERMOSTABILITY

Membranes of cell and cell organelles are highly ordered structures, mostly composed of lipid bilayers and proteins distributed in a mosaic pattern. High and low temperatures both have a profound effect on cell membrane thermostability (CMTS). Due to high temperature, kinetic energy across the membrane rises, resulting in the movement of molecules across the membrane, hence, loosening the cell membrane and making the cell membrane more fluid. Rise in temperature also denatures proteins, alters the tertiary and quaternary structure of the protein, and increases unsaturated fatty acid, which further increases membrane fluidity and permeability. Ultimately, increased permeability gives rise to electrolyte leakage and alteration of microdomains of proteins which further relay ambient cues to downstream signaling pathways.

6.2.3 MOLECULAR RESPONSES

ROS (e.g., $O_2^{\bullet -}$, H_2O_2, $^{\bullet}OH$ $_1O^2$) are considered unavoidable by-products of aerobic respiration. Higher organisms have intricately evolved mechanisms to protect themselves from ROS toxicity and also use it as a signaling molecule. The concentration of ROS should be kept checked in the cell, otherwise, it leads to oxidative damage of biomolecules, and ultimately entire cell may be damaged and it results in oxidative stress. Plants have developed several enzymatic and nonenzymatic pathways for the neutralization of ROS. Enzymes, such as superoxide dismutase (SOD), catalase (CAT), ascorbate peroxidase (APX), glutathione peroxidase (GPX), peroxiredoxin (PRX), guaiacol peroxidases (POD), and antioxidants like, ascorbic acid and glutathione (GSH) are responsible for scavenging of ROS in almost every cellular compartment. In addition to the enzyme and antioxidant, plants have also developed metabolic adaptations to check the accumulation of free transient ions, and ultimately, the production of toxic hydroxyl radical ($HO^{\cdot}$) via the Fenton reaction (Halliwell and Gutteridge, 2015). Production and scavenging of ROS is a continual process operated in every cellular compartment separately to maintain ROS homeostasis based on their redox potential, and hence, gene networks are activated according to it. ROS have a different signature with respect to cell type, cellular compartments, developmental stage, type, and intensity of stress (Choudhury et al., 2017). During abiotic stress, ROS is produced in apoplast, mitochondria, peroxisome, and chloroplast. In apoplast, NADPH oxidases, also known as RBOHs, catalyze the production of superoxide radical $O^{-2\cdot}O^{2-\cdot}$. RBOH-dependent ROS also act as signaling molecules to activate downstream stress-responsive genes (Gilroyet al., 2014, 2016). In addition to RBOH, peroxidase and oxalate oxidase also produce ROS in apoplast, which is scavenged by CuZn-SODs, APXs, cell wall-bound peroxidases, and low levels of ascorbate and GSH. The general mechanism involves the dismutation of superoxide radical to H_2O_2. H_2O_2 is subsequently detoxified into H_2O by CAT, APX, or POD. In chloroplast, stomatal closure due to heat stress produces $O^{2\cdot -}$, $_1O^2$ ions which subsequently affect the photosystem stoichiometry, chlorosis, and programmed cell death. Thylakoid-bound proteins EXECUTER1 (EX1) and EX2 are nuclear-encoded proteins responsible for sensing ROS in chloroplast and induction of various responses (Kleine and Leister, 2016). Fe- and CuZn-SODs and the Asada–Foyer–Halliwell pathway are mainly responsible for ROS scavenging in the chloroplast. In addition to that a high antioxidant concentration of ascorbic acid and GSH is also responsible for chloroplastic ROS scavenging. $O^{2\cdot -}$ production due to electron leakage in complex I and III in main response is due to heat stress in mitochondria which is scavenged by Mn-SOD. Type II NAD(P)H dehydrogenase, alternative oxidase (AOX), and inner membrane uncoupling protein are the enzymes of the electron transport chain of mitochondria. These enzymes and proteins are also responsible for the ROS mitigation in mitochondria (Rasmusson and Wallstrom, 2010).

Stress protein expressions are produced in profound amounts to cope with various environmental stress. Mostly, water-soluble proteins act as stress proteins which prevent cell and cellular integrity via hydration. In addition, molecular chaperones, mainly HSPs, are highly upregulated during heat stress, whose activation is controlled by heat shock transcription factors (Hsfs) (Discussed in Section 6).

6.3 BREEDING FOR HEAT TOLERANCE

6.3.1 SCREENING TECHNIQUES FOR HT BREEDING

Heat tolerance is a complex trait, and hence efficient screening techniques are needed to identify the key traits and heat-tolerant genotypes which can be used as a donor in the breeding program. Efficient screening techniques are required due to the complexity of the effect of heat stress assessments resulting from the interactions between genotype and their environment. Hence, multilocation screening resulted in the near selection of tolerant genotypes. For the success of screening techniques, a suitable screening environment is needed where an adequate amount of stress can be given either naturally or artificially. Therefore, screening techniques can be classified based on the environment, that is, natural environment or field condition and controlled environment or lab condition.

6.3.1.1 NATURAL ENVIRONMENT OR FIELD CONDITION

Natural heat stress is utilized in this screening technique to assess the tolerant or susceptible genotypes. Several traits are available in different crops, which help plant breeders in the selection of tolerant genotypes. In the field condition, the main advantage is that it allows natural selection for the traits. In the case of the wheat crop, the northern Gangetic plain, particularly in West Bengal, where the winter season is short and the heat stress prevailing at early stage coincides with the reproductive phase, is the best natural environment for heat tolerance assessment of wheat crops.

6.3.1.2 CONTROLLED ENVIRONMENT OR LAB CONDITION

In this screening technique, artificially, heat stress or optimum growth temperature, that is, the controlled condition, is given to the crop for selecting the tolerant lines. Under controlled conditions, the natural selections do not operate. The development of controlled environment infrastructure is very expensive. These factors limit the application of these techniques.

Heat tolerance can be evaluated by several methodologies like visual screening, measuring the thermostability of membrane, chlorophyll content, photosynthetic loss, and canopy temperature depression (CTD).

6.3.1.2.1 Visual Screening Methods

Screening is done on the basis of agronomical traits in different crops. In wheat (days to heading, GFD (grain filling duration)), maize (pollen viability, leaf firing, tassel sterility, tassel blasting, silk receptivity), rice (grain sterility) are the key agronomical traits for the selection of tolerant genotype under heat stress.

6.3.1.2.2 Thermo stability of Membrane

This is a destructive method of screening. Under normal conditions, there is the maintenance of equilibrium between the intercellular and intracellular environment. The heat stress affected this equilibrium and caused cell membrane dysfunction. Therefore, leakage of ions increases, and these leaked electrolytes can be measured by a conductivity meter from the heat stress plant leaf. According to Abrahim et al. (2001), the formula of CMTS measurement is as follows: CMTS = $(1 - T1/T2) \times 100$, where $T1$ = conductivity value after heat treatment and $T2$ = conductivity value after autoclaving. This method has been utilized in different crops, such as wheat, soybean, pearl millet, maize, and chickpea to screen heat-tolerant genotypes.

6.3.1.2.3 Chlorophyll Content and Chlorophyll Fluorescence

Chlorophyll is the key component of photosynthesis. The heat stress affected the chlorophyll synthesis mechanism, resulting in low chlorophyll content, and hence there is a reduction in the photosynthetic efficiency, leading to a lower crop yield. The measurement of chlorophyll content can be done by destructive as well as non-destructive methods (Soil Plant Analysis Development (*SPAD*)). The high temperature damages the photosynthetic tissues, and this damage can be measured by chlorophyll fluorescence. This method is inexpensive and can be used to screen large genotypes quickly. It is in the practice of several crops, such as wheat, soybean, chickpea, and groundnut.

6.3.1.2.4 Canopy Temperature Depression (CTD)

CTD is measured by a hand-held infrared thermometer. The healthy microclimate of a plant is essential for growth and development. Under high temperatures, the rate of transpiration increases to maintain the microclimate of the plant by nullifying the effect of high temperature. CTD measures the depression of canopy temperature as compared with the air temperature. The higher CTD value indicates that the plants have a greater adaptive capacity and higher leaf water status under high temperatures. In wheat crops, it is widely utilized for screening heat-tolerant genotypes.

6.3.2 BREEDING APPROACHES FOR HTST

6.3.2.1 CONVENTIONAL BREEDING APPROACH

Keeping an eye on climate change, several efforts have been made to develop heat-tolerant varieties of the different crops in the different corners of the world. In the conventional breeding approach, selection and hybridization are key tools to develop high-yielding varieties with heat tolerance. The success of any crop improvement for heat tolerance depends on the variability available in the cultivated and wild crop species. Several heat stress adaptive varieties have been developed in many crops. In rice, wild species like *Oryza meridionalis* and the CG14 genotype of *Oryza glaberrima* (Jagadish et al., 2008) exhibit heat tolerance and can be used as a donor for the trait. Several heat-tolerant varieties, that is, N22 and NH219 (Poli et al., 2013), Nipponbare and Akitakomachi (Matsui and Omasa, 2002), have been developed. In wheat, PBW550 is a heat-tolerant variety developed by the backcross breeding method, where *A. tauchii* is used as a donor parent for transferring heat-tolerant traits such as cell membrane stability and chlorophyll content (Sehgal, 2011). New varieties of wheat such as WH1021 and WH730 have been developed, which exhibit high yield under heat stress (Dhanda and Munjal, 2012).

6.3.2.2 NEW BREEDING TECHNIQUES (NBT) FOR HEAT TOLERANCE

In the present era, conventional breeding program is assisted by biotechnological tools and techniques collectively called new breeding techniques (NBT) for more precise and rapid breeding. The NBT includes marker-assisted selection (MAS), genome-wide selection (GWAS), genome editing (CRISPR/CAS), transgenics, and other techniques assisted by modern biotechnology. MAS, including MAS-backcrossing with foreground selection and background selection is mainly used to identify the gene/QTLs of interest or traits and track their inheritance and leads to enhance the heat tolerance breeding efficiency. Several genes and QTLs have been identified to regulate the different traits at high-temperature stress in different crops. In rice genotype HT54, a dominant candidate gene locus *OsHTAS* regulating heat tolerance at the seedling and grain filling stage was identified (Wei et al., 2013). With the help of marker-assisted backcross breeding (MABB), the major QTL identified in rice cultivar "Kokoromachi" regulating better grain quality under high temperature had been transferred to Tohoku 168 (Shirasawa et al., 2013). The major QTL "*thermo-tolerance 1* (*TT1*)" were identified in African rice (*Oryza glaberrima* Steud.), regulating the thermotolerance by the nullification of the toxic effect of denatured protein due to high-temperature stress.

The genomic selection (GS) or genome-wide association studies (GWAS) is a potential and promising NBT, which is used for genetic evaluation by narrowing down the candidate genomic region regulating the traits in the natural and wild species population (Verdeprado et al., 2018). Genome editing is nowadays burning NBT in which the most commonly used technique is CRISPR-Cas9, which modifies the targeted genome of different crops. CRISPR-Cas9 has been exploited in several crops, such as wheat, rice, maize, sorghum, and soybean. In rice, genome editing has been utilized for targeting the gene *OsMPK2*, *OsDEP1* (grain yield under stress conditions) (Shan et al., 2014), *GS3*, *Gn1a* (grain size and number increase under stress conditions) (Shen et al., 2017) for heat stress tolerance. The transgenic approach is applicable when the desired variability for heat stress is not available within the cultivated varieties and wild relatives. In this situation, the gene of interest is transferred to the desired cultivar. Under high-temperature stress, the gene OsDREB2B in rice (Matsukura et al., 2010) and ZmDREB2A in maize (Qin et al., 2007) regulate and induce the heat-responsive gene to overcome the stress. In tobacco, high germination has been observed due to the expression of the gene CAP2 under high temperatures (Shukla et al., 2009). HSP and AtHSP101 in cotton regulate pollen sterility and germination under high-temperature stress.

6.4 APPLICATION OF OMICS IN HEAT TOLERANCE BREEDING

Conventional breeding after the green revolution is mainly biased on achieving yield stability independent of environmental stress, which resulted in the accumulation of germplasm with increasing susceptibility toward changing climatic conditions, and therefore, loss of information underlying the mechanism of tolerance toward environmental stress. However, understanding the molecular and biochemical basis of tolerance has become crucial for the present day, and progress in omics technologies will tremendously increase our understanding of it. Comparative transcriptome analysis of crops, such as rice (Sharma et al., 2021), wheat (Rangan et al., 2020), maize (Qian et al., 2019), soybean (Wang et al., 2018), and tomato (Balyan et al., 2020) has highlighted the role of HSP and Hsfs in heat tolerance by upregulation of these genes under heat stress condition. Differentially expressed genes due to heat stress mainly belong to pathways, such as secondary metabolite production, signal transduction, sugar and lipid signaling, phytohormones, response to calcium ions, and protein phosphorylation (Janni et al., 2020). Genes related to ROS production are also found in abundance in heat stress-imposed tissues. ROS production is associated with basal heat tolerance (Bhattacharjee, 2012) and also has a role in intra and intercellular signal transduction (Petrov Van Breusegem, 2012). As global transcriptome analysis highlights the key role of *HSP* and *Hsfs* in heat stress response, the structure of these gene family has been investigated in many crops (Mittal et al., 2009; Lin et al., 2011; Liu et al., 2012; Scharf et al., 2012; Chung et al., 2013; Jin et al., 2013; Xue et al., 2014; Yang et al., 2016; Lavania et al., 2018) Hsfs gene family are structurally and functionally highly conserved in eukaryotes and are the chief regulator of HSP function. Expanding gene annotation data has led to the development of a database "HEATSTER" (Berz et al., 2019) for identifying and annotating Hsfs genes in plants. Transcriptome profile does not provide a true picture of the cellular state during stress due to post-transcriptional and translational regulation of RNA and proteins, respectively. Hence, the proteomic profile may display a slightly different picture of the cellular state than that of the transcriptomic profile. Techniques, such as 2D-gel electrophoresis, SDS–polyacrylamide gel electrophoresis, MALDI-TOF, mass spectrometry (MS), chromatography techniques coupled with MS (GC-MS, GC-MS-MS, LC-MS, LC-MS-MS), and iTRAQ coupled with MS are used in proteomic profiling. In proteomic analysis too, HSPs are found to be the most prominent proteins in the stressed cells. Although studies from different plant parts and different stages of growth showed that the GO category of proteins differs widely, but a major category still belongs to chaperones (Janni et al., 2020). LEA protein, oxidative stress-related proteins, TCA cycle protein sugar and energy metabolism, and regulatory proteins are other major proteins expressed in heat-stressed cells. In wheat, HSP17, calcium-dependent protein kinase (CDPK), oxygen-evolving extrinsic protein, catalase superoxide (SDO), Rubisco and Rubisco activase (RCA) were found as abundant proteins during heat stress (Kumar et al., 2019).

6.5 ENGINEERING THERMOTOLERANCE THROUGH HEAT SENSOR AND SIGNAL TRANSDUCTION PATHWAY

Temperature changes in the cells were perceived by the various sensors which are followed by a cascade of signal transduction to the nucleus to reprogram the transcriptional network. Temperature is a physical parameter having a simple thermodynamic effect on various molecules and supramolecules, which is rapid in nature. In general, each molecule in the cell can sense temperature, but only a few can induce a signaling cascade downstream to it. These molecules are considered temperature sensors, and hence temperature perception or sensing is the most upstream event of heat stress response (HSR) (Ruelland and Zachowski, 2010). Membranes are the most important temperature sensors due to their lipid bilayer structure, fluidity of the lipid layer varies with temperature, and it also depends upon the ratio of saturated and unsaturated fatty acids. At high temperatures, the fluidity of the membrane increases, which as a result, initiates the MAPK cascade downstream. Conformational changes or unfolding in protein due to heat stress is another signal sensor. Studies showed that replacing proline with its analogue l-azetidine-2-carboxylic acid (AZC) (causes unfolding of protein) activates various heat stress-related genes (Sugio et al., 2009). Transcription factor, AtHsfA1d, a master regulator of HSR (Discussed in Section 6) remains in an inactive state due to its binding with HSP90 protein in normal condition, but upon heat exposure, HSp90 gets denatured, releasing AtHsfA1d in an inactive state, and hence the activation of downstream heat response genes (Yamada et al., 2007). In addition to the effect of heat stress on individual protein, multimeric peptides like cytoskeleton can get disassembled due to heat stress and act as a sensor for signal transduction. Microtubule disruption and depolymerized actin microfilaments have been observed in tobacco cells and Arabidopsis roots at a temperature above 40°C (Smertenko et al., 1997; Müller et al., 2007; Malerba et al., 2010). The use of cytoskeleton stabilizer and destabilizer in the various experiments (Ruelland and Zachowski, 2010) has proved its upstream role in temperature stress as HAMK activation at 37°C is prevented by the cytoskeleton stabilizer jasplakinolide or taxol. However, destabilizing the microfilaments or microtubules activates the HAMK at 25°C (Sangwan et al., 2002). The activity of an enzyme governs various metabolic cycles, cascade, and redox chains. A smooth interplay between all the enzymes in these cascades and cycles is necessary. However, enzyme activity is temperature-dependent because temperature influences its catalytic activity and inactivation of the enzyme. The temperature dependence of an enzyme is defined by its Q_{10} value (rate of increase in enzyme activity per 10° rise in temperature). The efficiency of any metabolic pathway or cycle depends upon its slowest enzymes (rate-limiting enzyme), and due to temperature change, the efficiency of the rate-limiting enzyme can get affected, or its function may be replaced by another enzyme resulting in the change of intermediate metabolite pool (Korn et al., 2010). The cell senses these changes in metabolite status as stress, resulting in the initiation of signal transduction. As a signaling response to heat stress, the Ca^{2+} ion sharply increases in the cytosol of the cell, which in turn transduces the signal to MAPK/HAPK. MAPK transduction pathway is important in plant and works ubiquitously for many external stimuli. Sangwan and Dhindsa (2002) have identified heat stress-induced MAPK which is triggered by a change in membrane fluidity and cytoskeleton remodeling. They also showed the relation between HSP activation and Ca^{2+} ion and the action of Ca-dependent protein kinases (CDPK). ROS induced by heat stress is also an important signaling molecule which also transmits a signal through MAPK. Some hormones and molecules, such as salicylic acid, ABA, calcium chloride, H_2O_2, and ACC induced due to heat stress may reduce oxidative damage and provide heat tolerance (Larkindale and Huang, 2004). H_2O_2 and NO are also emerging as essential players in signaling during stressful conditions in plants (Dat et al., 2000).

TABLE 6.1 Heat Sensor and Mode of Heat Perception in Crop Plants.

Sensor	Mode of perception	Crop	References
Cell membrane	Membrane fluidization	-	Suri and Dhidsa (2008); Königshofer et al. (2008); Saidi et al. (2009)
HSP90.2	Protein conformational change	*Arabidopsis*	Yamada et al. (2007)
HAMK	Protein conformational change	Alfa alfa	Sangwan and Dhindsa (2002)
Thioredoxin and a Thioredoxin-like protein	Protein oligomerization	*Arabidopsis*	Lee et al. (2009); Park et al. (2009)
FAD8	Protein conformational change	*Arabidopsis*	Matsuda et al. (2005)
Cytoskeleton disassembly	Disruption of microtubules	*Arabidopsis*	Smertenko et al. (1997);
		Tobacco	Müller et al. (2007);
	Actin microfilament depolymerization		Malerba et al. (2010)

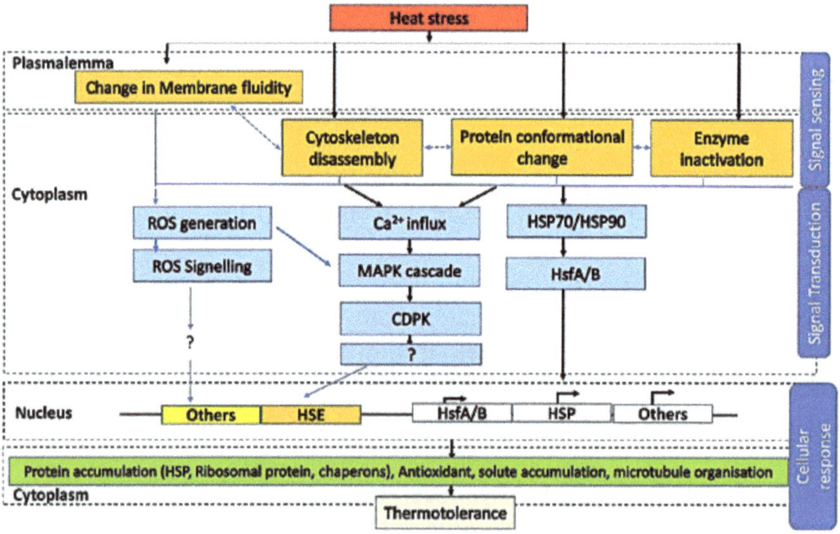

FIGURE 6.1 Mechanism of heat stress sensing and signal transduction.

The cell senses heat stress by a change in membrane fluidity, protein conformational changes, cytoskeleton disassembly, and enzyme inactivation. The ROS produced the above changes, act as a signaling molecule and relay the signal downstream either via MAPK cascade or some unknown pathway. Calcium flux due to the above changes also transduces the signal to the nucleus through the MAPK cascade. Conformational changes in proteins are stabilized by HSP70/HSP90, which also activates the master regulator HsfA, which binds to the HSE in the nucleus to activate the heat-responsive gene.

6.6 ENGINEERING THERMOTOLERANCE THROUGH HEAT SHOCK TRANSCRIPTION FACTOR AND HEAT SHOCK PROTEIN NETWORKS

6.6.1 HEAT STRESS TRANSCRIPTION FACTORS (HSFS)

During heat stress, plant tries to maintain its cellular activity by activating the HSR. The core of the HSR is heat shock transcription factors (Hsfs). Hsfs regulate transcription activation and repression of genes related to HSR, the most important of which is HSPs. The basic feature of HSR is well conserved across phylogeny, although plants being sessile in nature, have developed complex and unique cellular response system. Complexity in plant HSR can be seen by a huge number of Hsfs (21 gene family members in *Arabidopsis,* 24 Hsf-like genes in tomato, and 52 in soybean), their distinct role in different stress, and different regulatory mechanisms (Fragkostefanakis et al., 2015). Understanding the interaction between Hsfs and HSPs is important to developing feature heat-tolerant crops. In *Arabidopsis thaliana,* more than 240 genes are under transcriptional control of Hsfs under heat stress (Liu and Charng, 2013). Hsfs have been divided into three major classes (A, B, C) based on their phylogeny. Although, the basic structure is similar in all Hsfs, which comprise N-terminal DNA-binding domain followed by an oligomerization domain and C-terminal activator motif. HsfA1 is the master regulator of HSR. In tomato, studies have shown that HsfA1 is the single master regulator, and mutation in it decreases the thermotolerance of plants (Mishra et al., 2002). HsfA1 also activates HsfA2 (a major transcription factor in the heat-stressed cell), plays a role in its nuclear retention. In addition to HsfA2, it also activates and stabilizes HsfB1; therefore, together with HsfA2 and HsfB1, HsfA1 plays a principal role in HSR in tomato (Heerklotz et al., 2001). In *Arabidopsis thaliana,* four HsfA1 gene paralogs (a, b, c, d) have overlapping functions; therefore, a mutation in one gene can be compensated by other genes (Liu et al., 2011). Hsfs other than HdfA1 is not much essential for plant survival. However, the abundance of these enhances plants' tolerance toward heat stress. HsfA2 is the most abundant transcription factor in a heat-stressed cell. Studies revealed that HsfA2 becomes a major Hsf after repeated heat stress in the plant. Hetero-oligomerization of HsfA1 and HsfA2 forms a supramolecular complex responsible for activating several heat stress downstream genes more strongly than the individual transcription factors (Chan-Schaminet et al., 2009). Although HsfA1 is essential for nuclear retention of HsfA2, studies in *Arabidopsis* showed that overexpression of HsfA2 in quadruple knock out mutant of HsfA1 enhances thermotolerance (Liu and Charng, 2013). In addition to this, ectopic expression of rice HsfA2a and wheat HsfA2d in *Arabidopsis* increased its thermotolerance (Yokotani et al., 2008; Chauhan et al., 2013). In tomato, the concentration and activity of HsfA2 are modulated by changing its solubility, intracellular localization, and activator function by various proteins. For example, Hsp17 × 4-CII and small HSP function as a co-repressor of HsfA2 by forming a protein aggregate together with HsfA2, which under normal conditions, ultimately become solubilized with the help of class CI sHSPs or retained in the nucleus by hetero-oligomerization with HsfA1a. However, in heat-stressed conditions, accumulated HsfA2 and a network of heat-inducible proteins form a cytoplasmic multichaperone complex called heat stress granule (HSG), which is solubilized during the recovery phase by the action of ATP-dependent HSP70/101(Baniwal et al., 2004). On the other hand, HsfA4 is an antiapoptotic factor which is negatively regulated by a proapoptotic factor, HsfA5. HsfA4 controls ROS levels probably by regulating the ascorbate peroxidase (APX1) gene. Loss of functional mutant of *HsfA4* in *Arabidopsis* showed defective oxidative response and production and low APX1 expression (Baniwal et al., 2007). Similarly, in rice, a missense mutation in the DNA binding site of HsfA4d (*Spl7*) resulted in the formation of necrotic spots and hypersensitivity toward mild stresses (Davletova et al., 2005). The role of the representatives of the HsfA4/A5 group in rice may reside in the control of cell death connected with stress-induced damages. Unlike the HsfA family, HsfB does not have its own transcriptional regulation activity due to a lack of the characteristic contact motifs (AHA motifs) to interact with components of the basal transcription machinery. Therefore, HsfB works only with the cooperation of HsfA family genes by forming an enhanceosome-like complex to activate the heat stress-responsive genes. HsfB acts as an activator (synergistic coactivator of HsfA) and a repressor of heat stress genes. The repressor function of HsfB is necessary for suppressing heat stress genes once the plant reaches the heat attenuation phase. Loss of functional mutation in *hsfb1-hsfb2b* in *Arabidopsis* enhanced the HS genes' expression during normal conditions (Ikeda et al., 2011). Anhistone-like motif in the C-terminal of tomato HsfB1 is important for its coactivator function. The amino acid sequence for this C-terminal domain is GRGKMMK with indispensable Lysin residue at the center. This motif is needed for the recruitment of plant CREB-binding protein (CBP) ortholog HAC1. Together with HsfA1and HsfB1, HAC1/CBP forms a ternary complex.

6.6.2 HEAT SHOCK PROTEINS (HSPS)

Molecular chaperones are the proteins that assist protein folding and quality control by preventing irreversible protein aggregation both under normal and stress conditions. HSPs are molecular chaperones produced in abundance during stress conditions like high temperature and ensure proper folding as well as stabilization of proteins. In addition to a stress condition, HSPs also play an important role during ordinary cellular growth and development by ensuring proper folding, assembly, translocation, and degradation of proteins. Transcriptional reprogramming of gene expression during heat shock response (HSR) also includes induction of the HSP gene to maintain protein homeostasis. HSPs were first identified as proteins strongly induced by heat shock (Ritossa, 1962), but later, their role in other stresses, such as cold, biotic stress, wounding, UV light, heavy metal, and oxidative stress, have also been established. HSP gene family in plants are complex, but it can be subdivided into five conserved evolutionary groups based on their approximate molecular weight; HSP100, HSP90, HSP70, HSP60, and small heat shock proteins (sHSPs).

HSP100 (clpB) family comprises HSPs having a molecular weight of 100–104 kDa. These are ATP-dependent chaperones which play roles in re-solubilizing nonfunctional protein aggregate and protein degradation (Boston et al., 1996). Its conserved hexametric ring structure with a central pore is responsible for similar functional mechanisms in different organisms (Xu et al., 2011). HSP100 plays an important role in acclimatization in high temperatures; for example, a positive feedback loop between HSP100 and heat stress-associated 32-KD protein (HSA32) is known to prolong the effects of heat acclimation in rice seedlings (Lin et al., 2014). In addition to its role during stress conditions, it also has a housekeeping function in normal chloroplast development (Constan et al., 2004).

HSP90 is the most abundant HSPs in eukaryotic cells. It is also an ATP-dependent HSP with a molecular weight ranging from 82 to 90 kDa. Unlike HSP100, HSP90 is substrate-specific and binds only to relatively well-folded proteins mostly involved in transcriptional regulation and signal transduction pathways (Zhao et al., 2005). It works to coordinate with other cochaperones (e.g., HSP70, HSP40) by forming a complex structure that assists it in protein folding. It forms a dimer with its C-terminal dimerization domain. ATP binding and hydrolysis in the N-terminal domain change conformation of the substrate-binding domain and thereby act as a molecular clamp for substrate binding and release (Pearl and Prodromou, 2006). HSP90 is a key regulator of normal growth and development. Evidence has suggested that HSP90 interacts with R (resistance) protein to give rise to biotic stress tolerance during pathogen attack (Qian et al., 2018). One of the important roles of HSP90 during heat stress is the regulation of Hsf (heat shock factors). Cytoplasmic Hsp90 negatively inhibited hsf in the absence of heat stress, but under heat stress, this role is suspended temporarily to release hsf in its active form (Yamada et al., 2007).

HSP70 (DnaK) is a large and highly conserved protein with a molecular weight of 68–75 kDa. Eukaryotic cells have various homologs of HSP70 localized in the cytosol, mitochondria, chloroplast, and endoplasmic reticulum and exhibit constitutive as well as regulated expression. HSC70 (Het shock cognate70) is an HSP70 homolog,which has an housekeeping function in the cytosol by stabilizing a nascent protein released from the ribosome (Hartl, 1996). BiP is an abundant HSP70 protein present in the endoplasmic reticulum and has a role in guiding complex import or translocon (nascent proteins having N-glycan terminal are directed to ER), the ER quality control (selective retaining of misfolded and unassembled protein into ER), and directs the permanently misfolded protein to ERAD (the complex process through which the misfolded proteins are selected and ultimately degraded by the ubiquitin-proteasome system) (Gething, 1999). In eukaryotes, HSP70 has versatile roles, but a similar structure comprises the N-terminal ATPase domain of 45 kDa and the C-terminal substrate-binding domain of 25 kDa. ATP and ADP bound states have different conformations and hence different affinity toward the substrate. Changes in HSP70 conformation and HSP70 cycle are facilitated by HSP40, a J-domain cochaperone. With the help of HSP40, HSP70 goes through cycles of substrate binding and releasing in an ATP-dependent manner (Xu et al., 2011). As discussed, HSP90 plays a role in the structural maturation regulation of signal transduction molecules. However, it works downstream to HSP70. A tetratricopeptide repeat (TPR) containing protein Hop plays a role in recognizing and transferring client proteins from Hsp70 to Hsp90 by binding simultaneously to both chaperones (Scheufler et al., 2000). The importance of HSP70 for heat stress tolerance has been explored in many studies. *Arabidopsis* plants overexpressing AtHsc70-1 were more tolerant to heat shock (44°C for 10 min), but it also resulted in pleiotropic lethal effects (Sung and Guy, 2003). Similarly, nuclear localization of ntHSP70 in *Nicotiana tobaccum* confers thermotolerance to overexpressing transgenic tobacco plants (Cho and Choi, 2009). Some studies suggested that the use of prokaryotic HSPs70 has resulted in thermotolerance in transgenic plants; for example, overexpression of HSP70 from *Aphanothece halophytica* in tobacco (Ono et al., 2001) and rice (Uchida et al., 2008) resulted in increased HS tolerance. Furthermore, as ectopic expression of *Trichoderma harzianum* Hsp70 *A. thaliana* showed increased tolerance to osmotic, salt, and oxidative stresses (Montero-Barrientos et al., 2010).

The HSP60 family is called chaperonins 60 (cpn60), having a molecular weight from 58 to 65 kDa and being the first molecular chaperones identified. HSP60 and HSP70 share overlapping functions, although, they are distinct in mechanism and structure. In bacteria, GroEL is the most studied HSP60; it forms a huge homo-oligomer with a central hydrophobic cavity for accommodating unfolded protein of 10–60 kDa. A 10 kDa co-chaperonin GroES (HSP10 in eukaryotes) acts as a functional regulator of GroEL (Fink, 1999). In eukaryotes, GroEL-like homologs of HSP60 are found in mitochondria and chloroplast but not in the cytosol.

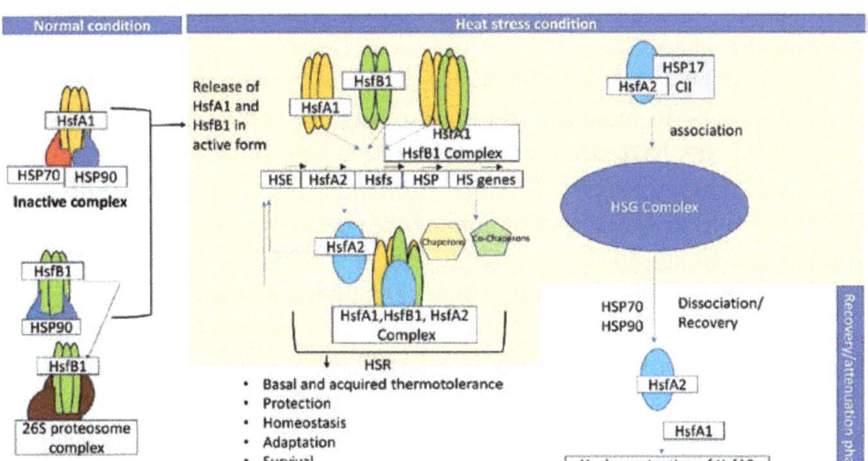

FIGURE 6.2 Mechanism of heat stress response (HSR).

In normal condition, constitutively expressing master regulator transcription factor, HsfA1 remains inactive by binding with HSP70 and HSP90. The concentration of another constitutively expressing transcription factor, HsfB1, also remains low due to its 26s proteasome-mediated degradation with the help of HSP90. During heat stress, HsfA/B is released in active form as a homodimer or heterodimer and activates other heat-responsive genes (HsfA2, HSPs, Chaperones, etc.) in the nucleus. During heat acclimatization, HsfA2 becomes the dominant TF which further activates the heat-responsive gene. A higher concentration of HsfA2 is regulated by HSP17 CII by forming an HSG complex. During attenuation, phage–HSG complex is dissociated with the help of HSP70/90, and free HsfA2 is retained in the nucleus with the help of HsfA1.

In mitochondria, these HSPs are induced due to heat shock and protect the existing proteins of the organelle (Martin et al., 1992). However, in the chloroplast (also called Cpn60 and Cpn10), its expression is constitutive and only escalates to some extent after exposure to the heat (Viitanen et al., 1995). Chloroplastic HSP60 is also known as "Rubisco large subunit binding protein" as it plays an essential role in the folding and assembly of Rubisco large subunit. Successful expression of Rubisco in the bacterial system was only achieved in the cells having a functional GroEL/GroES system (Hemmingsen et al., 1988). Although HSP70 and HSP 60 have some overlapping functions, the main difference between their mechanisms of folding is that Hsp60 can accommodate the entire domain or protein to their hydrophobic substrate-binding site. However, HSP70 can only recognize short peptide chains.

The small HSPs (sHSPs), most ubiquitous HSPs, have molecular weights of 12–342 kDa. In monocot and dicot plants, 10 conserved gene families encoding sHSPs have been recognized where four of them code cytosolic sHSPs and six families code for organelle sHSP. The unique feature of these HSP is a diverse N-terminal domain followed by a common a-crystallin domain containing 80–100 amino acid residues located in the C-terminal region. sHSPs have larger binding stoichiometry than other HSPs, and hence act as a reservoir of denatured protein due to stress, and later, they recruit these denatured proteins to HSP70 and HSP60 for proper folding. Another characteristic function of this class is the degradation of the proteins that have unsuitable folding (like sHsp, ubiquitin with its bounded enzymes) Land plants have distinct sHSP subfamilies. Most are highly upregulated in response to heat and other stress. Others are selectively expressed in seeds and pollen, and a few are constitutively expressed.

6.7 CONCLUSION

Rapid and accelerated temperature rise in recent years have imposed challenges in all forms of life on earth. Decline in crop production due to heat stress is threatening the food security worldwide. Heat stress affects every stage of plant growth and development from germination to seed setting by triggering phenological, morphological, physiological, biochemical, and molecular responses. Understanding the metabolic pathways, signaling mechanisms, QTLs, genes, molecular players, and interaction among them for conferring heat tolerance and integrations of this knowledge to develop climate-smart genotype. Several cellular and molecular machanisms have been deciphers in crop plants which enable them to thrive in stress conditions. Some examples are cellular protection mechanism by osmoregulation, ROS scavenging by enzymes and antioxidants. Similarly HSP protects proteins and enzymes from denaturation. Induction of MAPK and CDPK cascade after temperature sensing by different sensor again leads to the activation of heat-responsive genes in the nucleus to generate heat stress response. Genetic strategies can be formulated based on this knowledge by integrating traditional breeding, molecular breeding, transgenic technologies, and genome editing.

KEYWORDS

- **heat stress**
- **climate change**
- **high temperature**
- **thermotolerance**
- **new breeding techniques**

REFERENCES

Allakhverdiev, S. I.; Kreslavski, V. D.; Klimov, V. V.; Los, D. A.; Carpentier, R.; Mohanty, P. Heat Stress: An Overview of Molecular Responses in Photosynthesis. *Photosyn. Res.* **2008,** *98* (1), 541–550.

Allen, M. R.; Dube, O. P.; Solecki, W.; et al. Framing and Context. In *Global Warming of 1.5°C. An IPCC Special Report on the Impacts of Global Warming of 1.5°C Above*

Pre-industrial Levels and Related Global Greenhouse Gas Emission Pathways, in the Context of Strengthening the Global Response to the Threat of Climate Change, Sustainable Development, and Efforts to Eradicate Poverty; Masson-Delmotte, V. P., Zhai, H. O., Pörtner, D., et al.; In Press, 2018.

Ashraf, M.; Harris, P. Eds.; In *Abiotic Stresses: Plant Resistance Through Breeding and Molecular Approaches*; CRC Press, 2005.

Balyan, S.; Rao, S.; Jha, S.; Bansal, C.; Das, J. R.; Mathur, S. Characterization of Novel Regulators for Heat Stress Tolerance in Tomato from Indian Sub-continent. *Plant Biotechnol. J.* **2020,** *18* (10), 2118–2132.

Baniwal, S. K.; Bharti, K.; Chan, K. Y.; Fauth, M. Ganguli, A.;Kotak, S.; Mishra, S. M.; Nover, L.; Port, M.; Scharf, K. D.; Tripp, J.; Weber, C.; Zielinski, D.; von Koskull-Doring, P. Heat Stress Response in Plants: A Complex Game with Chaperones and More than Twenty Heat-Stress Transcription Factors. *J. Biosci.* **2004,** *29*, 471–487.

Baniwal, S. K.; Chan, K. Y.; Scharf, K. D.; Nover, L. Role of Heat Stress Transcription Factor HsfA5 as Specific Repressor of HsfA4. *J. Biol. Chem.* **2007,** *282* (6), 3605–3613.

Bañon, S.; Fernandez, J. A.; Franco, J. A.; Torrecillas, A.; Alarcón, J. J.; Sánchez-Blanco, M. J. Effects of Water Stress and Night Temperature Preconditioning on Water Relations and Morphological and Anatomical Changes of Lotus creticus Plants. *Sci. Hortic.* **2004,** *101* (3), 333–342.

Barnabás, B.; Jäger, K.; Fehér, A. The Effect of Drought and Heat Stress on Reproductive Processes in Cereals. *Plant Cell Environ.* **2008,** *31* (1), 11–38.

Bassirirad, H. Kinetics of Nutrient Uptake by Roots: Responses to Global Change. *New Phytol.* **2000,** *147* (1), 155–169.

Berz, J.; Simm, S.; Schuster, S.; Scharf, K. D.; Schleiff, E.; Ebersberger, I. HEATSTER: A Database and Web Server for Identification and Classification of Heat Stress Transcription Factors in Plants. *Bioinform. Biol. Insights* **2019,** *13*, 1177932218821365.

Bhattacharjee, S. An Inductive Pulse of Hydrogen Peroxide Pretreatment Restores Redox-Homeostasis and Oxidative Membrane Damage Under Extremes of Temperature in Two Rice Cultivars. *Plant Growth Regul.* **2012,** *68* (3), 395–410.

Bi, H.; Zhao, Y.; Li, H.; Liu, W. Wheat Heat Shock Factor TaHsfA6f Increases ABA Levels and Enhances Tolerance to Multiple Abiotic Stresses in Transgenic Plants. *Int. J. Mol. Sci.* **2020,** *21* (9), 3121.

Boston, R. S.; Viitanen, P. V.; Vierling, E. Molecular Chaperones and Protein Folding in Plants. *Plant Mol. Biol.* **1996,** *32* (1–2), 191–222.

Calleja-Cabrera, J.; Boter, M.; Oñate-Sánchez, L.; Pernas, M. Root Growth Adaptation to Climate Change in Crops. *Front. Plant Sci.* **2020,** *11*, 544.

ČERNÝ, M.; Jedelský, P. L.; Novák, J. A. N.; Schlosser, A.; Brzobohatý, B. Cytokinin Modulates Proteomic, Transcriptomic and Growth Responses to Temperature Shocks in A Rabidopsis. *Plant Cell Environ.* **2014,** *37* (7), 1641–1655.

Chan-Schaminet, K. Y.; Baniwal, S. K.; Bublak, D.; Nover, L.; Scharf, K. D. Specific Interaction Between Tomato HsfA1 and HsfA2 Creates Hetero-oligomeric Superactivator Complexes for Synergistic Activation of Heat Stress Gene Expression. *J. Biol. Chem.* **2009,** *284* (31), 20848–20857.

Chauhan, H.; Khurana, N.; Agarwal, P.; Khurana, J. P.; Khurana, P. A Seed Preferential Heat Shock Transcription Factor from Wheat Provides Abiotic Stress Tolerance and Yield Enhancement in Transgenic Arabidopsis Under Heat Stress Environment. *PLoS One* **2013,** *8* (11), e79577.

Cho, E. K.; Choi, Y. J. A Nuclear-Localized HSP70 Confers Thermoprotective Activity and Drought-Stress Tolerance on Plants. *Biotechnol. Lett.* **2009,** *31* (4), 597–606.

Choudhury, F. K.; Rivero, R. M.; Blumwald, E.; Mittler, R. Reactive Oxygen Species, Abiotic Stress and Stress Combination. *Plant J.* **2017,** *90* (5), 856–867.

Chung, E.; Kim, K. M.; Lee, J. H. Genome-Wide Analysis and Molecular Characterization of Heat Shock Transcription Factor Family in Glycine Max. *J. Genet. Genomics* **2013,** *40* (3), 127–135.

Constan, D.; Froehlich, J. E.; Rangarajan, S.; Keegstra, K. A Stromal Hsp100 Protein is Required for Normal Chloroplast Development and Function in Arabidopsis. *Plant Physiol.* **2004,** *136* (3), 3605–3615.

Dat, J. F.; Foyer, C. H.; Scott, I. M. Changes in Salicylic Acid and Antioxidants During Induced Thermotolerance in Mustard Seedlings. *Plant Physiol.* **1998,** *118* (4), 1455–1461.

Dat, J.; Vandenabeele, S.; Vranova, E. V. M. M.; Van Montagu, M.; Van Breusegem, F. Dual Action of the Active Oxygen Species During Plant Stress Responses. *Cell. Mol. Life Sci.* **2000,** *57* (5), 779–795.

Davletova, S.; Rizhsky, L.; Liang, H.; Shengqiang, Z.; Oliver, D. J.; Coutu, J.; ... & Mittler, R. Cytosolic Ascorbate Peroxidase 1 is a Central Component of the Reactive Oxygen Gene Network of Arabidopsis. *Plant Cell* **2005,** *17* (1), 268–281.

Dhanda, S. S.; Munjal, R. Heat Tolerance in Relation to Acquired Thermotolerance for Membrane Lipids in Bread Wheat. *Field Crops Res.* **2012,** *135*, 30–37.

Fahad, S.; Bajwa, A. A.; Nazir, U.; Anjum, S. A.; Farooq, A.; Zohaib, A.; ... Huang, J. Crop Production Under Drought and Heat Stress: Plant Responses and Management Options. *Front. Plant Sci.* **2017,** *1147*.

Fahad, S.; Hussain, S.; Saud, S.; Khan, F.; Hassan, S.; Nasim, W.; ... & Huang, J. Exogenously Applied Plant Growth Regulators Affect Heat-Stressed Rice Pollens. *J. Agron. Crop Sci.* **2016,** *202* (2), 139–150.

Feraru, E.; Feraru, M. I.; Barbez, E.; Waidmann, S.; Sun, L.; Gaidora, A.; Kleine-Vehn, J. In PILS6 is a Temperature-Sensitive Regulator of Nuclear Auxin Input and Organ Growth in Arabidopsis thaliana. *Proc. Natl. Acad. Sci.* **2019,** *116* (9), 3893–3898.

Fink, A. L. Chaperone-Mediated Protein Folding. *Physiol. Rev.* **1999,** *79* (2), 425–449.

Fragkostefanakis, S.; Roeth, S.; Schleiff, E., Scharf, K. D. Prospects of Engineering Thermotolerance in Crops Through Modulation of Heat Stress Transcription Factor and Heat Shock Protein Networks. *Plant Cell Environ.* **2015,** *38* (9), 1881–1895.

Gavelienė, V.; Jurkonienė, S.; Jankovska-Bortkevič, E.; Švegždienė, D. Effects of Elevated Temperature on Root System Development of Two Lupine Species. *Plants* **2022,** *11* (2), 192.

Gilroy, S.; Białasek, M.; Suzuki, N.; Górecka, M.; Devireddy, A. R.; Karpiński, S.; Mittler, R. ROS, Calcium, and Electric Signals: Key Mediators of Rapid Systemic Signaling in Plants. *Plant Physiol.* **2016,** *171* (3), 1606–1615.

Gilroy, S.; Suzuki, N.; Miller, G.; Choi, W. G.; Toyota, M.; Devireddy, A. R.; Mittler, R. A Tidal Wave of Signals: Calcium and ROS at the Forefront of Rapid Systemic Signaling. *Trends Plant Sci.* **2014,** *19* (10), 623–630.

Halliwell, B.; Gutteridge, J. M. In *Free Radicals in Biology and Medicine*; Oxford University Press, USA, 2015.

Hartl, F. U. Molecular Chaperones in Cellular Protein Folding. *Nature* **1996,** *381* (6583), 571–580.

Gething MJ. Role and Regulation of the ER Chaperone BiP. *Semin. Cell Dev. Biol.* **1999,** *10* (5), 465–472. DOI: 10.1006/scdb.1999.0318. PMID: 10597629.

Hasanuzzaman, M.; Nahar, K.; Alam, M.; Roychowdhury, R.; Fujita, M. Physiological, Biochemical, and Molecular Mechanisms of Heat Stress Tolerance in Plants. *Int. J. Mol. Sci.* **2013,** *14* (5), 9643–9684.

Heerklotz, D.; Döring, P.; Bonzelius, F.; Winkelhaus, S.; Nover, L. The Balance of Nuclear Import and Export Determines the Intracellular Distribution and Function of Tomato Heat Stress Transcription Factor HsfA2. *Mol. Cell. Biol.* **2001,** *21* (5), 1759–1768.

Hemmingsen, S. M.; Woolford, C.; van der Vies, S. M.; Tilly, K.; Dennis, D. T.; Georgopoulos, C. P.; ... Ellis, R. J. Homologous Plant and Bacterial Proteins Chaperone Oligomeric Protein Assembly. *Nature* **1988,** *333* (6171), 330–334.

Hurkman, W. J.; McCue, K. F.; Altenbach, S. B.; Korn, A.; Tanaka, C. K.; Kothari, K. M.; ... & DuPont, F. M. Effect of Temperature on Expression of Genes Encoding Enzymes for Starch Biosynthesis in Developing Wheat Endosperm. *Plant Sci.* **2003,** *164* (5), 873–881.

Ibrahim, A. M.; Quick, J. S. Genetic Control of High Temperature Tolerance in Wheat as Measured by Membrane Thermal Stability. *Crop Sci.* **2001,** *41* (5), 1405–1407.

Ikeda, M.; Mitsuda, N.; Ohme-Takagi, M. Arabidopsis HsfB1 and HsfB2b Act as Repressors of the Expression of Heat-Inducible Hsfs but Positively Regulate the Acquired Thermotolerance. *Plant Physiol.* **2011,** *157* (3), 1243–1254.

Jagadish, S. V. K.; Craufurd, P. Q.; Wheeler, T. R. Phenotyping Parents of Mapping Populations of Rice for Heat Tolerance During Anthesis. *Crop Sci.* **2008,** *48* (3), 1140–1146.

Janni, M.; Gullì, M.; Maestri, E.; Marmiroli, M.; Valliyodan, B.; Nguyen, H. T.; Marmiroli, N. Molecular and Genetic Bases of Heat Stress Responses in Crop Plants and Breeding for Increased Resilience and Productivity. *J. Exp. Bot.* **2020,** *71* (13), 3780–3802.

Jin, G. H.; Gho, H. J.; Jung, K. H. A Systematic View of Rice Heat Shock Transcription Factor Family Using Phylogenomic Analysis. *J. Plant Physiol.* **2013,** *170* (3), 321–329.

Karthik, L.; Kumar, G.; Keswani, T.; Bhattacharyya, A.; Chandar, S. S.; Bhaskara Rao, K. V. Protease Inhibitors from Marine Actinobacteria as a Potential Source for Antimalarial Compound. *PloS One* **2014,** *9* (3), e90972.

Kaya, H.; Takeda, S.; Kobayashi, M. J.; Kimura, S.; Iizuka, A.; Imai, A.; et al. Comparative Analysis of the Reactive Oxygen Species-Producing Enzymatic Activity of Arabidopsis NADPH Oxidases. *Plant J.* **2019,** *98*, 291–300. DOI: 10.1111/tpj.14212

Kinnersley, A. M.; Turano, F. J. Gamma Aminobutyric Acid (GABA) and Plant Responses to Stress. *Crit. Rev. Plant Sci.* **2000,** *19* (6), 479–509.

Kleine, T.; Leister, D. Retrograde Signaling: Organelles go Networking. *Biochimica Et Biophysica Acta (BBA)-Bioenergetics* **2016,** *1857* (8), 1313–1325.

Königshofer, H.; Tromballa, H. W.; Löppert, H. G. Early Events in Signalling High-Temperature Stress in Tobacco BY2 Cells Involve Alterations in Membrane Fluidity and Enhanced Hydrogen Peroxide Production. *Plant Cell Environ.* **2008,** *31* (12), 1771–1780.

Korn, M.; Gärtner, T.; Erban, A.; Kopka, J.; Selbig, J.; Hincha, D. K. Predicting Arabidopsis Freezing Tolerance and Heterosis in Freezing Tolerance from Metabolite Composition. *Mol. Plant* **2010,** *3* (1), 224–235.

Kumar, R. R.; Singh, K.; Ahuja, S.; Tasleem, M.; Singh, I.; Kumar, S.; ... & Praveen, S. Quantitative Proteomic Analysis Reveals Novel Stress-Associated Active Proteins (SAAPs) and Pathways Involved in Modulating Tolerance of Wheat Under Terminal Heat. *Funct. Integr. Genomic.* **2019,** *19* (2), 329–348.

Küpers, J. J.; Oskam, L.; Pierik, R. Photoreceptors Regulate Plant Developmental Plasticity Through Auxin. *Plants*, **2020,** *9* (8), 940.

Larkindale, J.; Huang, B. Thermotolerance and Antioxidant Systems in Agrostis Stolonifera: Involvement of Salicylic Acid, Abscisic Acid, Calcium, Hydrogen Peroxide, and Ethylene. *J. Plant Physiol.* **2004,** *161* (4), 405–413.

Larkindale, J.; Knight, M. R. Protection Against Heat Stress-Induced Oxidative Damage in Arabidopsis Involves Calcium, Abscisic Acid, Ethylene, and Salicylic Acid. *Plant Physiol.* **2002,** *128* (2), 682–695.

Lavania, D.; Dhingra, A.; Grover, A. Analysis of Transactivation Potential of Rice (Oryza sativa L.) Heat Shock Factors. *Planta* **2018,** *247* (6), 1267–1276.

Lee, J. R.; Lee, S. S.; Jang, H. H.; Lee, Y. M.; Park, J. H.; Park, S. C.; ... & Lee, S. Y. Heat-Shock Dependent Oligomeric Status Alters the Function of a Plant-Specific Thioredoxin-like Protein, AtTDX. *Proc. Natl. Acad. Sci.* **2009,** *106* (14), 5978–5983.

Lin, M. Y.; Chai, K. H.; Ko, S. S.; Kuang, L. Y.; Lur, H. S.; Charng, Y. Y. A Positive Feedback Loop Between HEAT SHOCK PROTEIN101 and HEAT STRESS-ASSOCIATED 32-KD PROTEIN Modulates Long-Term Acquired Thermotolerance Illustrating Diverse Heat Stress Responses in Rice Varieties. *Plant Physiol.* **2014,** *164* (4), 2045–2053.

Lin, Y. X.; Jiang, H. Y.; Chu, Z. X.; Tang, X. L.; Zhu, S. W.; Cheng, B. J. Genome-Wide Identification, Classification and Analysis of Heat Shock Transcription Factor Family in Maize. *BMC Genomic.* (**2011,** *12* (1), 1–14.

Liu, H. C.; Charng, Y. Y. Common and Distinct Functions of Arabidopsis Class A1 and A2 Heat Shock Factors in Diverse Abiotic Stress Responses and Development. *Plant Physiol.* **2013,** *163* (1), 276–290.

Liu, H. C.; Liao, H. T.; Charng, Y. Y. The Role of Class A1 Heat Shock Factors (HSFA1s) in Response to Heat and Other Stresses in Arabidopsis. *Plant Cell Environ.* **2011,** *34* (5), 738–751.

Liu, G. T.; Wang, J. F.; Cramer, G.; Dai, Z. W.; Duan, W.; Xu, H. G.; ... Li, S. H. Transcriptomic Analysis of Grape (Vitis vinifera L.) Leaves During and After Recovery from Heat Stress. *BMC Plant Biol.* **2012,** *12* (1), 1–10.

Llusià, J.; Penuelas, J.; Munné-Bosch, S. Sustained Accumulation of Methyl Salicylate Alters Antioxidant Protection and Reduces Tolerance of Holm Oak to Heat Stress. *Physiol. Plant.* **2005,** *124* (3), 353–361.

Macková, H.; Hronková, M.; Dobrá, J.; Turečková, V.; Novák, O.; Lubovská, Z.; ... Vanková, R. Enhanced Drought and Heat Stress Tolerance of Tobacco Plants with Ectopically Enhanced Cytokinin Oxidase/Dehydrogenase Gene Expression. *J. Exp. Bot.* **2013,** *64* (10), 2805–2815.

Maestri, E.; Klueva, N.; Perrotta, C.; Gulli, M.; Nguyen, H. T.; Marmiroli, N. Molecular Genetics of Heat Tolerance and Heat Shock Proteins in Cereals. *Plant Mol. Biol.* **2002,** *48* (5), 667–681.

Malerba, M.; Crosti, P.; Cerana, R. Effect of Heat Stress on Actin Cytoskeleton and Endoplasmic Reticulum of Tobacco BY-2 Cultured Cells and its Inhibition by Co2+. *Protoplasma* **2010,** *239* (1), 23–30.

Martin, J.; Horwich, A. L.; Hartl, F. U. Prevention of Protein Denaturation under Heat Stress by the Chaperonin Hsp60. *Science* **1992,** *258* (5084), 995–998.

Matsuda, O.; Sakamoto, H.; Hashimoto, T.; Iba, K. A Temperature-Sensitive Mechanism that Regulates Post-Translational Stability of a Plastidial ω-3 Fatty Acid Desaturase (FAD8) in Arabidopsis Leaf Tissues. *J. Biol. Chem.* **2005,** *280* (5), 3597–3604.

Matsui, T.; Omasa, K. Rice (Oryza sativa L.) Cultivars Tolerant to High Temperature at Flowering: Anther Characteristics. *Ann. Bot.* **2002,** *89* (6), 683–687.

Matsukura, S.; Mizoi, J.; Yoshida, T.; Todaka, D.; Ito, Y.; Maruyama, K.; ... Yamaguchi-Shinozaki, K. Comprehensive Analysis of Rice DREB2-Type Genes that Encode Transcription Factors Involved in the Expression of Abiotic Stress-Responsive Genes. *Mol. Genet. Genomic.* **2010,** *283* (2), 185–196.

Mishra, S. K.; Tripp, J.; Winkelhaus, S.; Tschiersch, B.; Theres, K.; Nover, L.; Scharf, K. D. In the Complex Family of Heat Stress Transcription Factors, HsfA1 has a Unique Role as Master Regulator of Thermotolerance in Tomato. *GenesDev.* **2002,** *16* (12), 1555–1567.

Mittal, D.; Chakrabarti, S.; Sarkar, A.; Singh, A.; Grover, A. Heat Shock Factor Gene Family in Rice: Genomic Organization and Transcript Expression Profiling in Response to High Temperature, Low Temperature and Oxidative Stresses. *Plant Physiol. Biochem.* **2009,** *47* (9), 785–795.

Montero-Barrientos, M.; Hermosa, R.; Cardoza, R. E.; Gutierrez, S.; Nicolas, C.; Monte, E. Transgenic Expression of the Trichoderma Harzianum hsp70 Gene Increases Arabidopsis Resistance to Heat and Other Abiotic Stresses. *J. Plant Physiol.* **2010,** *167* (8), 659–665.

Morales, D.; Rodríguez, P.; Dell'Amico, J.; Nicolas, E.; Torrecillas, A.; Sánchez-Blanco, M. J. High-Temperature Preconditioning and Thermal Shock Imposition Affects Water Relations, Gas Exchange and Root Hydraulic Conductivity in Tomato. *Biol. Plant.* **2003,** *47* (2), 203–208.

Müller, J.; Menzel, D.; Šamaj, J. Cell-Type-Specific Disruption and Recovery of the Cytoskeleton in Arabidopsis Thaliana Epidermal Root Cells Upon Heat Shock Stress. *Protoplasma* **2007,** *230* (3), 231–242.

Ono, K.; Hibino, T.; Kohinata, T.; Suzuki, S.; Tanaka, Y.; Nakamura, T.; Takabe, T. Overexpression of DnaK from a Halotolerant Cyanobacterium Aphanothece halophytica Enhances the High-Temperature Tolerance of Tobacco During Germination and Early Growth. *Plant Sci.* **2001,** *160* (3), 455–461.

Oosterhuis, D. M. Day or Night High Temperatures: A Major Cause of Yield Variability. *Cotton Grow.* **2002,** *46* (9), 8–9.

Osmond, C. B.; Austin, M. P.; Berry, J. A.; Billings, W. D.; Boyer, J. S.; Dacey, J. W. H.; ... Winner, W. E. Stress Physiology and the Distribution of Plants. *BioScience* **1987,** *37* (1), 38–48.

Park, S. K.; Jung, Y. J.; Lee, J. R.; Lee, Y. M.; Jang, H. H.; Lee, S. S.; ... Lee, S. Y. Heat-Shock and Redox-Dependent Functional Switching of an h-type Arabidopsis Thioredoxin from a Disulfide Reductase to a Molecular Chaperone. *Plant Physiol.* **2009,** *150* (2), 552–561.

Pearl, L. H.; Prodromou, C. Structure and Mechanism of the Hsp90 Molecular Chaperone Machinery. *Ann. Rev. Biochem.* **2006,** *75*, 271–294.

Perdomo, J. A.; Capó-Bauçà, S.; Carmo-Silva, E.; Galmés, J. Rubisco and Rubisco Activase Play an Important Role in the Biochemical Limitations of Photosynthesis in Rice, Wheat, and Maize Under High Temperature and Water Deficit. *Front. Plant Sci.* **2017,** *8*, 490.

Perrotta, C.; Treglia, A. S.; Mita, G.; Giangrande, E.; Rampino, P.; Ronga, G.; ... Marmiroli, N. Analysis of mRNAs from Ripening Wheat Seeds: The Effect of High Temperature. *J. Cereal Sci.* **1998,** *27* (2), 127–132.

Petrov, V. D.; Van Breusegem, F. Hydrogen Peroxide—A Central Hub for Information Flow in Plant Cells. *AoB Plants* **2012,** *2012.*

Poli, Y.; Basava, R. K.; Panigrahy, M.; Vinukonda, V. P.; Dokula, N. R.; Voleti, S. R.; ... Neelamraju, S. Characterization of a Nagina22 Rice Mutant for Heat Tolerance and Mapping of Yield Traits. *Rice* **2013,** *6* (1), 1–9.

Prasad, P. V.; Boote, K. J.; Allen Jr, L. H. Adverse High Temperature Effects on Pollen Viability, Seed-Set, Seed Yield and Harvest Index of Grain-sorghum [Sorghum bicolor (L.) Moench] are More Severe at Elevated Carbon Dioxide due to Higher Tissue Temperatures. *Agric. For. Meteorol.* **2006,** *139* (3-4), 237–251.

Prasad, P. V.; Djanaguiraman, M. Response of Floret Fertility and Individual Grain Weight of Wheat to High Temperature Stress: Sensitive Stages and Thresholds for Temperature and Duration. *Funct. Plant Biol.* **2014,** *41* (12), 1261–1269.

Qian, Y.; Ren, Q.; Zhang, J.; Chen, L. Transcriptomic Analysis of the Maize (Zea mays L.) Inbred Line B73 Response to Heat Stress at the Seedling Stage. *Gene* **2019,** *692*, 68–78.

Qian, L.; Zhao, J.; Du, Y.; Zhao, X.; Han, M.; Liu, Y. Hsp90 Interacts With Tm-22 and Is Essential for Tm-22-Mediated Resistance to Tobacco Mosaic Virus. *Front. Plant Sci.* **2018,** *9*, 411.

Qin, F.; Kakimoto, M.; Sakuma, Y.; Maruyama, K.; Osakabe, Y.; Tran, L. S. P.; ... Yamaguchi-Shinozaki, K. Regulation and Functional Analysis of ZmDREB2A in Response to Drought and Heat Stresses in Zea Mays L. *Plant J.* **2007,** *50* (1), 54–69.

Quan, R.; Shang, M.; Zhang, H.; Zhao, Y.; Zhang, J. Engineering of Enhanced Glycine Betaine Synthesis Improves Drought Tolerance in Maize. *Plant Biotechnol. J.* **2004,** *2* (6), 477–486.

Rangan, P.; Furtado, A.; Henry, R. Transcriptome Profiling of Wheat Genotypes Under Heat Stress During Grain-Filling. *J. Cereal Sci.* **2020,** *91*, 102895.

Rasmusson, A. G.; Wallström, S. V. Involvement of Mitochondria in the Control of Plant Cell NAD (P) H Reduction Levels. *Biochem. Soc. Trans.* **2010,** *38* (2), 661–666.

Rezaul, I. M.; Baohua, F.; Tingting, C.; Weimeng, F.; Caixia, Z.; Longxing, T.; Guanfu, F. Abscisic Acid Prevents Pollen Abortion Under High-Temperature Stress by Mediating Sugar Metabolism in Rice Spikelets. *Physiol. Plant.* **2019,** *165* (3), 644–663.

Ritossa, F. A New Puffing Pattern Induced by Temperature Shock and DNP in Drosophila. *Experientia* **1962,** *18* (12), 571–573.

Ruelland, E.; Zachowski, A. How Plants Sense Temperature. *Environ. Exp. Bot.* **2010,** *69* (3), 225–232.

Saidi, Y.; Finka, A.; Muriset, M.; Bromberg, Z.; Weiss, Y. G.; Maathuis, F. J.; Goloubinoff, P. The Heat Shock Response in Moss Plants is Regulated by Specific Calcium-Permeable Channels in the Plasma Membrane. *Plant Cell* **2009,** *21* (9), 2829–2843.

Sangwan, V.; Dhindsa, R. S. In Vivo and In Vitro Activation of Temperature-Responsive Plant Map Kinases. *FEBS Lett.* **2002,** *531* (3), 561–564.

Sangwan, V.; Örvar, B. L.; Beyerly, J.; Hirt, H.; Dhindsa, R. S. Opposite Changes in Membrane Fluidity Mimic Cold and Heat Stress Activation of Distinct Plant MAP Kinase Pathways. *Plant J.* **2002,** *31* (5), 629–638.

Scharf, K. D.; Berberich, T.; Ebersberger, I.; Nover, L. The Plant Heat Stress Transcription Factor (Hsf) Family: Structure, Function and Evolution. *Biochim. Biophys. Acta Gene Regul. Mech.* **2012,** *1819* (2), 104–119.

Scheufler, C.; Brinker, A.; Bourenkov, G.; Pegoraro, S.; Moroder, L.; Bartunik, H.; ... & Moarefi, I. Structure of TPR Domain–Peptide Complexes: Critical Elements in the Assembly of the Hsp70–Hsp90 Multichaperone Machine. *Cell* **2000,** *101* (2), 199–210.

Schöffl, F.; Prandl, R.; Reindl, A. Molecular Responses to Heat Stress. In *Molecular Responses to Cold, Drought, Heat and Salt Stress in Higher Plants*, 1999; vol *83* (93).

Schwacke, R.; Grallath, S.; Breitkreuz, K. E.; Stransky, E.; Stransky, H.; Frommer, W. B.; Rentsch, D. LeProT1, A Transporter for Proline, Glycine Betaine, and γ-amino Butyric Acid in Tomato Pollen. *Plant Cell* **1999,** *11* (3), 377–391.

Sehgal, S. K.; Kaur, S.; Gupta, S.; Sharma, A.; Kaur, R.; Bains, N. S. A Direct Hybridization Approach to Gene Transfer from Aegilops tauschii Coss. to Triticum aestivum L. *Plant Breed.* **2011,** *130* (1), 98–100.

Shan, Q.; Wang, Y.; Li, J.; Gao, C. Genome Editing in Rice and Wheat Using the CRISPR/Cas System. *Nat. Protoc.* **2014,** *9* (10), 2395–2410.

Sharma, E.; Borah, P.; Kaur, A.; Bhatnagar, A.; Mohapatra, T.; Kapoor, S.; Khurana, J. P. A Comprehensive Transcriptome Analysis of Contrasting Rice Cultivars Highlights the Role of Auxin and ABA Responsive Genes in Heat Stress Response. *Genomics* **2021,** *113* (3), 1247–1261.

Shelp, B. J.; Bown, A. W.; McLean, M. D. Metabolism and Functions of Gamma-aminobutyric Acid. *Trends Plant Sci.* **1999,** *4* (11), 446–452.

Shen, L.; Hua, Y.; Fu, Y.; Li, J.; Liu, Q.; Jiao, X.; ... & Wang, K. Rapid Generation of Genetic Diversity by Multiplex CRISPR/Cas9 Genome Editing in Rice. *Sci. China Life Sci.* **2017,** *60* (5), 506–515.

Shi, Q.; Bao, Z.; Zhu, Z.; Ying, Q.; Qian, Q. Effects of Different Treatments of Salicylic Acid on Heat Tolerance, Chlorophyll Fluorescence, and Antioxidant Enzyme Activity in Seedlings of Cucumis sativa L. *Plant Growth Regulat.* **2006,** *48* (2), 127–135.

Shirasawa, K.; Sekii, T.; Ogihara, Y.; Yamada, T.; Shirasawa, S.; Kishitani, S.; ... Nishio, T. Identification of the Chromosomal Region Responsible for High-Temperature Stress Tolerance During the Grain-Filling Period in Rice. *Mol. Breed.* **2013,** *32* (1), 223–232.

Shukla, R. K.; Tripathi, V.; Jain, D.; Yadav, R. K.; Chattopadhyay, D. CAP2 Enhances Germination of Transgenic Tobacco Seeds at High Temperature and Promotes Heat Stress Tolerance in Yeast. *FEBS J.* **2009,** *276* (18), 5252–5262.

Smertenko, A.; Dráber, P.; Viklický, V.; Opatrný, Z. Heat Stress Affects the Organization of Microtubules and Cell Division in Nicotiana Tabacum Cells. *Plant Cell Environ.* **1997,** *20* (12), 1534–1542.

Sugio, A.; Dreos, R.; Aparicio, F.; Maule, A. J. The Cytosolic Protein Response as a Subcomponent of the Wider Heat Shock Response in Arabidopsis. *Plant Cell* **2009,** *21* (2), 642–654.

Sung, D. Y.; Guy, C. L. Physiological and Molecular Assessment of Altered Expression of Hsc70-1 in Arabidopsis. Evidence for Pleiotropic Consequences. *Plant Physiol.* **2003,** *132* (2), 979–987.

Suri, S. S.; Dhindsa, R. S. A Heat-Activated MAP Kinase (HAMK) as A Mediator of Heat Shock Response in Tobacco Cells. *Plant Cell Environ.* **2008,** *31* (2), 218–226.

Tsukaguchi, T.; Kawamitsu, Y.; Takeda, H.; Suzuki, K.; Egawa, Y. Water Status of Flower Buds and Leaves as Affected by High Temperature in Heat-Tolerant and Heat-Sensitive Cultivars of Snap Bean (Phaseolus vulgaris L.). *Plant Prod. Sci.* **2003,** *6* (1), 24–27.

Uchida, A.; Hibino, T.; Shimada, T.; Saigusa, M.; Takabe, T.; Araki, E.; ... & Takabe, T. Overexpression of DnaK Chaperone from a Halotolerant Cyanobacterium Aphanothece Halophytica Increases Seed Yield in Rice and Tobacco. *Plant Biotechnol.* **2008,** *25* (2), 141–150.

Verdeprado, H.; Kretzschmar, T.; Begum, H.; Raghavan, C.; Joyce, P.; Lakshmanan, P.; ... Collard, B. C. Association Mapping in Rice: Basic Concepts and Perspectives for Molecular Breeding. *Plant Prod. Sci.* **2018,** *21* (3), 159–176.

Viitanen, P. V.; Schmidt, M.; Buchner, J.; Suzuki, T.; Vierling, E.; Dickson, R.; ... & Soll, J. Functional Characterization of the Higher Plant Chloroplast Chaperonins. *J. Biol. Chem.* **1995,** *270* (30), 18158–18164.

Wahid, A.; Close, T. J. Expression of Dehydrins Under Heat Stress and Their Relationship with Water Relations of Sugarcane Leaves. *Biol. Plant.* **2007,** *51* (1), 104–109.

Wahid, A.; Gelani, S.; Ashraf, M.; Foolad, M. R. Heat Tolerance in Plants: An Overview. *Environ. Exp. Bot.* **2007,** *61* (3), 199–223.

Wang, L.; Liu, L.; Ma, Y.; Li, S.; Dong, S.; Zu, W. Transcriptome Profiling Analysis Characterized the Gene Expression Patterns Responded to Combined Drought and Heat Stresses in Soybean. *Comput. Biol. Chem.* **2018,** *77*, 413–429.

Wang, X.; Zhuang, L.; Shi, Y.; Huang, B. Up-regulation of HSFA2c and HSPs by ABA Contributing to Improved Heat Tolerance in Tall Fescue and Arabidopsis. *Int. J. Mol. Sci.* **2017,** *18* (9), 1981.

Watanabe, C. K.; Yamori, W.; Takahashi, S.; Terashima, I.; Noguchi, K. Mitochondrial Alternative Pathway-Associated Photoprotection of Photosystem II is Related to the Photorespiratory Pathway. *Plant Cell Physiol.* **2016,** *57* (7), 1426–1431.

Wei, H.; Liu, J.; Wang, Y.; Huang, N.; Zhang, X.; Wang, L.; ... Zhong, X. A Dominant Major Locus in Chromosome 9 of Rice (Oryza sativa L.) Confers Tolerance to 48 C High Temperature at Seedling Stage. *J. Hered.* **2013,** *104* (2), 287–294.

Xu, Q.; Huang, B. Seasonal Changes in Root Metabolic Activity and Nitrogen Uptake for Two Cultivars of Creeping Bentgrass. *HortScience* **2006,** *41* (3), 822–826.

Xu, Y.; Huang, B. Effects of Foliar-Applied Ethylene Inhibitor and Synthetic Cytokinin on Creeping Bentgrass to Enhance Heat Tolerance. *Crop Sci.* **2009,** *49* (5), 1876–1884.

Xu, Y.; Zhan, C.; Huang, B. Heat Shock Proteins in Association with Heat Tolerance in Grasses. *Int. J. Proteomics* **2011,** *2011*.

Xue, G. P.; Sadat, S.; Drenth, J.; McIntyre, C. L. The Heat Shock Factor Family from Triticum Aestivum in Response to Heat and Other Major Abiotic Stresses and their Role in Regulation of Heat Shock Protein Genes. *J. Exp. Bot.* **2014,** *65* (2), 539–557.

Yadav, S. S.; Redden, R.; Hatfield, J. L.; Lotze-Campen, H.; Hall, A. J. W. Ed.; Crop Adaptation to Climate Change. JohnWiley & Sons: Hoboken, 2011. DOI: 10.1002/9780470960929.ch24

Yamada, K.; Fukao, Y.; Hayashi, M.; Fukazawa, M.; Suzuki, I.; Nishimura, M. Cytosolic HSP90 Regulates the Heat Shock Response that is Responsible for Heat Acclimation in Arabidopsis Thaliana. *J. Biol. Chem.* **2007,** *282* (52), 37794–37804.

Yamada, M.; Hidaka, T.; Fukamachi, H. Heat Tolerance in Leaves of Tropical Fruit Crops as Measured by Chlorophyll Fluorescence. *Sci. Hortic.* **1996,** *67* (1-2), 39–48.

Yamamoto, Y. Quality Control of Photosystem II: The Mechanisms for Avoidance and Tolerance of Light and Heat Stresses are Closely Linked to Membrane Fluidity of the Thylakoids. *Front. Plant Sci.* **2016,** *7*, 1136.

Yang, X.; Zhu, W.; Zhang, H.; Liu, N.; Tian, S. Heat Shock Factors in Tomatoes: Genome-Wide Identification, Phylogenetic Analysis and Expression Profiling Under Development and Heat Stress. *PeerJ*, **2016,** *4*, e1961.

Yokotani, N.; Ichikawa, T.; Kondou, Y.; Matsui, M.; Hirochika, H.; Iwabuchi, M.; Oda, K. Expression of Rice Heat Stress Transcription Factor OsHsfA2e Enhances Tolerance to Environmental Stresses in Transgenic Arabidopsis. *Planta*, **2008,** *227* (5), 957–967.

Zhang, J. H.; Huang, W. D.; Liu, Y. P.; Pan, Q. H. Effects of Temperature Acclimation Pretreatment on the Ultrastructure of Mesophyll Cells in Young Grape Plants (Vitis vinifera

L. cv. Jingxiu) Under Cross-Temperature Stresses. *J. Integr. Plant Biol.* **2005,** *47* (8), 959–970.

Zhao, R.; Houry, W. A. Hsp90: A Chaperone for Protein Folding and Gene Regulation. *Biochem. Cell Biol.* **2005,** *83* (6), 703–710.

Zhao, C.; Liu, B.; Piao, S.; Wang, X.; Lobell, D. B.; Huang, Y.; ... & Asseng, S. Temperature Increase Reduces Global Yields of Major Crops in Four Independent Estimates. *Proc. Natl. Acad. Sci.* **2017,** *114* (35), 9326–9331.

CHAPTER 7

Breeding and Molecular Approaches for Drought-Resilient Crops

H. G. KENCHARADDI,[1] SURESHA P. G.,[2] C. D. SOREGAON,[3] and AMARJEET KUMAR[4]

[1]*College of Agriculture, CAU(I), Kyrdemkulai, Meghalaya, India*

[2]*College of Agriculture, KSNUAHS, Shivamogga, Karnataka, India*

[3]*College of Agriculture, UASD, Vijayapura, Karnataka, India*

[4]*MTTC & VTC, Selesih, CAU (I), Mizoram, India*

ABSTRACT

Effects of climate change such as increase in temperature, altered/prolonged cycles of winter and summer, unseasonal rains, and global warming are causing huge threat to the food security during the 21st century. Drought stress is one of the most destructive among the abiotic stresses and more than 50% of the world's arable land will be affected due to drought stress in the year 2050. For achieving food and nutrition security, drought stress management is very important as it affects crop plants more frequently and severely. Plant breeders must be prepared for the substitution of species and varieties that can be adapted to the long-term and short-term effects of climate change. Hence, breeding for drought tolerance is imperative by identifying and utilizing resilient cultivars with drought-tolerant traits such as well-developed root system, early maturity, stay green nature, high harvest index, smaller and thicker leaves, smaller and denser stomata, more epidermal trichomes, stomata conductance, relative water content, etc., along with higher grain yield. In recent past, several conventional and molecular breeding techniques have been developed and successfully

Climate-Resilient Agriculture: A Molecular Perspective. Ashutosh Singh, Saurabh Pandey (Eds.)

applied to develop climate resilient crops. Several drought-tolerant varieties were released by using conventional breeding methods such as pureline selection, pedigree, bulk, recurrent selection, backcross breeding approach, etc., by utilizing various kinds of germplasms. In addition to the conventional breeding, modern molecular approaches namely in vitro selection methods, marker-assisted selection (MAS), transgenic approach, next-generation sequencing, genomic selection, and genome-editing techniques are also being followed to develop drought-tolerant varieties. The recent advances in tools and techniques of molecular biology and biotechnology required more consciousness to connect the genetics, physiology, and molecular signaling of drought tolerance. This information's would facilitate targeted breeding for drought tolerance in crops.

7.1 INTRODUCTION

The most important challenges of the present time are to produce sufficient food grains to meet the demands of increasing populations (Farooq et al., 2009; Tardieu et al., 2018). Abiotic stresses such as heat, cold, drought, flooding, salinity, deficiency/toxicity of mineral, variations in pH, heavy metal toxicity are frequent in occurrence due to change in climate and continue to affect crop growth in all stages individually or in combination (IPCC, 2014). Climate does not only mean fluctuation of temperature, but it also causes change in the magnitude of rainfall and their distribution pattern (Feng et al., 2013). Among the abiotic stresses, drought stress is one of the most critical and it has been predicted that more than half of the world's arable land will be affected due to drought stress by the year 2050 (Singhal, 2016). According to the IPCC, drought is "abnormal phase of dry weather to cause a serious hydrological imbalance" (IPCC, 2012). It is a slowly occurring disaster with less amount or lack of precipitation, resulting in a water shortage for the crop growth. Impact of drought is intensifying in near future as the world is moving toward being 2°C warmer (Ming, 2009). If not properly managed, drought is one of the major drivers of desertification, land degradation, increasing fragility of ecosystems, and social instability, especially in rural communities of developing counties (GARSRD, 2021). Impact of drought on health, economies, agriculture, energy, and the environment is also severe as every year passes. Drought is going to affect an estimated 55 million people around the globe. It may threaten people's livelihoods by increasing the risk of death, disease, malnutrition, and mass migration. As per the present climate change projections, many areas will experience droughts more frequently and in more severe form, which is affecting all kinds of crops plant growth, physiology, and reproduction level leading to the heavy economic losses in many countries (Yordanov et al., 2000; Barnabas et al., 2008).

As per the detailed study in the different regions of the world, several drought hotspots (areas particularly often or severely affected) have been identified as follows: The Mediterranean region (Hoerling et al., 2012), Southern Australia (Van Dijk et al., 2013), Sub-Saharan Africa (Greve et al., 2014), Southern South America (Penalba et al., 2014), some areas in China (Xu et al. 2015), South-western United States of America (Diffenbaugh et al., 2015), North-eastern Brazil (Marengo et al., 2017). As the world will continuously get warmer, the magnitude of the temperature rise during drought period is going to impact the severity of these projected droughts (GAR Report, 2021).

Plant behavior during drought stress is complex and depends on several environmental factors such as stage of the occurrence and duration of the stress, kind of germplasm, and the physiological stage of the plant at the time of the stress (Khan et al. 2016). Therefore, understanding the effects of drought in plants is critical for identifying and developing drought-resilient crops/cultivars with stable yield performance under drought and non-drought situations. Drought resilience is an ability of a plant to survive and recover from the effects of drought stress. To proceed under this direction, identification of drought-resilient native neglected and underutilized species/tolerant germplasm, development of tolerant cultivars through conventional, mutation breeding, participatory plant breeding, and application of advanced biotechnological approaches such as marker-assisted selection, identification and introgression of trait specific QTL's, genomic selection, and advanced genome editing technologies combined with other drought stress mitigating strategies like standardization of agronomic practices, seed quality enhancement techniques, water management at critical time of the crop growth, utilization of the osmoprotectants, etc., are imperative to address drought-associated problems and produce climate-resilient crop varieties.

7.2 IMPACT OF DROUGHT STRESS ON GROWTH AND DEVELOPMENT

Plant's behavior is species and genotype specific under drought stress. Crops' responses vary to drought stress, but final yield is significantly decreased due to occurrence of drought. Drought stress can affect plant development at all stages during crop growth. Plant establishment will be severely affected due to drought stress during seedling stage. The drought stress at flowering or grain-filling stages leads to complete loss of yields. Researchers have classified effect of drought on plant stages broadly as pre- and post-flowering stress (Nadew et al., 2021). Drought stress affects all levels including physiological, biochemical, morphological, and molecular levels by altering normal process leading to the poor crop performance and total crop failure. The drought stress causes disturbance in cellular processes at the physiological, molecular, and biochemical levels leading to morphological changes at crop growth stages.

At morphological level drought stress affects starting from germination, plant stand, root and shoot development, flowering time, flower development, pollen and embryo development, and normal fertilization followed by seed development and final yield. The negative impact on seed germination due to drought stress varies with species and cultivars. Numerous research findings in the past have revealed that drought stress during germination leads to poor development of root and shoot, reduced vigor, and poor crop stand ultimately reduces final crop yield (Kaya et al., 2006). Vibhuti et al. (2015) observed in rice, germination and seedling growth reduction by inhibiting water uptake and ultimately reduced the strength of seedling. During germination moisture stress alters essential enzymatic activities required for the normal germination and ultimately reduces the rate of germination. This is reported in soybean (Heatherly, 1993) and chickpea (Awari and Mate, 2015). Overall, moisture stress during germination, seedling stage will severely affect plant establishment, reduces leaf expansion (Alves and Setter. 2004), and elevated the leaf senescence (De Souza et al., 1997) and it results in the death of leaf tissue. In case of maize, typical morphological symptoms under drought stress include conversion of leaf color from green to green-gray and lower leaf rolling. Occurrence of stress 1 week before flowering leads to the slower growth of the ear than tassel growth and delay in silk emergence compared to pollen shed, giving rise to the problem in uniform pollination due to difference in anther extrusion and silk exposure (Edmeades, 2013).

Flowering and reproductive stages of the plant are highly sensitive to moisture stress (Robertson, 1993; Tuberosa and Salvi, 2006) which may lead to male sterility by inhibiting pollen grain germination and further development and embryo abort soon after pollination, leading to the decreased reproductive progenies for many plant species (Lopes et al., 2011). Drought-affected cobs typically had poorly filled grains (Edmeades et al., 2000). Matsuura (2012) observed the effect of moisture stress before and after flowering of four millets species namely proso, foxtail, little, and wild little millet. Reduction in final yield was reported in all four millets when the drought stress was imposed before flowering stage when compared with normal treatment. In rice, moisture stress during booting, flowering, and grain filling stages can lead to spikelet sterility, which results in lowering grain weight, grain size, seed setting rate, and poor grain yield (Kamoshita et al., 2004; Botwright et al., 2008). Severe drought in sunflowers caused a reduction in the head diameter, number of rows per head, and number of achenes per head, and these traits were highly correlated with yield components (Fereres et al., 1986; Gimenez and Fereres, 1986; Rauf and Sadaqat, 2007). José et al. (2018), in soybean, reported that drought stress at reproductive stage decreases flowering period and a smaller number of flowers, pods, and fewer number of seeds per plant. Other scientists have also reported earlier that drought stress during the seed formation period reduces number of seeds per plot (Meckel, 1984; Rotundo and Westgate, 2009) and ultimately reduces final yield.

At physiological level, numerous changes are induced in plant during drought stress by altering various activities such as reduction in plant water content, decreasing growth and elongation of cell, closure of stomata, gaseous exchange reduction, disruption of enzyme-catalyzed reactions viz., photosynthesis, respiration, ion uptake, translocation, carbohydrates and nutrient metabolism, imbalance between source–sink relationship, and other essential growth promoters (Farooq et al., 2009). Efficiency of photosynthesis in plants depends on plant stage and environmental stresses (Andrianasolo et al., 2016). Severe drought condition leads to decrease in the CO_2 diffusion and decrease in the chloroplast and leads to complete stop of the photosynthesis; furthermore, it closes the stomata and conserves the available water (Pinheiro and Chaves, 2011). Water content decrease in plant system damages chloroplast and photosynthetic output because of decreased light interception. The above mechanisms finally lead to the reduction in yield and death of the plant (Hussain et al., 2018; Chowdhury et al., 2016).

Drought stress hinders plant growth by decreasing water potential, cell division leading to the loss of relative water content and water potential of the leaves (Correia et al., 2006; Kiani et al., 2007) which creates problem in water supply to xylem and nutrient supply in phloem. During drought stress, weak absorption of plant nutrients and minerals is reported due to impaired active transport and membrane permeability (Kramer and Boyer, 1995). This affects the overall uptake of essential nutrients for optimum growth.

In Chickpea, Mafakheri et al. (2010) revealed decrease in chlorophyll content and photosynthesis. In addition to optimizing water use during drought stress, stomatal control is also a deciding factor for reducing excess water loss under severe drought situation (Makbul et al., 2011). For example in soybean, compared to the control plot, stomatal conductance is reduced by 60% during drought stress condition (Hao et al., 2013; Mak et al., 2014; Mutava, 2014). Atti et al. (2004) observed a reduction in photosynthetic rate by 78.4%, transpiration rate by 85%, and stomatal conductance by 92% under drought stress and also noticed a strong correlation between stomatal conductance with transpiration rate than with the photosynthetic rate. Hao et al. (2013) observed decrease in transpiration rate by 53% and Mark et al. (2014) noticed decrease by 57% during drought stress. This decline in stomatal conductance is due to the increase in the accumulation of root-originated ABA and severe stress is reported due to the increased concentration of ABA (Liu et al., 2004).

Moisture stress has a major effect on plant cellular and biochemical processes. At the cellular level drought stress leads to free radicles production causing shrinkage of cells, changing the fatty acids composition of plasma membrane and membranes apparatus of other cellular organelles (Terbea et al., 1995; Sgherri et al., 1996; Kang and Zhang, 1997). Drought stress promotes enhanced production of various ROS species (Cechin et al., 2015), causing damage to most of the cellular components. Reactive oxygen species (ROS) are oxygen-free radicals produced constantly in plants when required even under normal condition because of their involvement in cell signaling. But overproduction of ROS leads to the oxidative stress.

Under drought stress level of proline also increases which plays a significant role in development and metabolism of the plant. Proline is a cell-compatible solute and is accumulated at higher levels under moisture stress condition. Proline plays numerous roles in plants such as a protectant of various enzymes and biological membranes subjected to desiccation, as a better osmolyte, metal chelator (Hayat et al., 2012). It also helps plants under moisture stress condition through turgor pressure balance through membrane stabilization and electron transport complex II in Mitochondria (Hamilton and Heckathorn, 2001) and maintaining the required concentration of reactive Oxygen Species (ROS) in plants (Paleg et al., 1984; Hare et al., 1998). Moisture stress also affects structure of cell organelles and hydration of membranes, structure of DNA, RNA, and proteins. The difference in pressure in membrane cell wall complex under moisture stress also affects the structure and functions of the cell organelles. Hence, better understanding of plant response under moisture stress at morphological, physiological, and molecular levels is very important to address the stress-related problems and development of novel genotypes which performs well in moisture stress situation.

7.3 DROUGHT TOLERANCE MECHANISMS

Plants during evolution to survive under drought stress condition have developed numerous mechanisms. The mechanisms of plant under drought stress condition are broadly divided into three main forms, namely (1) drought escape, (2) drought avoidance, and (3) drought tolerance (Mitra, 2001).

Often plants use more than one mechanism synergistically to minimize the ill-effects of drought stress. Drought escape is a condition in which drought-susceptible plants perform well even in the drought environment by completing the life cycle early; thus avoiding the drought condition. For instance, early maturing trait is an important trait for escaping especially terminal droughts (May and Milthrope, 1962). This is particularly useful in the condition under which drought is a recurrent occurrence during terminal crop growth stage. In such cases, early maturing varieties can escape terminal drought. It is the simplest and most well-adopted mechanism for survival under drought condition by completing life cycle early by the plants. Hence, selection and development of early maturing cultivars under drought conditions have been used successfully while developing cultivars for drought stress (Hall and Patel, 1985; Singh, 1987, 1994). The major drought escape characters are rapid phenological adjustment, developmental plasticity, and high photosynthetic capacity (Kooyers, 2015). The study conducted on pearl millet in West Africa revealed that, based on the amount of rainfall and distribution, plant changes/adjusts its vegetative and reproductive phases during life cycle (Sivakumar, 1992). If main panicle development coincides with good rain, then it avoids drought situation during the terminal stage of the crop plant.

The second mechanism adopted by plant is drought avoidance; it is an ability of plant to retain considerable higher concentration of water/hydration under the soil water stress. It helps the plant by protecting crucial physiological, metabolic processes essential for the normal growth and development of a plant during moisture stress (Ludlow and Muchow, 1990). Plants develop several mechanisms such as development of efficient root system, including enhanced root biomass, deep rooting system, better water uptake, reduced transpiration due to closure of the stomata especially in xerophytes and CAM plants (Lawan, 1983; Boyer, 1996). The third mechanism is drought tolerance; here genotypes perform well compared to other genotypes by maintaining the crucial mechanisms like maintenance of membrane stability, integrity of the tissues, osmotic adjustment, proline accumulation, gaseous exchange essential for photosynthesis and respiration. The important traits conferring drought tolerance are thicker leaves, dense and small stomata, increased epidermal trichomes, and properly developed vascular bundle sheath (Wang et al., 2021).

Besides the above morphological traits, other significant traits associated with drought tolerance are: (1) Leaf rolling is an interesting physiological phenomenon under drought stress. Loss of water due to transpiration is less in moderate rolling leaves compared to full open leaves. Leaf rolling in plant is induced by loss of turgor; therefore, it is delayed by osmotic adjustment (Zhang, 2021). (2) Stay-green character of plant, support-efficient photosynthesis, and other vital processes of the plant during grain formation and grain maturity time under drought stress. Kencharaddi et al. (2015) studied stay green character in *G hirsutum* and developed a heterotic group involving the genotypes with stay green character. These stay green cultivars plants are present still green condition, even at the time of maturity such stay green type plants were identified in cotton, sorghum, and other plants (Patil et al., 2011; Kencharaddi et al., 2015). This is an ideal character in cotton harvesting through machines, as the trace content on the lint is very less. (3) Root system of the plant is a very important indirectly associated drought-tolerant trait. Important attributes associated with root system are number of roots, length and depth of the root system, density of the root, etc. (Rauf and Sadaqat, 2008). Comas et al. (2013), in pearl millet, observed the role of root biomass in producing final yield. Such genotypes are able to extract water from deep layers of the soil in the absence of water near the top portion of the root zone. Despite the significant positive association of root traits in influencing final yield under stress situation, root traits are not utilized often as the selection criteria in breeding, because of the difficulty in phenotyping of root traits. However, detailed understanding of development and architecture of roots in most of the crop plants is very important for the manipulation of root traits as per the requirement. Detailed characterization of germplasm for the above tolerant traits with high genetic variability, high heritability must be selected to harness the potentiality in drought stress situation.

7.4 MAJOR TRAITS ASSOCIATED WITH DROUGHT TOLERANCE

A wide range of traits of crop plants are associated with drought tolerance mechanisms. These inherent tolerance properties of plants are present in agronomic, morphological, and physiological levels. Traits associated at each level are described as mentioned below.

7.4.1 MORPHOLOGICAL AND ANATOMICAL TRAITS

These are simply observable phenotypic traits of plants that can have potential to provide indication for drought tolerance of the crop. For instance, seed germination rate, seedling vigor, high relative growth rate (RGR), leaf color, smaller and thicker leaves, plant stature, small plant height, wax content on the plant and leaves, early maturity, enhanced root biomass, root-to-shoot ratio above-ground biomass, harvest index, and final grain yield.

7.4.2 AGRONOMY AND YIELD-RELATED TRAITS

These traits are directly and indirectly contributing traits to the final yield of the crop plants. Among these are number of tillers per plant, early flowering, delayed flowering, size and number of panicle, boldness of the seed, seed weight, photosensitivity, etc. However, it is difficult to draw conclusions solely based on these traits, but these traits will give some indication about the plant response to the stress (Salekdeh et al., 2009).

7.4.3 PHYSIOLOGICAL AND BIOCHEMICAL TRAITS

The physiological and biochemical traits in relation to drought stress are smaller and denser stomata, thicker cuticle epidermis, stomata closure, cell wall integrity, osmotic adjustment, turgur pressure, relative water content, transpiration efficiency, antioxidant defence, hormone level, and enzyme activity (Sreeman et al., 2018). Other important traits are osmotic adjustment; ABA concentration, Cuticular wax, leaf characteristics like leaf rolling, leaf pubescence, delayed leaf rolling, stay greenness, increased water uptake, stomatal conductance, and feed forward response to stress.

Among all above-mentioned physiological traits, Osmotic Adjustment (OA) is one of the important drought-tolerant traits as it positively affects growth and final yield under drought stress. It is one of the most significant components of drought tolerance (Ribaut et al., 2009). Oxidative stress is a common phenomenon under drought stress. Reactive oxygen species (ROS) under stress may affect the membrane protein and enzymes (Gill et al., 2010). Abscisic acid (ABA) is known as stress hormone and its concentration increases in response to drought stress. It is considered the most potential endogenous phytohormones and is very closely associated with drought stress response. Under drought stress plants accumulate more intracellular ABA, especially in the root cap region which helps in water stress avoidance by affecting stomata closure, reduction of leaf area, promoting root growth, and activating the drought-related genes (Wang et al., 2016).

7.5 GENETICS OF DROUGHT RESISTANCE

It is a very complex phenomenon as many factors contributing to drought resistance have an interaction effect with several factors. Complexity of the trait discouraged many researchers to work in several crops (Sashidhar et al., 2013). The genetics of drought resistance characters ranged from oligogenic to polygenic in nature. Generally, leaf characters such as waxyness, glaucousness, and glabrous leaves are under oligogenic in nature. Some other resistance as traits like ABA accumulation in wheat, proline accumulation in mutants of barley, ability to produce pod formation in French bean are found to be under control of oligogenes. However, remaining traits associated with drought resistance are found to be controlled by polygenes in nature with the involvement of additive and dominance gene action. Plant breeders have identified several QTLs for traits associated for drought resistance with large to small phenotypic effects. Concerted efforts are still going on in many crops to elucidate the mode of action, understanding the complexity of the trait, identifying the key genes governing the trait, the promoters, QTLs, through advanced biotechnological techniques on genetics of drought resistance.

7.6 SELECTION ENVIRONMENTS FOR BREEDING DROUGHT RESISTANCE

The identification and use of suitable drought tolerance associated traits and suitable screening techniques are the important criteria while developing a genotype against moisture stress. The donor for drought tolerant requires integration of water-deficit response of plants at the molecular levels. Breeding of crops under drought must be conducted with due concern for the selection of target environment. Screening germplasm against drought stress can be broadly categorized into two groups namely screening germplasm under controlled environmental condition (greenhouse, growth chambers, and phytotrons) and screening in field condition as per the target environment. Screening under field condition is very complex as many traits contributing to drought tolerance have interaction effect with other factors. However, screening techniques under control condition have their own limitation of the absence of interacting environmental factors present in natural conditions. This leads to failure of genotypes selected from the control conditions to perform under natural environmental conditions.

Adequate care must be given during selection of parents and segregating genotypes at each segregating generation by using homogenous soil condition, along with proper experimental design and statistical analysis together to enhance the precision of genotypic means (Ribaut et al., 2009). However, achieving the above parameters under field condition is a very difficult task as most often target environment is highly variable in many factors like soil, water, and climate. Every cubic meter of soil condition is different from the other. Even plants under target environment also face variation with respect to climate during day and night under different season and locations. Any difference in the above-mentioned conditions of target environment might lead to the failure of performance of released variety for target location; hence, proper care must be taken to avoid such problems while conducting screening techniques and practising participatory plant breeding by involving the farmers of the target location is an ideal approach for breeding cultivars for the target environments (Sashidhara et al., 2013).

7.7 DIFFICULTIES IN BREEDING FOR DROUGHT RESISTANCE

- Drought-resistant variety should be developed with combination of drought resistance traits. However, usually, it is very difficult to achieve due to variation in moisture regime from season to season and location to location.
- Conventional drought-screening techniques are involved in more manual assessment process. These processes are labor intensive, time-consuming, error prone, and involve destructive sampling.
- Screening under field condition is very complex as many traits contributing to drought tolerance have interaction effect with other factors and creation of controlled condition under field is difficult.

- Screening techniques under control condition has its own limitation of absence of interacting environmental factors present in natural conditions. This may lead to the poor performance of genotypes selected from the control conditions to perform under natural environmental conditions.
- Drought resistance in plant usually is the result of combination of traits; hence, selection based on the single trait is not ideal and an integrated selection index has to be developed based on crop and prevailing condition.
- Measurement of most of the drought-associated resistance trait is difficult.
- Most of the drought-associated resistance traits have negative association with final yield. For instance, incorporation of earliness, stomatal conductivity traits may reduce yield in some genotypes.
- The use of wild relatives of the crop plant in drought resistance breeding affects the other traits of improved variety due to linkage drag.

7.8 BREEDING STRATEGIES FOR DESIGNING CLIMATE-RESILIENT DROUGHT-TOLERANT CROP PLANTS

Breeding for drought is an interdisciplinary approach involving the knowledge and techniques of agronomists, crop physiologists, plant breeders, and molecular biologists to understand the various factors and complex responses while identifying and developing a drought-tolerant cultivar. Despite many years of research in several crops on drought tolerance, little progress has been achieved with respect to genetic enhancement of crop yield under moisture-deficit environments due to association of so many factors and problems associated with bringing several drought tolerance traits with higher yield potentiality under normal conditions.

Genetic modification of crop plants by modifying the plant morphology or incorporating some traits through various breeding and modern biotechnology approaches is usually the most successful and cheapest approach toward managing drought stress (Rauf, 2008). However, availability of variation at various levels such as inter-specific, intra-specific, and intra-varietal levels has major importance for the selection of donor and helps to enhance breeding for resistance drought stress (Blum, 1985; Ashraf and Sharif, 1998; Serraj et al., 2005). Combining genes from both intra- and inter-species level is ideal to develop the drought-tolerant genotypes, but till now very limited efforts were made in most of the crops to study genotypic variation for major drought tolerance associated traits. Physiological traits leading toward drought tolerance should be considered ideal if it shows strong positive correlation with final yield, moderate to high heritability, cheaply and easily measured, and hopefully determined before flowering so that undesirable plants may be rouged before crossing. Hence, more focused efforts are required to screen and identify drought germplasm with drought-associated traits contributing the final yield (Ceccarelli, 1996; Cleveland, 2001), to discover new genes and pyrimidining these genes in crop plant to induce enhanced degree of drought tolerance.

Various management strategies such as the use of germplasms as a traits donor, incorporation of NUS species for the diversification of the agriculture, conventional breeding methods to develop novel gene combinations, induced mutation and novel breeding approaches such as rapid generation achievement, doubled haploid, in vitro selection and Somaclonal variations, marker-assisted selections, identification and introgression of QTLs, genomic selections, etc., have been opted by different scientists to enhance the performance of crop plants under drought stress. The details of each strategy is discussed in the following section.

7.9 GERMPLASM AS A SOURCE FOR DROUGHT STRESS TOLERANCE

Germplasm of crop plants has shown varied genetic responses for drought tolerance. Accelerated research efforts are required at conventional and molecular levels to harness the genetic potential of germplasm by utilizing prebreeding followed by interventions from modern genomics approaches. Prebreeding approach is mostly used for transferring traits from unadapted material and for identification of desirable traits from wild relatives. Prebreeding activity is a bridge for linking the desirable traits from wild relatives to modern cultivars by providing essential traits (Valkoun, 2001; Sharma et al., 2013). The drought-resistant traits may be found in following kinds of germplasm namely elite cultivars, landraces, wild relatives, and unrelated organisms. Generally, ample genetic variability for the traits contributing to drought resistance is present in elite germplasm of most of the crops. In some crops even cultivated varieties may contain usable genetic variations for these traits. Few important kinds of germplasm utilized for drought tolerance breeding are discussed as follows.

7.9.1 LANDRACES

Landraces are primitive cultivars which are selected and cultivated by farmers for many generations and highly adopted to the local environments and could be potential source for drought stress. There are many reports available for genetic variability for drought-tolerant traits such as osmotic adjustments, root characteristics, gas exchange parameters, seedling establishment, drought susceptibility index, etc. For example, Ethiopian sorghum landraces exhibit native genetic variation for drought resistance even though yet there is no development of sorghum cultivars with resistance to these important stresses.

In pearl millet, landraces adopted to the dry regions acts as a novel source for breeding drought tolerance (Kusaka et al., 2005; Yadav, 2010). The yield potentiality of most of the traditional landraces is bit low compared to modern varieties, but traditional landraces acts a reservoir for genes against drought stress (Yadav et al., 2011). Scientists from the IRRI, Philippines, screened nearly 1000 germplasm collected from Gene banks of 47 countries for drought tolerance (Bin Rahman and Zhang, 2016). More than 65 drought-tolerant accessions were identified, which are either aus or indica (Torres et al, 2013). The greater number of drought-tolerant aus accessions, that is, 19 are from Bangladesh, followed by 7 from India, while 16 drought-tolerant indica accessions are originated from India followed by 3 each from Bangladesh and Sri Lanka. Utilization of landraces in breeding programmes may be somewhat problematic due to undesirable linkages. But since these materials have evolved both under the influence of natural and artificial sections, the magnitude of such problems would be limited.

7.9.2 INTROGRESSION OF TRAITS FROM CROP WILD RELATIVES: CRADLE OF DROUGHT-TOLERANT GENES

The wild relatives, particularly those adapted to arid habitats of many crop plants, are known to be rich in drought resistance attributes and are regarded as valuable genetic sources for these traits. Screening a large number of CWR of the cultivated crops could help in identifying drought-tolerant relevant traits. Several wild species acted as a donor while developing drought-resistant varieties. For example, drought-tolerant chickpea variety BG1103 was released by introgression of genes from *Cicer reticulatum* Ladiz. The variety BG1103 showed more pod filling and higher yield under drought stress condition (Hajjar and Hodgkin, 2007). Genes from *Oryza longistaminata* Chev. Roehr were introgressed to cultivated rice cultivar in Philippines for the development of drought-tolerant rice cultivars (Brar, 2005). In tomato, extensive use has been made of the genetic variation present in wild species (Bai and Lindhout, 2007; Robertson and Labate, 2007) in developing today's commercial varieties. Helianthus *anomalus* is crop wild relative of sunflower which is successfully used in breeding programmes and transferred drought-tolerant genes to cultivated sunflower (Ludwig et al., 2004) and also in drought-tolerant genes were transferred from wild emmer wheat case of *Triticum dicoccoides* to cultivated Emmer wheat (Budak et al., 2013). Crop wild relatives are one of the potential sources for resistance against drought stress. However, utilization of such material is subject to the two serious limitations. First, it is likely that the various attributes of such wild species, which enable them to survive the stresses of their habitats, may be of little advantage to a crop variety. Second, even if this were not so, the transfer of genes governing drought resistance traits from these species into the crop plant would more often than not present crossability and linkage drag problems, which may not be easy to overcome.

7.9.3 INCORPORATION OF DROUGHT STRESS-TOLERANT NUS SPECIES

The incorporation of NUS into the present cropping systems could help to diversify the cropping system and reduce the global food and nutritional insecurity and scarcity (Mayes, 2012). NUS are neglected by policy makers, plant breeders, and agricultural scientists (Padulosi et al., 2013). Several neglected and underutilized species are climate resilient in nature. Underutilized fruits and vegetables, leafy vegetables, minor millets are also important members of NUS crops which have potential to be incorporated to present cropping systems to diversify (Table 7.1). NUS crops have not just met the food and nutritional security of man especially in marginal and degraded land and also during drought, but they have ensured the survival of mankind and livestock's during drought periods especially in many African countries. Except a few, genetic improvement and conservation work is still in initial stage in other NUS species. Hence, it needs an attention as many NUS crops are considered to be rich source of abiotic and biotic stress-resistant traits. Lastly, NUS crop species are preferred in dry land or under restricted irrigation because of its wider adaptability. We need to evaluate all our native

NUS crop landraces under controlled irrigation to harness their real yield potential. Wisdom of the past combined with science for the future shall ensure millets for sustainable agriculture.

TABLE 7.1 Potential NUS Crops for Crop Diversifications.

Cereals	Pulses	Fruits	Vegetables	Root and tuber crops
Finger millet	Faba bean	Jackfruit	Spiny gourd	Taro
Little millets	Cowpea	Bread fruit	Chow Chow	Yams
Proso millet	Horsegram	Wood apple	Snack gourd	Sweet potato
Buck wheat	Rice bean	Beal	Jack bean	Elephant foot yam
Amaranths	Moth bean	Jewish plum	Fenugreek	-
Quinoa	Lima bean	Drumstik	Coccinia	-
Barnyard millet	Adzuki beans	Tamrind	Lettuce	-

7.9.4 CONVENTIONAL BREEDING APPROACHES

The genetic control of drought resistance is polygenic in nature, very complex and influenced by environment. Understanding inheritance pattern of genes governing drought resistance is very important to design appropriate breeding approaches. Conventional breeding methods are commonly used for crop improvements such as mass selection, pure line selection, pedigree, bulk, single seed descent, and backcross breeding methods used for the improvement of self-pollinated crops while recurrent selection, disruptive selection, diallelic selective mating, bi-parental mating, synthetics, composites, heterosis used for cross-pollinated crops and mutation breeding, back cross breeding used for both self- and cross-pollinated crops. Basically, conventional breeding methods utilize existing variability, creation of variability through hybridization between contrasting parents and selection/development of novel genetic traits. Conventional breeding programmes for drought tolerance involve the following steps, namely: (1) characterization of accessions drought tolerance associated traits by using various methods. (2) Characterization and classification of germplasms for morphological and physiological traits. (3) Hybridization and selection in segregating generations under drought stress condition. (4) Continuing the selection till the population becomes completely homozygous and homogeneous. (5) Potential tolerant homozygous and homogeneous lines will release a variety or it will be used as a donor during future breeding programs.

Over the past four decades, breeders of many institutes have developed a series of potential cultivars that are drought stresses using mass selection and pureline selection methods (Kumar et al., 2008). For instance, IRRI Philippines developed and released several drought-resistant cultivars in countries through conventional breeding (Raman et al., 2012). Conventional bred Sagbhagi Dhan variety has made huge impact in the farming community due to its drought tolerance and maximum yielding ability under fovorable condition (Manzoor, 2020). The pedigree breeding method is the most widely used and oldest breeding method in many self-pollinated crops. The term pedigree means keeping all the relevant records like generation and year, etc., related to particular plant/genotype. In this method pedigree of all selected genotypes is maintained. Moreover, this method is more commonly used for the improvement of quantitative traits governed by polygenes than oligogenic traits as this method provides opportunity to identify transgressive segregants. Ali (2011) reported grain yield improvement of 25.00% and 25.54% compared to bulk selection under drought stress condition. It is a simple and most widely used method for identifying potential genotypes in self-pollinated crops, and also this approach provides an opportunity to isolate the lines performing well under drought situation. The disadvantages of pedigree selections can be rectified by using a recurrent selection approach. It is a population improvement strategy involving multiple parents and from each generation of selection, selfing and inter-mating will increase the concentration of the favorable alleles in the population. This method is mostly used in cross-pollinated crops like maize to increase the proportion of favorable alleles (Bolaños and Edmeades, 19193, 1996); Foroughi-Wehr and Wenzel (1990); rice (Veillet et al., 1996); barley (Parlevliet and van Ommeren, 1988); soybean (Posadas, 2014), and pearl millet (Kountche, 2013). This approach is used in wheat for grain protein improvement (Loffler et al., 1983; Delzer, 1995), weight of the kernel (Busch and Kofoid, 1982; Wiersma et al., 2001), and final yield (Olmeda-Arcega et al., 1995). In wheat genotypes are identified by using recurrent selection for nutrient efficiency and increased water (Zhang et al., 2014). The backcross breeding method is commonly used in many crops to rectify one or two defects and introgress gene of interest from donor parent. After crossing with donor parent repeated backcrossing needs to be done to reduce the genome content of donor species. This is one of the most widely used conventional breeding methods for rectifying the defects of the popular varieties and transferring desirable monogenic or oligogenic traits. Many drought-tolerant varieties were released through this method (Busch and Kofoid, 1982). Laffitte et al. (2004) conducted a large-scale backcross breeding experiment and observed that the backcross breeding method along with direct selection for yield is an important breeding method to improve drought tolerance in rice.

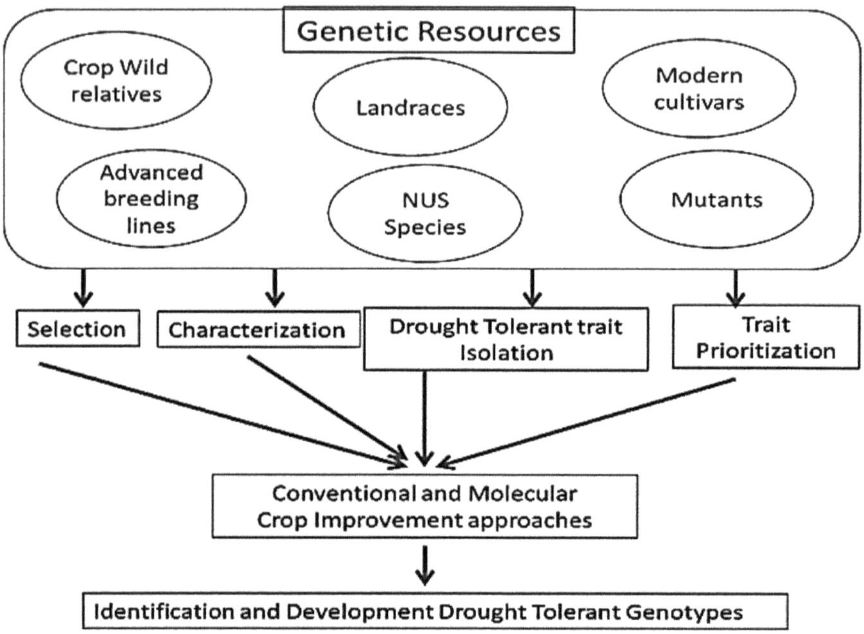

FIGURE 7.1 Breeding approaches for drought-tolerant traits.

7.9.5 PARTICIPATORY PLANT BREEDING

In case of Participatory Plant Breeding, the farmers, researchers, processors, and consumers participate in crop improvement programme and contribute to the selection of varieties as per their requirement. This method is ideally suited for addressing complexity agricultural system by developing specifically adapted cultivars to a magnitude of target environment (Annicchiarico et al., 2005; Singh et al., 2006). The programme broadly involves two steps: (1) participatory varietal evaluation and (2) participatory plant breeding. In participatory varietal selection, popular local varieties, advanced varieties developed by breeders, advanced generation elite breeding lines are grown in farmer's field. The farmers identify and select the genotypes most suitable as per their needs. This approach has been successfully used in several breeding programmes. However, in case of participatory plant breeding, all the stakeholders participate in the selection process from the segregating materials. Thus, plants selected in segregating generations as per their needs will be advanced to next generation in bulk till the line attains near homozygosity stage and later superior pureline will be released as a new variety. This method is ideally suited for developing cultivars in marginal field condition and stress situations as per the needs of the farmers of that location (Table 7.2). Several programs are in progress in many parts of the world to develop location-specific climate-resilient varieties, for instance, in maize, in western India (Witcombe et al., 2005) and in eastern India (Virk et al., 2005) the traits that farmers wanted in new maize genotypes is a variety having white grain (endosperm), with early maturity and tolerance to most common abiotic constraints of drought encountered in these regions was determined from interviews and from the results of participatory trials. Through participatory plant breeding approach, Ceccarelli et al. (2007) identified barley with superior yield (7–47 %) under drought condition compared to local varieties grown previously.

TABLE 7.2 Key Differences Between Conventional and Participatory Plant Breeding Approaches.

Conventional breeding	Participatory plant breeding
Conducted by plant breeders of public and private organizations	Conducted by involving all stake holders namely breeders, farmers, consumers, and other end users
Centralized approach	Decentralized approach
Focuses on broad objectives suitable for all kinds of environment	Focuses on selection and breeding as per the need of stake holders
Finally released as varieties after development	The overall output is improved and sometimes material for own use or release varieties
Elite cultivars and advanced breeding lines often used as base material	Landraces and locally adapted genotypes often used as base material
Ideally suited for yield improvement	Ideally suitable for quality traits improvement
cultivars with unacceptable traits by the farmer may be released	Farmer-unacceptable traits can be identified early
No participatory involvement of the farmers	Farmers' awareness about cultivar
The farmer's involvement is late, even several years taken between final product and exposure to farmers	Early farmer-to-farmer spread of material occurs

7.9.6 *MUTATION BREEDING*

Mutation breeding is another potential breeding technique highly used in many crops when there is no variability available in the population for particular traits. This method complements other conventional breeding methods, to improve oligogenic and polygenic traits such as flowering, ripening time, adaptability, plant type, growth habit, resistance to lodging, tolerance to temperature, drought, heat, salinity, and finally yield of plants (Aetsveit et al., 1997). The performance of mutant lines showed more tolerance to the concerned stresses as compared to their parents and check lines. These mutants will be used either (1) directly as commercial varieties in the target environments (2) or indirectly could be used as donor parents (Roy Chowdhury and Tah, 2013). Several drought-resistant varieties have been developed in many crops in most of the countries in the world. In India, drought varieties were developed in several crops such as rice, cotton, sugarcane, potato, and many in ornamental plants. Drought-tolerant mutants have been identified in sugar beet (*Beta vulgaris* L.) (Sen and Alikamanoglu, 2012) and several other crops. In cotton, drought-resistant varieties namely MCU-1 and MUC-10 have been developed and through induced mutations released for cultivation in India. The group of scientists from BARC and UAS Dharwad produced a bold seeded variety of groundnut. These bold seeded varieties with early maturing and drought-tolerant early maturing of groundnuts are frequently grown in desert areas in Rajasthan (Parthasarathy, 2013). Oladosu et al. (2016) summarized the success stories of the use of induced mutation while developing several potential rice varieties. Till today, the mutation breeding method is commonly practiced by many breeders especially asexually propagated crops for creating variability and identification of novel traits in various crop improvement programs. Present progress in mutation breeding for drought tolerance in plants clearly indicates that mutation breeding is a very useful tool in developing variation and donors or cultivars with higher yields, and these varieties could be grown in the cultivation of different plant species in drought periods or less rainy areas of the world.

7.10 MOLECULAR TECHNIQUES IN BREEDING FOR DROUGHT TOLERANCE

Though the conventional breeding approaches have helped in identifying and developing several drought-tolerant varieties (Table 7.3), most of the interesting traits with low heritability due to G × E interaction were difficult

TABLE 7.3 Crop Varieties Suitable for Cultivation under Drought Stress in India.

Variety	Crop	Seed availability
Ashoka-200F	Rice	RSSC, Rajasthan
Gujarat Nagli-4	Rice	DRNAU, Gujarat
Kalinga, Sahbhagi dhan, IR-36	Rice	BCKV, Kalyani
Kalinga-2, Kalyani-3, Narendar dhan 97	Rice	BCKV, Kalyani
Pant Dhan 16, Barani, Dhan-1, Aditya, Kalinga-3	Rice	GBPUA&T, Pantnagar; IGKV, Jabalpur
Pradhan, Poornima	Rice	IGKV, Jabalpur
HD2888, K8027	Wheat	BAU, Sabour
Sujata, C-306, JWS-17, HI-8627, HI-1531	Wheat	IGKV, Jabalpur
HPW-155, HPW-236	Wheat	CSK HPKVV, Palampur
Suwan	Maize	BAU, Sabour/RAU, Pusa
HQPM-5, HQPM-1	Maize	IGKV, Raipur
PMH-2, Parkash	Maize	PAU, Ludhiana
GHB-538 and GHB-719	Pearl Millet	JAU, Jamnagar
WCC-75	Pearl Millet	GKVK, UAS Bengalore
Raj-171	Pearl Millet	RSSC, Rajasthan
PL-419	Barley	PAU, Ludhiana
K603	Barley	CSAUAT, Kanpur
RD2660	Barley	RARS, Durgapura, Jaipur
RS-118, K-211-1, PS-4, SIA-326	Foxtail Millet	GKVK, UAS(B)
HHB-67	Foxtail Millet	CAZRI, Jodhpur
MPMH-17	Foxtail Millet	CAZRI, Jodhpur
VR-708 (Padmavati), HR-374	Finger millet	IGKV/NRC millets Bangalore
Jawahar Kutki 4 (JK 4)	Little millet	JNKVV, Rewa, Madhya Pradesh
Jawahar Kodo 137	Kodo millet	IGKV, COA, and ARS, Jagdalpur, Bastar
MR-1, MR-6, GPU-66, GPU28, KMR-301, ML-365	Kodo millet	GKVK, UAS(B)/KSSC
Phule Nachani	Kodo millet	MPKV, Rahuri; ZARS, Kolhapur, Maharashtra

Source: Adapted from Maheshwari et al. 2015

to exploit through conventional breeding (Prohens, 2011). Molecular markers help plant breeders to select the parents with different traits as the basis for the improvement of the trait of interest (Nadeem et al., 2018). Identification of the markers that are closely linked to the traits and relationships are varying on the type's markers used. The molecular marker is a chromosomal landmark of genes that target genes of desirable traits (Collard et al., 2005; Singh and Singh, 2015). These markers are not directly related to the morphology of the trait because they are found in the vicinity of target genes (O'Boyle et al., 2007; Jiang, 2013). Successful application of marker-assisted selection along with conventional breeding has led to the significant improvement in many crops such as in rice (Jena and Mackill, 2008), wheat (Miedaner and Korzun, 2012), maize (Prasanna et al., 2010), and sorghum (Mohamed et al., 2014); hence, marker-assisted selection is expected to increase the speed of crop improvement programme especially for the complex traits.

7.10.1 QTL MAPPING FOR DROUGHT TOLERANCE

The initial QTL tagging for root-associated traits was done by IIRI Philippines in 1995. Ever since a large number of QTLs were discovered for many traits predicted to contribute to drought tolerance in many crops. A large number of QTLs identified for traits associated with drought resistance have been tagged by using different kinds of markers namely RFLPs, RAPDs, CAPS, PCR indels, AFLPs, microsatellites (SSRs), SNPs, etc. Due to the complexity of drought tolerance-associated traits and the problems of G X E interaction during phenotypic-based selection, identification of QTLs associated with drought tolerance using marker-assisted selection (MAS) is the most widely followed approach in several crops (Farooq et al., 2014). Identification of QTLs controlling drought-tolerant traits under drought stress condition involves a series of aspects starting from the development of mapping populations with contrasting parents, polymorphic markers identification, phenotyping and genotyping of the mapping populations used in the study, genetic maps construction, phenotyping based on drought tolerance associated traits, and, finally QTL mapping based on genotypic and phenotypic data and finally identification of drought-responsive genes using molecular biology (Collard et al., 2005).

Many QTLs were identified and exploited in the past for drought-tolerant yield and yield-attributing traits. For example, increased root diameter and grain yield in sorghum are due to PSTOL1 and Sb07g02840 genes

(Bernardino et al., 2019). In sorghum, the stay-green trait is strongly associated with terminal drought. A mapping population developed between B35 x Tx7000 was tested over the environment for four stay green QTLs Stg1, Stg2, Stg3, Stg4, and Stg2 mainly controlling stay green trait (Xu et al. 2000; Subudhi et al. 2000). Guo et al. (2008), in maize, characterized QTL for traits such as flowering time, yield-attributing traits, and plant height using RIL population under both water-deficit and well-watered conditions and 51 QTL for 10 traits were identified on 10 different chromosomes. Under water-deficit conditions, 22 QTL were identified for seven traits. In rice, a large number of drought-related QTL have been identified in the past for different growth and physiological traits involved in drought tolerance. The major achievement has been made recently to introgress the major drought-tolerant QTLs such as qDTY1.1, qDTY2.1, qDTY2.2, qDTY3.1, qDTY4.1, qDTY12.1 qDTY6.3. into the background of the popular rice varieties like Swarna IR64 and TDK-1, which led to the development of several high-yielding drought-tolerant rice varieties (Venuprasad et al., 2009; Mishra et al., 2013; Bernier et al., 2007; Vikram et al., 2011; Yadaw et al., 2013).

7.10.2 MAKER-ASSISTED BACKCROSSING

Marker-assisted backcrossing is the most widely used and effective method for introgressing gene of interest. The marker alleles linked to gene(s)/QTL of interest is identified easily through MABC compared to the conventional backcross breeding method (Jiang, 2013). It has been mostly applied in several crops to transfer gene of interest into the lines that lack particular gene. The use of MAS in backcrossing (BC) has three major advantages: (1) It increases the efficiency of selection of the traits difficult to select through conventional breeding. (2) It helps in introgression of a desirable gene to a recipient parent from donor parent while maintaining the other essential traits of the recipient parent; (3) It prevents the excessive selfing step while transferring recessive gene.

Several QTLs introgressed in many crops by using MABC, for instance, Stg3A and Stg3B QTLs mapped in sorghum using SNPs markers. QTLs were linked for delaying leaf senescence during post-flowering drought stress and successfully introgressed to the local varieties in Tanzania that led to the yield improvement by 30% and stay green trait under moisture stress condition (Mwamahonje et al., 2021). Through MABC, an upland rice variety namely PY 84 has been released recently by selecting multiple QTL with better root growth under drought conditions (Sashidhar et al., 2013). PY 84 is the first rice variety developed by using marker-assisted selection and client-oriented breeding approach. 10% yield improvement was observed under rain-fed conditions compared to the recurrent parent (Steele et al., 2006).

7.10.3 PYRAMIDING OF QTLS FOR DROUGHT TOLERANCE

Assembling two or more genes from different donors into single genotypes for specific trait is known as marker-assisted gene pyramiding. Pyramiding of marker-assisted gene/QTLs is a powerful tool for introgressed multiple genes/QTLs for inducing deploy broad-spectrum resistance to abiotic and biotic stresses. Generally, the QTLs with large effect of phenotypic expression for the trait will be used for introgression of QTL's pyrimidining. For example, gene pyramiding of sub-1 gene in rice for developing submergence tolerance in rice. Most of the QTLs may not produce sufficiently large phenotypic effects so that they are unable to generate useful improvement in the trait of interest. Under such a situation, introgression of QTLs may fail to produce the expected improvement in the trait of interest. Efforts are being made to develop breeding and selection strategies for enhancing the effectiveness of QTL introgressions. There are several successful examples of QTL introgression, and some of these have led to the development of improved versions of a popular variety. For instance, in Malaysian high-quality rice cultivar viz., MRQ74, by using marker-assisted breeding (MAB) three drought yield QTLs (qDTY2.2, qDTY3.1, and qDTY12.1) were pyramided to induce resistance (Shamsudin et al., 2016).

In rice, major QTLs linked to drought stress viz., tolerance (qDTY1.1, qDTY2.1, qDTY3.1, and qDTY6.1 in Apo (Liu et al., 2020; Tsai et al., 2019) identified and deployed the above QTLs through molecular breeding and developed the drought-tolerant version of IR64, namely DRR Dhan 42, drought-tolerant MR219, drought-tolerant Sabitri, through the marker-assisted back cross breeding (MABB) approach. For example, a major QTL (Qyld.csdh.7AL) in wheat was introgressed into other cultivars to develop a better performing genotype for Rainfed situation (Quarrie, 2006). Pyramiding of QTLs situated on different chromosomes related to root traits was attempted by using an Indian indica rice as recipient parent and deep rooting and grain aroma producing cultivar Azucena, a japonica from Philippines used as donor parent.

7.10.4 GENOMIC SELECTION

Marker-assisted selection has been applied in several crops, but its efficiency is limited with respect to the number of molecular markers used. Most of the drought stress-tolerant traits and gronomic traits are quantitative in nature and governed by polygenes with additive effects. Under such a situation, genomic selection approach is well suited to address complex traits (Meuwissen et al., 2001) by involving advanced statistical methods and high-throughput genotyping methods. The genome selection scheme uses selection information derived from genotypic data of a large number of markers distributed over the whole genome irrespective of their associations with the traits being nonsignificant. This method is preferred over traditional marker-assisted selection as an effect of minor genes also captured in genomic selection (Hayes et al., 2009). All types of markers can be used in genomic selections for predicting the performance (Jannink et al., 2010). In cassava, genomic selection has been successfully applied for improvement of quantitative traits (Heffner et al., 2009; Jannink et al., 2010). Genomic selection has been shown to increase the efficiency of breeding in several crops such as in rice (Gurgul et al., 2019), maize (Yuan, et al., 2019), and wheat (Ray et al., 2014). In wheat, by using Genomic selection, 77 SNPs were identified related to many physiological traits such as root development, stomatal closure, and photosynthesis (Rai et al., 2018).

7.10.5 GENETIC IMPROVEMENT THROUGH GENOME EDITING TECHNIQUES

The recent development in phenotyping and genotyping platforms have paved the way for better understanding of drought stress tolerance in crops. The other dimension in the omics era is reverse genetics that provides an opportunity for genome editing in plants to induce moisture stress tolerance. Zhang et al. (2019) defined genome editing as an advanced molecular technique used to make precise, targeted modification at exact genomic loci. Three main genome editing technologies namely Transcription Actor like effector Nucleases (TALENs), Zinc Finger Nucleases (ZFNs), Clustered Regularly interspersed Short Palinromic Repeats (CRISPR) are commonly used to induce drought tolerance in crop plants. These genome editing techniques are used to study the functions of genes and modify them to enhance the functions as per the trait of interest. Among three, the commonly used genome editing tool is CRISPR because of its simplicity in use, flexibility, accuracy, and requirement of less time as compared to the other genome editing techniques. Although this technique is widely used in plant science only few reports are available on its utilization in abiotic stress tolerance breeding (Vats et al., 2019). CRISPR/Cas is a multiediting system and it has the potential to modify multiple genes and complex metabolic pathways essential to induce drought tolerance in crop plants. Drought-tolerant variety in maize was developed by editing ARGO58 gene through the CRISPR editing approach (Jinruishi et al., 2017).

7.10.6 SOMACLONAL VARIATION AND IN VITRO SELECTION

The plant regenerated from plant cells cultured in vitro often shows heritable genetic variation for both monogenic and polygenic traits; such variations are known as Somaclonal variation. Heritable variation is observed in plants regenerated from tissue and cell cultures. These variations lead to the production of aneuploids, sterile plants, morphological variants, and sometimes involving the traits of economic importance of crop plants. The usefulness of variations was first reported in potato by recovering disease-resistant plants. Somaclonal variations are likely to arise due to chimeral rearrangements and as a result of chromosomal structural and numerical aberrations, gene mutations, gene amplification, and altered expression of multi-gene families. There are several successful examples of the isolation of useful drought-tolerant mutants. The occurrence of genetic variation in the plants regenerated from tissue culture also offers opportunities to isolate disease-resistant Somaclonal variants. Several drought stress-tolerant genotypes were identified in the past by suing induction of selective pressure in culture conditions (Sakhanokho et al., 2009). In vitro tissue culture-based tools are ideal for understanding behavior of plant under drought stress and various mechanisms at the physiological and biochemical levels (Benderradji et al., 2012) as it is difficult to analyze the plant response to drought stress under field condition due to variable and complexity of these stresses under natural environment. In vitro selection pressure technique is most widely used to screen the germplasm under in vitro condition in a medium by inducing stress with selective agents for desirable traits. Level of tolerance for drought can be also assessed through this approach (Purohit et al., 1998). Polyethylene glycol (PEG) mannitol is commonly used as a selective agent for screening drought situation. Several selection criteria are available for screening the cultivars under in vitro condition, but seedling growth under PEG stress, change in plant phenology, and grain filling by translocated stem reserve are relatively easy to assay and are being mostly in some crop species. During past years, in vitro selection for plants exhibiting increased drought tolerance was reported in several crop plants (Barakat and Abdel-Latif, 1995; Hassan et al., 2004; Errabii et al., 2006).

7.11 CONCLUSIONS AND THE FUTURE PROSPECTS

Among the abiotic stresses, drought stress is one of the most devastating and by 2050 around 50% of the world's arable portion of land will be affected due to drought. Projections on climate change predict that many areas of the world are going to experience droughts that are more recurrent and more rigorous, which is affecting all kinds of crops' plant growth, physiology, and reproduction level leading to the heavy economic losses in many countries. Plant responses to drought stress are complex depending on several environmental factors such as frequency and duration of the stress and other climatic factors. Therefore, understanding physiological, biochemical, and molecular responses toward drought stress and tolerance mechanisms and the effects of drought on plants are critical for identifying and developing drought-resilient varieties with stable yield performance. Germplasms of plant species have a wide range of genes and traits associated with drought tolerance and that can be used in future breeding programmes to induce drought tolerance in crop plants. For this detailed characterization of all available germplasm is imperative to harness they offer. The other research gaps that need attention for developing drought-tolerant cultivars are development of simple and efficient screening techniques to identify drought-tolerant germplasm, large number of screening landraces, and crop wild relatives and characterization of available germplasms for draught-associated traits and understanding the molecular basis of drought stress and identification of QTLs will accelerate the future breeding programs. With the advent of plant genomics tools, many novel genes related to drought tolerance have been identified and are being used to improve drought tolerance with the help of innovative approaches. Genomic selection, genome-wide association studies, next-generation sequencing technologies, and genome editing techniques will play a vital role in future plant breeding programmes. The advanced tools will provide novel insights into drought-tolerant mechanisms and help in better designing of drought-tolerant high-yielding climate-resilient crops and finally the challenges of drought stress are not insurmountable; they can be addressed with sufficient understanding of the genetics, physiological, biochemical, and molecular responses of the plant under stress and the available resources/techniques to scientific community.

KEYWORDS

- **food and nutritional security**
- **drought**
- **wild relatives**
- **conventional breeding**
- **MAS**
- **QTL**
- **genomic selection**

REFERENCES

Aetsveit, K.; Kawai, T.; Sigurbjornson, B.; Scarascia Mugnozza, G. T.; Gottschalk, W. Plant Traits to be Improved by Mutation Breeding. In *Manual on Mutation Breeding*, 2nd Ed.; IAEA: Vienna, USA, 1997; p 219.

Ali, M. A. Pedigree Selection for Gratin Yield in Spring What (Triticum aestivum L.) under Drought Stress Condtions. *Asian J.* **2011,** 158–168.

Alves, A. A. C.; Setter, T. L. Abscisic Acid Accumulation and Osmotic Adjustment in Cassava Under Water Deficit. *Env. Exp. Bot.* **2004,** *51*, 259–271.

Andrianasolo, F. N.; Casadebaig, P.; Maza, E.; Champolivier, L.; Maury, P.; Debaeke, P. Prediction of Sunflower Grain Oil Concentration as a Function of Variety, Crop Management and Environment Using Statistical Models. *Eur. J. Agron.* **2014,** *54*, 84–96.

Annicchiaric, P.; Bellah, F.; Chiari, T. Defining Sub Regions and Estimating Benefits for a Specific Adaptation Strategy by Breeding Programs: A Case Study. *Crop Sci.* **2005,** *45*, 1741–1749.

Ashraf, M.; Sharif, R. Assessment of Inter-cultivar/line Variation of Drought Resistance in a Potential Oil-Seed Crop, Ethiopian Mustard (Brassica carinata). *J. Plant Nutr.* **1998,** *43*, 251–265.

Atti, S.; Bonnell, R.; Smith, D.; Prasher, S. Response of an Indeterminate Soybean {Glycine Max (L.) Merr} to Chronic Water Deficit During Reproductive Development Under Greenhouse Conditions. *Can. Water Resour. J.* **2004,** *29*, 209–222.

Awari, V. R.; Mate, S. N. E_ect of Drought Stress on Early Seedling Growth of Chickpea (Cicer arietinum L.) Genotypes. *Life Sci. Int. Res. J.* **2015,** *2*, 356–361.

Bai, Y.; Lindhout, P. Domestication and Breeding of Tomatoes: What Have We Gained and What Can We Gain in the Future. *Ann. Bot.* **2007,** *100*, 1085–1094.

Barakat, M. N.; Abdel-Latif, T. H. In Vitro Selection for Drought Tolerant Lines in Wheat. I. Effect of PEG on the Embryonic Cu0ltures. *Alex. J. Agric. Res.* **1995,** *40*, 97–112.

Barnabas, B.; Jager, K.; Feher, A. The Effect of Drought and Heat Stress on Reproductive Processes in Cereals. *Plant Cell Environ.* **2008,** *31*, 11–38. DOI: 10.1111/j.1365-3040.2007.01727.x

Benderradji, L.; Brini, F.; Kellou, K.; Ykhelf, N.; Djekoun, A.; Masmoudi, K.; Bouzerour. H. Callus Induction, Proliferation, and Plantlets Regeneration of Two Bread Wheat (Triticum aestivum L.) Genotypes Under Saline and Heat Stress Conditions. **2012,** *2012* ISRN Agronomy, Article ID 367851.

Bernardino, K. C.; Pastina, M. M.; Menezes, C. B.; de Sousa, S. M.; Maciel, L. S.; Carvalho, G. Jr.; Guimarães, C. T.; Barros, B. A.; Silva1, L. D.; Carneiro, P. C. S.; et al. The Genetic Architecture of Phosphorus Efficiency in Sorghum Involves Pleiotropic QTL for Root Morphology and Grain Yield Under Low Phosphorus Availability in the Soil. *BMC Plant Biol.* **2019,** *19*, 87

Bernier, J.; Kumar, A.; Ramaiah, V.; Spaner, D.; Atlin. G. A Large-Effect QTL for Grain Yield Under Reproductive-Stage Drought Stress in Upland Rice. *Crop Sci.* **2007,** *47*, 507–516. https:// doi. org/ 10. 2135/ crops ci2006. 07. 0495.

Bin Rahman, A. N. M. R.; Zhang, J. H. Flood and Drought Tolerance in Rice: Opposite but may Coexist. *Food Energy Secur.* **2016,** *5* (2), 76–88.

Blum, A. Breeding for Crop Varieties for Stress Environments. *Crit. Rev. Plant Sci.* **1985,** *2*, 199–238.

Bolaños, J.; Edmeades, G. O. Eight Cycles of Selection for Drought Tolerance in Lowland Tropical Maize. I. Responses in Grain Yield, Biomass, and Radiation Utilization. *Field Crops Res.* **1993,** *31*, 233–252.

Bolaños, J.; Edmeades, G. O. The Importance of the Anthesis-Silking Interval in Breeding for Drought Tolerance in Tropical Maize. *Field Crops Res.* **1996,** *48*, 65–80.

Botwright Acuna, T. L.; Latte, H. R.; Wade, L. J. Genotype and Environment Interactions for Grain Yield of Upland Rice Backcross Lines in Diverse Hydrological Environments. *Field Crops Res.* **2008,** *108*, 117–125.

Boyer, J. S. Advances in Drought Tolerance in Plants. *Adv. Agron.* **1996,** *56*, 189–218.

Brar, D. Broadening the Genepool and Exploiting Heterosis in Cultivated Rice, In *Rice is Life: Scientific Perspectives for the 21st Century*; Toriyama, K.; Heong, K. L.; Hardy, B., Eds.; Proceedings of the World Rice Research Conference, Nov 4–7, 2004; Tokyo and Tsukuba, Japan, 2005.

Budak, H.; Kantar, M.; Kurtoglu, K. Y. Drought Tolerance in Modern and Wild Wheat. *Sci. World J.* **2013,** *2013*, 1–16.

Busch, R. H.; Kofoid, K. Recurrent Selection for Kernel Weight in Spring Wheat. *Crop Sci.* **1982,** *22*, 568–572.

Ceccarelli, S. Adaptation to Low/High Input Cultivation. *Euphytica* **1996,** *92*, 203–214.

Ceccarelli, S.; Grando, S.; Baum, M. Participatory Plant Breeding in Water-Limited Environments. *Exp. Agric.* **2007,** *43*, 411–435.

Cechin, I.; Rossi, S. C.; Oliveira, V. C.; Fumis, T. F. Photosynthetic Responses and Proline Content of Mature and Young Leaves of Sunflower Plants Under Water Deficit. *Photosynthetica* **2006,** *44* (1), 143–146.

Chowdhury, J. A.; Karim, M. A.; Khaliq, Q. A.; Ahmed, A. U.; Khan, M. S. A. Effect of Drought Stress on Gas Exchange Characteristics of Four Soybean Genotypes. *Bangladesh J. Agr. Res.* **2016,** *41*, 195–205.

Cleveland, D. A. Is Plant Breeding Science Objective Truth or Social Construction? The Case of Yield Stability. *Agric. Human Values* **2001,** *18*, 251– 270.

Collard, B. C. Y.; Jahufer, M. Z. Z.; Brouwer, J. B.; Pang, E. C. K. An Introduction to Markers, Quantitative Trait Loci (QTL) Mapping and Marker-Assisted Selection for Crop Improvement: The Basic Concepts. *Euphytica* **2005,** *142* (1–2), 169–196.

Comas, L. H.; Becker, S. R.; Cruz, V. M. V.; Byrne, P. F.; Dierig, D. A. Root Traits Contributing to Plant Productivity Under Drought. *Front. Plant Sci.* **2013,** *4*, 442. DOI: 10.3389/fpls.2013.00442

Correia, M. J.; Osorio, M. L.; Osorio, J.; Barrote, I.; Martins, M.; David, M. M. Influence of Transient Shade Periods on the Effects of Drought on Photosynthesis, Carbohydrate Accumulation and Lipid Peroxidation in Sunflower Leaves. *Environ. Exp. Bot.* **2006,** *58*, 75–84.

Dar, M. H.; Waza, S. A.; Shukla, S.; Zaidi, N. W.; Nayak, S.; Hossain, M.; Kumar, A.; Ismail, A. M.; Singh, U. S. Drought Tolerant Rice for Ensuring Food Security in Eastern India. *Sustainability* **2020,** *12*, 2214. DOI: 10.3390/su12062214

Delzer, B. W.; Busch, R. H.; Hareland, G. A. Recurrent Selection for Grain Protein in Hard Red Spring Wheat. *Crop Sci.* **1995,** *35*, 730–735.

De Souza, P. I.; Egli, D. B.; Bruening, W. P. Water Stress During Seed Filling and Leaf Senescence in Soybean. *Agron. J.* **1997,** *89*, 807–812.

Diffenbaugh, N. S.; Swain, D. L.; Touma, D. Anthropogenic Warming has Increased Drought Risk in California. *Proc. Natl. Acad. Sci.* **2015,** *112* (13), 3931–3936.

Edmeades, G. O. Progress in Achieving and Delivering Drought Tolerance in Maize: An Update. ISAAA, Ithaca, 2013.

Edmeades, G. O.; Bolaños, J.; Elings, A.; Ribaut, J. M.; Bänziger, M.; Westgate, M. E. The Role and Regulation of the Anthesis-Silking Interval in Maize. In *Physiology and Modeling Kernel Set in Maize*; Westgate, M. E., Boote, K. J., Eds.; CSSA Special Publication: Wisconsin, 2000; Vol. 29, pp 43–73.

El-Haris, M. K.; Barakat, M. N. Evaluation of the In Vitro Selected Drought Tolerant Wheat Lines Under Drought Stress Conditions. *Alex. J. Agric. Res.* **1998,** *43*, 293–302.

Errabii, T.; Gandonou, C. B.; Essalmani, H.; Abrini, J.; Idomar, M.; Senhaji, N. S. Growth, Proline and Ion Accumulation in Sugarcane Callus Cultures Under Drought-Induced Osmotic Stress and its Subsequent Relief. *Afri. J. Biotechnol.* **2006,** *5*, 1488–1493.

Farooq, M.; Hussain, M.; Siddique, K. H. M. Drought Stress in Wheat During Flowering and Grainfilling Periods. *Crit. Rev. Plant Sci.* **2014,** *33*, 331–349.

Farooq, M.; Wahid, A.; Kobayashi, N.; Fujita, D.; Basra, S. M. A. Plant Drought Stress: Effects, Mechanisms and Management. *Agron. Sustain. Dev.* **2009,** *29*, 185–212.

Farooq, M.; Wahid, A.; Kobayashi, N.; Fujita, D.; Basra, S. M. A. Plant Drought Stress: Effects, Mechanisms and Management. In *Sustainable Agriculture*; Springer: Dordrecht, 2009; pp 153–188.

Feng, X.; Porporato, A.; Rodriguez- Iturbe, I. Changes in Rainfall Seasonality in the Tropics. *Nat. Clim. Change* **2013**. DOI: 10.1038/NCLIMATE1907

Fereres, E.; Gimenez, C.; Fernandez, J. M. Genetic Variability in Sunflower Cultivars Under Drought. I. Yield Relationships. *Australian J. Agric. Res.* **1986,** *37*, 573–582.

Foroughi-Wehr, B.; Wenzel, G. Recurrent Selection Alternating with Haploid Steps—A Rapid Breeding Procedure for Combining Agronomic Traits in Inbreeders. *Theor. Appl. Genet.* **1990,** *80*, 564–568.

Gimenz, C.; Fereres, E. Genetic Variability in Sunflower Cultivars Under Drought II. Growth and Water Relations. *Aust. J. Agric. Res.* **1986,** *37*, 573–582.

Greve, P.; Orlowsky, B.; Mueller, B.; Sheffield, J.; Reichstein, M.; Seneviratne, S. I. Global Assessment of Trends in Wetting and Drying Over Land. *Nat. Geosci.* **2014,** *7* (10), 716–721.

Guo, O.; Zhang, J.; Gao, Q.; Xing, S.; Li, F.; Wang, W. Drought Tolerance Through Over-Expression of Mono Ubiquitin in Transgenic Tobacco. *J. Plant Physiol.* **2008,** *165*, 1745–1755.

Gurgul, A.; Miksza-Cybulska, A.; Szmatoła, T.; Jasielczuk, I.; Piestrzynska-Kajtoch, A.; Fornal, A.; Semik-Gurgul, E.; Bugno-Poniewierska, M. Genotyping-by-Sequencing Performance in Selected Livestock Species. *Genomics* **2019,** *111*, 186–195.

Hall, A. E.; Patel, P. N. Breeding for Resistance to Drought and Heat. In *Cowpea Research, Production and Utilization*; Singh, S. R., Rachie, K. O., Eds.; Wiley: New York, 1985; pp 137–151.

Hamilton, E. W.; Heckathorn, S. A. Mitochondrial Adaptationsto NaCl. Complex I is Protected by Anti-oxidants and Small Heats Hock Proteins, Whereas Complex II is Protected by Proline and Betaine. *Plant Physiol.* **2001,** *126*, 1266–1274. DOI: 10.1104/pp.126.3.1266

Hao, X.; Li, P.; Feng, Y.; Han, X.; Gao, J.; Lin, E.; Han, Y. Effects of Fully Open-Air [CO2] Elevation on Leaf Photosynthesis and Ultrastructure of Isatis indigotica Fort. *PLoS One* **2013,** *8*, e74600.

Hare, P. D.; Cress, W. A.; Van Staden, J. Dissecting the Roles of Osmolyte Accumulation During Stress. *Plant Cell Environ.* **1998,** *21*, 535–553. DOI: 10.1046/j. 1365-3040.1998.00309.x

Hassan, N. M.; Serag, M. S.; El-Feky, F. M. Changes in Nitrogen Content and Protein Profiles Following In Vitro Selection of NaCl Resistant Mung Bean and Tomato. *Acta Physiol. Plant.* **2004,** *26*, 165–175.

Hayat, S.; Hayat, Q.; Alyemeni, M. N.; Wani, A. S.; Pichtel, J.; Ahmad, A. Role of Proline Under Changing Environments: A Review. *Plant Signal. Behav.* **2012,** *7*, 1456–1466. DOI: 10.4161/psb.21949

Heatherly, L. G. Drought Stress and Irrigation Effects on Germination of Harvested Soybean Seed. *Crop Sci.* **1993,** *33*, 777–781.

Heffner, E. L.; Sorrells, M. E.; Jannink, J. L. Genomic Selection For Crop Improvement. *Crop Sci.* **2009,** *49*, 1–12.

Hoerling, M.; Eischeid, J.; Perlwitz, J.; Quan, X.; Zhang, T.; Pegion, P. On the Increased Frequency of Mediterranean Drought. *J. Clim.* 2012, 25 (6), 2146–2161.

Hussain, M.; Farooq, S.; Hasan, W.; Ul-allah, S.; Tanveer, M. Drought Stress in Sunflower: Physiological Effects and its Management Through Breeding and Agronomic Alternatives. *Agric. Water Manag. J.* **2018,** *201*, 152–166.

IPCC (Intergovernmental Panel on Climate Change. Managing the Risks of Extreme Events and Disasters to Advance Climate Change Adaptation. Special Report of Working Groups I and II of the Intergovernmental Panel on Climate Change, Field, C. B., Barros, V., Stocker, T. F., Qin, D., Dokken, D. J., Ebi, K. L., Mastrandrea, M. D., Mach, K. J., Plattner, G. K., Allen, S. K., Tignor, M., Midgley, P. M., Ed.; Cambridge University Press: Cambridge and New York, 2012.

IPCC. In *Climate Change 2014: Impacts, Adaptation, and Vulnerability*; Field, C. B., et al., Ed.; Cambridge University Press, 2014. http://www.ipcc.ch/report/ar5/wg2/

Jannink, J. L.; Lorenz, A. J.; Iwata, H. Genomic Selection in Plant Breeding: From Theory to Practice. *Brief Funct Genom.* 2010, *9*, 166–177.

Jena, K.; Mackill, D. Molecular Markers and Their Use in Marker Assisted Selection in Rice. *Crop Sci.* **2008,** *48*, 1266–1276.

Jiang, G. L. Molecular Markers and Marker Assisted Breeding in Plants. In *Plant Breeding from Laboratories to Fields*; Anderson, S. B., Eds.; InTech: Croatia, 2013; pp 45–83.

José, A.; Rodrigues, J.; Filho, M.; Rodrigues, C.; Sales, G.; Célia, R.; Pires, D. M.; Machado, E. C. Source-Sink Relationships in Two Soybean Cultivars with Indeterminate Growth Under Water Deficit. *Bragantia* **2018,** *77*, 23–35.

Kamoshita, A.; Rofriguez, R.; Yamauchi, A.; Wade, L. J. Genotypic Variation in Response of Rainfed Lowland to Prolonged Drought and Re-watering. *Plant Prod. Sci.* **2004,** *7*, 406–420.

Kang, S. Z.; Zhang, J. H. Hydraulic Conductivities in Soil-Root System and Relative Importance at Different Soil Water Potential and Temperature. *Trans. Chinese Soc. Agric. Eng.* **1997,** *13* (2), 76–81.

Kaya, M. D.; Okçu, G.; Atak, M.; Çıkılıc, Y.; Kolsarıcı, Ö. Seed Treatments to Overcome Salt and Drought Stress During Germination in Sunflower (Helianthus annuus L.). *Europ. J. Agron.* **2006,** *24*, 291–295.

Kencharaddi, H. G.; Hanchinal, R. R.; Patil, S. S. Studies on Combining Ability in Inter Heterotic Group Derived Cotton Hybrids for Lint Yield and its Components. *Res. Environ. Life Sci.* **2015,** *8* (3), 451–456.

Khan, A.; Sovero, V.; Gemenet, D. Genome-Assisted Breeding for Drought Resistance. *Curr. Genomics* **2016,** *17* (4), 330–342.

Kiani, S. P.; Grieu, P.; Maury, P.; Hewezi, T.; Gentzbittel, L.; Sarrafi, A. Genetic Variability for Physiological Traits Under Drought Conditions and Differential Expression of Water Stress-Associated Genes in Sunflower (Helianthus annuus L.). *Theor. Appl. Gene.* **2007,** *114* (2), 193–207.

Kooyers, N. J. The Evolution of Drought Escape and Avoidance in Natural Herbaceous Populations. *Plant Sci.* **2015,** *234*, 155–162.

Kountche, B. A.; Tom Hash, C.; Dodo, H.; Laoualy, O.; Timbeli, A.; Vigouroux, Y.; et al. Development of a Pearl Millet Striga-Resistant Genepool: Response to Five Cycles of Recurrent Selection Under Strigainfetsed Field Conditions in West Africa. *Field Crops Res.* **2013,** *154*, 82–90.

Kramer, P. J.; Boyer, J. S. In *Water Relations of Plants and Soils*; Academic Press: New York, 1995.

Kumar, A.; Bernier, J.; Verulkar, S.; Lafitte, H. R.; Atlin, G. N. Breeding for Drought Tolerance: Direct Selection for Yield, Response to Selection and use of Drought Tolerant Donors in Upland and Lowland Adapted Populations. *Field Crops Res.* **2008,** 107, 221–231.

Kusaka, M.; Lalusin, A. G.; Fujimura, T. The Maintenance of Growth and Turgor in Pearl Millet (Pennisetum glaucum [L.] Leeke) Cultivars With Different Root Structures and Osmo-regulation under Drought Stress. *Plant Sci.* **2005,** *168*, 1–14. DOI: 10.1016/j.plantsci.2004.06.021

Lafitte, H. R.; Price, A. H.; Courtois, B. Yield Response to Water Deficit in an Upland Rice Mapping Population: Associations Among Traits and Genetic Markers. *Theor. Appl. Genet.* **2004,** *109*, 1237–1246.

Lawan, R. J. Responses of Four Grain Legumes to Water Stress in Higher Plants. *Annu. Rev. Plant Physiol.* **1983,** *35*, 299–319.

Li, R.; Liu, C.; Zhao, R.; Wang, L.; Chen, L.; Yu, W.; Zhang, S.; Sheng, J.; Shen, L. CRISPR/Cas9-Mediated SlNPR1 Mutagenesis Reduces Tomato Plant Drought Tolerance. *BMC Plant Biol.* **2019,** *19*, 38.

Liu, F.; Andersen, M. N.; Jacobsen, S.; Jensen, C. R. Stomatal Control and Water Use E_ciency of Soybean (Glycine max L. Merr.) During Progressive Soil Drying. *Environ. Exp. Bot.* **2005,** *54*, 33–40.

Liu, F.; Jensen, C. R. Andersen, M. N. Pod Set Related to Photosynthetic Rate and Endogenous ABA in Soybeans Subjected to Di_erent Water Regimes and Exogenous ABA and BA at Early Reproductive Stages. *Ann. Bot.* **2004,** *94*, 405–411.

Liu, M.; Yu, H.; Ouyang, B.; Shi, C.; Demidchik, V.; Hao, Z.; Yu, M.; Shabala, S. NADPH Oxidases and the Evolution of Plant Salinity Tolerance. *Plant Cell Environ.* **2020,** *43*, 2957–2968.

Loffler, C. M.; Busch, R. H.; Wiersma, J. V. Recurrent Selection for Grain Protein Percentage in Hard Red Spring Wheat. *Crop Sci.* **1983,** *23*, 1097–1101.

Lopes, M. S.; Araus, J. L.; Van Heerden, P. D. R.; Foyer, C. H. Enhancing Drought Tolerance in C4 Crops. *J. Exp. Bot.* **2011,** *62*, 3135–3153.

Ludlow, M. M.; Muchow, R. C. A Critical Evaluation of Traits for Improving Crop Yield in Water-Limited Environments. *Advan. Agron.* **1990,** *43*, 107–153.

Ludwig, F.; Rosenthal, D. M.; Johnston, J. A.; Kane, N. C.; Gross, B. L.; Lexer, C.; Dudley, S. A.; Rieseberg, L. H.; Donovan, L. A. Selection on Leaf Ecophysiological Traits in a Desert Hybrid Helianthus Species and Early-Generation Hybrids. *Evolution* **2004,** *58*, 2682–2692.

Mafakheri, A.; Siosemardeh, A.; Bahramnejad, B.; Struik, P. C.; Sohrabi, Y. E.; et al. Drought Stress on Yield, Proline and Chlorophyll Contents in Three Chickpea Cultivars. *Aust. J. Crop Sci.* **2010,** *4*, 580–585.

Maheswari, M.; Sarkar, B.; Vanaja, M.; Srinivasa Rao, M.; Prasad, J. V. N. S.; Prabhakar, M.; Ravindra Chary, G.; Venkateswarlu, B.; Ray Choudhury, P.; Yadava, D. K.; Bhaskar, S.; Alagusundaram, K. Eds.; Climate Resilient Crop Varieties for Sustainable Food Production under Aberrant Weather Conditions; ICAR-Central Research Institute for Dryland Agriculture: Hyderabad, 2019; p 64.

Mak, M.; Babla, M.; Xu, S.; Carrigan, A. O.; Liu, X.; Gong, Y.; Holford, P.; Chen, Z. Leaf Mesophyll K+, H+ and Ca2+ Fluxes are Involved in Drought-Induced Decrease in Photosynthesis and Stomatal Closure in Soybean. *Environ. Exp. Bot.* **2014,** *98*, 1–12.

Makbul, S.; Güler, N. S.; Durmu,¸S. N.; Güven, S. Changes in Anatomical and Physiological Parameters of Soybean Under Drought Stress. *Turk. J. Bot.* **2011,** *35*, 369–377.

Marengo, J. A.; Torres, R. R.; Alves, L. M. Drought in Northeast Brazil—Past, Present, and Future. *Theor. Appl. Climatol.* (**2017,** *129* (3), 1189–1200.

Matsuura, A. Effect of Pre- and Post-heading Water Deficit on Growth and Grain Yield of Four Millets. *Plant Prod. Sci.* **2012,** *15* (4), 323–331.

May, L. H.; Milthrope, F. L.; Drought Resistance in Crop Plants. *Field Crops Abstr.* **1962,** *15*, 171–179.

Mayes, S.; Massawe, F. J.; Alderson, P. G. P.; Roberts, J. A.; Azam-Ali, S. N.; Hermann, M. The Potential for Underutilized Crops to Improve Security of Food Production. *J. Exp. Bot.* **2012,** *63*, 1075–1079. https:// doi.org/10.1093/jxb/err396

Meckel, L.; Egli, D. B.; Phillips, R. E.; Radcli, E. D.; Leggett, J. E.; et al. Moisture Stress on Seed Growth in Soybeans. *Agron. J.* **1984,** *76*, 647–650.

Meuwissen, T. H. E.; Hayes, B. J.; Goddard, M. E. Prediction of Total Genetic Value Using Genome Wide Dense Marker Maps. *Genetics* **2001,** *157*, 1819–1829.

Miedaner, T.; Korzun, V. Marker-Assisted Selection for Disease Resistance in Wheat and Barley Breeding. *Phytopathology* **2012,** *102*, 560–566.

Ming, X. World Wide Fund for Nature: Yangtze River Basin Climate Change Vulnerability and Adaptation Report, WWF-China Program Office, Beijing, 2009.

Mishra, K. K.; Vikram, P.; Yadaw, R. B.; Swamy, B. M.; Dixit, S.; Sta Cruz, M. T.; Paul, M.; Marker, S.; Kumar, A. qDTY12.1: A Locus with a Consistent Effect on Grain Yield Under Drought in Rice. *BMC Genet.* **2013,** *14*, 12. https:// doi. org/ 10. 1186/ 1471-2156-14-12.

Mitra, J. Genetics and Genetic Improvement of Drought Resistance of Crop Plants. *Curr. Sci.* **2001,** *80*, 758–763.

Mohamed, A.; Ali, R.; Elhassan, O.; et al. First Products of DNA Marker-Assisted Selection in Sorghum Released for Cultivation by Farmers in Sub-saharan. *Afr. J. Plant Sci. Mol. Breed.* **2014,** *3*, 1–10.

Mutava, R. N.; Jebakumar, S.; Prince, K.; Hasan, N.; Song, L.; Valliyodan, B.; Chen, W.; Nguyen, H. T. Understanding Abiotic Stress Tolerance Mechanisms in Soybean: A Comparative Evaluation of Soybean Response to Drought and Flooding Stress. *Plant Physiol. Biochem.* **2014,** 109–120.

Mwamahonje, A.; Eleblu, J. S. Y.; Ofori, K.; Deshpande, S.; Feyissa, T.; Tongoona, P. Drought Tolerance and Application of Marker-Assisted Selection in Sorghum. *Biology* **2021,** *10*, 1249. https://doi.org/10.3390/ biology10121249

Nadew, D.; Bejiga, T.; Teressa, T. Breeding of Sorghum Crop for Resistance and Tolerance to Drought. *Int. Res. Plant Crops Sci.* (IRJPCS) **2021,** *6* (2), 188–195.

Nadeem, M. A.; Nawaz, M. A.; Shahid, M. Q.; et al. DNA Molecular Markers in Plant Breeding: Current Status and Recent Advancements in Genomic Selection and Genome Editing. *Biotechnol. Biotechnol. Equip.* **2018,** *32*, 261–285.

Neha, R.; Amasiddha, B.; Sinha, N.; et al. Physiological and Morphological Evaluation of MABB Derived Lines under Drought Stress in Bread Wheat (Triticum Aestivum L. Em. Thell.). *Indian J. Genet. Plant Breed.* **2018,** *78* (04), 417–425. https://doi.org/10.31742/IJGPB.78.4.3.

Shamsudin, N. A. A.; Swamy, B. P. M.; et al. Marker Assisted Pyramiding of Drought Yield QTLs into a Popular Malaysian Rice Cultivar, MR219. *BMC Genet.* **2016,** *17*, 30.

O'Boyle, P. D.; James, D.; Kelly, J. D.; Kirk, W. W. Use of Marker-Assisted Selection to Breed for Resistance to Common Bacterial Blight in Common Bean. *J. Am. Soc. Hortic. Sci.* **2007,** *132* (3), 381–386.

Oladosu, Y.; Rafii, Mohd Y.; Abdullah, N.; Hussin, G.; Ramli, A.; Rahim, H. A.; Miah, G.; Usman, M. Principle and Application of Plant Mutagenesis in Crop Improvement: A Review. *Biotechnol. Biotechnol. Equip.* **2016,** *30*, 11–16, DOI: 10.1080/13102818.2015.1087333

Olmeda-Arcega, O. B.; Elias, E. M.; Cantrell, R. G. Recurrent Selection for Grain Yield in Durum Wheat. *Crop Sci.* **1995,** *35*, 714–719.

Padulosi, S.; Thompson, J.; Rudebjer, P. Fighting Poverty, Hunger and Malnutrition with Neglected and Underutilized Species (NUS): Needs, Challenges and the Way Forward: Neglected and Underutilized Species; Bioversity International: Rome, Italy, 2013; ISBN 978-92-9043-941-7.

Paleg, L. G.; Stewart, G. R.; Bradbeer, J. W. Prolineandglycine Betaine Influence Protein Solvation. *Plant Physiol.* **1984,** *75*, 974–978. DOI: 10.1104/pp.75.4.974

Parlevliet, J. E.; van Ommeren, A. Accumulation of Partial Resistance in Barley to Barley Leaf Rust and Powdery Mildew Through Recurrent Selection Against Susceptibility. *Euphytica* **1988,** *37*, 261–274.

Parthasarathy, K. S. Mutation Breeding of Oil Seeds, Pulses and Cereals, 2013 [Online]. http://www. thehindu.com/sci-tech/agriculture/mutation-breeding-of-oil-seeds-pulses-and-cereals/article 5142401.ece

Patil, S. S.; Ramakrishna, V.; Manjula, S. M.; Swathi, P.; Ranganatha, H. M.; Kencharaddi, H. G. Deploying Reciprocal Selection for Combining Ability for Improving Performance of Hybrids in Cotton (Gossypium hirsutum). *Indian J. Genet. Plant Breed.* **2011,** *71* (2), 180–184.

Penalba, O. C.; Rivera, J. A.; Pántano, V. C. The CLARIS LPB Database: Constructing a Long-Term Daily Hydro-Meteorological Dataset for La Plata Basin, Southern South America. *Geosci. Data J.* **2014,** *1* (1), 20–29.

Pinheiro, C.; Chaves, M. M. Photosynthesis and Drought: Can we Make Metabolic Connections from Available Data? *J. Exp. Bot.* **2011,** *62*, 869–882. DOI: 10.1093/jxb/erq340

Posadas, L. G.; Eskridge, K. M.; Specht, J. E.; Graef, G. L. Elite Performance for Grain Yield from Unadapted Exotic Soybean Germplasm in Three Cycles of a Recurrent Selection Experiment. *Crop Sci.* **2014,** *54*, 2536–2546.

Prasanna, B.; Pixley, K.; Warburton, M. L.; Xie, C. X. Molecular Marker-Assisted Breeding Options for Maize Improvement in Asia. *Mol. Breed.* **2010,** *26*, 339–356.

Prohens, J. Plant Breeding: A Success Story to be Continued Thanks to the Advances in Genomics. *Front. Plant Sci.* **2011,** *2*, 1–3.

Purohit, M.; Srivastava, S.; Srivastava, P. S. Stress Tolerant Plants Through Tissue Culture. In *Plant Tissue Culture and Molecular Biology: Application and Prospects*; Srivastava, P. S., Ed.; Narosa Publishing House: New Delhi, 1998; pp 554–578.

Quarrie, S. A.; Pekic Quarrie, S.; Radosevic, R. Dissecting a Wheat QTL for Yield Present in a Range of Environments: From the QTL to Candidate Genes. *J. Exp. Bot.* **2006,** *57*, 2627–2637.

Raman, A.; Verulkar, S. B.; Mandal, N. P.; Variar, M.; et al. Drought Yield Index to Select High Yielding Rice Lines Under Different Drought Stress Severities. *Rice* **2012,** *5*, 1–12.

Rauf, S. Breeding Sunflower (Helianthus annuus L.) for Drought Tolerance. *Commun. Biometry Crop Sci.* **2008,** *3* (1), 29–44.

Rauf, S.; Sadaqat, H. A. Sunflower (Helinathus annuus L.) Germplasm Evaluation for Drought Tolerance. *Commun. Biometry Crop Sci.* **2007,** *2* (1), 8–16.

Ray, S.; Satya, P. Next Generation Sequencing Technologies for Next Generation Plant Breeding. *Front. Plant Sci.* **2014,** *5*, 367.

Reem Hajjar, A. E.; Hodgkin, T. The Use of Wild Relatives in Crop Improvement: A Survey of Developments Over the Last 20 Years. *Euphytica* **2007,** *156*, 1–13.

Ribaut, J. M.; Betran, J.; Monneveux, P.; Setter, T. Drought Tolerance in Maize, In *Handbook of Maize: Its Biology*; Bennetzen, J., Hake, S., Eds.; Springer: New York, 2009; pp 311–344.

Robertson, M J.; Fukai, S.; Ludlow, M. M.; Hammer, G. L. Water Extraction by Grain Sorghum in a Sub-Humid Environment. II. Extraction in Relation to Root Growth. *Field Crops Res.* **1993,** *33*, 99–112.

Robertson, L. D.; Labate, J. A. Genetic Resources of Tomato (Lycopersicon esculentum var. esculentum) and Wild Relatives, In *Genetic Improvement of Solanaceous Crops*, 2007; vol 2, pp 25–75.

Rotundo, L.; Westgate, M. E. Meta-analysis of Environmental Effects on Soybean Seed Composition. *F. Crop. Res.* **2009,** *110*, 147–156.

Roy Chowdhury, R.; Tah, J. Chapter 4, Mutagenesis—A Potential Approach for Crop Improvement, In *Crop Improvement: New Approaches and Modern Techniques*; Hakeem, K. R., Ahmad, P., Öztürk, M., Eds.; 2013; pp 149–187.

Sakhanokho, H. F.; Kelley, R. Y. Influence of Salicylic Acid on In Vitro Propagation and Salt Tolerance in Hibiscus Acetosella and Hibiscus moscheutos (cv 'Luna Red'). *Afr. J. Biotechnol.* **2009,** *8*, 1474–1481.

Salekdeh, G. H.; Reynolds, M.; Bennett, J.; Boyer, J. Conceptual Framework for Drought Phenotyping During Molecular Breeding. *Trends Plant Sci.* **2009,** *14*, 488–496.

Sen, A.; Alikamanoglu, S. Analysis of Drought-Tolerant Sugar Beet (Beta vulgaris L.) Mutants Induced with Gamma Radiation Using SDS-PAGE and ISSR Markers. *Mutat. Res.* **2012,** *738–739*, 38–44.

Serraj, R.; Hash, T. C.; Buhariwalla, H. K.; Bidinger, F. R.; Folkertsma, R. T.; Chandra, S.; Gaur, P. M.; Kashiwagi, J.; Nigam, S. N.; Rupakula, A.; Crouch, J. H. In *Marker-Assisted Breeding for Crop Drought Tolerance at ICRISAT: Achievements and Prospects*, Tuberosa, R., Phillips, R. L., Gale, M., Eds.; Proceedings of the International Congress "In the Wake of the Double Helix: From the Green Revolution to the Gene Revolution; Avenue Media: Bologna, Italy, 2005; pp 217–238.

Sgherri, C. L. M.; Pinzino, C.; Navari-Izzo, F. Sunflower Seedlings Subjected to Increasing Stress by Water Deficit: Changes in O2 Production Related to the Composition of Thylakoid Membranes. *Plant Physiol.* **1996,** *196* (3), 446–452.

Sharma, S.; Upadhyaya, H. D.; Varshney, R. K.; Gowda, C. L. Pre-breeding for Diversification of Primary Gene Pool and Genetic Enhancement of Grain Legumes. *Front. Plant Sci.* **2013,** *4*, 309.

Shashidhar, H. E.; Kanbar, A.; Toorchi, M.; Raveendra, G. M.; Kundur, P.; Vimarsha, H. S.; Soman, R.; Kumar, N. G.; Bekele, B. D.; Bhavani, P. Breeding for Drought Resistance Using Whole Plant Architecture Conventional and Molecular Approach, 2013. DOI: 10.5772/54983.

Shi, J.; Gao, H.; Wang, H.; Lafitte, H. R.; Archibald, R. L.; Yang, M.; Hakimi, S. M.; Mo, H.; Habben, J. E. ARGOS8 Variants Generated by CRISPR-Cas9 Improve Maize Grain Yield Under Field Drought Stress Conditions. *Plant Biotechnol. J.* **2018,** *15*, 207–216.

Singh, B. B. Breeding Cowpea Varieties for Drought Escape. In *Food Seed Production in Semi Arid Africa*; Menyonga, J. M., Bezuneh, T., Youdeowei, A. Eds.; OAU/STRCSAFGRAD: Ouagadougou, 1987; pp 299–306.

Singh, B. B. Breeding Suitable Cowpea Varieties for West and Central African Savanna. In *Progress in Food Grains Research and Production in Semiarid Africa*; Menyonga, J. M., Bezuneh, J. B., Yayock, J. Y., Soumana, I., Eds.; OAU/STRCSAFGRAD: Ouagadougou, 1994; pp 77–85.

Singh, M.; Malhotra, R. S.; Ceccarelli, S.; Sarker, A.; Grando, S.; Erskine, W. Spatial Variability Models to Improve Dry Land Field Trials. *Exp. Agric.* **2003,** *39*, 1–10.

Singh, B. D.; Singh, A. K. Linkage Mapping of Molecular Markers and Oligogenes. In *Marker- Assisted Plant Breeding: Principles and Practices*; Springer Nature, 2015; pp 151–183.

Singhal, P.; Jan, A. T.; Azam, M.; Haq, Q. M. R. Plant Abiotic Stress: A Prospective Strategy of Exploiting Promoters as Alternative to Overcome the Escalating Burden. *Front. Life Sci.* **2016,** *9*, 52–63.

Sivakumar, M. V. K. Empirical-Analysis of Dry Spells for Agricultural Applications in West Africa. *J. Climate* 1992, *5* (5), 532–539.

Sreeman, S. M.; Vijayaraghavareddy, P.; Sreevathsa, R.; Rajendrareddy, S.; Arakesh, S.; Bharti, P.; Dharmappa, P.; Soolanayakanahally, R. Introgression of Physiological Traits for a Comprehensive Improvement of Drought Adaptation in Crop Plants. *Front. Chem.* **2018,** *6*, 92.

Steele, K. A.; Price, A. H.; Shashidhar, H. E.; Witcombe, J. R. Marker-Assisted Selection to Introgress QTLs Controlling Root Traits into an Indian Upland Rice Variety. *Theor. Appl. Genet.* **2006,** *112*, 208–221.

Stone, P. The Effects of Heat Stress on Cereal Yield and Quality. In *Crop Responses and Adaptations to Temperature Stress*; Basra, A. S., Ed.; Food Products Press: Binghamton, NY, 2001; pp 243–291.

Subudhi, P.; Rosenow, D.; Nguyen, H. Quantitative Trait Loci for the Stay Green Trait in Sorghum (Sorghum bicolor L. Moench): Consistency Across Genetic Backgrounds and Environments. *Theor. Appl. Genet.* **2000,** *101*, 733–741. https://doi.org/10.1007/s001220051538

Tardieu, F.; Simonneau, T.; Muller, B. The Physiological Basis of Drought Tolerance in Crop Plants: A Scenario-Dependent Probabilistic Approach. *Annu. Rev. Plant Biol.* **2018,** *69*, 733–759. DOI: 10.1146/annurev-arplant-042817-040218

Terbea, M.; Vranceanu, A. V.; Petcu, E.; Craiciu, D. S.; Micut, G. Physiological Response of Sunflower Plants to Drought. *Rom. Agric. Res.* **1995,** *3*, 61–67.

Torres, R. O.; McNally, K. L.; Cruz, C. V.; Serraj, R.; Henry, A. Screening of Rice Genebank Germplasm for Yield and Selection of New Drought Tolerance Donors. *Field Crops Res.* **2013,** *147*, 12–22.

Tsai, Y. C.; Chen, K. C.; Cheng, T. S.; Lee, C.; Lin, S. H.; Tung, C. W. Chlorophyll Fluorescence Analysis in Diverse Rice Varieties Reveals the Positive Correlation Between the Seedlings Salt Tolerance and Photosynthetic Efficiency. *BMC Plant Biol.* **2019,** *19*, 403.

Tuberosa, R.; Salvi, S. Genomics-Based Approaches to Improve Drought Tolerance of Crops. *Trends in Plant Sci.* **2006,** *11*, 405–412.

United Nations Office for Disaster Risk Reduction. GAR Special Report on Drought 2021, Geneva, 2021.

Valkoun, J. J. Wheat Pre-breeding Using Wild Progenitors. *Euphytica* **2011,** *119*, 17–23.

Van Dijk, A. I.; Beck, H. E.; Crosbie, R. S.; de Jeu, R. A.; Liu, Y. Y.; Podger, G. M.; Timbal, B.;Viney, N. R. The Millennium Drought in Southeast Australia (2001– 2009): Natural and Human Causes and Implications for Water Resources, Ecosystems, Economy, and Society. *Water Res. Res.* **2013,** *49* (2), 1040–1057.

Vats, S.; Kumawat, S. V.; Patil, G. B.; Joshi, T.; Sonah, H.; Sharma, T. R.; Deshmukh, R.; Genome Editing in Plants : Exploration of Technologies, Advancement and Challenges. *Cells* **2019,** *8* (11), 1–39.

Veillet, S.; Filippi, M. C.; Gallais, A. A Combined Genetic Analysis of Partial Blast Resistance in an Upland Rice Population and Recurrent Selection for Line and Hybrid Values. *Theor. Appl. Genet.* **1996,** *92*, 644–653.

Venuprasad, R.; Dalid, C. O.; Del Valle, M.; Zhao, D.; Espiritu, M.; Sta Cruz, M. T.; Amante, M.; Kumar, A.; Atlin, G. N. Identification and Characterization of Large Effect Quantitative Trait Loci for Grain Yield Under Lowland Drought Stress in Rice Using Bulk-Segregant Analysis. *Theor. Appl. Genet.* **2009,** *120* (1), 177–190. DOI: https:// doi. org/10. 1007/s00122-009-1168-1

Vibhuti, V.; Shahi, C.; Bargali, K.; Bargali, S. S. Seed Germination and Seedling Growth Parameters of Rice (Oryza sativa L.) Varieties as Affected by Salt and Water Stress. *Ind. J. Agric. Sci.* **2015,** *85* (1), 102–108.

Vikram, P.; Swamy, B. M.; Dixit, S.; Ahmed, H. U.; Sta Cruz, M. T.; Singh, A. K.; Kumar, A. qDTY1.1, A Major QTL for Rice Grain Yield Under Reproductive Stage Drought Stress with a Consistent Effect in Multiple Elite Genetic Backgrounds. *BMC Genet.* **2011,** *12*, 89. https:// doi. org/ 10. 1186/ 1471-2156-12-89

Virk, D. S.; Chakraborty, M.; Ghosh, J.; Prasad, S. C.; Witcombe, J. R. Increasing the Client-Orientation of Maize Breeding Using Farmer Participation in Eastern India. *Exp. Agric.* **2005,** *41*, 413–426.

Wang, L.; Chen, L.; Li, R.; Zhao, R.; Yang, M.; Sheng, J.; Shen, L. Reduced Drought Tolerance by CRISPR/Cas9-mediated SlMAPK3 Mutagenesis in Tomato Plants. *J. Agric. Food Chem.* **2017,** *65*, 8674–8682.

Wang, J.; Li, C.; Li, L.; Reynolds, M.; Mao, X.; Jing, R. Exploitation of Drought Tolerance-Related Genes for Crop Improvement. *Int. J. Mol. Sci.* **2021,** *22*, 10265.

Wang, J. Y.; Li, Q.; Mao, X.G.; Li, A.; Jing, R. L. Wheat Transcription Factor TaAREB3 Participates in Drought and Freezing Tolerancesin Arabidopsis. *Int. J. Biol. Sci.* **2016,** *12*, 257–269.

Wiersma, J. J.; Busch, R. H.; Fulcher, G. G.; Hareland, G. A. Recurrent Selection for Kernel Weight in Spring Wheat. *Crop Sci.* **2001,** *41*, 999–1005.

Witcombe, J. R.; Joshi, K. D.; Gyawali, S.; Musa, A. M.; Johansen, C.; Virk, D. S.; Sthapit, B. R. Participatory Plant Breeding is Better Described as Highly Client-Oriented Plant Breeding. I. Four Indicators of Clientorientation in Plant Breeding. *Exp. Agric.* **2005,** *41*, 299–319.

Xu, W. W.; Subudhi, P. K.; Crasta, O. R.; Rosenow, D. T.; Mullet, J. E.; Nguyen, H. T. Molecular Mapping of QTLs Conferring Stay-Green in Grain Sorghum (Sorghum bicolor L. Moench). *Genome* **2000,** *43*, 461–469.

Xu, K.; Yang, D.; Yang, H.; Li, Z.; Qin, Y.; Shen, Y. Spatio-temporal Variation of Drought in China During 1961–2012: A Climatic Perspective. *J. Hydrol.* **2015,** *526*, 253–264.

Yadav, O. P. Drought Response of Pearl Millet Landrace-Based Populations and Their Crosses with Elite Composites. *Field Crops Res.* **2010,** *118*, 51–57. DOI: 10.1016/j.fcr.2010.04.00

Yadav, R. S.; Sehgal, D.; Vadez, V. Using Genetic Mapping and Genomics Approaches in Understanding and Improving Drought Tolerance in Pearl Millet. *J. Exp. Bot.* **2011,** *62*, 397–408. DOI: 10.1093/jxb/erq265

Yadaw, R. B.; Raman, A.; Kumar, A.; Swamy, B. P. M.; Mishra, K. K.; Sta Cruz, M. T.; Pandey, M.; Maturan, P. T.; Vikram, P.; Dixit, S. A QTL for High Grain Yield Under Lowland Drought in the Background of Popular Rice Variety Sabitri from Nepal. *Field Crops Res.* **2013,** *144*, 281–287. https:// doi. org/ 10. 1016/j. fcr. 2013. 01.019

Yordanov, I.; Velikova, V.; Tsonev, T. Plant Responses to Drought, Acclimation, and Stress Tolerance. *Photosynthetica* **2000,** *38*, 171–186. DOI: 10.1023/ A:100720141

Yuan, Y.; Cairns, J. E.; Babu, R.; Gowda, M.; et al. Genome-Wide Association Mapping and Genomic Prediction Analyses Reveal the Genetic Architecture of Grain Yield and Flowering Time Under Drought and Heat Stress Conditions in Maize. *Front. Plant Sci.* **2019,** *9*, 9.

Zhang, G. H.; Hou, X.; Wang, L.; Xu, J.; Chen, J.; Fu, X.; Shen, N. W.; Nian, J. Q.; Jiang, Z. Z.; Hu, J.; et al. Photo-Sensitive Leaf Rolling 1 Encodes a Polygalacturonase that Modifies Cell Wall Structure and Drought Tolerance in Rice. *New Phytol.* **2021,** *229*, 890–901.

Zhang, L.; Richards, R. A.; Condon, A. G.; Liu, D. C.; Rebetzke, G. J. Recurrent Selection for Wider Seedling Leaves Increases Early Biomass and Leaf Area in Wheat (TriticumaestivumL.) *J. Exp. Bot.* **2014**. DOI: 10.1093/jxb/eru468

Zhang, Y.; Malzahn, A. A.; Sretenovic, S.; Qi, Y. The Emerging and Uncultivated Potential of CRISPR Technology in Plant Science. *Nat. Plants* **2019,** *5*, 778–794. DOI: 10.1038/s41477-019-0461-5

CHAPTER 8

Pollen-Based Selection Strategy for Biotic and Abiotic Stress Tolerance

SURESH ANTRE,[1] SAUMYA KUMARI,[1] APURBA PAL,[2] SAURABH PANDEY,[3] and ASHUTOSH SINGH[4]

[1]*University of Agricultural Sciences, GKVK Campus, Bangalore, India*

[2]*Horticulture College, Khuntpani, Birsa Agricultural University, Ranchi, Jharkhand, India*

[3]*Department of Agriculture, Guru Nanak Dev University, Amritsar, Punjab, India*

[4]*Centre for Advance Studies on Climate Change, RPCAU, Pusa, Samastipur, India*

ABSTRACT

Pollen grain could be used to screen the traits controlled by oligogenic and complex polygenic traits. The pollen performance against the biotic and abiotic stresses is an indicator of sporophytic performance. Apart from this, the genetic overlap of structural genes between sporophyte and gametophyte stages provides the possibility of selection of superior pollen genotypes. Since more pollen can be exposed for selection pressure at the haploid level, it makes pollen-based selection more efficient compared to sporophytic selection. This approach channelizes the variability in very less time for improvement of both biotic and abiotic stresses. Therefore, it can reduce the number of selection cycles to help the homozygosity during crop improvement programs. The integration of pollen-based selection with MAS and speed breeding will lead to a synergistic effect on the efficiency of pollen-based selection as well as speed breeding. This book chapter will

Climate-Resilient Agriculture: A Molecular Perspective. Ashutosh Singh, Saurabh Pandey (Eds.)

discuss the concept, strategies, and case study of the pollen-based selection for the improvement of stress tolerance. Its integration with existing crop improvement tools will help to attain more efficient outcomes on major food crop varietal improvement.

8.1 INTRODUCTION

Plant growth mainly consists of two stages *viz.* vegetative and reproductive stages. Both biotic and abiotic stresses affect all stages of crop growth from germination to harvesting. Flowering stage of crops is most susceptible to stresses leading to a huge loss in crop yield (Singh et al., 2021). The life cycle of crop plants consists of gametophytic and sporophytic stages. There is genetic overlap between these two stages. The overlapping of genes provides an opportunity to introduce early selection into the breeding program *via* pollen-based selection. The sporophytic phase is the dominant phase in the angiosperm and it develops by fertilization of gamete and forms the zygote. But meiosis of spores produces the gametophyte generation. These gametophyte and sporophyte generations allow early selection that can be applied to both male and female gametes. The male gametes are more sensitive to stresses; thus, identification of vigorous/resistant pollens to the stresses can produce more resistant crops (Hormaza and Herrero, 1992). Plants produce a large number of pollen grains; thus, it produces competition during development in the anther, germination of pollen grains, and pollen tube growth in the style. Further, the independent phase shedding and stylar deposition create a good opportunity for manipulation using the male gametes.

The scientist mainly focused on the sporophytic phase while the influence of the gametophytic phase was considered "passive" (Ottaviano and Sari Gorla, 1993). However, the sporophytic selection requires a large population size and needs to be repeated over several seasons. It requires more resources to obtain the essential combination of attributes needed to improve a crop which increases the cost. Many woody perennial species have very few new cultivar developments due to long juvenile phases and larger physical size (Hormaza and Herrero, 1996). Therefore, male gametes could be used for selection against various stresses for crop improvement. Small size and very simple life cycle of pollen attracted the attention of artificial selection programs seeking increased sporophyte yield in crops (Ottaviano et al., 1980). Despite nearly a century of research, quantifying the extent of pollen selection and its fitness effects in sporophytes remains challenging. Pollen has drawn the most attention to research efforts and early thoughts focused on pollen and sporophytes' potential for conflict (Haldane, 1932). The selection during the gametophyte stage could increase sporophyte fitness in crop plants (Mulcahy, 1979). There are many pieces of evidence supporting that pollen competition resulted in differences in offspring quality; it created widespread enthusiasm (Davis et al., 1987). But the evidence for significant genetic variation in pollen performance required for pollen selection, as well as the nature of pleiotropic effects caused by this variation was highly contested (Charlesworth et al., 1987; Charlesworth, 1988; Walsh and Charlesworth, 1992).

The gametophytic selection cannot possibly influence the sporophyte unless there is a correlation between pre- and post-zygotic qualities. In *Petunia hybrida*, the quantity of pollen placed on the stigma seems to influence the quality of the resultant generation (Mulcahy et al., 1975). The varying degrees of gametophytic selection, applied in the production of the F_1 generation, have resulted in detectable changes in the genetic composition of that generation. These changes appear to be expressed also in the F_2 generation. This indicates that gametophytic selection can modify the genetic composition of the sporophytic generation (Mulcahy et al., 1978). The evidence suggests that some alleles which are transcribed in the gametophyte are transcribed also in the sporophyte. Alleles that play this dual role have been characterized as showing "haplodiploid" transcription (Mulcahy et al., 1975). Therefore, populations improved through pollen-based selection emerged as more promising because pollen-based selection allowed the selection of minor positive allele combinations that would barely detect at the sporophytic level. A combination of pollen-based selection and traditional sporophytic selection could be an effective tool in population improvement programs to achieve higher levels of resistance in a relatively short time (Shobharani and Ravikumar, 2006).

8.2 POLLEN SELECTION STRATEGY FOR BIOTIC STRESS IMPROVEMENT

The purpose of any crop improvement program is boosting all characteristics/traits that have economic value. These traits may increase yield by improving the tolerance level of plants to biotic and abiotic stresses (Fehr, 1991). The advancement of breeding and biotechnological tools/techniques provides new options to plant scientists interested in improving different plant responses and characteristics.

The selection pressure at the gametophyte stage leads to tolerance to several biotic factors at the sporophytic level. Biotic stress is an important area in which pollen response is parallel to sporophytic performance. There are many reports available, where indirect evidence is obtained by the analysis of the behavior of gametophytic and sporophytic phases against the different biotic agents (Sacher et al., 1983; Sari-gorla et al., 1988). The yield improvement was observed in the inbred lines by pollen selection than in the control plants (Melian, 1994) when pollen was selected against toxin applied to the stigmatic tissue. In the case of *Brassica – Alternaria* interaction, destruxin B, a pathotoxin of *Alternaria brassica,* could distinguish resistance and susceptibility differences in pollen grain bioassay (Sharma and Tiwari, 1996). Further, the pollen selection for Alternaria leaf blight resistance in sunflowers demonstrated a corresponding increase in progeny resistance (Chikkodi and Ravikumar, 2000). They also found that pollen selection was effective in increasing the proportions of resistant individuals in sunflower for alternaria leaf and stem blight disease after two generations of selection (Chikkodi and Ravikumar, 2003).

Ravikumar and Patil (2004) studied the genetic effect of pollen-based selection for Fusarium wilt in chickpeas. They selected JG 62 (susceptible) control by homozygous dominant *H1* gene and WR315 (tolerant) control by homozygous recessive *h1* is resistant to fusarium wilt. The F_1 plant was produced by crossing JG62 and WR315, the pollen selection was applied in the F_1 generation, and the segregating pattern was checked by the ASAP (Allele-specific amplification polymorphism) marker in the F_2 population. The Chi-square test showed a significant deviation from the expected monohybrid ratio (3:1) in the selected population. The observed number of plants with ASAP amplification was significantly less than expected, and the plants without ASAP (linked to H1 locus) amplification were more than that expected. The number of the resistant plants was more than that of the susceptible plants in the case of the selected population as compared to the control. Similarly, this study was confirmed by more number of linked and unlinked markers and backcross and segregating populations (Ravikumar et al., 2007; Babu and Ravikumar, 2010; Ravikumar et al., 2013) (Table 8.1). Pollen selection channelizes the tolerant alleles, but there are chances that it may result in low productivity due to linkage drag. It was also observed that the effect of pollen selection also leads to variability of non-target traits in the F_2 generation (Choukimath and Ravikumar, 2011). The result clearly showed that the early selection for fusarium wilt result improved the sporophytic performance. It clearly indicates that pollen selection can be easily combined with a classical crop improvement program. The pollen-based selection is more promising as the pollen selection allowed the selection of rare positive allelic combinations that would barely be detected at the sporophytic level (Shobharani and Ravikumar, 2006). We can say that the studies of pollen selection for biotic stress improvement are very limited. But these studies clearly indicate the pollen-based selection potential technique for biotic stress improvement. It also allows selection in favor of recessive alleles at the haploid level, with the advantage that they are not masked by dominant alleles. Thus, there is a potential benefit of employing pollen selection for recessive alleles in crop improvement.

TABLE 8.1 The Improvement of Biotic Stress Tolerance by Pollen-Based Selection in Different Crops.

Crop	Disease	Traits	Improvement (%)	Reference
Brassica juncea	*Alternaria brassicae*	Pollen tube length	40	Shivanna and Sawhney (1993)
Brassica nigra	*Alternaria brassicae*	Pollen tube length	38	Shivanna and Sawhney (1993)
Dianthus spp.	*Fusarium oxysporum* f. sp. Dianthi	Solute leakage	600	Tejaswini (2002)
Helianthus annuus	*Alternaria helianthi* Hanf.	Seed yield per plant	110.60	Rani and Ravikumar (2006)
Cicer arietinum	*Fusarium oxysporum* f.sp. cicero	Pollen tube length	50	Babu and Ravikumar (2010)

8.3 POLLEN-BASED SELECTION STRATEGY FOR IMPROVEMENT OF ABIOTIC STRESS TOLERANCE

There are many approaches available for the improvement of abiotic stress tolerance in plants. But there is a lack of comprehensive screening methods for abiotic stress, and screening at the field level is very difficult because of fluctuations in the environment and differences in field conditions. So, we need to develop methods by which we can screen the crops under control conditions. The gametophytic/pollen selection could be a potential option. There are three ways of gene expression gametophytic and sporophytic phases: (1) diploid expression (genes expressed only in the sporophyte), (2) haploid expression (genes expressed only in the gametophyte), (3) haplodiploid expression (genes expressed in both gametophyte and sporophyte). The haplodiploid expression is suitable for gametophytic/pollen selection. Genes expressed only in sporophytes can also be used, but this should be closely linked to genes expressed in gametophytes in addition to the overlaps of gene expression; there is a need for correlation between the two phases at the phenotypic level. The studies on the possibility of correlated response between the two generations are hinted (Hormaza and Herrero, 1996; Ravikumar and Chikkodi, 1998) and proved by subjecting to selection against different abiotic stresses.

TABLE 8.2 The Improvement of Abiotic Stress Tolerance by Pollen-Based Selection in Different Crops.

Crop	Abiotic stress	Traits	Improvement (%)	Reference
Lycopersicon esculentus	Salt	Dry weight	25	Sacher et al. (1983)
Zea mays	Temperature	Grain yield	35	Landi et al. (1989)
Zea mays	Chlorsulfuron	Tolerant	89-100	Frascaroli and Songstad (2001)
Cicer arietinum	Chilling	Pod yield	42.10	Clarke et al. (2004)
Helianthus annus	Drought (PEG)	Seed germination	38.80	Totsky and Lyakh (2015)
Zea mays	Heat	Seed yield	95.85	Mohapatra et al. (2020)

8.3.1 DROUGHT STRESS

Drought is a complex trait and there is no single trait that was proven to be effective for screening drought stress. Hence, identification of alternative technology, pollen assay, has the potential to solve the purpose. The moisture stress-tolerant genotype produces more tolerant pollen grains in tomatoes (Mulcahy et al., 1988). Further, PEG-induced stress in vitro conditions improve the performance at the field level in sunflower genotypes. It indicates the correlated responses between the gametophytic and sporophytic stages (Keshavamurthy, 1994). Similarly, a correlated response between gametophytic and sporophytic was observed in wheat (Morgan, 1999). Further, these studies extended to F_1 and F_2 generation and the result clearly showed that the PEG-treated pollen grains improve the performance at the field level for different parameters such as seedling tolerance, seed set, etc., in sorghum (Ravikumar et al., 2003). The pollen selection for moisture stress tolerance using PEG in F_1 generation showed that the pollen selection treatment significantly increased the drought tolerance of F_2 populations (Table 8.2) (Totsky and Lyakh, 2015).

8.3.2 *HEAT STRESS*

The pollen-based selection can be capitalized which offers the opportunity to introduce early selection in breeding programs (Singh et al., 2017). The plants derived from the heat-treated flowers were more heat-tolerant than plants derived from normal temperature-treated flowers in maize exposed to high temperatures (Petolino et al., 1990). Further studies proved that intensive pollen selection for heat stress in F1 generation improved the yield-related traits in F2 generation in maize (Table 8.2) (Landi et al., 1989). Milatovic and Nikolic (2014) showed differential response of sweet cherry cultivar to the temperature during the pollination period. Burke and Chen (2015) provided proof of the concept of pollen assay for heat tolerance by the transgenic approach. They selected *AtHSP101 genes and* overexpressed them in cotton. The in vitro pollen germination and pollen tube length were improved in transgenic plants as compared to wild type. In addition, sporophytic performances such as boll set and seed numbers were significantly improved in transgenic cotton lines exposed to high day and night temperatures in field and greenhouse conditions. Further, studying pollen-based selection progress, Mohapatra et al. (2020) clearly showed that the application of pollen selection in the F_1 generation improved the sporophytic performance in the F2 generation (Table 8.2). The genetic evidence of pollen-based selection showed that the F2 population derived from pollen selection shifted toward the tolerant parent.

8.3.3 *OTHER ABIOTIC STRESSES*

There are many reports available where pollen-based selections improved the abiotic stress. The competitive ability of pollen is higher for those tomato genotypes that are grown at high altitudes, compared to that of *Lycopersicon esculentum* grown under cold conditions. Further work concluded that this effect was due to the haploid genome. Furthermore, they showed that some genes responsible for cold tolerance were located in the haploid genome linked to sporophytic isozymes (Zamir et al., 1981, 1982). Further, this study progressed in chickpeas; it was clearly shown that chilling tolerant pollen fertilizes the ovule at low temperatures. It was tracked by a morphological marker such as flower color. They also observed that the chilling tolerance improved in those progenies which are derived from pollen selection (Table 8.2) (Clarke et al., 2004). The first indirect evidence of salinity tolerance was observed in coastal plants that produced more tolerant pollen grains than those plants grown away from the sea coast (Eisikowitch and Woodell, 1975). It means those plants/pollen exposed to saline conditions produce more tolerant pollen grains. Later studies showed that the genotypes grown under salt stress produced pollen grains tolerant to salt stress (Sari-Gorla et al., 1988). The contribution of gametophyte in the inheritance of salt tolerance in barley is notable. The application of salt reduces performance in pollen germination and yield in nonsaline conditions. It shows the slow reduction in all phenotypic parameters at high salinity. These results have indicated the possibility of application of stress at pollen grain level to achieve transmission of alleles of selected traits preferentially from gametophytic generation to succeeding sporophytic progeny levels (Gul and Ahmad, 2006). There are many studies available that showed that the heavy metal reduced the pollen germination and pollen tube length resulting in a reduction in the grain yield (Gur and Topdemir, 2005). The screening of the plants for heavy metal toxicity by pollen germination and pollen tube length appears suitable. But further progress is required to prove the positive effect of pollen-based selection on sporophytic performance. Apart from the above abiotic factors, other stress factors like ozone also cause reduced pollen germination and pollen tube growth, but the response was variable depending on the origin of the pollen grains: pollen from sporophytes susceptible to ozone was more affected than those from less susceptible sporophytes. A similar correlation in the tolerance to ozone between sporophytes and male gametophytes was obtained in several fruit crops also (Hormaza et al., 1996).

We can conclude that abiotic stress had an effect on pollen germination, pollen tube growth, and fertilization. It directly affects the seed set. The pollen-based selection in the F_1 generation of populations and subsequent generations increases the number of heat-tolerant plants under natural abiotic stress conditions. The pollen-based selected progenies showed improvement in the pollen quality and quantity, which resulted in improved sporophytic performance than non-pollen selected progenies under different abiotic stress conditions. Molecular genetic evidence results confirmed the positive effect of pollen selection on parental allele frequencies for the abiotic stress, as the population skewed toward the abiotic stress-tolerant alleles. In addition, the continuous pollen-based selection for abiotic stress tolerance in the early generations has resulted in the improved tolerance of subsequent generations' progenies. Thus, we can combine this pollen-based (gametophyte) selection strategy with sporophytic selection for the improvement of abiotic stress tolerance in crops.

8.4 CASE STUDIES ON PBS-BASED HEAT STRESS TOLERANCE

Heat stress is a polygenic trait and different types of screenable traits/characters are already identified. However, screening for heat stress under field conditions has many challenges due to the environment and its interaction with other factors. Alternatively, pollen grains can be independently exposed to heat stress and afterward tested for their performance in pollen germination, viability, fertilization success, and finally seed set to determine. Pollen could be used to screen for abiotic stress such as heat stress. The pollen grains of 10 inbred lines were subjected to heat stress 36°C-2 h and 4 h duration and cultured on a pollen germination medium. Heat stress showed a significant variation in pollen germination and pollen tube length. It is clearly suggesting the genetic variability of pollen performance against heat stress tolerance. At the same time heat stress treated (36°C-3 h) pollen grains were also used to self-fertilize the field-grown plants of these 10 inbred lines. Heat stress significantly reduced the seed set in all the 10 inbred lines. The lines which showed tolerance of pollen grains to heat stress under in vitro conditions produced more seed sets per cob after heat treatment to pollen grains. The pollen performance in vivo and in vitro indicated the potentiality of using pollen bioassay as an efficient and reliable technique to screen the maize inbred lines for heat tolerance (Singh et al., 2017).

The pollen-based selection is used not only in the screening of polygenic traits but also used for the improvement of sporophytic performance for abiotic stress such as heat stress. There is a genetic overlap of structural genes expressed in both sporophytic and gametophytic generations. This overlap makes it possible to select superior pollen genotype. In this context, we have chosen BTM6 (heat stress-tolerant) and BTM4 (heat stress susceptible) (Singh et al., 2017). The heat stress pollen was used to self-pollinate the plants and produce pollen-selected F_2 progenies and control F_2 progenies were produced by selfing the pollen grains without exposing pollen grains for heat stress. The number of seeds per cob of pollen selected F_2 progenies showed 412.82% more number of seed per plant over the control F_2. Similarly, F_3 progenies were also produced by pollen selection and without selection. Further, these F_3 progenies were screened for heat stress tolerance and the result clearly showed the two-generation pollen-selected progenies showed 122.74% more number of seeds per cob than the control.

Similarly, screening at F_4 progenies at the seedling stage showed the same trend (Singh et al., 2020). The F_2 populations were screened with the polymorphic molecular marker for parents and the result clearly showed that the frequency of tolerant parents increased significantly in the pollen selected population. Therefore, the pollen selection in the F_1 generation has resulted in the improved heat stress tolerance of F_3 and F_4 progenies. In addition, the generation-wise repeated pollen selection for heat stress tolerance has resulted in the improved tolerance of F_4 progenies (Fig. 8.1). Thus, we can combine this pollen-based selection strategy with sporophytic selection to improve heat-stress tolerance in crops.

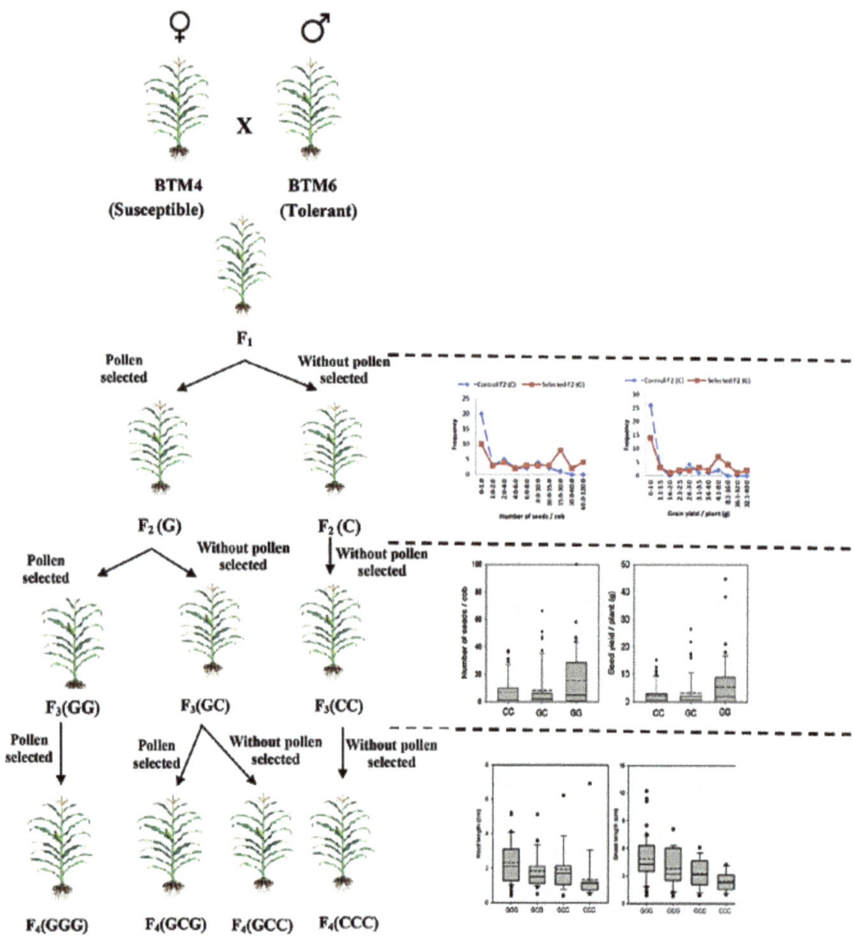

FIGURE 8.1 Pollen selection mediated improvement of heat stress tolerance in maize.

8.5 CONCLUSION AND FUTURE PROSPECTS

The major limitation of conventional breeding is the lack of comprehensive screening methods for complex polygenic traits such as drought, heat, etc. Therefore, pollen-based selection/response has been proposed as an alternate inclusive technique (Sakata and Higastitani, 2008). Pollen-based selection improves stress tolerance by channelizing the variation in less time. The cyclic pollen-based selection for desirable traits will lead to homozygosity, reducing the breeding cycle. However, this process needs at least 4–5 generations/seasons of crops (Singh et al., 2020). Speed breeding platforms could be used to increase the efficiency of pollen-based selection. The pollen-based selection at the F_1 generation increases the tolerant plants in the F_2 generation by channelizing the variation. Further, the F_2 population can be forwarded by speed breeding to produce tolerant genotypes in significantly less time. Pollen-based selection may also be employed as a foreground selection and a molecular marker used for background selection and forwarded with the SB platform. Combining pollen-based selection, marker-assisted selection, and speed breeding is the potential approach for incorporating desirable alleles in significantly less time, even for the complex polygenic trait. Overall, the pollen-selected progeny is more vigorous and tolerant to adverse conditions (Singh et al., 2020; Mohapatra et al., 2020). Therefore, pollen-based selection will also overcome the problem in speed breeding, such as less seed set and poor germination. That can help to shape our future breeding programs at the field level.

KEYWORDS

- **pollen-based selection**
- **gametophytic generation**
- **sporophytic generation**
- **marker-assisted selection**

REFERENCES

Babu, D. R.; Ravikumar, R. L. Parallel Response Between Gametophyte and Sporophyte for Fusarium Wilt Resistance in the Recombinant Inbred Lines of Chickpea (Cicer arietinum L.). *Curr. Sci.* **2010,** 513–518.

Burke, J. J.; Chen, J. Enhancement of Reproductive Heat Tolerance in Plants. *PLoS One* **2015,** *10* (4), 122933.

Chikkodi, S. B.; Ravikumar, R. L. Influence of Pollen Selection for Alternaria helianthi Resistance on the Progeny Performance Against Leaf Blight in Sunflower (Helianthus annuus L.). *Sex Plant Reprod.* **2000,** *12*, 222–226.

Chikkodi, S. B.; Ravikumar, R. L. Effect of Pollen Selection for Alternaria Blight Resistance in Sunflower on F2 Generation. In *Extended summaries, National Seminar on Stress Management in Oil Seeds for Attaining Self-Reliance in Vegetable Oils*; Hyderabad, India, 2003; pp 29–31.

Choukimath, A. M.; Ravikumar, R. L. Influence of Gamete Selection for Wilt Resistance in the F1 Generation on the Variability of Non-target Quantitative Traits in F2 Generation of Chickpea. *Indian J. Genet. Plant Breed.* **2011,** *71* (3), 269.

Clarke, S. M.; Mur, L. A.; Wood, J. E.; Scott, I. M. Salicylic Acid Dependent Signaling Promotes Basal Thermotolerance but is not Essential for Acquired Thermotolerance in Arabidopsis thaliana. *Plant J.* **2004,** *38* (3), 432–447.

Eisikowitch, D.; WoodelL, S. R. Some Aspects of Pollination Ecology of Armeria maritima (Mill.) Willd. in Britain. *New Phytol.* **1975,** *74* (2), 307–322.

Fehr, W.R. In *Principles of Cultivar Development*; Iowa State University, Macmillian Publishing Company: New York, 1991; p 593.

Frascaroli, E.; Songstad, D. D. Pollen Genotype Selection for a Simply Inherited Qualitative Factor Determining Resistance to Chlorsulfuron in Maize. *Theor. Appl. Genet.* **2001,** *102* (2-3), 342–346.

Gul, H.; Ahmad, R. Effect of Salinity on Pollen Viability of Different Canola (Brassica napus L.) Cultivars as Reflected by the Formation of Fruits and Seeds. *Pak. J. Bot.* **2006,** *38* (2), 237.

Gür, N.; Topdemir, A. Effects of Some Heavy Metals on in vitro Pollen Germination and Tube Growth of Apricot (Armenica vulgaris Lam.) and Cherry (Cerasus avium L.). *World Appl. Sci. J.* **2008,** *4* (2), 195–198.

Haldane, J. B. S. The Time of Action of Genes, and its Bearing on Some Evolutionary Problems. *Am. Nat.* **1932,** *66* (702). 5–24.

Hormaza, J. I.; Herrero, M. Pollen Selection. *Theor. Appl. Genet.* **1992,** *83*, 663–672.

Hormaza, J. I.; Herrero, M. Male Gametophytic Selection as a Plant Breeding Tool. *Sci. Hortic.* **1996,** *65* (4), 321–333.

Hormaza, J. I.; Pinney, K.; Polito, V. S. Correlation in the Tolerance to Ozone Between Sporophytes and Male Gametophytes of Several Fruit and Nut Tree Species (Rosaceae). *Sex. Plant Reprod.* **1996,** *9* (1), 44–48.

Keshavamurthy, M. N. Male Gametophytic Selection in Identification and Development of Moisture Stress Tolerant Lines in Sunflower. M.Sc. (Agri.) Thesis, University of Agricultural Sciences, Bangalore, 1994.

Landi, P.; Frascaroli, E.; Tuberosa, R.; Conti, S. Comparison Between Responses to Gametophytic and Sporophytic Recurrent Selection in Maize (Zea mays L.). *Theor. Appl. Genet.* **1989,** *77* (6), 761–767.

Melian, L. G. Method of Pollen Selection in Plants for Resistance to Phytopathogens (Exemplified by Tomato). *Sel'skokho zyaistvennaya Biol.* **1994,** *1*, 121–129.

Milatovic, D.; Nikolic, D. In *The Effect of Temperature on Pollen Germination and Pollen Tube Growth In Vitro of Sweet Cherry Cultivars*, VII International Cherry Symposium, 2014; vol *1161*, pp 401–404.

Mohapatra, U.; Singh, A.; Ravikumar, R. L. Effect of Gamete Selection in Improving of Heat Tolerance as Demonstrated by Shift in Allele Frequency in Maize (Zea mays L.). *Euphytica* **2020,** *216*, 76.

Morgan, J. M. Pollen Grain Expression of a Gene Controlling Differences in Osmoregulation in Wheat Leaves: A Simple Breeding Method. *Aust. J. Agric. Res.* **1999,** *50*, 953–962.

Mulcahy, D. L. The Rise of the Angiosperms: A Genecological Factor. *Science* **1979,** *206* (4414), 20–23.

Mulcahy, D. L.; Mulcahy, G. B.; Popp, R.; Fong, N.; Pallais, N.; Kallinowski, A.; Marien, J. N. Pollen Selection for Stress Tolerance or the Advantage of Selecting Before Pollination. *Sex. Plant Reprod.* **1988,** 41–50.

Ottaviano, E.; Sari Gorla, M. Gametophytic and Sporophytic Selection. In *Plant Breeding: Principles and Prospects*; Hayward, M. D., Bosemark, N. O., Romagosa, I., Eds.; 1993; pp 332–352.

Ottaviano, E.; Sari-Gorla, M.; Mulcahy, D. L. Pollen Tube Growth Rates in Zea mays: Implications for Genetic Improvement of Crops. *Science* **1980,** *210* (4468), 437–438.

Petolino, J. F.; Cowen, N. M.; Thompson, S. A.; Mitchell, J. C. Gamete Selection for Heat-Stress Tolerance in Maize. *J. Plant Physiol.* **1990,** *136*, 219–224.

Ravikumar, R. L.; Chaitra, G. N.; Choukimath, A. M.; Soregaon, C. D. Gametophytic Selection for Wilt Resistance and its Impact on the Segregation of Wilt Resistance Alleles in Chickpea (Cicer arietinum L.). *Euphytica* **2013,** *189* (2), 173–181.

Ravikumar, R. L.; Chikkodi, S. B. Association Between Sporophytic Reaction to Alternaria helianthi and Gametophytic Tolerance to Pathogen Culture Filtrate in Sunflower (Helianthus annuus L.). *Euphytica* **1998,** *103* (2), 173–180.

Ravikumar, R. L.; Patil, B. S. Effect of Gamete Selection on Segregation of Wilt Susceptibility-Linked DNA Marker in Chickpea. *Curr. Sci.* **2004,** *86*, 642–643.

Ravikumar, R. L.; Patil, B. S.; Salimath, P. M. Drought Tolerance in Sorghum by Pollen Selection Using Osmotic Stress. *Euphytica* **2003,** (3), 371–376.

Ravikumar, R. L.; Patil, B. S.; Soregaon, C. D.; Hegde, S. G. Genetic Evidence for Gametophytic Selection of Wilt Resistant Alleles in Chickpea. *Theor. Appl. Genet.* **2007,** *114* (4), 619–625.

Sacher, R. F.; Mulcahy, D. L.; Staples, R. C. Developmental Selection During Self Pollination of Lycopersicon x Solanum F1 for Salt Tolerance of F2. In *Pollen: Biology and Implication for Plant Breeding*; Mulcahy, D. L., and Ottaviano, E., Eds.; Elsevier Press: New York, NY, 1983; pp 329–334.

Sakata, T.; Higashitani, A. Male Sterility Accompanied with Abnormal Anther Development in Plants–Genes and Environmental Stresses with Special Reference to High Temperature Injury. *Int. J. Plant Dev. Biol.* **2008,** *2* (4).

Sari-Gorla, M.; Mulcahy, D. L.; Gianfrancheschi, L.; Ottaviano, E. Gametophytic Selection for Salt Tolerance. *Genet. Agrar.* **1988,** *42*, 92–93.

Sharma, T. R.; Tewari, J. P. Flow Cytometric Analysis of Brassica Junceacell and Pollen Cultures Treated with Destruxin B, A Toxin Produced by Alternaria brassicae. *Physiol. Mol. Plant Pathol.* **1996,** *48* (6), 379–387.

Shobha Rani, T.; Ravikumar, R. L. Sporophytic and Gametophytic Selection for Improvement of Partial Resistance to Alternaria Leaf Blight in Sunflower (Helianthus annuus L.). *Euphytica* **2006,** *147*, 421–431.

Singh, A.; Antre, S. H.; Ravikumar, R. L.; Kuchanur, P. H.; Lohithaswa, H. C. Genetic Evidence of Pollen Selection Mediated Phenotypic Changes in Maize Conferring Transgenerational Heat-Stress Tolerance. *Crop Sci.* **2020,** *60* (4), 1907–1924.

Singh, A.; Kuchanur, P. H.; Ravikumar, R. L. Identification of Heat Tolerant Inbred Lines using TIR Technique and its Association with Field Tolerance. *Bioscan* **2017,** *12*, 2053–2058.

Totsky, I. V.; Lyakh V. A. Pollen Selection for Drought Tolerance in Sunflower. *Helia* **2015,** *63* (38), 152–166.

Zamir, D.; Tanksley, S. D.; Jones, R. A. Low Temperature Effect on Selective Fertilization by Pollen Mixtures of Wild and Cultivated Tomato Species. *Theor. Appl. Genet.* **1981,** *59* (4), 235–238.

Zamir, D.; Tanksley, S. D.; Jones, R. A. Haploid Selection for Low Temperature Tolerance of Tomato Pollen. *Genetics* **1982,** *101* (1), 129–137.

CHAPTER 9

From Genetics to Genomics: Connecting Conventional to Modern Approaches for Designing Climate-Resilient Underutilized Orphan Crops

AJAY KUMAR CHANDRA[1], AMARJEET KUMAR[2], ALKA BHARTI[3], LOVE KUMAR[4], and ASHUTOSH SINGH[5]

[1]*ICAR-Indian Agricultural Research Institute, New Delhi, Delhi, India*

[2]*MTTC & VTC, Selesih, Central Agricultural University, Imphal, Mizoram, India*

[3]*ICAR-Central Agro-Forestry Research Institute (CAFRI), Jhansi, Uttar Pradesh, India*

[4]*Centre for Protected Cultivation Technology, IARI, New Delhi, India*

[5]*Centre for Advanced Studies on Climate Change, RPCAU, Pusa, Samastipur, India*

ABSTRACT

The unprecedented alteration in global climatic conditions, especially rise in average temperature due to excessive release of greenhouse gasses, has resulted in low production of major cereals. Under such global food and nutrient insecurity condition, millet crops can be an alternative that can supply an adequate amount of nutritionally enriched food to all growing mouths. Due to its innate ability to withstand such adverse conditions without affecting its productivity and nutritional contents, the crop is considered to be "super cereals" and the "crop of the future." But due to some neglect and knowledge

Climate-Resilient Agriculture: A Molecular Perspective. Ashutosh Singh, Saurabh Pandey (Eds.)

gaps, most millet crops were not completely explored and were underutilized by our ancestors since long. However, in past few years, several conventional breeding programs have been developed to improve these millet crops. But comprehensive success has not been achieved by these scientific groups. It is either because of low genetic and molecular inputs or low responsiveness of crops under stress-induced environments; although the crop has innate ability. Therefore, it needs to take a U-turn from conventional to modern molecular breeding technologies to resolve these issues. It is necessary to adopt newly developed modern tools and techniques that can make it easy to explore the genetics and genomics behind the targeted improvement of these underutilized millet crops.

9.1 INTRODUCTION

Millets represent an assorted group of small-grained minor cereal crops widely grown for feed, forage, and food. It has been predominantly cultivated in regions of central, south-east Asia, North Africa, and East Africa since the prehistoric era and is probably considered one of the oldest known food to humans. Although the crop is underutilized for long period, millets are designated as a "super cereal" because of their rich nutritional values and unique nutraceutical properties, often much higher than that of major cereals. Besides these, the millets are organic by default as they can capitalize on low fertilizer input, which indicates that millets can be cultivated easily in nutrient-poor soil and are widely adaptable (Chandra et al., 2020a). It is probably because of its C_4 nature that makes them a "climate-smart" crop. Moreover, the crop has a good resource for genetic variations that can strengthen the stress resilience of other model crops in the era of unprecedented climate and nutrient-resilient conditions. Further, earlier studies suggested that breeding for climate resilience through conventional approaches owe to be more challenging probably because of the complex genome, ploidy level, occurrence of allele-specific stress regulatory factors, etc. Therefore, it is an urgent need to take a U-turn from conventional to modern genomics approaches, which could be the only feasible strategy to sustain these underlying concerns (Chandra et al., 2021). Recently, several cutting-edge genetics to genomics approaches have been developed such as molecular breeding tools, diversity analysis, genomic selection, gene discovery, expression profiling, and transgenic development that could serve as sustainable and efficient strategies for the development of abiotic stress tolerance of millet crops (Ceasar et al., 2018). Moreover, the advancement of next-generation genomic biotechnologies, such as high-throughput sequencing technology, TILLING, EcoTILLING, genome editing, transcriptome sequencing, RNA editing, system biology, etc., will certainly complement the knowledge gaps and future genomic research perspective for the development of more climate-resilient millet crops. Hence, with this, this chapter aimed to present and boost the innate ability of millets to develop tolerance to various abiotic stresses through bridging the genetics to modern genomics approaches so that the orphan crops can sustain future global food and nutrient security concerns despite the growing climate and nutrient-resilient condition.

9.2 CLIMATE-SMART AGRICULTURE

To address the sustainable and increased millet production under altered climate changes, it requires a transition of agricultural systems. Sustainable agricultural systems boost the overall millet production and in turn address shocks created by climate risks. The so-called climate-smart agriculture (CSA) is a type of sustainable agricultural system that can contribute to the preservation and mitigation of climate change. The recently evolved concept of climate-smart agriculture is the inbuilt method developed based on the real-time experience of the grounds. The on-ground sustainable agricultural practices can address various agricultural challenges evolved under changing climate. Climate-smart agriculture by adopting various sustainable agricultural practices can achieve three major dimensions, (1) increased agricultural productivity by conservation of biodiversity, (2) building resilience and adapting to climate change, and (3) reducing the emission of greenhouse gas (Amin et al., 2015). The available plant genetic resources as a major constituent of sustainable agricultural can also contribute to crop improvement so that it can resist climate-induced various biotic and abiotic stresses (Wambi et al., 2021). In addition to this, different genome-aided tools assisted by genomic resources can also contribute to the sustainable development of climate-smart crop varieties (Ceasar et al., 2018). Earlier studies also reported the success of sustainable agriculture achieved by the effective utilization of genetic and genomic resources. Since the millet is widely grown on marginal land by substantial farmers, this could lead to the implementation of different mitigation and adaptation options to ensure food and nutritional security while preserving the available agroecosystem.

9.3 MILLET CROPS: AN UNDERUTILIZED ORPHAN CROP

Over 50% of the world's population relies upon major cereals. Rice, wheat, and maize are the most preferred cereals, while millets have been largely neglected even after the green revolution. In broad terms, millets represent small-seeded grasses that are widely grown for food, feed, and fodder (Chandra et al., 2020b; Lata et al., 2015). Millets comprise crop species that largely originated and domesticated in third-world countries. The crop is widely cultivated by marginal farmers across the globe, mainly in semi-arid, arid, and tropical regions. The millet is considered a "super" and/or "stress-resilient" crop because of the following reasons—(1) nutritionally enriched crop, (2) adaptive to adverse climatic conditions, and (3) requirement of minimal inputs (Lata et al., 2015; Chandra et al., 2020b). Besides, the millets also offer huge genetic resources necessary for the breeding of food-based crops that can ensure future food and nutritional security concerns. Additionally, all-season growth and early maturation characteristics make them an ideal candidate crop for more intensive cropping. The millet can also be utilized as a relay or catch crop in combination with other slow-maturing crops. Despite the fact, millet is sometimes considered as "lost crops" or "orphan crops." The crops are not lost but the terminology indicates their relatively low production and productivity compared to major cereal crops. However, in many terms, these neglected crops contributed significantly to conserving biodiversity and the livelihood of the poor farmers. Millets are composed of six major small-seeded crops, namely, foxtail millet (*S. italica*), finger millet (*E. coracana*), kodo millet (*P. scrobiculatum*), barnyard millet (*Echinochloa spp.*), proso millet (*P. miliaceum*), and little millet (*P. sumatrense*) (Table 9.1). Millet is known to be enriched with nutritional values and has unique nutraceutical properties (Kumar et al., 2016). The millets require very low water and fertilizer inputs for production and even can be cultivated under rain-fed (non-irrigated) conditions. Compared to major cereal crops, millets are considered a "crop for food security" because of their many-fold richness in nutritional contents and low-cost production.

9.4 MILLETS AS POTENTIAL CLIMATE-RESILIENT CROPS: TOLERANCE VS YIELD LOSS

Over the past few decades, climate alteration and its negative impact on agriculture is of great interest (Chandra et al., 2021). So, it has been extensively studied for the possible effect on yield and productivity of major crops. The rate of climate-changing is alarmingly very high. This alarming situation can trigger various abiotic stresses to the standing crops and in turn, can affect directly the global food and nutritional security. The major abiotic stresses faced by millet crops include lodging, waterlogging stress, flooding, heat stress, and drought stress. Most developing countries, especially arid and semi-arid regions of the world, are facing water scarcity problems that directly affect the performance of different crops. Several studies have been carried out on foxtail millet, little millet, finger millet, proso millet, and wild millet and reported a significant decrease in yield performance when grown under drought-prone conditions (Satyavathi et al., 2019; Numan et al., 2021). Likewise, up to 80–90% yield loss was recorded in studied finger millet subjected to drought after 3–4 weeks of sowing (Maqsood and Ali, 2007). In the same way, up to 60% yield loss was reported in studied millets because of terminal drought (Hadebe et al., 2017). The estimated yield loss in pearl millet that occurred due to severe drought was recorded as up to 51% across various germplasm (Ashok et al., 2018). Similarly, undesirable effects of various abiotic stresses including drought, lodging, waterlogging stress, flooding, and heat stress in different millet crops are presented in Table 9.2. It was also reported that undesirable effects of various abiotic stresses not only affect the productivity of millet crops but also affect overall nutritional qualities like grain minerals, protein, fibers, etc., (Saleh et al., 2013). However, millet is resistant to abiotic stresses including heat and drought that can induce various physiological, biochemical, and molecular alterations. In addition, respiration and photosynthesis are the two most sensitive biochemical processes to heat and drought which can dramatically affect the yield performance of millet crops (Sood et al., 2019).

9.5 AVAILABLE GENETIC RESOURCES FOR CLIMATE-SMART BREEDING

Diversity in the crop is one of the most essential criteria for the sustainable development of agriculture. Genetic resources like germplasm, cultivars, landraces, etc., provide the requisite genetic variability for the improvement of the crop. The contracted genetic base of cultivars resulted in an increased risk of vulnerability to crop failure. It might occur due to disease epidemics, insect-pest infestation, and/or unpredictable climatic alterations. Globally, about 135,000 cultivated germplasm of millets including more than 30,627

TABLE 9.1 Details of Different Millet Crops Widely Cultivated Across the Arid and Semi-Arid Regions of the World.

S. No.	Millet crops	Species	Ploidy	Chromosome number	Domestication	Cultivated germplasm in India	Regions of cultivation
	Finger millet	*E. coracana*	Tetraploid	2n = 4X = 36	East Africa and India	22,583	India, Nepal, Kenya, Zimbabwe, Uganda, Rwanda, Sudan, Eritrea, Malawi, Zambia, Madagascar, Burundi, Sri Lanka, Myanmar, and China
	Pearl millet	*P. glaucum*	Diploid	2n = 2x = 14	West Africa	7980	India, Sudan, Nigeria, Niger, Mali, Sudan, and Namibia
	Foxtail millet	*S. italica*	Diploid	2n = 2X = 18	Eastern Asia	8377	India, Japan, Nepal, Korea, and China
	Kodo millet	*P. scrobiculatum*	Tetraploid	2n = 4X = 40	India	3937	India and West African countries
	Proso millet	*P. miliaceum*	Tetraploid	2n = 4X = 36	Egypt and Arabia	849	India, Sri Lanka, Nepal, China, Pakistan, and southeast African countries
	Indian barnyard millet	*E. frumentacea*	Hexaploid	2n = 6X = 36	India	743	India, Nepal, Pakistan, and central African countries
	Japanese barnyard millet	*E. esculenta*	Hexaploid	2n = 6X = 36	Japan	NA	Japan, India, China, and Korea
	Little millet	*P. sumatrense*	Tetraploid	2n = 4X = 36	South east Asia	473	India, Sri Lanka, Nepal, Myanmar, China, Pakistan, and southeast Asian countries

TABLE 9.2 Effect of Different Abiotic Stresses on Yield Loss (Percent or, kg.ha^{-1}) of Millet Crops.

S. No.	Millet crops	Drought	Heat	Water lodging	Lodging	References
	Finger millet	61%	75–84% at 38/28°C	42.14 kg.ha^{-1}	–	Hittalmani et al. (2017)
	Foxtail millet	20.3%	60% at 38/28°C	–	41.2–51.1%	Matsuura et al. (2012)
	Kodo millet	–	–	42.84 kg.ha^{-1}	–	Jency et al. (2020)
	Barnyard millet	63.5%	–	–	–	Hittalmani et al. (2017)
	Pearl millet	60.1%	70–75% at 36/26°C	–	–	Yadav et al. (2020)
	Little millet	80.5%	–	18.14 kg.ha^{-1}	–	Hittalmani et al. (2017)
	Proso millet	64%	–	–	–	Matsuura et al. (2012)

accessions of these genera are conserved in gene banks (Table 9.1). Region-wise, about 13.5% of accessions of small millets are conserved in Europe, 13.8% in Africa, and the majority of accessions (64.4%) are conserved in Asia. However, the figures might vary slightly because some germplasm can be duplicated between and/or within the genebanks. Indian and African countries, such as India, Nepal, Pakistan, Ethiopia, Zambia, Uganda, and Kenya, are commonly known as "home of finger millet" and have the majority of finger millet collections. Similarly, India, Japan, China, France, and some East African countries have conserved the majority of foxtail millet germplasm. Correspondingly, barnyard millet is in India and Japan; proso millet is in India, China, Russia, and Ukraine; Kodo millet is in the USA and USA; and the major collections of little millet germplasm is conserved in India (Numan et al., 2021). Institute-wise, the largest collection center for small millets conserved across the countries are little millet (1253), Kodo millet (2180), and finger millet (9522) at NBPGR, India; foxtail millet (26,233) at ICS-CAAS, China; barnyard millet (3671) at NIAS, Japan; proso millet (8778) at NRSRIPI, Russia. Except for foxtail millet, proso millet, and finger millet, germplasm conserved across various collection center or gene banks are very low for other small millet crops. This could be because of low research priority or most landraces are already been lost. This can occur due to shifting from a traditional cropping system to cash cropping for the improvement of hybrids and varieties. Therefore, conservation and collection of existing genetic resources are very crucial before complete genetic erosion, particularly in small millets like barnyard millet, Kodo millet, and little millet where very little (<4000) germplasm is conserved globally.

9.6 PROSPECTS FOR CLIMATE-SMART BREEDING USING GENOME-AIDED TOOLS

9.6.1 MARKER-ASSISTED BREEDING

Application of genomic tools and techniques is one of the prerequisites of any breeding programs that have been proven to be useful for mining novel QTLs, alleles, or genes governing traits of interest like stress tolerance and other agronomic importance. Genomic tools like marker-assisted selection (MAS) and introgression of the desired gene into particular genetic backgrounds have an immense potential to assist the improvement of crops by minimizing the lengthy selection and phenotypic evaluation process. Recently, several high-throughput molecular breeding tools have been developed that can assist the improvement of different crops by innovation in population genotyping (Fig. 9.1). In the last few decades, PCR-based markers have mostly been used for marker-assisted selection, genotyping, and diversity analysis programs. For the identification of desirable genes, functional gene markers have readily been used for the construction of linkage maps and genome-wide association analysis (GWAS) (Sehgal et al., 2012). This approach was proven practical for the identification of candidate genes putatively involved in drought tolerance in major crops like rice, wheat, maize, and barley (Diab et al., 2008; Serba and Yadav, 2016). In the last two decades, sizable progress has only been made in orphan crops like pearl millet for drought tolerance using the marker-assisted selection (MAS) method (Vadez et al., 2012). This study proved the efficiency of marker-assisted breeding for QTL detection, functional marker discovery, and genotyping in the orphan crop. However, in most orphan crops, breeding through marker-assisted breeding are still in a nascent phase.

9.6.2 GENOMIC SELECTION

The MAS uses the linkage disequilibrium principle with major QTLs using molecular markers and inherited together. Recently, the genomic selection method is developed as a modern extension of the marker-assisted selection (MAS) tool (Serba and Yadav, 2016). In genomic selection, genome-wide markers were employed in a large breeding population for targeted improvement of desirable quantitative traits. Genomic selection resulted in the selection of promising individuals based on higher genomic estimated breeding values (GEBV). It combines phenotype with marker genotype data from different environments and kinship or pedigree was performed to estimate breeding values. This selection method based on GEBV was anticipated to speed up the genetic gain which was substantially higher than conventional marker-assisted selection (Zhao et al., 2015). The method has been successfully employed in several crops for the improvement of desirable traits. Being a minor and neglected crop, the millets have not yet been evaluated using this genomic selection method. Similarly, none of the reports are available that can pinpoint its application in crop improvement, particularly for drought, heat, water lodging, cold, etc., in these orphan crops. However, as the earlier reports stated in other major crops, this method can also be efficiently used for the selection and high-throughput improvement of these orphan crops.

9.6.3 TRANSGENIC

Transgenic technology or transgenics is a functional genomics-based tool that provides valuable molecular insights into the metabolic and gene regulatory networks underlying biotic and abiotic stress tolerance in plants (Lata et al., 2015). In recent years, it has been one of the most widely used technology for genetic improvement of crop plants, in particular major cereals. However, the application of transgenomics in these orphan crops is still in the nascent phase despite it having huge nutritional and economic importance. Although some studies have been successfully conducted for the identification and validation of candidate genes in these millet crops (Chandra et al., 2020b). Some of the identified functional genes were successfully introgressed into tobacco and arabidopsis. Most of the candidate genes studied in the millet crops are stress-responsive genes or transcription factors (TFs) that can impart an important role in stress tolerance (Table 9.3). Stress-responsive candidate genes or TFs like MYB, MYC, ABRE, DRE, NAC, AP2/ERFs, LEA, PHGPX, APX, bHLH, WRKY, and C_2H_2 zinc finger were extensively studied in foxtail millet (Muthamilarasan et al., 2014; Lata et al., 2015; Zhang et al., 2017). In addition, several other genes like NAC, apx, ZIP, bHLH, WRKY, AP, MYB, SOD, NAM, HSP, GBF, HYP, and Dehydrin family genes were comprehensively studied in finger millet (Ramegowda et al., 2012; Singh et al., 2014; Parvathi et al., 2019). Likewise, stress-responsive candidate genes or TFs including WRKYs, HSP, and LEA were studied in pearl millet (Reddy et al., 2015; Chanwala et al., 2020). However, these candidate genes or TFs require further characterization using suitable functional genomics approaches so that they can be used for engineering stress resistance. Despite several advancements made in millet functional genomics, the development of efficient protocols for stress-responsive transgenic orphan crops is still in its premature stage.

9.6.4 GENE EXPRESSION AND TRANSCRIPTOME ANALYSIS

The "transcriptome" can be defined as the total of all transcripts present in a cell/tissue or organism and their extensive investigation in terms of relative expression is termed "transcriptomics." Consequently, genome-wide identification and expression analysis serve as powerful functional biology tools for the identification of candidate genes. In the past few years, gene expression and transcriptome analysis targeting candidate genes putatively involved in stress regulatory networks and various biological processes have been increased. Plants exposed to salinity, heat, and drought generally undergo extreme cellular dehydration and therefore most of the genes/TFs expressed

TABLE 9.3 Functional Genomics for Identification of Stress-Responsive Genes or TFs for Improving Stress Tolerance in Millet Crops.

S. No.	Candidate genes/TFs	Source	Abiotic stress tolerance	References
1.	*Late embryogenesis abundant (LEA), phospholipid hydroperoxide glutathione peroxidase (PHGPX), ascorbate peroxidase (APX), and catalase 1 (CAT1)*	Foxtail millet	Tolerance against drought and salinity stress	Lata et al. (2015)
2.	*EcNAC1*	Finger millet	Overexpression induces salinity, drought, and osmotic stress tolerance	Ramegowda et al. (2012)
3.	*Late embryogenesis abundant (LEA) protein*	Pearl millet	Tolerance against heat and salinity stress	Reddy et al. (2015)
4.	*Ec-apx1*	Finger millet	Overexpression of *Ec-apx1* induces drought stress tolerances	Bhatt et al. (2013)
5.	*Metallothionein, Farnesylated protein ATFP6, Farnesyl pyrophosphate Synthase, Protein phosphatase 2A,* and *RISBZ4*	Finger millet	Induced drought stress tolerance	Parvathi et al. (2019)
6.	*ZIP, WRKY29, AP2, MYB, NAM,* and *Vascular ATP synthase family*	Finger millet	Overexpression induces salinity, drought, and osmotic stress tolerance	Rahman et al. (2016)
7.	*C2H2 type zinc finger transcription factors (TFs)*	Foxtail millet	Salinity, cold, and dehydration stress tolerance	Muthamilarasan et al. (2014)
8.	*EcDehydrin7*	Finger millet	Overexpression of *EcDehydrin7* induces drought stress tolerance	Singh et al. (2014)
9.	*HSP70* and *HSP90*	Pearl millet	Multiple-stress tolerance like heat and salinity stress	Reddy et al. (2015)
10.	*Monodehydroascorbate reductase*	Finger millet	Overexpression induced salt, UV radiation, and drought tolerance	Sudan et al. (2015)
11.	*EcbHLH57, SOD, rd29A, rd29B, ADH1, APX, HSP70, LEA14,* and *PP2C*	Finger millet and sorghum	Overexpression induces multiple-stress tolerance	Babitha et al. (2015)

TABLE 9.3 *(Continued)*

S. No.	Candidate genes/TFs	Source	Abiotic stress tolerance	References
12.	*EcNAC 67*	Finger millet	Tolerance against drought and salinity stress	Rahman et al. (2016)
13.	*EcCIPK31*	Finger millet and Foxtail millet	Tolerance against drought stress	Nagarjuna et al. (2016)
14.	*EcJAZ*	Finger millet	Overexpression induces abiotic stress tolerance	Sen et al. (2016)
15.	*EcGBF3, RGAP2, DBH, RSLP*, and *HYP*	Finger millet	Tolerance to salinity, drought, and osmotic stress	Ramegowda et al. (2012)
16.	*bHLH* and *WRKY*	Foxtail millet	Tolerance to salinity and drought stress	Zhang et al. (2017)
17.	*EcbZIP17*	Finger millet	Overexpression induces salinity and heat stress tolerance	Ramakrishna et al. (2018)
18.	*EcTAF6*	Finger millet	Tolerance against drought and salinity stress	Parvathi et al. (2019)
19.	*WRKY TFs*	Pearl millet	Tolerance to salinity and drought stress	Chanwala et al. (2020)

under these stresses are almost common (Reddy et al., 2015; Lata et al., 2015). In general, stress-responsive genes can be categorized into early- and late-responsive genes in terms of their duration of response. Early responsive genes respond within minimal duration after plants are exposed to stress whereas the late responsive genes take even weeks for their relative expression. Early responsive genes like protein kinases involve in various signal transductions and it provides initial protection. In contrast, late responsive genes, like heat shock proteins (HSPs), late embryogenesis abundant (LEA) proteins, reactive oxygen species (ROS) scavengers, etc., are speculated to provide stress adaptation (Table 9.3). Notably, it has been revealed that more than 11% of stress-responsive genes are potential transcription factors (TFs) that regulate the expression of downstream genes putatively involved in stress tolerance (Ramanjulu and Bartels, 2002). Several extensive studies have been carried out in monocots like rice, wheat, and maize for gene expression and transcriptomics studies of stress-inducible genes. The results could also be employed for comparative and functional genomics in these orphan crops. Though, these monocots are probably to be quite different from most millets in terms of morphology, physiology, metabolism, and stress responsiveness because they are physiologically different in terms of C_3 and C_4 crops. Therefore, developing the C_4 millet model by taking C_3 model crops as a reference has become more important under present changing climatic conditions. However, significant progress has been made in millets functional genomics research, but mostly is concentrated on foxtail millet, finger millet, and pearl millets (Table 9.3). Therefore, under the current scenario, there is an urgent need for a paradigm shift in millet functional genomics research to other minor millets. So, it can help to sustain food and nutritional security concerns under altered climatic conditions shortly.

9.6.5 GENE OR GENOME EDITING

In the era of modern molecular biology, gene or genome editing represents a group of advanced molecular tools that can facilitate genome modification more precisely at a particular locus. It is one of the most recently developed molecular biology tools and is grouped into four different classes based on engineered nucleases (EN). It includes clustered regularly interspaced short palindromic repeats (CRISPR)/CRISPR associated 9 (CAS9), zinc finger nucleases (ZFNs), and transcription activator-like effector nucleases (TALENs), and meganucleases (MNs) (Sood et al., 2019; Chandra et al., 2020a). Since its development, application for genetic improvement of major crops like rice, wheat, and maize is being of great interest. Encouraged by the success of genome editing in particular CRISPR/Cas9 in major crops, it has also been considered for improvement of millet crops. For example, *the less shattering1* (*SvLes1*) gene was modified using the CRISPR/Cas9 system and speculated to result in controlled seed shattering in *Setaria viridis* (Mamidi et al., 2020). In other studies, CRISPR/Cas9 system was used for targeted modification of phosphate transporters and resulted in significant down-regulation of *SiPHT1-4* genes in shoot and root tissues of foxtail millet (Ceasar et al., 2018). Like that, several other genome editing events have been successfully performed in millets targeting quality and quantitative traits governing genes. However, to the best of my knowledge, there are least or no reports have been available for editing stress-inducible genes in these millet crops. Therefore, these findings might help in possible CRISPR/Cas9 editing of stress-inducible genes, so that the scientist can engineer stress-resilient millet crops in near future.

9.6.6 THE NEXT-GEN WHOLE GENOME SEQUENCING PROJECT

The genomic sequences are considered as a genetic blueprint of a species that provides valuable genetic information about the structural and functional annotation of the whole genome or a gene. This data can be useful to understand the structure, sequence variation, evolution, diversity, and genome mapping for traits of interest (Serba and Yadav, 2016). The sequence data also facilitates the development of functional genic markers useful for genomics-assisted breeding. Due to the availability of different versatile and high-throughput sequencing platforms, it not only lowers sequencing costs but also offers higher reliability, depth, and long sequence coverage. So far, the whole-genome sequence of the small millets, like finger millet, foxtail millet, Japanese barnyard millet, and proso millet are available (Zhang et al., 2012; Varshney et al., 2017; Hatakeyama et al., 2018; Zou et al., 2019). In addition, chloroplast sequencing of proso millet, barnyard millet, foxtail millet, and little millet have been sequenced. Among them, foxtail millet was the first millets crop whose genome is fully sequenced. It also serves as a C_4 model crop because of its self-pollinating nature, short duration, and small diploid genome (Zhang et al., 2012). Finger millet has the largest genome size (1.5 Gb), followed by barnyard millet (1.27 Gb) while foxtail millet has the smallest one (423–510 Mb) (Table 9.4). Despite the availability of various sequencing platforms, most of the minor millet genomes are under the draft sequence stage which requires further re-annotation and re-sequencing to fill the mis-annotations, chromosomal assignments, and gaps. However, these drafts are adequate for providing sufficient sequence information useful for gene mining and large-scale genotyping. It was reported that genome sequencing of millets is somewhat difficult because of a higher percentage of repetitive DNA and ploidy levels (Serba and Yadav, 2016). However, with the emergence of high-throughput third-generation sequencing platforms, it is expected to resolve all these limitations in near future. Recently ICRISAT, a leading institute in India genotyped six small millets (proso millet, barnyard millet, finger millet, foxtail millet, little millet, and Kodo millet) using a genotyping-by-sequencing (GBS) approach.

9.6.7 SYSTEM-BIOLOGY AND GENOMIC DATABASES

Understanding biological principles underlying complex metabolic and gene regulatory networks requires sound knowledge of system biology assisted with biological databases. Several databases are available that mostly provide gene expression datasets for model plants like rice and arabidopsis. Analysis of gene expression data of more complex genomes like millets requires different bioinformatics tools. These databases provide information like genome sequences, structural organization of genes, expression datasets, presence of SNPs, linked molecular markers, mapping, synteny, etc.; this information can boost the improvement of millet crops for the trait of interest. Until now no or little work has been done for the development of millet databases using systems biology tools (Table 9.4). Millet databases deciphering metabolic and gene regulatory networks underlying stress-responsiveness in different millet crops are not yet developed. Therefore, under current environment-changing conditions, it is very critical to put rigorous efforts into the development of such databases.

9.7 CONCLUSION AND PROSPECTS

Under current un-predictive and changing climatic conditions, millets are considered to be the future crop. Recent developments in molecular biology with the availability of huge genetic and genomic resources for these orphan crops put it over the major cereals like wheat, rice, and maize. Since millet crops are nutritional at par with most cereals under stress conditions so it has been considered food and nutritional-secure crop. Despite its huge importance under stressful conditions, it was neglected for long by our scientific communities. However, it deserves attention from germplasm

TABLE 9.4 Next-Gen Whole Genome Sequencing Project and Genomic Databases in Millet Crops.

S. No.	Millet crops	Genome size	No. of genes	Sequencing platform	Genotypes used	Freely available genomic databases	References
1	Finger millet	1.5 GB	85,243	Illumina NextSeq500	PR202 and ML365	NA	Hatakeyama et al. (2018)
2	Pearl millet	1.76 GB	38,579	Illumina HiSeq 2000	Tift23D$_2$B$_1$-P1-P5	Pearl Millet Drought Transcriptome database (PMDTDb) Giga Science Database	Varshney et al. (2017)
3	Foxtail millet	423–510 Mb	38,801	Illumina Genome Analyzer II and HiSeq 2000	Zhang gu and Yugu 1	Setria italic genome database (SiGDB) Foxtail millet Marker Database (FmMDb) Foxtail millet Transcription Factor Database (FmTFDb) Foxtail millet miRNA Database (FmMiRNADb) Foxtail millet Transposable Elements-based Marker Database (FmTEMDb)	Zhang et al. (2012)
4	Kodo millet	1.91–1.98 GB	NA	NA	NA	NA	NA
5	Proso millet	~923 Mb	55,930	Illumina HiSeq 2000	Landrace accession number: 00000390	NA	Zou et al. (2019)
6	Indian barnyard millet	1.27 GB	1,08,771	NA	STB08	NA	NA
7	Japanese barnyard millet	NA	NA	NA	NA	NA	NA
8	Little millet	NA	NA	NA	NA	NA	NA

conservationists, geneticists, breeders, biotechnologists, etc., for the improvement of global food and nutritional security. Amidst this, almost all millets have intrinsic mechanisms to withstand certain environmental stresses likely waterlogging, lodging, cold, heat, and drought, however, they threaten millet production with variable impact. Various genetics to genomics techniques, such as marker-assisted breeding, genomic selection, transgenic, gene profiling, transcriptomics, genome editing, etc., can bridge the conventional to modern technology. These integrated tools integrated with system biology and computational biology are needed to expedite future needs. Therefore, these high-throughput technologies can help scientists and plant breeders to develop millet crops with improved stress tolerance so that they can sustain food and nutritional insecurity threats under altered environmental conditions.

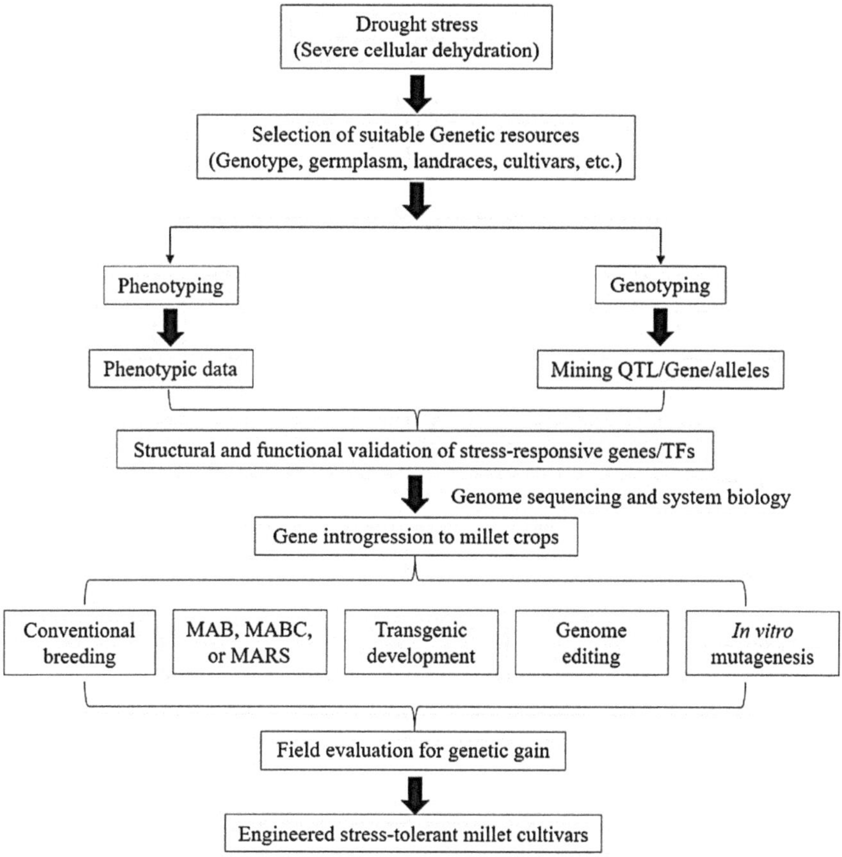

FIGURE 9.1 Integrated genetic to genomics approaches for designing stress-tolerant millet cultivars.

KEYWORDS

- **millets**
- **genomics**
- **climate-smart breeding**
- **climate-resilience**
- **underutilized crops**

REFERENCES

Amin, A.; Mubeen, M.; Hammad, H. M.; Nasim, W. Climate-Smart Agriculture: An Approach for Sustainable Food Security. *Agric. Res. Commun. Centres* **2015,** *2*, 13–21.

Babitha, K. C.; Vemanna, R. S.; Nataraja, K. N.; Udayakumar, M. Overexpression of *EcbHLH57* Transcription Factor from *Eleusinecoracana* L. in Tobacco Confers Tolerance to Salt, Oxidative and Drought Stress. *PLoS One* **2015,** *10* (9), e0137098.

Bhatt, D.; Saxena, S. C.; Jain, S.; et al. Cloning, Expression, and Functional Validation of Drought Inducible Ascorbate Peroxidase (Ec-apx1) from *Eleusinecoracana. Mol. Biol. Rep.* **2013,** *40*, 1155–1165.

Ceasar, S. A.; Maharajan, T.; Ajeesh, T. P. K.; et al. "Finger Millet [*EleusineCoracana* (L.) Gaertn.] Improvement: Current Status and Future Interventions of Whole Genome Sequence. *Front. Plant Sci.* **2018,** *9*, 1–16.

Chandra, A. K.; Kumar, A.; Bharati, A.; et al. Microbial-Assisted and Genomic-Assisted Breeding: A Two-way Approach for the Improvement of Nutritional Quality Traits in Agricultural Crops. *3Biotech* **2020a,** *2* (10), 1–15.

Chandra, A. K.; Pandey, D.; Tiwari, A.; et al. An Omics Study of Iron and Zinc Homeostasis in Finger Millet: Biofortified Foods for Micronutrient Deficiency in an Era of Climate Change? *OMICS J. Int. Biol.* **2020b**. DOI: 10.1089/omi.2020.0095.

Chandra, A. K.; Pandey, D.; Tiwari, A.; et al. Metal Based Nanoparticles Trigger the Differential Expression of Key Regulatory Genes Which Regulate Iron and Zinc Homeostasis Mechanism in Finger Millet. *J. Cer. Sci.* **2021,** *100*, 103235.

Chanwala, J.; Satpati, S.; Dixit, A.; et al. Genome-Wide Identification and Expression Analysis of WRKY Transcription Factors in Pearl Millet (*Pennisetumglaucum*) Under Dehydration and Salinity Stress. *BMC Gen.* **2020,** *21*, 231.

Hadebe, S. T.; Modi, A. T.; Mabhaudhi, T. Drought Tolerance and Water use of Cereal Crops: A Focus on Sorghum as a Food Security Crop in Sub-Saharan Africa. *J. Agron. Crop Sci.* **2017,** *203* (3), 177–191.

Hatakeyama, M.; Aluri, S.; Balachadran, M. T.; et al. Multiple Hybrid De Novo Genome Assembly of Finger Millet, an Orphan Allotetraploid Crop. *DNA Res.* **2018,** *2*, 39–47.

Hittalmani, S.; Mahesh, H.; Shirke, M. D.; et al. Genome and Transcriptome Sequence of Finger Millet (*Eleusinecoracana* (L.) Gaertn.) Provides Insights into Drought Tolerance and Nutraceutical Properties. *BMC Gen.* **2017,** *18*, 46.

Jency, J. P.; Rajasekaran, R.; Singh, R. K.; et al. Induced Mutagenesis Enhances Lodging Resistance and Photosynthetic Efficiency of Kodomillet (*Paspalumscrobiculatum*). *Agronomy* **2020,** *10*, 227.

Kumar, A.; Metwal, M.; Kaur, S.; et al. Nutraceutical Value of Finger Millet [EleusineCoracana (L.) Gaertn.], And Their Improvement Using Omics Approaches. *Front. Plant Sci.* 2016, *7*, 1–14.

Lata, C. Advances in Omics for Enhancing Abiotic Stress Tolerance in Millets. *Proc. Indian Natl. Sci. Acad.* **2015,** *81* (2), 397–417.

Mamidi, S.; Healey, A.; Huang, P.; et al. Genome Resource for Green Millet *Setariaviridis* Enables Discovery of Agronomically Valuable Loci. *Nat. Biotechnol.* **2020,** *38*, 1203–1210.

Maqsood, M.; Ali, S. N. A. Effects of Drought on Growth, Develompent, Radiation Use Efficiency and Yield of Finger Millet (*EleucineCoracana* (L.) Gaertn." *Pak. J. Bot.* **2007,** *39*, 123–134.

Matsuura, A.; Tsuji, W.; An, P.; et al. Effect of Pre- and Post-heading Water Deficit on Growth and Grain Yield of Four Millets. *Plant Prod. Sci.* **2012,** *15*, 323–331.

Muthamilarasan, M.; Bonthala, V. S.; Mishra, A. K.; et al. C2H2 -Type of Zinc Finger Transcription Factors in Foxtail Millet Define Response to Abiotic Stresses *Funct. Integr. Genomics* **2014**. DOI: 10.1007/ s10142-014-0383-2.

Nagarjuna, K. N.; Parvathi, M. S.; Sajeevan, R. S.; et al. Full-Length Cloning and Characterization of Abiotic Stress Responsive CIPK31-like Gene from Finger Millet, a Drought-tolerant Crop. *Curr. Sci.* **2016,** *111*, 890–894.

Numan, M.; Khan, A. L.; Asaf, S.; et al. From Traditional Breeding to Genome Editing for Boosting Productivity of the Ancient Grain Tef [*Eragrostistef* (Zucc.) Trotter]. *Plants* **2021,** *10*, 628.

Parvathi, M. S.; Nataraja, K. N.; Reddy, Y. A. N.; et al. Transcriptome Analysis of Finger Millet (*EleusineCoracana* (L.) Gaertn.) Reveals Unique Drought Responsive Genes. *J. Genet.* **2019,** *98*, 1–12.

Rahman, H.; Ramanathan, V.; Nallathambi, J.; et al. Over-Expression of a NAC 67 Transcription Factor from Finger Millet (*EleusineCoracana* L.) Confers Tolerance against Salinity and Drought Stress in Rice. *BMC Biotechnol.* **2016,** *16*, 8–20.

Ramakrishna, C.; Singh, S.; Raghavendrarao, S.; et al. The Membrane Tethered Transcription Factor *EcbZIP17* from Finger Millet Promotes Plant Growth and Enhances Tolerance to Abiotic Stresses. *Sci. Rep.* **2018,** *8*, 2148.

Ramanjulu, S.; Bartels, D. Drought- and Desiccation Induced Modulation of Gene Expression in Plants. *Plant Cell Environ.* **2002,** *25*, 141–151.

Ramegowda, V.; Senthil-Kumar, M.; Nataraja, K. N.; et al. Expression of a Finger Millet Transcription Factor, *EcNAC1*, in Tobacco Confers Abiotic Stress-Tolerance. *PLoS One* **2012,** *7*, e40397.

Reddy, P. S.; Tata, S.; Rao, R. B.; et al. Genome-Wide Identification and Characterization of the Aquaporin Gene Family in *Sorghum bicolor* (L.). *Plant Gene.* **2015,** *1*, 18–28.

Saleh, A. S.; Zhang, Q.; Chen, J.; Shen, Q. Millet Grains: Nutritional Quality, Processing, and Potential Health Benefits. *Compr. Rev. Food Sci. Food Saf.* **2013,** *12*, 281–295.

Satyavathi, C.; Tara, R. K.; Solanki, R. K.; et al. Genomics Assisted Breeding for Abiotic Stress Tolerance in Millets, Genomics Assisted Breeding of Crops for Abiotic Stress Tolerance. Rajpal, V. R., et al., Eds.; Sustainable Development and Biodiversity; Springer Nature: Switzerland, 2019; vol II, pp 241–255.

Sehgal, D.; Rajaram, V.; Armstead, I. P.; et al. Integration of Gene-Based Markers in a Pearl Millet Genetic Map for Identification of Candidate Genes Underlying Drought Tolerance Quantitative trait Loci. *BMC Plant Biol.* **2012,** *12*, 9.

Sen, S.; Kundu, S.; Dutta, S. K. Proteomic Analysis of JAZ Interacting Proteins Under Methyl jasmonate Treatment in Finger Millet. *Plant Physiol. Biochem.* **2016,** *108*, 79–89.

Serba, D. D.; Yadav, R. S. Genomic Tools in Pearl Millet Breeding for Drought Tolerance: Status and Prospects. *Front. Plant Sci.* **2016,** *7*, 1724.

Singh, R. K.; Phanindra, L. V.; Singh, M.; et al. Isolation and Characterization of Drought Responsive *EcDehydrin7* Gene from Finger Millet *Eleusinecoracana* (L.) Gaertn.)". *Ind. J. Genet. Plant Breed.* **2014,** *74* (4), 456–462.

Sood, S.; Joshi, D. C.; Chandra, A. K.; Kumar, A. Phenomics and Genomics of Finger Millet: Current Status and Future Prospects. *Planta* **2019,** *250* (3), 731–751.

Subiramaniyan, A.; Alagarswamy, S.; Senthil, N.; Punitha, S.**;** et al. Yield Potential of Small Millets under Drought Condition. *Madras Agric. J.* **2018,** 105.

Sudan, J.; Negi, B.; Arora, S. Oxidative Stress Induced Expression of Monodehydroascorbate Reductase Gene in *Eleusinecoracana. Physiol. Mol. Biol. Plants* **2015,** *21*, 551–558.

Varshney, R. K.; Shi, C.; Thudi, M., et al. Pearl Millet Genome Sequence Provides a Resource to Improve Agronomic Traits in Arid Environments. *Nat. Biotechnol.* **2017,** *35*, 969–976.

Wambi, W.; Otienno, G.; Tumwesigye, W.; Mulumba, J. Genetic and Genomic Resources for Finger Millet Improvement: Opportunities for Advancing Climate-Smart Agriculture. *J. Crop Imp.* **2021,** *35* (2), 204–233.

Yadav, T.; Kumar, A.; Yadav, R.; et al. Salicylic Acid and Thiourea Mitigate the Salinity and Drought Stress on Physiological Traits Governing Yield in Pearl Millet-Wheat. *Saudi J. Biol. Sci.* 2020, *27*, 2010–2017.

Zhang, L.; Geng, X.; Zhang, H.; et al. Isolation and Characterization of Heat-Responsive Gene TaGASR1 from Wheat (*Triticumaestivum* L.). *J. Plant Biol.* **2017,** *60*, 57–65.

Zhang, G.; Liu, X.; Quan, Z.; et al. Genome Sequence of Foxtail Millet (*Setariaitalica*) Provides Insights into Grass Evolution and Biofuel Potential. *Nat. Biot.* **2012,** *30*, 549–554.

Zhao, Y.; Mette, M. F.; Reif, J. C. Genomic Selection in Hybrid Breeding. *Plant Breed.* **2015,** *134*, 1–10.

Zou, C.; Li, L.; Miki, D.; et al. The Genome of Broomcorn Millet. *Nat. Commun.* **2019,** *10*, 436.

PART III

Climate-Resilient Agriculture: Advanced Techniques

CHAPTER 10

Speed Breeding: Rapid Generation Tool for Breeding Climate Smart Crops

SNEHA ADHIKARI[1], AMARJEET KUMAR[2], ANJALI JOSHI[3], ALKA BHARATI[4], PRAMOD PRASAD[1], O. P. GANGWAR[1], CHARU LATA[1], SUBODH KUMAR[1], and S. C. BHARDWAJ[1]

[1]*ICAR-IIWBR, Regional Station Flowerdale, Shimla, Himachal Pradesh, India*

[2]*Department of Genetics and Plant Breeding, MTTC & VTC, Selesih, CAU, Imphal, India*

[3]*Genetics and Tree Improvement Division, Arid Forest Research Institute, Jodhpur, Rajasthan, India*

[4]*ICAR- Central Agroforestry Research Institute, Jhansi, India*

ABSTRACT

The world's population is increasing at an alarming rate and about 60–80% more food would be required to feed 9 billion people by 2050. Meeting this mammoth challenge is becoming increasingly difficult from limited arable land under rapidly changing climatic conditions. Hence, there is a need to focus on smart agriculture, which is based on cost-effective novel technologies. The speed breeding, a potential technique of breeding cops in shortest possible duration, has the potential to accelerate crop research through a rapid generation advancement. Under speed breeding, artificial environment is created with longer daylight regimes to accelerate breeding cycles of photo-insensitive crops. Earlier, this technique was largely used for research purposes only, but currently, it is being adopted by the industries as well. There is a need to design breeding programs in consideration with

Climate-Resilient Agriculture: A Molecular Perspective. Ashutosh Singh, Saurabh Pandey (Eds.)

the changing climate conditions, and breeding for climate-resilient crops is the immediate challenge which can be accomplished through new concepts like speed breeding. In the absence of integrated pre-breeding program, the development of new high performing cultivars with market-oriented traits may take more than 10 years. As during early phase of breeding after initial crossing between parental genotypes for selection and genetic advancement stages, huge time, space, and resources were invested. Hence, speed breeding is helping by accelerating up the breeding and research program through shortening of generation's time which will consolidate the genetic gains in the plant breeding programs. Repeated and detailed studies on plant–pathogen interactions, plant anatomy, flowering time, etc., can be carried out using this technology. Therefore, speed breeding plays a potential role in reducing the time required for cultivar development, release, commercialization, and thus ensuring global food security. In addition, speed breeding can also be combined with several other technologies, such as marker-assisted selection, genomic selection, and CRISPR gene editing, which holds promise to relieve the long-standing bottleneck of lengthy crop breeding cycles to enable plant breeders to keep up a pace with an evolving climate and ever-increasing human population. In this chapter, we briefly discuss the traditional as well as advancement method of accelerated breeding in different plant species with special focus on speed breeding system.

10.1 INTRODUCTION

The fast-increasing population and unprecedented climate change are two major challenges for sustainable food production. The present world population is about 7.8 billion and by 2050, it is assumed to reach nearly 9.9 billion which represents an increase of > 25% from 2020 (IISD, 2020). To maintain the surplus food supply, we require at least 0.5 hectare arable land/person but the availability is only half of the required, that is, 0.27 hectare/capita which will further reduce to 0.14 hectare/person within next 40 years due to drastically increasing population. Hence, it is impossible to feed world population by relying on limited arable land (Raju and Sagar, 2020).

Climatic fluctuations broadly include rapid temperature changes, elevated CO_2 concentration, ice melting, frequent floods, natural calamities, and drought. These may cause novel diseases and more frequent disease and pest outbreaks, which may adversely affect crop productivity thereby challenges to food security (Prasad et al., 2021). Therefore, food security under adverse climatic conditions can be achieved by developing climate-resilient crop cultivars. Hence, sustainable food production system requiring an agile plant breeding response is needed (Hussain, 2015). Since the beginning of human civilization, the plant breeding has been performed in various forms. For fulfilling increasing food demand continuous evolution in plant breeding procedure is needed. Cultivar development through conventional plant breeding procedure is time-taking that may take 10–12 years or more depending upon plant habit, reproductive behavior, traits complexity, etc. Transition in breeding work is needed to obtain self-sufficiency under climate change scenario. Varietal development in a shorter span of time is necessary for countering the unpredictable weather for which the concept of accelerated plant breeding is considered best. It can shorten the breeding cycle and thereby improves the efficiency of conventional plant breeding strategies by combining them with modern molecular tools.

Accelerated crop breeding leads to rapid release of improved crop varieties, leading to improved rate of genetic gains. According to breeders' equation the speed of plant breeding relied on the breeding cycle time (t) which is a most critical factor that significantly affects per unit time genetic gain (Samantara et al., 2022). Since 1940s, breeders are trying to fasten the plants lifecycle turnover by using some classical techniques such as a single-seed descent (SSD) (Goulden, 1939; Brim, 1966) and shuttle breeding (Borlaug, 1968).

Recently, for further reduction in plant lifecycle duration, researchers have manipulated environment, under which crops are grown, to fasttrack flowering and seed set, thereby advancement of breeding generation which in turns resulted in valuable plant varieties which helped in gaining global food security. Modern techniques that enhance the generation turnover are broadly clustered under "speed breeding" and includes (1) Rapid development of homozygous lines through accelerated SSD method, (2) Rapid generation cycling using DNA marker technology, (3) fast generation cycling using stressed conditions and in vitro embryo rescue, and (4) rapid generation turnover through immature seed harvest and photoperiod response (Samantara et al., 2022). There are several successful outcomes of speed breeding including rapid phenotyping in bread wheat, multitrait phenotyping in durum wheat for rapid selection of early filial generations or off-season characterization of fixed lines, simultaneous evaluation of traits governing disease resistance in European two-rowed barley, breeding for herbicide tolerant chickpea, allelic introgression of *Aphanomyces* root rot resistance from wild form to cultivated lentil, etc. (Alahmad et al., 2018; Atieno et al., 2017; Croser et al., 2021;

Hickey et al., 2017; Lulsdorf and Banniza, 2018). Likewise, speed breeding coupled with molecular breeding techniques resulted in the development of novel salt-tolerant rice line YNU31-2-4 which serves as a potential source for breeding of salt-tolerant rice varieties (Rana et al., 2019). Various protocols have been tested and found suitable for speed breeding system, which facilitates faster characterization of huge segregating populations for several diseases and disorders, such as pod shattering, wheat preharvest sprouting, wheat rusts, yellow spot of wheat, Fusarial head blight of wheat, and crown rot under SB methodology (Ghosh et al. 2018; Voss-Fels et al., 2019). These outcomes established the potentiality of speed breeding system for sustainable and accelerated genetic gain.

10.2 IMPACT OF CHANGING CLIMATIC SCENARIO ON AGRICULTURE AND FOOD PRODUCTION

Comparison to global climatic variability, local climates pattern affects agricultural activities. Climate change adversely affects crop yields both directly through reducing plant growth and production and indirectly through accelerating biotic and abiotic stresses. Variation in temperature, CO_2 level, greenhouse gases, rainfall pattern, etc., which are the major cause of extreme calamities including water deficiency-drought, or water excess-flood, hail, and hurricanes that are the major hurdles in achieving global food security (Leisner, 2020). Extreme weather conditions specially drought and floods prevent crops from growing and result in significant yield reduction. In 2010 and 2012 across U.S. corn belt, the high night time temperature had severely affected corn yield, and warm winter in Michigan resulted in premature budding in cherries which led to 220 million USD loss in the year 2012 (Hatfield et al., 2014). The impact of drought is more prevalent and become more challenging where soils become drier due to rising summer temperature. However, by increased frequency of irrigation, it can be managed at some places but limited water availability is another problem. According to an estimate of Inter-governmental Panel on Climate Change (IPCC) very high rise in global temperature, that is, 1.7–4.8°C may be observed and simultaneously rainfall pattern will also be altered (IPCC, 2014). In China, the impact of rising temperature can be seen as the Yangtze river basin is becoming hotter, and its temperature will be expected to increase up to 2°C by 2050 in comparison to the temperature prevailed during year 1950. This will also lead to reduced rice and maize production by 41% and 50%, respectively at the end of 21st century (Bakala et al., 2020).

In fourth assessment report of IPCC, it has been predicted that in India by 2080, 2.7–4.3°C temperature rise will be seen (IPCC, 2007). According to the government of India's economic survey (2018) report, the annual loss of 9–10 billion USD due to adverse climatic condition was estimated (Rao et al., 2019). For assessing the impact of climate change on Indian agriculture, the government of India has initiated a project "NICRA" (National Innovations in Climate Resilient Agriculture). It was projected that by 2050 and 2080, the rice yield will be reduced marginally by <2.5%, whereas the yield of irrigated rice will be reduced by 7% and 10% in the year 2050 and 2080, respectively. Similarly, the wheat and maize yield is predicted to reduce by 6–25% and 18–23% sequentially by the year 2100 (*Effect of Climate Change on Agriculture*, n.d.).

Elevated CO_2 levels are also affecting crop yields. There are some experimental evidences that support the positive association between increased CO_2 levels and plant growth. But the other environmental factors, namely, varying temperature regime, O_3 level, and moisture along with nutrients factors may hinder the potential improvement in crop yield. Likewise, if temperature surpassed crop's optimum level, water, and nutrients are not available at sufficient level, in such case, yield increase may be reversed. Though rising CO_2 can stimulate plant growth, in some crop species such as alfalfa, elevated CO_2 level may reduce protein and nitrogen content, resulting in the loss of crop quality (Hatfield et al., 2014). Likewise, concentration of protein and essential minerals may also be reduced in most of the plant species, such as cereals (wheat, rice) and oilseed (soybean) under increased concentration of atmospheric CO_2. Therefore, it will possess threat to human and animal health as the nutritional value of most of the food crops is directly affected under elevated CO_2 level.

With changing climatic scenario, the frequency of occurrence of abiotic and biotic stresses in crop plants increase and pests and diseases which were minor previously, can become economically important. During the past three decades, due to severe impact of variable climatic conditions on biotic and abiotic stresses, global agricultural production is severely affected (1–5%) (Newbery et al., 2016). Globally, in unprotected crops, 16% yield loss has been recorded by the occurrence of crop diseases (Oerke, 2006). Therefore, for ensuring future food security, obtaining higher yield via disease management is very much critical. Warmer temperature, humid climates, and elevated atmospheric CO_2 level are some factors that favor many weeds to grow and pests as well as fungi to reproduce and multiply exponentially. Therefore, for fighting with weeds annually, the U.S. farmers spend more than 11 billion USD, which ultimately competes with other agriculture inputs, including nutrients, light, and water (Doney et al., 2014). With ever-changing climatic conditions, the coverage and distribution pattern of weeds and pests are likely to increase that could cause new problems for crops, which were previously unexposed to these threads. Among abiotic stresses, drought is a major threat, and in the next five decades, by covering almost half of the global arable land, it is expected to reduce crop productivity by 50% (Dhankher and Foyer, 2018). Climate shift may also have associated adverse effects on soil properties and processes, therefore, by affecting soil productivity, it may severely threaten food security (Karmakar et al., 2016). The nutrient level and their bioavailability in the soil are directly and indirectly altered under variable climatic conditions. Like under elevated CO_2 concentration, the wider C:N ratio of crop residues leads to reduced rate of decomposition and nutrient supply further. Hence, the development of climate-resilient varieties would ensure food security under adverse climatic conditions.

10.3 ACCELERATED BREEDING APPROACHES TOWARDS CLIMATE-RESILIENT AGRICULTURE

Development of new crop variety via traditional breeding approaches required selection of desirable parental genotypes followed by crossing among selected parents, thereby selection and generation advancement of superior progenies for release of cultivar that fitted best with market, and thereafter registration of cultivar, seed multiplication, and ultimately seed distribution to growers (Shimelis and Laing, 2012). The prime goal behind breeding of novel cultivar includes higher yield potential, quality, and biotic as well abiotic stress tolerance (Breseghello and Coelho, 2013). In traditional breeding methods, breeders make selection under specific environmental condition. Therefore, most of the varieties will not perform well across the wider range of environment. For handling of segregating generation/generation advancement pedigree and backcross are the most commonly used traditional methods. A majority of traditionally bred varieties are based on pedigree method. Pedigree method depends on growing all of the field's plant populations, and for attaining homozygosity fixation of lines, it takes a longer period. Whereas for the incorporation of desirable traits in a popular variety which is lacking in some traits but otherwise agronomically superior, another commonly used method is backcrossing. Variability in crop stages upon which trait evaluation is carried out, the need of repeated screening of progenies for the selection of desired variant of recurrent parent are the main factors that increase the time required for development of a variety by this method. Moreover, these methods require prolonged time period, that is, 10–14 years, and in some cases, time duration may exceed depending upon plant habit, reproductive behavior, and complexity of traits involved in the absence of an integrated pre-breeding program (Ahmar et al., 2020). Traditional varietal development program is resource, space, and time-intensive which results in slow release of cultivar and their commercialization further. Conventional breeding needs entire season for field selection, and there is a slow rate of generation advancement. In cereals and legumes, the crop cycle is 3–6 months/crop/generation/year. One breeding generation takes 15–18 months in cassava. Under some agro-ecological conditions, variable weather attributes including extreme temperature fluctuation, erratic precipitation, day length, etc., may permit completion of only 1 cycle/year. Even under some tropical agroecologies with maximum efforts, only two generation/year can be obtained (Laux et al., 2010). Besides this, for achieving continuous/uninterrupted field selection and rapid generation advancement (RGA), there is need of skills, expertise, and production resources for phenotyping and management of huge-sized segregating populations before evaluation for yield and economic traits in homozygous or genetically stable breeding populations (Mobini et al., 2015).

For catering growing food demand, evolution in plant breeding method is continuously needed. Since the start of human civilization, crop breeding has been performed in various forms. To counter the effect of unpredictable weather parameters, varietal development needs to be complete in shortest time period. Therefore, smart breeding strategies which involve coupling of conventional breeding methodologies with advance molecular, genomic, and phenomic tools need to be performed for efficient and faster breeding of high-yielding resilient crop cultivars. The duration of breeding cycle which is the most critical factor in deciding pace of cultivar development can be reduced by the integration of modern technologies, such as doubled haploid (DH), shuttle breeding, rapid generation advance (RGA), marker-assisted backcrossing (MABC) and recurrent selection (MARS), genomic selection (GS), speed breeding. These strategies not only help in shortening of breeding cycle but also lead to systematic screening for various biotic and abiotic stresses. Therefore, in the recent past, accelerated breeding has gained popularity among researchers and is considered an efficient tool for rapid development of new varieties tailored to environmental fluctuations for reducing the impact of rapid climate change.

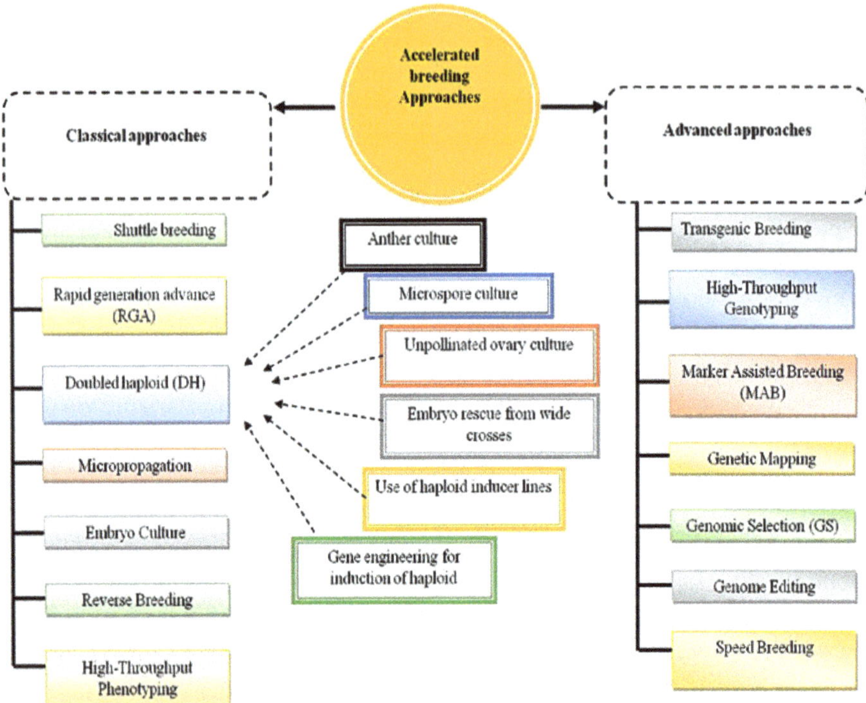

FIGURE 10.1 Schemetic representation of accelerated breeding approaches.

10.4 SPEED BREEDING CONCEPT AND HISTORY

Speed breeding is a combination of techniques that aims at quick advancement of breeding generations by accelerated flowering and seed setting via manipulation of environment under which crop genotypes will be grown. During earlier days, in various parts of world, uncertain climatic fluctuation, such as inadequate lighting during crop growth stages particularly during winter season could lead to yield reduction and in some cases, complete crop fails. For preventing crop failure and productivity improvement in these regions, protected cultivation in green house having controlled environment with artificial light came into existence. From long time back, artificial lighting has been used successfully for supplementing inadequate sunlight under protected cultivation in greenhouses and in tissue culture laboratories for maintaining plant tissue cultures (Mpelkas, 1980). Sequential development in speed breeding concept has been depicted in Figure 10.2. During 1860s, Mangon (1861) and Prilleux (1869) reported the possible role of artificial electric lamps for achieving desirable plant growth and development.

Thereafter, for speeding up plant growth and development, the role of incandescent lamps, and electric arc lamps was established by Siemens (1880). In 1900s, the significant influence of artificial lighting period on phytoconstituents of plants was recorded by Pfeiffer (1926). During the same period, various researchers have demonstrated the possible role of light-emitting diodes particularly blue diodes for attaining required plant growth (Gosal et al., 2020).

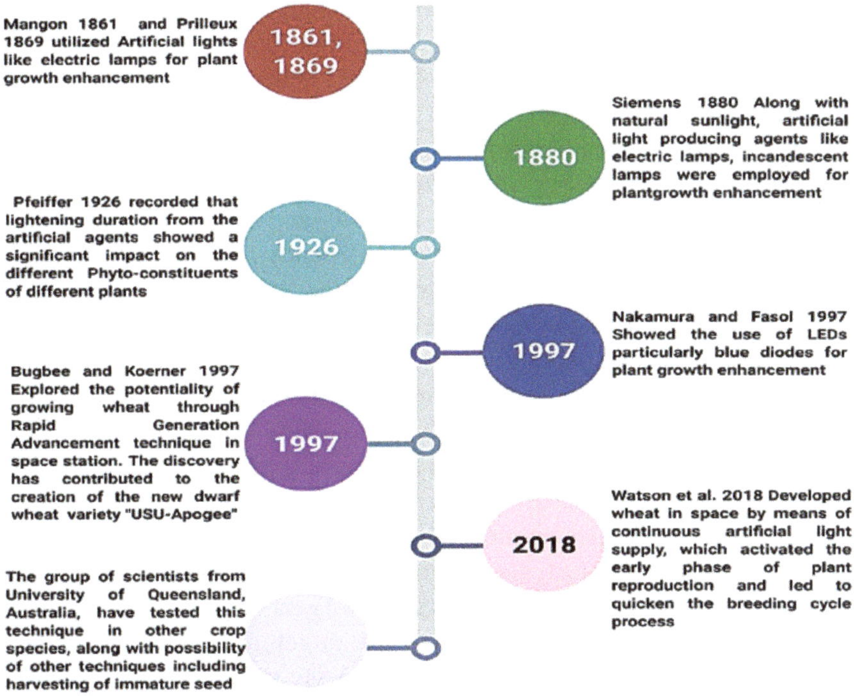

FIGURE 10.2 Sequential development in speed breeding concept.

During late 1900s NASA, USA, in collaboration with Utah State University, designed an experiment with wheat crop to know the effect of constant light on food production in space stations. Based on their experiment they have established the possibilities of rapid cycling/generation advancement of wheat via manipulation of environmental conditions and this experiment yielded a new dwarf wheat variety "USU-Apogee" in the year 1996 (Bugbee and Koerner, 1997). This variety was also selected for calcium-induced leaf tip chlorosis resistance which was generally reported under controlled environments. Later on, various scientists from the

University of Sydney, University of Queensland, and John Innes Center, Australia, got inspired from this technique and they worked on these methodologies and had made significant improvement further and devised the term "speed breeding." By the help of this technique, it is possible to grow wheat in space by using constant light which triggered the establishment of reproductive stage and thereby shortening of wheat breeding cycle (Watson et al., 2018). Light-dependent speed breeding protocols have provided benefits of sustainable photosynthesis for entire light period throughout the years (Bhatta et al., 2021). Thereafter, the team of scientists from University of Queensland, Australia, have tested this technique in other crop species, along with possibility of other techniques including harvesting of immature seed.

For growing plants under speed breeding techniques, there is no need of in vitro set-up, whereas they can be grown under greenhouses or controlled growth chamber by manipulating the intensity and quality of light, photoperiod, and temperature. Speed breeding depends on planned/intentional manipulation of various plant growing conditions that are enlisted in Figure 10.3; Table 10.1. These parameters help in acceleration of plant physiological processes, such as photosynthesis, flowering hence leads to shortening of generation time. In addition to these genetic engineering targeted components of flowering pathway, grafting juvenile plants to mature rootstock are some other promising techniques for rapid cycling (Chiurugwi et al., 2019). The possible reduction in generation time via speed breeding in different crop species has been compiled in Table 10.2. As compared with the conventional generation advancement systems that permits only up to 2–3 generations/year in a regular glasshouse or 1–2 generations/year in the field. Under speed breeding system it is possible to get up to nine generations/year (Ghosh et al., 2018; Jamali et al., 2020). By targeting speed breeding protocols for various crop species, researchers were able to produce up to six generations/year for bread wheat, durum wheat, barley, chickpea, pea, and 4 generations/year for canola (Ghosh et al., 2018; Watson et al., 2018). To accelerate research activities, speed breeding protocol was designed for minor crops as well (Chiurugwi et al., 2019). Improvement in annual generation turnover by speed breeding techniques in various crop species is briefly compiled in Table 10.3. Around three-fold increased improvement in annual generation turnover has been achieved by application of speed breeding approaches. To date, speed breeding protocols have been designed successfully for various plant species including the members of Poaceae, Fabaceae, and Brassicaceae families. In these crop species, speed breeding in conjunction with various studies including phenotyping for traits of interest, genotyping by sequencing (GBS), mutant analysis, transformation studies have led to acceleration of improved varietal development (Rai and Rai, 2022). Under fully enclosed, controlled environment growth chamber, speed breeding not only helps to accelerate pace of varietal development but also beneficial in research, such as phenotyping of adult plant traits, mutation, and transformation studies. Various traditional breeding methods such as SSD (single-seed descent), single pod descent (SPD), single plant selection (SPS), clonal selection can integrate well with speed breeding for achieving better resource use efficiency along with shortening of breeding cycle (Wanga et al., 2021). In conjunction with other modern techniques, such as high-throughput genotyping, GS, and genome, editing speed breeding techniques can be utilized in other crop species for achieving accelerated crop improvement. Likewise, speed breeding in combination with MAS and high-throughput phenotyping facilitates multiple traits selection simultaneously.

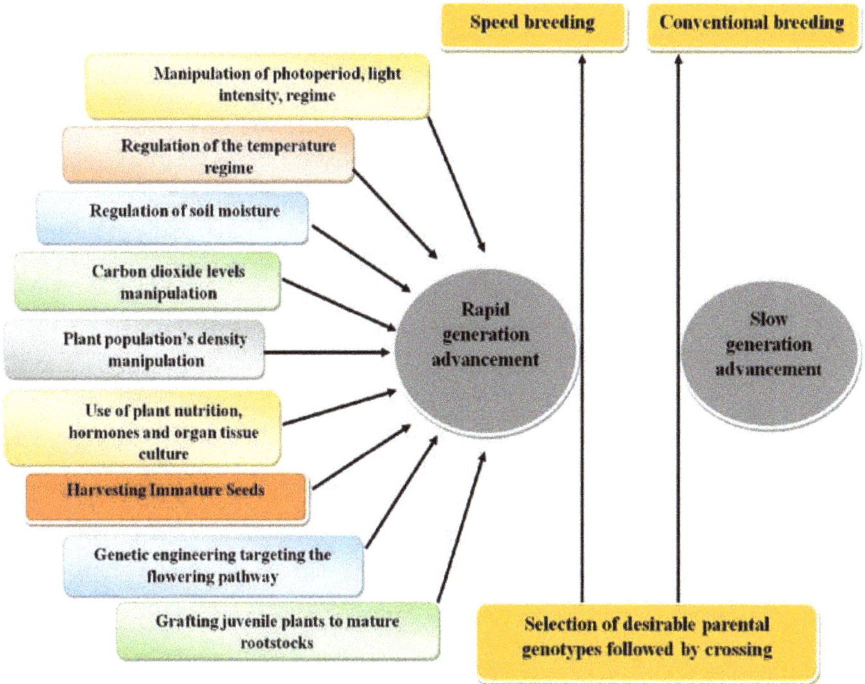

FIGURE 10.3 Speed breeding relies on deliberate manipulation of enlisted growing conditions and other promising techniques.

TABLE 10.1 Crop-specific Deliberate Manipulation of Growing Conditions Under Speed Breeding.

Environmental factor	Manipulation	Crop	References
Photoperiod and light intensity	Photosynthetic active radiation (PAR) within the range of 400–700 nm with an intensity of 360–650 μmol $m^{-2}s^{-1}$	Wheat, barley, chickpea, pea, and canola	Watson et al. (2018)
	22/2 h light/dark under PAR of 150–190 μE $m^{-2}s^{-1}$	Wheat	Dubcovsky et al. (2006)
	8/16 h light/dark for 6 weeks before being transferred to long day (LD) conditions (16/8 h light/dark) at a light intensity of 200–270 μmol $m^{-2}s^{-1}$	Wheat	Wanga et al. (2021)
	16/8 h light/dark with a light intensity of 500 μmol $m^{-2}s^{-1}$	Barley	Zheng et al. (2013)
	12/12 h light/dark using standard incandescent bulb of 60 W with a light intensity of 870 Lm	Chickpea	Samineni et al. (2020)
	24 h light with 450 W PAR	Groundnut	O'Connor et al. (2013)
	14/10 h light/dark period post-germination up to 30 days to support vegetative development followed by 10/14 h light/dark regime were followed to induce flowering and enabling 4–5 generations	Rice	Collard et al. (2017), Rana et al. (2019)
	LD conditions (16/8 h light/dark) followed by SD conditions (8/16 h light/dark) using a light intensity of 150 mmol	Grain amaranth	Stetter et al. (2016)
	28–32°C temperature and relative humidity at 50–60%	Pigeon pea	Saxena et al. (2019)
	10 h photoperiod with blue light enrichment and far-red deprivation	Soybean, rice, and amaranth	Jähne et al. (2020)
	20 h photoperiod, 21/16 h ligh/dark, hydroponic system	Pea	Mobini and Warkentin (2016)
Temperature	20–22C temperature for the germination of immature seed derived from embryo culture	Wheat, barley	Zheng et al. (2013)
	25 ± 1°C temperature for germination of direct sown immature seed	Chickpea	Samineni et al. (2020)
	17/32°C temperature, facilitated rapid plant growth, flower induction, and seed set	Groundnut	O'Connor et al. (2013)
	30/23°C day/night temperature	Sorghum	Yang et al. (2014)

TABLE 10.1 *(Continued)*

Environmental factor	Manipulation	Crop	References
Soil moisture (stress)	Drought stress	Wheat, barley, and pearl millet	Shavrukov et al. (2017)
	Drought stress	Pearl millet	Vadez et al. (2012)
	Drought stress	Cowpea	Goufo et al. (2017)
	Reducing the watering frequency from daily to twice/week, from 4 to 6 weeks after flowering followed by non-watering in the last week before harvesting	Wheat, barley, canola, and chickpea	Watson et al. (2018)
CO_2 level elevation	Increased levels of CO_2 of 400/700, 350/700, and 350/650/100 ppm reduced days to flowering	Soybean, rice, and cowpea	Springer and Ward (2007)
	CO_2 supplementation >400 ppm	Soybean	Nagatoshi and Fujita (2019)
	Increased CO_2 (600 ppm)	Rice	Tanaka et al. (2016)
High population density	16–38 plant m^{-2}	Sorghum	Villar et al. (1989)
	High-density planting of 400 plants m^{-2}	Rice	Rahman (2019)
	11 plant m^{-2}	Cotton	Khan et al. (2019)
Plant nutrition, harmone	Embryo culture supplemented with 4% of sugar solution and 10 times increased concentration of KH_2PO_4 triggered flowering and seed set	Wheat	Yao et al. (2017)
	Combination of auxin and cytokinin hormones, in the form of flurprimidol (0.3 μM), indole-3-acetic acid (5.7 μM) and zeatin (2.3 μM) promoted in vitro flowering and seed setting	Faba bean	Mobini et al. (2015)
	Combination of flurprimidol (0.9 μM) and 4- chloroindole- 3- acetic acid (0.05 μM) promoted in vitro flowering and seed setting	Lentil	Mobini and Warkentin (2016)
	Combination 0.175 mg/L indole- 3- acetic acid (IAA) and 0 mg/L 6- benzylamino-purine (BAP) was found to be suitable for embryo culture	Lentil	Bermejo et al. (2016)
	22 h light at 25–28°C temperature and 2 h dark at 12–18°C temperature in a growth chamber	Lentil	Idrissi (2020)

TABLE 10.2 Reduction in Generation Time via Speed Breeding Efforts in Different Food and Industrial Crops.

Crop	Method	Duration in field	Duration of accelerated generation	References
Wheat	Speed breeding	105	62	Watson et al. (2018)
	Comprehensive speed breeding	105	46	Song et al. (2022)
	Speed breeding	105–132	67–73	Vikas et al. (2021)
	Speed breeding	105	60	Mukade (1974)
	Speed breeding	105	80	Mukade (1974)
Rice	Biotron breeding	122	80	Ohnishi et al. (2011)
	Modified Biotron breeding	135	100	Rana et al. (2019)
	Rapid generation advancement (RGA)	110–121	95–105	Collard et al. (2017)
Barley	Speed breeding	132	68	Watson et al. (2018)
Sorghum	Speed breeding	119	88	Hickey et al. (2019)
Chickpea	Speed breeding	115	82	Watson et al. (2018)
	Rapid generation advancement (RGA)	115–125	80–85	Fikre et al. (2021)
	Speed breeding	90–160	43–60, 44–64, and 52–79	Samineni et al. (2020)
Pea	Speed breeding	84	51	Watson et al. (2018)
	Speed breeding	143 ± 3	67.2 ± 4.6	Ochatt et al. (2002)
	Speed breeding	98.4–114.4	68.4	Mobini and Warkentin (2016)
	Speed breeding	142–152	70–81	Cazzola et al. (2020)
Canola	Speed breeding	171	98	Watson et al. (2018)
	Comprehensive speed breeding	113	67	Song et al. (2022)
	Speed breeding	150–180	62–71, 48–51	Yao et al. (2016)
Amaranthus sp.	Speed breeding	70–120		Stetter et al. (2016)
Soybean	Speed breeding	120	77	Jähne et al. (2020)
	Speed breeding	130–140	65–70	Roumet and Morin (1997)
	Speed breeding	102–132	70	Nagatoshi and Fujita (2019)
	Speed breeding	120	63 (mid-maturity), 81 (late maturity)	Harrison et al. (2021)
Oat	Speed breeding	114	98	González-Barrios et al. (2021)
	Speed breeding	70–80	49–59	Liu et al. (2016)

TABLE 10.2 *(Continued)*

Crop	Method	Duration in field	Duration of accelerated generation	References
Triticale	Speed breeding	68–81	48–61	Liu et al. (2016)
Feba bean	Speed breeding	104	54	Mobini et al. (2015)
Lentil	Speed breeding	102–107	45	Mobini et al. (2015)
	Speed breeding	105	56	Lulsdorf and Banniza (2018)
Groundnut	Speed breeding	145	89	O'Connor et al. (2013)
Arabidopsis	Speed breeding	49–56	20–26	Ochatt and Sangwan (2008)

TABLE 10.3 Improvement in Annual Generation Turnover by Speed Breeding Techniques.

Crop	Techniques used	Generations/ year	References
Rice (*Oryza sativa*)	Photoperiod, temperature manipulation, and embryo rescue	4–5	Rana et al. (2019)
	Restricted root growth and canopy thinning with high CO_2 concentration, followed by early harvest and embryo rescue	4	Tanaka et al. (2016)
	Photoperiod, temperature manipulation, and high-density planting	4	Collard et al. (2017)
Wheat (*Triticum aestivum*)	Acceleration of seed maturation, germination of immature seeds, green, and seed green vernalization	5.7–6	Mukade (1974)
	Photoperiod, temperature, soil fertility, immature seed germination, and embryo rescue	8	Zheng et al. (2013)
	Photoperiod, temperature, harvesting of immature spikes and drying them in an oven/ dehydrator	5.7	Watson et al. (2018)
	Photoperiod manipulations	5	Vikas et al. (2021)
	Vernalization of germinated seeds, high-density seedling culture, and accelerated flowering and maturation with optimized light regime	4.5	Song et al. (2022)
Barley (*Hordeum vulgare*)	Photoperiod and temperature manipulation	6	Hickey et al. (2017)
	Photoperiod, temperature, soil fertility, immature seed germination, and embryo rescue	9	Zheng et al. (2013)
Sorghum (*Sorghum bicolor*)	Photoperiod, temperature, and immature seed germination	6	Forster et al. (2014)
	Embryo rescue		Rizal et al. (2014)
Oat (*Avena sativa*)	Stress, embryo rescue	7	Liu et al. (2016)

TABLE 10.3 *(Continued)*

Crop	Techniques used	Generations/ year	References
Triticale	Stress, embryo rescue	7	Liu et al. (2016)
Clover	Plant hormones, light intensity, and photoperiod	3–4	Castello et al. (2016)
(*Trifolium subterraneum*)	Plant hormones, photoperiod, temperature	2.7–6.1	Pazos-Navarro et al. (2017)
French serradella (*Ornithopus sativus* Brot.)	Plant hormones, light intensity, and photoperiod	3–4	Castello et al. (2016)
Lentil (*Lens culinaris*)	Plant hormones, photoperiod manipulation, light intensity, and embryo rescue	8	Mobini et al. (2015)
	Photoperiod and immature seed germination	5	Lulsdorf and Banniza (2018)
	Plant hormones, embryo rescue	4	Bermejo et al. (2016)
	Photoperiod and temperature manipulation, plant harmones	5–6	Bermejo et al. (2020)
Faba bean (*Vicia faba*)	Plant hormones, photoperiod manipulation, light intensity, and embryo rescue	7	Mobini et al. (2015)
Chickpea (*Cicer arietinum*)	Photoperiod manipulation and immature seed germination	7 (early maturing accessions), 6.2 (mid maturing accessions), 6 (late maturing accessions	Samineni et al. (2020)
Pea (*Pisum sativum*)	Embryo rescue	6.9	Ochatt et al. (2002)
	Plant hormones, photoperiod, and Embryo rescue	5.3	Mobini and Warkentin (2016)
	Light intensity, photoperiod, temperature manipulation, embryo rescue	5	Ribalta et al. (2017)
	Plant hormones, photoperiod, temperature manipulation, early grain harvest	5	Cazzola et al. (2020)
Lupin (*Lupinus angustifolius*)	Light intensity	5	Croser et al. (2016)
Pigeon pea (*Cajanus cajan*)	Photoperiod, temperature, immature seed germination	4	Saxena et al. (2017)
Soybean (*Glycine max*)	Photoperiod, temperature, and immature seed germination	5	Jähne et al. (2020)

TABLE 10.3 *(Continued)*

Crop	Techniques used	Generations/ year	References
	Embryo rescue	5	Roumet and Morin (1997)
	CO_2 supplementation, light intensity, temperature, and immature seed germination	5	Nagatoshi and Fujita (2019)
	Red and blue light (RB), full-spectrum white light (FS), and photoperiod	5	Harrison et al. (2021)
	off-site nursery, fresh-seeding method, and marker-assisted selection	4	Fang et al. (2021)
Bambara groundnut (*Vigna ubterranean*)		4	Ochatt et al. (2002)
Groundnut (*Arachis hypogaea*)	Photoperiod and temperature manipulation	4	O'Connor et al. (2013)
Canola (*Brassica napus*)	Photoperiod, light intensity, temperature, immature seed germination, and soil moisture	3.8	Watson et al. (2018)
	High temperature, water stress, long lighting hours, natural seed maturation, and embryo rescue	6.5–7.6	Yao et al. (2016)
	Vernalization of germinated seeds, high-density seedling culture, and accelerated flowering and maturation with optimized light regime	5.5	Song et al. (2022)
Linseed (*Linum usitatissimum*)	Photoperiod	3	Sysoeva et al. (2010)
Sun flower	Embryo culture	4	Dagustu et al. (2012)
Grain amaranthus (*Amaranthus* spp.)	Photoperiod and temperature manipulation	6	Stetter et al. (2016)
Arabidopsis thaliana	Plant hormones application, immature seed germination, and photoperiod manipulation	10	Ochatt and Sangwan (2008)
Pepper (*Capsicum annuum* L.)	Immature seeds germination	3–4	Manzur et al. (2014)
Tomato (*Solanum lycopersicum*)	Immature seeds germination	5	Bhattarai et al. (2009)

10.5 OPPORTUNITIES OF SPEED BREEDING TECHNIQUES

10.5.1 RAPID DEVELOPMENT OF HOMOZYGOUS LINES FOR ACCELERATED BREEDING AND MAPPING POPULATION DEVELOPMENT

The duration of variety development is largely dependent on the number of years needed for homozygous line development after initial crossing among complementary parents. Generally, after initial hybridization, it takes about 7–9 years if only one crop cycle or generation/year is produced. Therefore, RGA is used to shorten the time required for achieving desired level of homozygosity. In several crop species, speed breeding systems have been utilized effectively for rapid development of homozygous lines which facilitates the pace of breeding cycles and breeding progress thereby varietal development program. As mentioned above, this technique is based on manipulation of various factors (Fig. 10.3) that facilitate early flowering, seed setting, and thereby reduces the overall generation time and results in the production of 3–9 generations/annum depending upon life cycle, photoperiod requirement. Hence, homozygous lines can be developed in a short span of time may be in 1–2 years in contrast >7 years in case of traditional breeding methods.

The mapping population is primarily needed to carry out linkage mapping studies for identifying linked molecular markers, major genes governing monogenic or oligogenic traits, and quantitative trait loci (QTLs) governing polygenic traits. There are several kinds of mapping populations depending upon their method of development, such as F_2, backcross inbred lines (BILs), double haploid (DH), recombinant inbred lines (RILs), and chromosome segment substitution lines (CSSLs) each of them having their own associated benefits and limitation. Majority of mapping studies are based on BILs and RILs population and the main limitation is time required for their development (5–7 generations) (Ghosh et al., 2018; Watson et al., 2018). Based on Biotron speed breeding method in rice involving regulated temperature, day length, restricting numbers of tillers, and embryo rescue, Rana et al. (2019) successfully developed a rice mapping population for the detection of salinity stress tolerance. They made three backcross generations and due to embryo rescue technique, each generation took only 80 days. In backcross breeding program for other than transferring desired traits genes to elite parental cultivars for attaining similar genetic proportion of elite cultivar, backcrossing is needed for 3–5 generations. However, through speed breeding, these results could be obtained in 1 year time period. Likewise, in rice, other mapping populations/

experimental lines, such as RILs, CSSLs, isogenic cultivars have been developed which accelerate rice genomics research (Tanaka et al., 2016). In the case of pea RILs (F_1–F_7 generation) population obtained from crossing of parental lines, CDC Amarillo × CDC Dakota have been developed in 422 days with an average of 68.4 days/generation (Mobini and Warkentin, 2016). In the case of wheat, the speed breeding protocol has been designed for development of RIL populations for mapping of novel APR genes within a time period of 12–18 months (Hickey et al., 2015). Through speed breeding protocol, the duration of mapping population development in pea can be shortened by 30–45 days than standard SSD method (Mobini and Warkentin, 2016). The speed breeding protocol coupled with MAS has worked as a powerful tool for the development of climate smart varieties of different crops. In rice, various genes including *hst1* associated with salinity tolerance, *dro1*, *pup1*, associated with root system architecture can be transferred into salinity susceptible but otherwise high-yielding elite cultivars in shortest time period using speed breeding.

10.5.2 HAPLOTYPE-BASED SPEED BREEDING

The main objective of plant breeding is cultivar improvement for desirable traits, in other words making improved version of existing cultivar by correcting undesired trait. For achieving these goals, there are various approaches including haplotype-based breeding. The concept of haplotype was given by (Bevan et al., 2017) and the group of genes that inherited collectively from single parent is referred as haplotype (Sinha et al., 2020). Haplotype breeding includes the identification of superior haplotypes and utilization of variation as a block from single parent that were tagged to a marker to tailor future cultivars. The aim behind haplotype-based breeding is the transfer of superior haplotypes associated with genetic variation that showed strong linkage disequilibrium with candidate genomic regions regulating trait of interest (Varshney et al., 2021). Haplotype strategy which is based on single or few markers is inefficient and does not give any idea about the surrounding sequences. However, these limitations could be overcome by using modern genomic technologies, which may result in identification of novel genetic variation and thereby leads to improved haplotype-based breeding. In various crop species, haplotypes related to adaptation under various biotic and abiotic stresses have been identified, it showed potential for trait improvement but their introgression into popular varieties is a time-taking process. Therefore, for attaining faster introgression of desirable haplotypes, there is a need of integration of haplotype breeding approach with speed breeding technology. As compared with earlier established MAS it showed high accuracy rate.

10.5.3 SPEED BREEDING COUPLING WITH TRANSGENIC BREEDING PROGRAM

Conventional breeding methods are not only lengthy but also exhibit linkage drag along with undirected mutagenesis and random recombinations (Wolter et al., 2019). Therefore, transgenic approaches are considered best for getting rid of these problems and there is possibility of rapid transgenic crop development by combining it with speed breeding strategies (Flachowsky et al., 2011). Because of protracted generation cycles, in apple, transgenic development is a slow process. In the case of apple, transgenic overexpression of *BpMADS4* gene from silver birch causes early flowering (Sharma et al., 2022). The over-expressing line was crossed with wild apple possessing fire blight tolerance resulted in tolerance introgression in progeny. Likewise, in rice, overexpression of *OsRFT1*, *OsmiR393* genes and constitutive expression of the Arabidopsis *LEAFY* gene associated with early flowering (Sharma et al., 2022) are considered most critical traits for the completion of rapid plant lifecycle. Therefore, speed breeding system is the most promising technique for obtaining the desired outcome of transgenic plants within a short span of time.

10.5.4 SPEED BREEDING AND GENOME EDITING

Ethical issues are associated with transgenic approaches which hinder the progress of breeding programs; hence to overcome this hurdle, genome editing is a very effective strategy. This technique specifically modifies the targeted genes either by insertion, deletion, or substitution process. ZFNs (zinc finger nucleases), transcription activator-like effector nucleases (TALENs), and CRISPR-Cas (clustered regularly interspaced short palindromic repeats) are three popular genome editing techniques that are successfully employed in crop improvement programs (Savadi et al., 2017). Among these three techniques, CRISPR/Cas is most recent and advantageous over rest of the methods in terms of economics, speed, and frequencies of mutagenesis (Vats et al., 2019). Simultaneous RNA-I (RNA interference)-mediated knockout of *O. sativa* ABA-responsive element binding factor 1 (*OsABF1*), and its homolog basic leucine zipper (*OsbZIP40*) results in significantly early flowering phenotype (Zhang et al., 2016). In maize, early flowering under long days was forced through CRISPR/Cas9-directed knockdown of *ZmCCT9* (Huang et al., 2018). Therefore, genes governing flowering in various crop species could be targeted by using genome editing techniques and these transgenics could be grown by using speed breeding protocol to ensure speedy generation advancement to ensure stable homozygous lines within 1 year.

10.6 CHALLENGES OF SPEED BREEDING

For improvement of genetic gain in a number of crop species speed breeding is considered a very powerful breeding tool with some associated limitations. It requires infrastructure, expertise, complementary plant phenomic facilities, and continuous financial support (Shimelis et al., 2019). The key limitation of speed breeding is the availability of controlled environmental conditions including photoperiod, temperature, humidity that is suitable for RGA of targeted crop species and diseases as well as pest infestation. Continuous electricity supply for maintaining temperature and other parameters is a costly affair. More than 50% cost of speed breeding is just for maintaining light and temperature controlled environment (O'Connor et al., 2013), which could be reduced by using LED which is a energy efficient lighting source, solar-oriented electricity system, and inverter-based AC (air conditioning) system (Ghosh et al., 2018). The number of crosses and size of population under evaluation is restricted by the availability of size and cost of infrastructure facility. There is a reduced access to reliable field facilities, trained manpower, and training opportunities as well as professional development. In addition, specialized equipments are needed for the selection of significant traits through rapid development stage of breeding cycle. In addition, through rapid development stage of breeding cycle, selection of significant traits needs specialized equipment which further increases the cost. However, these extra breeding costs could to be recovered as speed breeding could result in a fast commercialization of new varieties (Collard et al., 2017).

Under intensive growth conditions, species or even genotypes within species exhibit varied response which often results in reduced seed yield which can affect further field evaluations (González-Barrios et al., 2021). For overcoming the problem of low seed number, next generation sparse phenotyping field trial designs could help up to some extent (Watson et al.,

2018). There are some negative effects associated with manipulation of environmental conditions such as excessive photoperiod which can cause photo-oxidation, which in turn produces a high level of starch as stress hormones. Likewise, when speed breeding is based on harvesting of immature seed in such cases it is impossible to evaluate for seed traits (Samantara et al., 2022). When we push the plants for speeding up their life cycle, then plants grow by utilizing their maximum physiological capability which reduces the ability of plant to defend itself and lack of proper management practices may lead to loss of beneficial breeding material. Management strategies involved the determination of suitable photoperiod and optimum temperature that supports maximum plant growth for each species and even some time within species (among genotypes). One of the possible strategies that prevent loss of breeding material by disease incidence, pest attack, loss of controlled environmental conditions, etc., is the conservation of few seeds of every individual in each generation. The phenotyping of agronomic traits can be performed under speed breeding system, but expression can be biased in controlled environmental condition; therefore, proper care should be taken. Artificial environment under speed breeding differs significantly from field conditions where crop is actually grown; therefore, selection for adaptability in target environment must be performed under field environment. In order to ensure unbiased phenotyping of field crops under speed breeding, the phenotyping should be validated under field environment. For example, in pea to ensure boron tolerance, evaluation under hydroponic condition is integrated with speed breeding condition and similarly in wheat phenotyping for disease resistance was combined with speed breeding condition (Bennett et al., 2017; Watson et al., 2019). A number of traits such as plant height and flowering duration in oat may not be consistently phenotyped due to genotypic interaction with growth system (González-Barrios et al., 2021). Disease and pest management under intensive growing conditions and tracking of individuals when we are dealing with mapping population development for gene discovery studies are other challenges of growing plants under speed breeding condition.

Under poorly resourced breeding program where access to suitable land, specialized equipment, trained personal is limited, speed breeding is still used at experimental level only for accelerated improvement of first-generation breeding resources (O'Connor et al., 2013). If sophisticated controlled environment facilities are not available, then additional resources and expertise are needed for employing additional techniques including embryo rescue, MAS. etc. By integrating modern improved techniques such as genomic-assisted breeding with speed breeding resource use efficiency could be maximized (Li et al., 2018). In comparison to developed countries, implementation of speed breeding methodologies is more challenging, due to limited infrastructure facilities and expertise along with little collaboration with international organizations (Wanga et al., 2021). For wider adaptation of speed breeding technologies, there is a need for integration of suitable skills, and expertise in plant breeding and biotechnology along with a long-term funding with government policies is needed.

10.7 APPLICATION OF SPEED BREEDING IN VARIOUS CROPS

There is a need of novel speed breeding approaches that facilitate reduced flowering time, seed setting, faster embryo development, etc. Over the years, speed breeding approaches have been evolved which are broadly categorized into three groups. Under first category, plants are grown with speed breeding specifications under controlled growth chamber; in the second category, glasshouse is used with speed breeding specifications, and plants are grown under tailored homemade growth room specifically developed for low-cost speed breeding programs under third category (Pandey et al., 2022). There are various successful applications of speed breeding including speedy development of advanced steady lines in addition to mapping populations, screening for identifying donor sources for character(s), pyramiding of multiple traits, phenotyping of adult plant traits, faster backcrossing, mutation experiments, transformation studies, and rapid improved cultivars development in crop plants (Watson et al., 2018). In combination with other breeding methods, speed breeding has ability to revolutionize plant breeding that could help to feed rapidly increasing world population. Different methods of speed breeding have been applied successfully in various crop species for the improvement of valuable traits. Various crop-specific deliberate manipulation of growing conditions/crop-specific protocols under speed breeding, such as photoperiod manipulation (wheat, rice, barley, grain amaranth, chickpea, pea, canola, groundnut, pigeon pea, soybean, chickpea, groundnut), temperature manipulation (wheat, barley, sorghum), soil moisture stress (wheat, barley, pearl millet, cowpea, canola, chickpea), CO_2 elevation (rice, cowpea, soybean), high population density (rice, cotton, sorghum), plant nutrients and hormonal manipulation (wheat, faba bean, lentil) was established by various researchers. Along with these, early harvesting coupled with artificial drying under oven/dehydrator up to 3 days further shortens the seed cycle as compared with normal seed ripening process which may take up to 15 days. However, it significantly reduced grain weight but the seed viability remains unaffected (Hickey et al., 2019).

Some of the crop varieties/breeding lines/transgenic developed by RGA and speed breeding (SB) are enlisted in Table 10.4. Crop wise, these protocols have been compiled in Table 10.1 and some of them are summarized species wise.

TABLE 10.4 Crop Varieties/Breeding Lines/Transgenic Developed by Rapid Generation Advance (RGA) and Speed Breeding (SB).

Crop	Method	Name of variety	Desirable features	References
Rice	RGA	IRRI142	Grain quality	Ha et al. (2011)
	RGA	IRRI165	Salinity tolerance	Gregorio et al. (2013)
	RGA	BINA dhan8	Salinity tolerance	Gregorio et al. (2013)
	RGA	BRRI dhan61	Salinity tolerance	BRRI (2013, 2015, 2017); http://www.brri.gov.bd/
	RGA	BRRI dhan62	First zinc-enriched variety	
	RGA	BRRI dhan67	Salinity tolerance	
	RGA	BRRI dhan72	Zinc-enriched variety	
	RGA	Luna Sankhi	Salinity tolerance	Gregorio et al. (2013)
	SB (Biotron breeding system)	YNU31-2-4	Salinity tolerance	Rana et al. (2019)
	Greenhouse-based RGA	Nipponbare		Koumura (1972)
Wheat	SB	USU-Apogee		Bugbee and Koerner (1997)
		DS Faraday	Preharvest sprouting tolerance, high protein content, and milling recovery	Shivakumar et al. (2018)
Apple	SB	TSB3, TSB15, TSB17 (Trasgenics)	Apple scab and fire blight resistance	Flachowsky et al. (2011)
		TSC3, TSC6 (Trasgenics)	Powdery mildew resistance	

10.7.1 SPEED BREEDING IN CEREALS

Researchers have successfully developed efficient speed breeding protocols for obtaining homozygous lines after hybridization of desirable genotypes in case of cereals. The speed breeding work was initiated with wheat crop itself. In an initial experiment Mukade et al. (1974) successfully obtained 4–6 generation/year of wheat by harvesting immature seeds (15–20 days after anthesis) followed by low temperature H_2O_2 treatment but germination was poor. Later on, for the improvement in immature seed, germination (De Pauw and Clarke, 1976) has increased the duration of low temperature (11°C) H_2O_2 treatment which also resulted in generation time reduction by 12–23 days. Similarly, >90% germination for air-dried immature seeds that were harvested 21 days after anthesis was also observed by Robertson and Curtis (1967). Multiple traits including plant height, flowering duration (Ghosh et al., 2018), drought adaptation-associated root traits (Richard et al., 2015), preharvest sprouting tolerance via grain dormancy (Hickey et al., 2009), and resistance to diseases, such as leaf rust (Riaz et al., 2016; Alahmad et al., 2018b), yellow spot (Dinglasan et al., 2016), fusarium head blight, drought tolerance in wheat imparted by stay green and root adoption traits (Christopher et al., 2015) were phenotyped efficiently with the help of speed breeding protocols. By using speed breeding technique wheat variety DS Faraday, was developed which showed preharvest sprouting tolerance along with high protein content and milling recovery (Shivakumar et al., 2018). Biotron breeding system is popular in rice which comprised of artificial environmental chamber where temperature, photoperiod, CO_2 level were regulated along with embryo rescue and tiller removal to shorten generation period (Ohnishi et al., 2011). The protocol was further refined by Rana et al. (2019) who have utilized modified Biotron breeding system coupled with backcrossing to transfer salt-tolerance gene "*hst1*" from rice variety "Kajin" to a high-yielding variety "Yukinko-mai" with the help of SNP-based marker-assisted selection which yielded a new salt-tolerant high-yielding rice line YNU31-2-4. By using generation advancement method, Japanese scientists have bred several rice cultivar including Nipponbare (Tanaka et al., 2016). Recently, in rice, Rahman (2019) observed possible role of high-density planting (400 plants m^{-2}) in shortening of generation cycle. Speed breeding has been successfully used for screening of drought tolerance, glaucousness, and pod-shattering traits in barley can be measured under SB conditions (Ghosh et al., 2018; Watson et al., 2018). Modified backcrossing in conjunction with speed breeding, yielded rust, and spot blotch resistant barley lines which were susceptible earlier (Hickey et al., 2017). In Argentina, scarlet is most popular barley variety which was susceptible to many diseases by combining modified backcrossing with speed breeding resistant lines were developed within 2 years (Hickey et al., 2017).

In case of sorghum photoperiod, temperature manipulation, and immature seed germination is performed for obtaining 6 generations/year (Forster et al., 2014) and embryo rescue was found effective in reducing life cycle from 17 to 11 weeks (Rizal et al., 2014). With an aim at accelerated purification of sorghum mutant line by using rapid cycling methods (embryo rescue) speed breeding experiment was performed by Dan Anisiyah (2020). They found rapid cycling technique could effectively shorten the purification of sorghum mutant lines in 4 weeks/generation than conventional methods. Likewise for ensuring higher generation turnover in oats stress, embryo rescue (Liu et al., 2016) and photoperiod manipulation (González-Barrios et al., 2021) were found effective. Speed breeding protocols have been developed for major cereals except maize and sugarcane at the end of this decade but suitable protocols are needed to be developed for minor cereals including pearl millet and other millets (Chiurugwi et al., 2019). Because of their high nutritional value, these are needed for combating hidden hunger (Barman et al., 2020) which is a major concern nowadays.

10.7.2 SPEED BREEDING IN OILSEEDS

Breeding of high-yielding oilseed varieties has been always challenging as it engages the improvement in the seed yield along with increased oil content. This resulted in slower varietal release program. By application of speed breeding techniques faster oilseed improvement can be achieved. Several protocols under speed breeding have been developed for oilseeds by various researchers, including soybean (Fang et al., 2021; Harrison et al., 2021; Jähne et al., 2020; Nagatoshi and Fujita, 2019), groundnut (O'Connor et al., 2013; Ochatt et al., 2002), canola (Song et al., 2022; Watson et al., 2018; Yao et al., 2016), linseed (Sysoeva et al., 2010), and sunflower (Dagustu et al., 2012). It is very much possible to produce viable seed via precocious germination in soybean and it was established well, based on which Roumet and Morin (1997) successfully established a protocol for shortening crop cycle from 130–140 days to 65–70 days. They have carried out precocious germination of immature, pretreated soybean pods. In a recent study, Nagatoshi and Fujita (2019) developed a standard speed breeding protocol for soybean Japanese cultivar "Enrei" which reduces generation time from 102 to 132 days to 70 days enables 5 generations/year instead of 1–2 generations in a year. They had manipulated photoperiod with the help of fluorescent lamps (14/10-h, light /dark at 30°C and 25°C, respectively) along with elevated CO_2 (> 400 ppm). In another study, Jähne et al. (2020) have obtained five generations/year in the case of soybean as plant matures within 77 days after sowing under controlled environmental condition comprised of 10 h photoperiod supplemented with LED. Besides, various other methods have been suggested by scientists for speed breeding procedure in oilseeds, Watson et al. (2018) suggested a method of growing canola in a controlled environment chamber for rapid generation turnover and evaluation of pod-shattering trait. Likewise in sunflower, embryo rescue supports shortening the generation time which resulted in four generations/year (Dagustu et al., 2012) in which they had cultured embryo on MS medium after 10–12 days of pollination. Photoperiod manipulations via using LEDs were found effective for reducing maturation time of groundnut (*Arachis hypogea*) up to 89 days from 145 days (O'Connor et al., 2013). In a recent experiment, Parmar et al. (2021) have designed a novel strategy, that is, single-seed chipping (SSC) coupled with high-throughput genotyping and speed breeding for accelerating groundnut genetics and breeding research. In this method, DNA is extracted from a small part of seed cotyledon, that is, posterior end of seed and rest chipped seed is used for generation advancement which showed 95–99% germination. This methodology can save up to 6–8 months in groundnut genetics and breeding research. For many major and minor oilseed crops, speed breeding protocols are under developing phase and will get success in near future to fulfil oil demand of ever-escalating population.

10.7.3 SPEED BREEDING IN PULSES

Pulses are a rich source of nutrients including protein and form important component of human diet. It provides animal feed, and via biological nitrogen fixation, it helps in replenishing soil fertility. With rapidly increasing population, the plant proteins demand for human consumption and animal feed is increasing and due to time-consuming conventional pulse breeding process, it is difficult to meet out demand. Therefore, rapid development of high-yielding pulse varieties is urgently needed. Temperature and photoperiod are the key factors for rapid transition of vegetative growth stage into flowering legumes (Nelson et al., 2010). In vitro culture methodology is successfully tested on various pulse species to accelerate genetic gain in breeding programs by rapid generation cycling. For the induction of in vitro flowering in peas, Ochatt et al. (2002) and Surma et al. (2013) have successfully designed a methodology, and the period was shortened by photoperiod manipulation and hormonal application in some cases. Subsequently, Ribalta et al. (2014) improved in vitro flowering protocol for wider range of pea genotypes using anti-gibberellin, flurprimidol, which help in internodes length reduction. Flurprimidol application is associated with compact and smaller plants which is essential for working in breeding chamber due to less space accommodation. Likewise various speed breeding methods have been designed in pulses. Pazos-Navarro et al. (2017) have obtained 2.7–6.1 generations/years across the clover genotypes by designing in vitro-assisted SSD technique combining temperature, photoperiod manipulation with curtailed seed-filling duration and embryo rescue technique. Similarly, generation acceleration methodologies have been established well in many pulse species, especially for temperate pulses because due to their facultative, long-day nature, they positively respond to photoperiod extension (Croser et al., 2016). For example, in pea, increased plant growth rate is observed under controlled glass house facilitated with LED-supplemented continuous light along with optimal temperature and humidity (Jähne et al., 2020). In addition to these, for in vitro flowering and RGA, the role of plant growth regulators is well proven which in turns gives up to 8 generations of lentil and 7 generations of faba bean within a year. In a recent study in chickpea, Samineni et al. (2020) have recovered up to 7 generations/year via photoperiod manipulation and immature seed germination. In pigeon pea, immature seeds harvested from 35-day-old plants give 100% germination which yielded 3–4 generations/year in association with photoperiod and temperature manipulation (Saxena et al., 2017). Speed breeding research has opened new avenue in pulse research by eradicating toxins, that is, BOAA (β-*N*-oxalyl-amino-L-alanine) present in *Lathyrus sativus* (white pea/chickling pea), which can cause neurological diseases, such as lathyrism in human and animals. Therefore, a group of scientists has initiated speed breeding program for rapid eradication of lythyrus toxins to make it safer for consumption in East Africa. Efforts are underway to transfer low-toxin trait developed at the John Innes Centre, UK, into high-yielding varieties that are adapted to East Africa. Due to hardy nature of white pea, it can thrive well under extreme weather conditions; hence researchers hope that if nontoxic white pea cultivars bred successfully, then it will contribute in food and nutritional security under ever-changing climatic condition especially in Ethiopia, where it is already widely consumed.

10.7.4 SPEED BREEDING IN FRUIT CROPS

Most of the fruit trees before entering into flowering undergo a long juvenile phase which may range from few years to 20 years depending upon plant species (Nocker and Gardiner, 2014). Therefore, the speed breeding in fruit crops is mainly focused on reduction of juvenile phase. One such example is speed breeding in apple leads to rapid vegetative growth and reduced flowering time in apple, that is, from 5 years to 10 months (Nocker and Gardiner, 2014). Likewise Baier et al. (2012) recorded reduced flowering period (2 years instead of 7 years) in the case of chestnut trees when grown in a growth chamber supplemented with a high dose and intensity of light. Similarly, olive plants enter into flowering stage 1.2–2.2 years by forcing seedling growth in greenhouse which normally takes >12 years in normal field condition (Santos-Antunes et al., 2005). In the case of apple, Flachowsky et al. (2011) have started a breeding program for juvenile phase reduction for successful and timely introgression of traits from wild species into pre-breeding material using speed breeding technology in association with transgenic, and marker-assisted selection. In this experiment, at first in glasshouse conditions, they evaluated early flowering transgenic apple lines, carrying *BpMADS4* gene from silver birch for tree morphology. Based on tree morphology line, T1190 was selected that was used for further study and application of speed breeding. The DNA sequence flanking T-DNA were isolated and the T-DNA integration site was mapped on linkage group 4. Then the line T1190 was crossed with the fire blight-resistant wild species of apple, that is, *Malus fusca.* Subsequently, early flowering, fire blight-resistant resistant F_1 transgenic seedlings were selected and backcrossed with "Regia" which carried apple scab resistance gene *Rvi2, Rvi4,* fire blight resistance QTL *FB-F7,* and 98/6-10 carrying powdery mildew resistance gene *Pl-1, Pl-2* to pyramid apple scab, powdery mildew, and fire blight resistance. They could successfully identify three transgenic BC_1 seedlings carrying both apple scab resistance gene and fire blight resistance QTL, along with three other BC_1 seedlings carrying both powdery mildew resistance genes. Therefore, this experiment led to a successful establishment of the first transgenic early flowering-based apple breeding program combined with MAS. Likewise in several of the clonally/asexually propagated crop species, including banana, speed breeding programs have been initiated to achieve increased flowering along with reduced flowering time (Samantara et al., 2022). The tree species become taller in height after certain period of time therefore, their maintenance under controlled greenhouse chamber is a cumbersome process which is a major challenge in the implementation of speed breeding procedures in case of tree crops.

10.7.5 SPEED BREEDING IN VEGETABLE CROPS

These crops generally lack sexual reproduction and are clonally propagated hence exhibit lack of diversity. Major fraction of world population depends on theses crops for fulfilling their daily energy demand. Therefore, for introducing desirable changes, including disease/pest resistance, modification in economically important part speed breeding under artificially controlled environment will be more advantageous and trouble-free. These crops are most suitable for controlled greenhouse chamber due to small plant stature. The seeds of these crops are not much important from farmer's point of view, and these species also show non-synchronized flowering. Whereas most of the root and tuber crops enter into flowering stage either rarely, or too late (after several years of vegetative growth), or under some specialized conditions, whereas leafy vegetables may enter in flowering between 35 and 50 days after sowing. Therefore, the possibility of successful crossing in these crops during conventional breeding is negligible. Hence, speed breeding will help to obtain synchronized flowering among diverse genotypes of leafy vegetables for utilization in breeding program (Barman et al., 2021). Even some late flowering or recalcitrant lines of leafy vegetables start flowering under speed breeding condition, therefore to accelerate the pace of varietal development, speed breeding is needed. Likewise, in the case of cassava, grafting leads to increased flowering (Silva Souza et al., 2018). A speed breeding protocol has been designed by Stetter et al. (2016) in amaranth which is a leafy vegetable and they have generated up to six generations/year via photoperiod and temperature manipulation. Likewise, the response of other vegetables was tested under extended photoperiod which resulted in shortened generation interval in pepper, tomato (Samantara et al., 2022), and earlier flowering in cassava (Pineda et al., 2020). Other technique that leads to accelerated genetic gains involves germination of immature seeds which in turn produces 3–5 generations/years in tomato (Bhattarai et al., 2009; Geboloğlu et al., 2011), 3–4 generation in pepper (Manzur et al., 2014). Further, integration of speed breeding with other modern technologies may lead to rapid genetic gain, for example, in pepper, higher generation turnover can be achieved by higher expression of flowering genes *CaFT-LIKE* (Borovsky et al., 2020). Tomato is very sensitive to constant light which makes it unsuitable for speed breeding environment, but scientists have identified a tomato gene (*CAB-13*) whose introgression leads to production of long photoperiod tolerent plants that could be grown successfully under speed breeding conditions (Velez-Ramirez et al., 2014). Significant improvement in breeding efficiency for heat tolerance in tomatoes could be achieved by combining speed breeding with MAS, GS, and high-throughput phenotyping (Ayenan et al., 2019).

10.8 CONCLUSION

To handle the present era of climate change along with ever-increasing world population, a major responsibility of crop researchers is to produce sufficient quantity of quality food to fulfill global food demand in short time frame. Availability of climate-resilient varieties will prepare human community to face unprecedented climate change effects. The speed at which the present varietal development programs are marching will not be enough to ensure self-sufficient food production. Climate change may cause multiple negative impacts on crop growth and development to handle the detrimental effects of climate change; there is need of comprehensive breeding strategies that accelerate crop improvement programs for achieving hunger-free world in future. Development of climate-resilient cultivar is an immediate challenge which can be fulfilled through novel breeding approaches such as "speed breeding." The speed breeding technique utilizes optimal light quality and intensity, day length, temperature, CO_2 level, and humidity control to accelerate photosynthesis and flowering. In addition to these, plant population density manipulation, plant nutrient manipulation, plant growth hormonal supplementation and organ tissue culture, harvesting of immature seeds, genetic engineering targeting the key elements of flowering pathway and grafting of juvenile plants to mature rootstocks are some other promising techniques for rapid cycling. Speed breeding can accelerate breeding of improved cultivars carrying marker preferred traits in a short span of time by investing limited spaces and resources during selection and generation advancement of superior crop cultivars. For effective and efficient integration of speed breeding technologies into crop improvement program reduced labor requirement and cost-effective controlled facilities are critical factors. Further, integration of speed breeding with modern technologies, such as marker-assisted selection, GS, and genome editing facilitate speedy detection of best genotypes/lines carrying novel traits, including higher yield, superior nutritional qualities, along with biotic and abiotic stress resiliency. However, in most of the developing countries, adoption rate of speed breeding in public plant breeding program is very much limited due to limited access to suitable land, specialized equipment, requisite infrastructure, trained personal, and interrupted water and electricity supplies. Henceforth, for sustainable and wider adoption of speed breeding system in public plant breeding programs, integration of suitable skills and expertise in plant breeding and biotechnology, along with government support at policy and financial level is needed.

KEYWORDS

- **speed breeding**
- **accelerated breeding**
- **climate change**
- **crop breeding**

REFERENCES

Ahmar, S.; Gill, R. A.; Jung, K. H.; Faheem, A.; Qasim, M. U.; Mubeen, M.; Zhou, W. Conventional and Molecular Techniques from Simple Breeding to Speed Breeding in Crop Plants: Recent Advances and Future Outlook. *Int. J. Mol. Sci.* **2020,** *21*, 2590.

Alahmad, S.; Dinglasan, E.; Leung, K. M.; Riaz, A.; Derbal, N.; Voss-Fels, K. P.; Able, J. A.; Bassi, F. M.; Christopher, J.; Hickey, L. T. Speed Breeding for Multiple Quantitative Traits in Durum Wheat. *Plant Methods* **2018,** *14*, 1–15.

Atieno, J.; Li, Y.; Langridge, P.; Dowling, K.; Brien, C.; Berger, B.; Varshney, R. K.; Sutton, T. Exploring Genetic Variation for Salinity Tolerance in Chickpea Using Image-Based Phenotyping. *Sci. Rep.* **2017,** *7*, 1–11.

Ayenan, M. A. T.; Danquah, A.; Hanson, P.; Ampomah-Dwamena, C.; Sodedji, F. A. K.; Asante, I. K.;Danquah, E. Y. Accelerating Breeding for Heat Tolerance in Tomato (*Solanum lycopersicum* L.): An Integrated Approach. *Agronomy* **2019,** *9*, 720.

Baier, K.; Maynard, C.; Powell, W. A. Early Flowering in Chestnut Species Induced under High Intensity, High Dose Light in Growth Chambers. *J. Am. Chestnut Found.* **2012,** *26*, 8–10.

Bakala, H. S.; Singh, G.; Srivastava, P. Smart Breeding for Climate Resilient Agriculture. In *Plant Breeding-Current and Future Views*; Abdurakhmonov, I. Y., Eds.; IntechOpen, 2021.

Barman, M.; Chandra, B.; Viswavidyalaya., K. Speed Breeding : Voyage in the Ocean of Innovative Crop Improvement Era. In Emerging Tools for Sustainable Agriculture and Food Security; Rajput, V. D., Singh, A., Singh, A. K., Minkina, T., Eds.; Deepika Book Agency: Delhi, India, 2021.

Barman, M.; Kumar Choudhary, V.; Kumar Singh, S.; Kumar Singh, M.; Parveen, R.; Wang, H. Y. Genetic Variability Analysis in Bread Wheat (*Triticum aestivum* L.) Genotypes for

Morpho-Physiological Characters and Grain Micronutrient Content. *Int. Res. J. Pure Appl. Chem.* **2020,** *21*, 1–8.

Bennett, R. G.; Ribalta, F. M.; Pazos-Navarro, M.; Leonforte, A.; Croser, J. S. Discrimination of Boron Tolerance in *Pisum sativum* L. Genotypes Using a Rapid, High-Throughput Hydroponic Screen and Precociously Germinated Seed Grown Under Far-Red Enriched Light. *Plant Methods* **2017,** *13*, 1–11.

Bermejo, C.; Gatti, I.; Cointry, E. *In Vitro* Embryo Culture to Shorten the Breeding Cycle in Lentil (*Lens culinaris* Medik). *Plant Cell Tiss. Organ Cult.* **2016,** *127*, 585–590.

Bevan, M. W.; Uauy, C.; Wulff, B. B. H.; Zhou, J.; Krasileva, K.; Clark, M. D. Genomic Innovation for Crop Improvement. *Nature* **2017,** *543*, 346–354.

Bhatta, M.; Sandro, P.; Smith, M. R.; Delaney, O.; Voss-Fels, K. P.; Gutierrez, L.; Hickey, L. T. Need for Speed: Manipulating Plant Growth to Accelerate Breeding Cycles. *Curr. Opin. Plant Biol.* **2021,** *60*, 101986.

Bhattarai, S. P.; De La Pena, R. C.; Midmore, D. J.; Palchamy, K. *In Vitro* Culture of Immature Seed for Rapid Generation Advancement in Tomato. *Euphytica* **2009,** *167*, 23–30.

Borlaug, N. In *Wheat Breeding and Its Impact on World Food Supply*, The Proceedings of the Third International Wheat Genetics Symposium; Finlay, K. W., Shephard, K. W., Eds.; Australian Academy of Sciences: Canberra, Australia, 1968; pp 1–36.

Borovsky, Y.; Mohan, V.; Shabtai, S.; Paran, I. *CaFT-LIKE* is a Flowering Promoter in Pepper and Functions as Florigen in Tomato. *Plant Sci.* **2020,** *301*, 110678.

Breseghello, F.; Coelho, A. S. G. Traditional and Modern Plant Breeding Methods with Examples in Rice (*Oryza sativa* L.). *J. Agric. Food Chem.* 2013, *61*, 8277–8286.

Brim, C. A. A Modified Pedigree Method of Selection in Soybeans. *Crop Sci.* **1966,** *6*, 220.

BRRI. In *Modern Rice Cultivation*, 17th Ed.; Bangladesh Rice Research Institute: Gazipur, 2013.

BRRI. In *Modern Rice Cultivation*, 18th Ed.; Bangladesh Rice Research Institute: Gazipur, 2015.

BRRI. In *Modern Rice Cultivation*, 20th Ed.; Bangladesh Rice Research Institute: Gazipur, 2017.

Bugbee, B.; Koerner, G. Yield Comparisons and Unique Characteristics of the Dwarf Wheat Cultivar "USU-Apogee." *Adv. Space Res.* **1997,** *20*, 1891–1894.

Castello, M.; Stefanova, K.; Nichols, P. G. H.; Nutt, B. J.; Revell, C. K.; Croser, J. S. *In vitro* Reproduction in the Annual Pasture Legumes Subterranean Clover (*Trifolium subterraneum* L.) and French Serradella (*Ornithopus sativus* Brot.). *Grass Forage Sci.* **2016,** *71*, 79–89.

Cazzola, F.; Bermejo, C. J.; Guindon, M. F.; Cointry, E. Speed Breeding in Pea (*Pisum sativum* L.), An Efficient and Simple System to Accelerate Breeding Programs. *Euphytica* **2020,** *216*, 1–11.

Chiurugwi, T.; Kemp, S.; Powell, W.;Hickey, L. T. Speed Breeding Orphan Crops. *Theor. Appl. Genet.* **2019,** *132*, 607–616.

Christopher, J.; Richard, C.; Chenu, K.; Christopher, M.; Borrell, A.; Hickey, L. Integrating Rapid Phenotyping and Speed Breeding to Improve Stay-Green and Root Adaptation of Wheat in Changing, Water-Limited, Australian Environments. *Procedia Environ. Sci.* **2015,** *29*, 175–176.

Collard, B. C. Y.; Beredo, J. C.; Lenaerts, B.; Mendoza, R.; Santelices, R.; Lopena, V.; Verdeprado, H.; Raghavan, C.; Gregorio, G. B.; Vial, L.; Demont, M.; Biswas, P. S.; Iftekharuddaula, K. M.; Rahman, M. A.; Cobb, J. N.; Islam, M. R. Revisiting Rice Breeding

Methods Evaluating the Use of Rapid Generation Advance (RGA) for Routine Rice Breeding. *Plant Prod. Sci.* **2017,** *20*, 337–352.

Collard, B.; Raghavan, C.; Islam, M. R. Critical Considerations for Molecular Breeding in Rice. *ORYZA Int. J. Rice* **2017,** *53*, 346–353.

Croser, J. S.; Pazos-Navarro, M.; Bennett, R. G.; Tschirren, S.; Edwards, K.; Erskine, W.; Creasy, R.; Ribalta, F. M. Time to Flowering of Temperate Pulses *in vivo* and Generation Turnover *in vivo–in vitro* of Narrow-Leaf Lupin Accelerated by Low Red to Far-Red Ratio and High Intensity in the Far-Red Region. *Plant Cell Tiss. Organ Cult.* **2016,** *127*, 591–599.

Croser, J.; Mao, D.; Dron, N.; Michelmore, S.; McMurray, L.; Preston, C.; Bruce, D.; Ogbonnaya, F. C.; Ribalta, F. M.; Hayes, J.; Lichtenzveig, J.; Erskine, W.; Cullis, B.; Sutton, T.; Hobson, K. Evidence for the Application of Emerging Technologies to Accelerate Crop Improvement- A Collaborative Pipeline to Introgress Herbicide Tolerance Into Chickpea. *Front. Plant Sci.* **2021,** *12*, 2839.

Dagustu, N.; Bayram, G.; Sincik, M.; Bayraktaroglu, M. The Short Breeding Cycle Protocol Effective on Diverse Genotypes of Sunflower (*Helianthus annuus* L.). *Turkish J. Field Crop.* **2012,** *17*, 124–128.

Dan Anisiyah, I. W. M. Accelerated Purification of Sorghum Mutant Line by Using Rapid Cycling Methods. *J. Phys. Conf. Ser.* **2020,** *1436*, 012028.

De Pauw, R. M.; Clarke, J. M. Acceleration of Generation Advancement in Spring Wheat. *Euphytica* **1976,** *25*, 415–418.

Dhankher, O. P.; Foyer, C. H. Climate Resilient Crops for Improving Global Food Security and Safety. *Plant Cell Environ.* **2018,** *41*, 877–884.

Doney, S.; Rosenberg, A. A.; Alexander, M.; Chavez, F.; Harvell, C. D.; Hofmann, G.; Orbach, M.;Ruckelshaus, M. In *Oceans and Marine Resources* Melillo, J. M., Richmond, T.,Yohe, G. W., Eds.; Climate Change Impacts in the United States: USA, 2014; pp 558–578.

Dubcovsky, J.; Loukoianov, A.; Fu, D.; Valarik, M.; Sanchez, A.;Yan, L. Effect of Photoperiod on the Regulation of Wheat Vernalization Genes *VRN1* and *VRN2*. *Plant Mol. Biol.* **2006,** *60*, 469–480.

Effect of Climate Change on Agriculture. [Online] (n.d.). Retrieved Mar 21, 2022. https://pib.gov.in/Pressreleaseshare.aspx?PRID=1696468

Fang, Y.; Wang, L.; Sapey, E.; Fu, S.; Wu, T.; Zeng, H.; Sun, X.; Qian, S.; Khan, M. A. A.; Yuan, S.; Wu, C.; Hou, W.; Sun, S.; Han, T. Speed-Breeding System in Soybean: Integrating Off-Site Generation Advancement, Fresh Seeding, and Marker-Assisted Selection. *Front. Plant Sci.* **2021,** *12*.

Fikre, A.; Degefu, T.; Geleta, T.; Thudi, M.; Gaur, P.; Ojiewo, C.; Hickey, L.; Varshney, R. K. Rapid Generation Advance in Chickpea for Accelerated Breeding Gain in Ethiopia. *Ethiop. J. Agric. Sci.* **2021,** *31*, 1–10.

Flachowsky, H.; Le Roux, P. M.; Peil, A.; Patocchi, A.; Richter, K.;Hanke, M. V. Application of a High-Speed Breeding Technology to Apple Based on Transgenic Early Flowering Plants and Marker-Assisted Selection. *New Phytol.* **2011,** *192*, 364–377.

Forster, B. P.; Till, B. J.; Ghanim, A. M. A.; Huynh, H. O. A.; Burstmayr, H.; Caligari, P. D. S. Accelerated Plant Breeding. *Cab Rev.* **2014,** *9*, 1–16.

Geboloğlu, N.; Bozmaz, S.; Aydin, M.;Çakmak, P. The Role of Growth Regulators, Embryo Age and Genotypes on Immature Embryo Germination and Rapid Generation Advancement in Tomato (*Lycopersicon esculentum* Mill.). *Afr. J. Biotechnol.* **2011,** *10*, 4895–4900.

Ghosh, S.; Watson, A.; Gonzalez-Navarro, O. E.; Ramirez-Gonzalez, R. H.; Yanes, L.; Mendoza-Suárez, M.; Simmonds, J.; Wells, R.; Rayner, T.; Green, P.; Hafeez, A.; Hayta,

S.; Melton, R. E.; Steed, A.; Sarkar, A.; Carter, J.; Perkins, L.; Lord, J.; Tester, M.; Hickey, L. T. Speed Breeding in Growth Chambers and Glasshouses for Crop Breeding and Model Plant Research. *Nat. Protoc.* **2018,** *13*, 2944–2963.

González-Barrios, P.; Bhatta, M.; Halley, M.; Sandro, P.; Gutiérrez, L. Speed Breeding and Early Panicle Harvest Accelerates Oat (*Avena sativa* L.) Breeding Cycles. *Crop Sci.* **2021,** *61*, 320–330.

Gosal, S. S.; Pathak, D.; Wani, S. H.; Vij, S.;Pathak, M. Accelerated Breeding of Plants: Methods and Applications. In *Accelerated Plant Breeding*; Gosal, S., Wani, S., Eds.; Springer: Cham, 2020; pp 1–29.

Goufo, P.; Moutinho-Pereira, J. M.; Jorge, T. F.; Correia, C. M.; Oliveira, M. R.; Rosa, E. A. S.; António, C.;Trindade, H. Cowpea (*Vigna unguiculata* L. Walp.) Metabolomics: Osmoprotection as a Physiological Strategy for Drought Stress Resistance and Improved Yield. *Front. Plant Sci.* **2017,** *8*, 586.

Goulden, C. H. In *Problems in Plant Selection,* The Proceedings of the 7th International Genetic Congress; Burnett, R. C., Eds.; Cambridge University Press: Cambridge, United Kingdom, 1939; pp 132–133.

Gregorio, G. B.; Islam, M. R.; Vergara, G. V.; Thirumeni, S. Recent Advances in Rice Science to Design Salinity and Other Abiotic Stress Tolerant Rice Varieties. *Sabrao J. Breed. Genet.* 2013, *45*, 31–41.

Ha, W. G.; Torollo, G. V.; Kang, K. H.; Lapiz, M.; Padolina, T. Breeding Temperate Japonica Rice MS11 for the Tropics. *Philipp. J. Crop. Sci.* **2011,** *36*, 111.

Harrison, D.; Da Silva, M.; Wu, C.; De Oliveira, M.; Ravelombola, F.; Florez-Palacios, L.; Acuña, A.;Mozzoni, L. Effect of Light Wavelength on Soybean Growth and Development in a Context of Speed Breeding. *Crop Sci.* **2021,** *61*, 917–928.

Hatfield, J.; Takle, G.; Grotjahn, R.; Holden, P.; Izaurralde, R. C.; Mader, T.; Marshall, E.; Liverman, D. Agriculture. In *Climate Change Impacts in the United States*; Melillo, J. M., Richmond, T., Yohe, G. W., Eds.; USA, 2014; pp 150–174.

Hickey, L. T.; Germán, S. E.; Pereyra, S. A.; Diaz, J. E.; Ziems, L. A.; Fowler, R. A.; Platz, G. J.; Franckowiak, J. D.;Dieters, M. J. Speed Breeding for Multiple Disease Resistance in Barley. *Euphytica* **2017,** *213*, 1–14.

Hickey, L. T.; Hafeez, N. A.; Robinson, H.; Jackson, S. A.; Leal-Bertioli, S. C. M.; Tester, M.; Gao, C.; Godwin, I. D.; Hayes, B. J.;Wulff, B. B. H. Breeding Crops to Feed 10 Billion. *Nat. Biotechnol.* **2019,** *37*, 744–754.

Hickey, L.; Rutkoski, J.; Riaz, A.; NG, W.; Singh, D.; Godwin, I.; Aitken, E.; Platz, G.;Dieters, M. Using 'Speed Breeding' to Harness Rust Resistance: Faster, Cheaper and Easier, 2015. https://globalrust.org/content/using-%E2%80%98speed-breeding%E2%80%99-harness-rust-resistance-faster-cheaper-and-easier.

Huang, C.; Sun, H.; Xu, D.; Chen, Q.; Liang, Y.; Wang, X.; Xu, G.; Tian, J.; Wang, C.; Li, D.; Wu, L.; Yang, X.; Jin, W.; Doebley, J. F.; Tian, F. *ZmCCT9* Enhances Maize Adaptation to Higher Latitudes. *PNAS* **2018,** *115*, 334–341.

Hussain, B. Modernization in Plant Breeding Approaches for Improving Biotic Stress Resistance in Crop Plants. *Turk. J. Agric. For.* **2015,** *39*, 515–530.

Idrissi, O. Application of Extended Photoperiod in Lentil: Towards Accelerated Genetic Gain in Breeding for Rapid Improved Variety Development. *Moroccan J. Agri. Sci.* **2020,** *1* (1).

IISD. (n.d.). Retrieved Mar 11, 2022 [Online]. https://sdg.iisd.org/news/world-population-to-reach-9-9-billion-by-2050/

IPCC. Climate Change 2007: Synthesis Report. (n.d.). Retrieved Mar 21, 2022 [Online] 2007. http://pure.iiasa.ac.at/id/eprint/8667/

IPCC. Climate Change: Impacts, Adaptation, and Vulnerability. Part A: Global and Sectoral Aspects. Contribution of Working Group II to the Fifth Assessment Report of the Inter Governmental Panel on Climate Change. Cambridge University Press, Cambridge, United Kingdom, 2014; p 32.

Jähne, F.; Hahn, V.; Würschum, T.; Leiser, W. L. Speed Breeding Short-Day Crops by LED-Controlled Light Schemes. *Theor. Appl. Genet.* **2020,** *133*, 2335–2342.

Jamali, S. H.; Cockram, J.; Hickey, L. T. Is Plant Variety Registration Keeping Pace with Speed Breeding Techniques? *Euphytica* **2020,** *216*, 1–13.

Karmakar, R.; Das, I.; Dutta, D.; Rakshit, A. Potential Effects of Climate Change on Soil Properties: A Review. *Sci. Int.* 2016, *4*, 51–73.

Khan, N.; Han, Y.; Xing, F.; Feng, L.; Wang, Z.; Wang, G.; Yang, B.; Fan, Z.; Lei, Y.; Xiong, S.; Li, X.; Li, Y. Plant Density Influences Reproductive Growth, Lint Yield and Boll Spatial Distribution of Cotton. *Agronomy* **2019,** *10*, 14.

Koumura, T. Breeding of New Rice Variety 'Nipponbare'. *Agric. Tech.* **1972a,** *27*, 112–116.

Laux, P.; Jäckel, G.; Tingem, R. M.;Kunstmann, H. Impact of Climate Change on Agricultural Productivity Under Rainfed Conditions in Cameroon-A Method to Improve Attainable Crop Yields by Planting Date Adaptations. *Agric. For. Meteorol.* **2010,** *150*, 1258–1271.

Leisner, C. P. Review: Climate Change Impacts on Food Security- Focus on Perennial Cropping Systems and Nutritional Value. *Plant Sci.* 2020, *293*, 110412.

Li, H.; Rasheed, A.; Hickey, L. T.; He, Z. Fast-Forwarding Genetic Gain. *Trends Plant Sci.* **2018,** *23*, 184–186.

Liu, H.; Zwer, P.; Wang, H.; Liu, C.; Lu, Z.; Wang, Y.; Yan, G. A Fast Generation Cycling System for Oat and Triticale Breeding. *Plant Breed.* **2016,** *135*, 574–579.

Lulsdorf, M. M.; Banniza, S. Rapid Generation Cycling of an F2 Population Derived from a Cross Between *Lens culinaris* Medik. and *Lens ervoides* (Brign.) Grande After Aphanomyces Root Rot Selection. *Plant Breed.* **2018,** *137*, 486–491.

Maglia, F.; Bermejo, C.; Palacios, L. T.; Cointry, E. Speed Breeding Para la Multiplicación de colecciones activas en lenteja (*Lens culinaris* Medik.). *J. Appl. Genet.* **2020,** *31*, 123–145.

Mangon, H. Production de la matière verte des feuilles sous l'influence de la lumièreélectrique. *Compt. Rend. Acad. Sci. Paris* 1861, *53*, 243–244.

Manzur, J. P.; Oliva-Alarcón, M.; Rodríguez-Burruezo, A. *In vitro* Germination of Immature Embryos for Accelerating Generation Advancement in Peppers (*Capsicum annuum* L.). *Sci. Hortic.* **2014,** *170*, 203–210.

Mobini, S. H.; Lulsdorf, M.; Warkentin, T. D.; Vandenberg, A. Plant Growth Regulators Improve *in vitro* Flowering and Rapid Generation Advancement in Lentil and Faba Bean. *In vitro* Cellular and Developmental Biology. *Plant* **2015,** *51*, 71–79.

Mobini, S. H.; Warkentin, T. D. A Simple and Efficient Method of *in vivo* Rapid Generation Technology in Pea (*Pisum sativum* L.). *In vitro* Cellular and Developmental Biology. *Plant* **2016,** *52*, 530–536.

Mpelkas, C. C. Light Sources for Horticultural Lighting. *Inst. Electr. Electron Eng. Trans. Ind. Appl.* **1980,** *16*, 557–565.

Mukade, K. New Procedures for Accelerating Generation Advancement in Wheat Breeding. *JARQ* **1974,** *8*, 1–5.

Nagatoshi, Y.; Fujita, Y. Accelerating Soybean Breeding in a CO2-Supplemented Growth Chamber. *Plant Cell Physiol.* **2019,** *60*, 77–84.

Nelson, G.C.; Rosegrant, M.; Palazzo, A.; Gray, I.; Ingersoll, C.; Robertson, R.;Tokgoz, S.; Zhu, T.; Sulser, T.; Ringler, C.; Msangi, S.; You, L. Food Security, Farming, and Climate Change to 2050: Scenarios, Results, Policy Options; International Food Policy Research Institute (IFPRI): Washington, DC, 2010.

Newbery, F.; Qi, A.; Fitt, B. D. Modelling Impacts of Climate Change on Arable Crop Diseases: Progress, Challenges and Applications. *Curr. Opin. Plant Biol.* 2016, *32*, 101–109.

Nocker, S. Van; Gardiner, S. E. Breeding Better Cultivars, Faster: Applications of New Technologies for the Rapid Deployment of Superior Horticultural Tree Crops. *Hortic. Res.* **2014,** *1*. 1–8.

O'Connor, D. J.; Wright, G. C.; Dieters, M. J.; George, D. L.; Hunter, M. N.; Tatnell, J. R.; Fleischfresser, D. B. Development and Application of Speed Breeding Technologies in a Commercial Peanut Breeding Program. *Peanut Sci.* **2013,** *40*, 107–114.

Ochatt, S. J.; Sangwan, R. S. *In vitro* Shortening of Generation Time in *Arabidopsis thaliana. Plant Cell Tiss. Organ Cult.* **2008,** *93*, 133–137.

Ochatt, S. J.; Sangwan, R. S.; Marget, P.; Assoumou Ndong, Y.; Rancillac, M.; Perney, P. New Approaches Towards the Shortening of Generation Cycles for Faster Breeding of Protein Legumes. *Plant Breed.* **2002,** *121*, 436–440.

Ochatt, S. J.; Sangwan, R. S.; Marget, P.; Ndong, Y. A.; Rancillac, M.; Perney, P.; Röbbelen, G. New Approaches Towards the Shortening of Generation Cycles for Faster Breeding of Protein Legumes. *Plant Breed.* **2002,** *121*, 436–440.

Oerke, E. C. Crop Losses to Pests. *J. Agric. Sci.* **2006,** *144*, 31–43.

Ohnishi, T.; Yoshino, M.; Yamakawa, H.; Kinoshita, T. The Biotron Breeding System: A Rapid and Reliable Procedure for Genetic Studies and Breeding in Rice. *Plant Cell Physiol.* **2011,** *52*, 1249–1257.

Pandey, S.; Singh, A.; Parida, S. K.; Prasad, M. Combining Speed Breeding with Traditional and Genomics-Assisted Breeding for Crop Improvement. *Plant Breed.* **2022**.

Parmar, S.; Deshmukh, D. B.; Kumar, R.; Manohar, S. S.; Joshi, P.; Sharma, V.; Chaudhari, S.; Variath, M. T.; Gangurde, S. S.; Bohar, R.; Singam, P.; Varshney, R. K.; Janila, P.;Pandey, M. K. Single Seed-Based High-Throughput Genotyping and Rapid Generation Advancement for Accelerated Groundnut Genetics and Breeding Research. *Agronomy* **2021,** *11*, 1226.

Pazos-Navarro, M.; Castello, M.; Bennett, R. G.; Nichols, P.; Croser, J.; Pazos-Navarro, M.; Castello, M.; Bennett, R. G.; Nichols, P.; Croser, J. *In vitro*-Assisted Single-Seed Descent for Breeding-Cycle Compression in Subterranean Clover (*Trifolium subterraneum* L.). *Crop Pasture Sci.* **2017,** *68*, 958–966.

Pfeiffer, N. E. Microchemical and Morphological Studies of Effect of Light on Plants. *Bot. Gaz.* **1926,** *81*, 173–195.

Pineda, M.; Morante, N.; Salazar, S.; Cuásquer, J.; Hyde, P. T.; Setter, T. L.; Ceballos, H. Induction of Earlier Flowering in Cassava Through Extended Photoperiod. *Agronomy* **2020,** *10* (9).

Prasad, P.; Bhardwaj, S.C.; Thakur, R. K.; Adhikari, S.; Gangwar, O. P.; Lata, C.; Kumar, S. Prospects of Climate Change Effects on Crop Diseases with Particular Reference to Wheat. *J. Cer. Res.* **2021,** *13*, 117–134.

Prilleux, E. De l'influence de la lumière artificielle sur la réduction de l'acide carbonique par les plantes. *Compt. Rend. Acad. Sci. Paris* **1869,** *69*, 408–412.

Rahman, M. A. Field Rapid Generation Advance : An Effective Technique for Industrial Scale Rice Breeding Program. *Experiment* **2019,** *47*, 2659–2670.

Rai, A. C.; Rai, K. K. Speed Breeding for Rapid Cycling of Crops for Stress Management and Global Food Security. In *Microbial Biocontrol: Food Security and Post Harvest Management*; Kumar, A., Eds.; Springer: Cham, 2022; pp 23–37.

Raju, C. S. N.; Sagar, C. K. Speed Breeding in Agriculture Future Prospects. *Int. J. Curr. Microbiol.* **2020,** *9*, 1059–1076.

Rana, M. M.; Takamatsu, T.; Baslam, M.; Kaneko, K.; Itoh, K.; Harada, N.; Sugiyama, T.; Ohnishi, T.; Kinoshita, T.; Takagi, H.; Mitsui, T. Salt Tolerance Improvement in Rice through Efficient SNP Marker-Assisted Selection Coupled with Speed-Breeding. *Int. J. Mol. Sci.* **2019,** *20*, 2585.

Rao, C. S.; Prasad, R. S.; Mohapatra, T. Climate Change and Indian Agriculture: Impacts, Coping Strategies, Programmes and Policy. Technical Bulletin/Policy Document 2019. Indian Council of Agricultural Research, Ministry of Agriculture and Farmers' Welfare and Ministry of Environment, Forestry and Climate Change, Government of India: New Delhi, 2019; p 25.

Ribalta, F. M.; Croser, J. S.; Erskine, W.; Finnegan, P. M.; Lulsdorf, M. M.; Ochatt, S. J. Antigibberellin-Induced Reduction of Internode Length Favors *in vitro* Flowering and Seed-Set in Different Pea Genotypes. *Biol. Plant.* **2014,** *58*, 39–46.

Ribalta, F. M.; Pazos-Navarro, M.; Nelson, K.; Edwards, K.; Ross, J. J.; Bennett, R. G.; Munday, C.; Erskine, W.; Ochatt, S. J.; Croser, J. S. Precocious Floral Initiation and Identification of Exact Timing of Embryo Physiological Maturity Facilitate Germination of Immature Seeds to Truncate the Lifecycle of Pea. *Plant Growth Regul.* **2017,** *81*, 345–353.

Rizal, G.; Karki, S.; Alcasid, M.; Montecillo, F.; Acebron, K.; Larazo, N.; Garcia, R.; Slamet-Loedin, I. H.;Quick, W. P. Shortening the Breeding Cycle of Sorghum, a Model Crop for Research. *Crop Sci.* **2014,** *54*, 520–529.

Robertson, L. D.;Curtis, B. C. Germination of Immature Kernels of Winter Wheat. *Crop Sci.* **1967,** *7*, 269–270.

Roumet, P.; Morin, F. Germination of Immature Soybean Seeds to Shorten Reproductive Cycle Duration. *Crop Sci.* **1997,** *37*, 521–525.

Samantara, K.; Bohra, A.; Mohapatra, S. R.; Prihatini, R.; Asibe, F.; Singh, L.; Reyes, V. P.; Tiwari, A.; Maurya, A. K.; Croser, J. S.; Wani, S. H.; Siddique, K. H. M.; Varshney, R. K. Breeding More Crops in Less Time: A Perspective on Speed Breeding. *Biology* **2022,** *11*, 275.

Samineni, S.; Sen, M.; Sajja, S. B.; Gaur, P. M. Rapid Generation Advance (RGA) in Chickpea to Produce up to Seven Generations Per Year and Enable Speed Breeding. *Crop J.* **2020,** *8*, 164–169.

Santos-Antunes, F.; Léon, L.; De La Rosa, R.; Alvarado, J.; Mohedo, A.; Trujillo, I.; Rallo, L. The Length of the Juvenile Period in Olive as Influenced by Vigor of the Seedlings and the Precocity of the Parents. *Hort. Sci.* **2005,** *40*, 1213–1215.

Saxena, K. B.; Saxena, R. K.; Hickey, L. T.; Varshney, R. K. Can a Speed Breeding Approach Accelerate Genetic Gain in Pigeonpea? *Euphytica* **2019,** *215*, 1–7.

Saxena, K.; Saxena, R. K.;Varshney, R. K. Use of Immature Seed Germination and Single Seed Descent for Rapid Genetic Gains in Pigeonpea. *Plant Breed.* **2017,** *136*, 954–957.

Sharma, S.; Kumar, A.; Dhakte, P.; Raturi, G.; Vishwakarma, G.; Barbadikar, K. M.; Das, B. K.; Shivaraj, S. M.; Sonah, H.;Deshmukh, R. Speed Breeding Opportunities and Challenges for Crop Improvement. *J. Plant Growth Regul.* **2022,** *15*, 1–14.

Shavrukov, Y.; Kurishbayev, A.; Jatayev, S.; Shvidchenko, V.; Zotova, L.; Koekemoer, F.; De Groot, S.; Soole, K.; Langridge, P. Early Flowering as a Drought Escape Mechanism in Plants: How Can It Aid Wheat Production? *Front. Plant Sci.* **2017,** *8*, 1950.

Shimelis, H.; Gwata, E. T.;Laing, M. D. Crop Improvement for Agricultural Transformation in Southern Africa. In *Transforming Agriculture in Southern Africa*; Sikora, R. A., Terry, E. R., Vlek, P. L.G., Chitja, J. Eds.; Taylor and Francis: London, New York, USA, 2019; pp 97–103.

Shimelis, H.;Laing, M. D. Timelines in Conventional Crop Improvement : Pre-breeding and Breeding Procedures. *Aust. J. Crop Sci.* **2012,** *6*, 1542–1549.

Shivakumar, M.; Nataraj, V.; Kumawat, G.; Rajesh, V.; Chandra, S.; Gupta, S.; Bhatia, V. S. Speed Breeding for Indian Agriculture:A Rapid Method for Development of New Crop Varieties. *Curr. Sci.* **2018,** *115*, 1241.

Siemens, C. W. On the Influence of Electric Light Upon Vegetation, and on Certain Physical Principles Involved. *Proc. R. Soc. Lond.* **1880,** *30*, 210–219.

Silva Souza, L.; Diniz, R. P.; Neves, R. de J.; Alves, A. A. C.;Oliveira, E. J. Grafting as a Strategy to Increase Flowering of Cassava. *Sci. Hortic.* **2018,** *240*, 544–551.

Sinha, P.; Singh, V. K.; Saxena, R. K.; Khan, A. W.; Abbai, R.; Chitikineni, A.; Desai, A.; Molla, J.; Upadhyaya, H. D.; Kumar, A.; Varshney, R. K. Superior Haplotypes for Haplotype-Based Breeding for Drought Tolerance in Pigeonpea (*Cajanus cajan* L.). *Plant Biotechnol. J.* **2020,** *18*, 2482.

Song, Y.; Duan, X.; Wang, P.; Li, X.; Yuan, X.; Wang, Z.; Wan, L.; Yang, G.; Hong, D. Comprehensive Speed Breeding: A High-Throughput and Rapid Generation System for Long-Day Crops. *Plant Biotechnol. J.* **2022,** *20*, 13.

Springer, C. J.; Ward, J. K. Flowering Time and Elevated Atmospheric CO2. *New Phytol.* **2007,** *176*, 243–255.

Stetter, M. G.; Zeitler, L.; Steinhaus, A.; Kroener, K.; Biljecki, M.; Schmid, K. J. Crossing Methods and Cultivation Conditions for Rapid Production of Segregating Populations in Three Grain Amaranth Species. *Front. Plant Sci.* **2016,** *7*, 816.

Surma, M.; Adamski, T.; Święcicki, W.; Barzyk, P.; Kaczmarek, Z.; Kuczyńska, A.; Krystkowiak, K.; Mikołajczak, K.; Ogrodowicz, P. Preliminary Results of *in vitro* Culture of Pea and Lupin Embryos for the Reduction of Generation Cycles in Single Seed Descent Technique. *Acta Soc. Bot. Pol.* 2013, *82*, 231–236.

Sysoeva, M.; Markovskaya, E.; Shibaeva, T. Plants Under Continuous Light: A Review. *Plant Stress* **2010,** *4*, 5–17.

Tanaka, J.; Hayashi, T.; Iwata, H. A Practical, Rapid Generation-Advancement System for Rice Breeding Using Simplified Biotron Breeding System. *Breed. Sci.* **2016,** *66*, 542.

Vadez, V.; Hash, T.; Bidinger, F. R.; Kholova, J. Phenotyping Pearl Millet for Adaptation to Drought. *Front. Physiol.* **2012,** *3*, 386.

Varshney, R. K.; Bohra, A.; Yu, J.; Graner, A.; Zhang, Q.;Sorrells, M. E. Designing Future Crops: Genomics-Assisted Breeding Comes of Age. *Trends Plant Sci.* **2021,** *26*, 631–649.

Vats, S.; Kumawat, S.; Kumar, V.; Patil, G. B.; Joshi, T.; Sonah, H.; Sharma, T. R.; Deshmukh, R. Genome Editing in Plants: Exploration of Technological Advancements and Challenges. *Cell J.* **2019,** *8*, 1386.

Velez-Ramirez, A. I.; Van Ieperen, W.; Vreugdenhil, D.; Van Poppel, P. M. J. A.; Heuvelink, E.; Millenaar, F. F. A Single Locus Confers Tolerance to Continuous Light and Allows Substantial Yield Increase in Tomato. *Nat. Commun.* **2014,** *5*, 1–13.

Vikas, V. K.; Sivasamy, M.; Jayaprakash, P.; Vinod, K. K.; Geetha, M.; Nisha, R.; Shajitha, P.; Peter, J. Customized Speed Breeding as a Potential Tool to Advance Generation in Wheat. *Indian J. Genet. Plant Breed.* **2021,** *81*, 199–207.

Villar, J. L.; Maranville, J. W.; Gardner, J. C. High Density Sorghum Production for Late Planting in the Central Great Plains. *J. Prod. Agric.* **1989,** *2*, 333–338.

Voss-Fels, K. P.; Herzog, E.; Dreisigacker, S.; Sukumaran, S.; Watson, A.; Frisch, M.; Hayes, B.; Hickey, L. T. In *"SpeedGS" to Accelerate Genetic Gain in Spring Wheat. Applications of Genetic and Genomic Research in Cereals*; Miedaner, T., **Korzun, V.,** Eds.; Elsevier: New York, USA, 2019; pp 303–327.

Wanga, M. A.; Shimelis, H.; Mashilo, J.; Laing, M. D. Opportunities and Challenges of Speed Breeding: A Review. *Plant Breed.* **2021,** *140*, 185–194.

Watson, A.; Ghosh, S.; Williams, M. J.; Cuddy, W. S.; Simmonds, J.; Rey, M. D.; Asyraf Md Hatta, M.; Hinchliffe, A.; Steed, A.; Reynolds, D.; Adamski, N. M.; Breakspear, A.; Korolev, A.; Rayner, T.; Dixon, L. E.; Riaz, A.; Martin, W.; Ryan, M.; Edwards, D.; Hickey, L. T. Speed Breeding is a Powerful Tool to Accelerate Crop Research and Breeding. *Nat. Plants* **2018,** *4*, 23–29.

Watson, A.; Hickey, L. T.; Christopher, J.; Rutkoski, J.; Poland, J.; Hayes, B. J. Multivariate Genomic Selection and Potential of Rapid Indirect Selection with Speed Breeding in Spring Wheat. *Crop Sci.* **2019,** *59*, 1945–1959.

Wolter, F.; Schindele, P.; Puchta, H. Plant Breeding at the Speed of Light: The Power of CRISPR/Cas to Generate Directed Genetic Diversity at Multiple Sites. *BMC Plant Bio.* **2019,** *19*, 1–8.

Yang, S.; Weers, B. D.; Morishige, D. T.; Mullet, J. E. *CONSTANS* is a Photoperiod Regulated Activator of Flowering in Sorghum. *BMC Plant Bio.* **2014,** *14*, 1–15.

Yao, Y.; Zhang, P.; Liu, H.; Lu, Z.; Yan, G. A Fully *in vitro* Protocol Towards Large Scale Production of Recombinant Inbred Lines in Wheat (*Triticum aestivum* L.). *Plant Cell Tiss. Organ Cult.* **2017,** *128*, 655–661.

Yao, Y.; Zhang, P.; Wang, H. B.; Lu, Z. Y.; Liu, C. J.; Liu, H.; Yan, G. J. How to Advance up to Seven Generations of Canola (*Brassica napus* L.) Per Annum for the Production of Pure Line Populations? *Euphytica* **2016,** *209*, 113–119.

Zhang, C.; Liu, J.; Zhao, T.; Gomez, A.; Li, C.; Yu, C.; Li, H.; Lin, J.; Yang, Y.; Liu, B.; Lin, C. A Drought-Inducible Transcription Factor Delays Reproductive Timing in Rice. *Plant Physiol.* **2016,** *171*, 334–343.

Zheng, Z.; Wang, H. B.; Chen, G. D.; Yan, G. J.; Liu, C. J. A Procedure Allowing up to Eight Generations of Wheat and Nine Generations of Barley Per Annum. *Euphytica* **2013,** *191*, 311–316.

CHAPTER 11

Recent Advances in CRISPR-Cas for Climate Resilient Agriculture in Cereals

SACHIN PHOGAT[1,2], ANKUR POUDEL[1,2,5], ALKA BHARATI[1,2,3], AMRESH KUMAR[2,4], and P. K. MANDAL[2]

[1]*ICAR-Indian Agricultural Research Institute, Pusa, New Delhi, India*

[2]*ICAR-National Institute of Plant Biotechnology, Pusa, New Delhi India*

[3]*ICAR-Central Agroforestry Research Institute, Jhansi, India*

[4]*Department of Biotechnology, Deenbandhu Chottu Ram University of Science and Technology, Murthal, Haryana, India*

[5]*Lamjung Campus, Institute of Agriculture & Animal Sciences, Tribhuvan University, Nepal*

ABSTRACT

Climate change has become a serious threat to agriculture. Plants are under continuous threat of biotic and abiotic stresses. Plant breeding approaches are currently in use for pyramiding preferred agricultural traits for stress tolerance from their wild lineages. However, these techniques are time consuming, and exclusive trait selection causes biodiversity loss. Although efforts have been directed toward addressing stress tolerance some successes have been achieved in biotic stress tolerance but still, abiotic stress tolerance is a major concern. The complex nature of inheritance of these multiple abiotic stress-tolerant genes and their environmental interaction is a bottleneck in enhancing stress tolerance in crops. Thus, the search continues for precise and reliable techniques to enhance stress tolerance in crops besides classical breeding. Recent advances in genome editing techniques such as

Climate-Resilient Agriculture: A Molecular Perspective. Ashutosh Singh, Saurabh Pandey (Eds.)

zinc finger nucleases (ZFNs), transcription activator-like effector nucleases (TALENs), and clustered regularly interspaced short palindromic repeats-Cas (CRISPR-Cas) systems have the potential to address this issue. Among these, the CRISPR-Cas technique is more robust, precise, easy, quick, and generates heritable changes. CRISPR-Cas has developed crop plants with desirable changes in their genomes by deleting negative genetic sections governing unwanted agronomic traits, the gain of function mutation, base editing, multiplex editing, etc. Cas9 protein-modified versions such as Nicknase (nCas9) and deactivated dCas9 have mutations in one or both HNH and RuvC active sites, which can perform different functions. Cereal crops such as wheat, rice, and maize have already been developed with sufficient tolerance to drought and salinity using CRISPR-Cas9 technique to cope with a constantly changing environment. Nowadays, Cas12a/Cpfl RNA-guided DNA nuclease—which is a different CRISPR-Cas system—is also available and it requires smaller guide RNA, T-rich PAM instead of G-rich, and produces staggered ends more appropriate for knock-in DNA fragments. Cas12b/C2c I class of nuclease has also been applied in rice for transcriptional activation and in Arabidopsis for multiplex editing. Another CRISPR effector variant Cas13/C2c2 has been applied in plants which targets single-stranded RNA in eukaryotes. Its RNase activity is governed by the HEPN (higher eukaryotes and prokaryotes nucleotide-binding) domain. The fusion of Cas13 with other Cas nucleases could provide new prospects for crop improvement by targeting both DNA and RNA in the same plant. Another class of CRISPR is Cas14a/Casl4f, which is an RNA-guided DNA nuclease but it cleaves single-stranded DNA. This unique property makes it special and useful in combating plant-infecting viruses with single-stranded DNA. Further, this CRISPR system is continuing to evolve and researchers expect its full potential is yet to come.

11.1 INTRODUCTION

The climate is changing. Patterns of different climatic events such as temperature, rainfall, etc. have changed significantly over a period of time. Climate change has been caused by anthropogenic activity during the last several decades. Indiscriminate human activities have resulted in an imbalance in the global atmosphere (Parry et al., 2007). People are being affected by the changing climate in a variety of ways. Thus, the impacts of changing climate on the environment have been a serious concern for people around the world (Watts and Scales, 2015). The IPCC's Special Report "Global Warming of 1.5 degrees Celsius" has revealed several worrisome facts about climate change. It is predicted that between 2030 and 2052, the temperature will rise by 1.5°C; global mean surface temperature has increased to around 0.87°C which is much higher as compared to the period 1850–1900 (IPCC, 2018). Moreover, much higher annual temperatures have been experienced with frequent heatwaves, droughts, floods, and persistent sea level rise in different parts of the world. Porter et al. (2014) stated that the level of GHGs (due to anthropogenic causes) is the highest in history. The atmospheric CO_2 concentration has risen from 280 to 400 mol^{-1} as a consequence of deforestation and reliance on fossil fuels. Also, the concentration of CO_2 is expected to double (800 μmol^{-1}) by the end of this century (Raza et al., 2019). Accumulation of greenhouse gases (GHG) such as CO_2 has resulted in the greenhouse effect (Vaughan et al., 2018). These extremes in climate have a severe detrimental influence on biodiversity. It is most likely that some crops/varieties native to the region might not adapt to the changing scenario of climate thus resulting in crop failure. In addition, interactions among agricultural ecosystem components such as pollinators, predator-prey balance, and so on would be disrupted. Resurgence and emergence of minor or new pests/diseases have been seen due to changes in climate. The trend and pattern of climatic events such as rainfall, summers, and winters have changed resulting in extreme weather. Therefore, the cumulative effect will result in the decline of productivity of major food crops particularly at lower altitudes where crops will suffer much (Shukla et al., 2019). Crop yields are anticipated to drop by about 20% by 2020 (Vaughan et al., 2018). On one hand, the climate continues to change while on the other hand, the population keeps on increasing, and arable land keeps on decreasing (Stamm et al., 2011). At present, the population is increasing at a rate of 1.10% which means around 83 million people are added annually. According to current trends, the global population will reach 8.6 billion in 2030, 9.8 billion in 2050, and 11.2 billion in 2100 (UN, 2017). Thus, agriculture must be ready to sustainably feed the ever-increasing population under pressure of changing climate and other impacts. Agriculture must be ready to meet the unprecedented demand for food in the upcoming decade. As a result, it is necessary to safeguard the long-term utilization of natural resources.

However, climate change is expected to increase in the next years. Extreme weather conditions and their frequency, intensity, duration, and area of occurrence will increase (Murray and Ebi, 2012). 2019 was Europe's warmest year, the worldwide increase in drought, glacier retreat, floods, heat waves, and precipitation patterns has altered dramatically (Ch et al., 2019). This has affected agriculture the most. Economic losses in agriculture due to these disasters are escalating as it is the most vulnerable sector. Between 1998 and 2017, Aronsson-Storrier (2021) calculated a direct economic loss of roughly US$ 2908 billion, of which 77% was attributable to climate-related disasters. So, if climate change is not addressed with caution, costs will climb, and the risk associated will worsen.

So, to adjust to the changing climate and protect the extremely vulnerable agriculture, different adaptations and mitigation strategies and actions are required. Climate resilient agriculture (CRA) is one such strategy in which landscapes, crops, land, livestock, forests, and fisheries are integrated to tackle the challenges of climate change and food security. Climate resilient agriculture investigates agricultural practices that can tolerate and withstand the stress of changing environments and extreme weather. It is the incorporation of mitigation, adaptation, and other strategies to resist the damage due to climate and immediately recover. CRA is an integrated approach that relies on increasing productivity, resilience to biotic and abiotic stresses, and reduction in emissions and carbon sequestration.

Climate change has been affecting the agriculture system in multiple ways. The vulnerability depends on the magnitude, type, and degree of impact of the agricultural component of the system. The major effect on crops is the shortening of crop duration. Increased temperature has led to shortening of crop duration by almost 2–3 weeks in annual crops. Similarly, in many winter crops high temperature during reproductive stages (anthesis, pollination, and fertilization) has been a key factor to reduce production. Similarly, erratic rainfall (inadequate/excessive) has also resulted in crop losses. Plant physiological processes have been harmed as a result of these stresses, resulting in yield reductions. Since 1982, climate change has resulted in yield losses of up to 70% (Boyer, 1982). High temperatures have resulted in a yield loss of around 6% for each degree increase in temperature in wheat (Asseng et al., 2015). Many important enzymes have been affected due to temperature stress in plants. When the temperature exceeds 35°C activity of the ribulose bisphosphate carboxylase oxygenase (RUBISCO) enzyme is disrupted which ceases photosynthesis in plants (Griffin et al., 2004). When two or more stresses are combined together to act on plants, the effects are more adverse. Under both heat and drought stresses, the function of photosystem II is found to be reduced in *Leymus chinensis* (Xu and Zhou, 2006). Crop performance has been observed to be affected by drought stress. In wheat drought stress is responsible for decreasing yield 1–30% under mild stress whereas almost 90% yield reduction is seen under prolonged drought stress during the anthesis and postanthesis stages (de Oliveira et al., 2012). In case of maize yield decline was seen beyond 29°C (Schlenker and Roberts, 2009) whereas maize yield declined 8.3% for every 1°C increase in temperature (Lobell and Field, 2007).

11.2 CLIMATE CHANGE ADAPTATION

Climate change adaptation measures are being used to reduce the impact of climate change on agriculture. Adaptation refers to the modification of an agricultural system to a changing environment or the preparation of an agricultural system to adapt to predicted climatic conditions. Adaptation is done either by improving the package of practices of cultivation, integrated farming, water/nutrient management, crop diversification, pest surveillance, integrated pest control, breeding of stress-tolerant varieties, etc. or by risk coverage via insurance, social networking, and community-based adaptation, agrometeorology advisory, international collaboration, etc. (Godfray and Garnett, 2014). On-farm practices for climate change adaptation include incorporation and maintenance of high soil organic matter, using local crop genetic diversity, integration of crop and livestock, mixed cropping, multiple cropping, minimizing topsoil erosion, and maintaining soil moisture (Altieri and Koohafkan, 2008). Variable weather and extreme weather such as heavy rain and wind accelerates soil erosion. So, to avoid this loss windbreaks, hedgerows, cover crops, etc. are planted. Similarly, microirrigation, water harvesting, terracing, and mulching help to reduce moisture losses and harsh surface temperatures. Similarly, changing tillage techniques reduces surface temperature by 1–4°C and improves irrigation water efficiency by 66–100%, while zero tillage with residue retention has boosted soil organic content by 4.66 tons per hectare over the last 7 years (Sapkota et al., 2017).

It is a proven fact that a diverse production system is more resilient in enhancing food and nutritional security by nutrient recycling, reducing GHG emissions, and controlling hydrological processes (Chivenge et al., 2015). Crop diversity, whether through crop substitution or crop rotation, will assist in making cropping systems more robust in both time and space (Lin, 2011). Crop diversification lowers risk of disease and pest and improves soil health.

One crucial component of a climate change adaptation plan is management practices. Another major technique is the creation of climate-resistant agricultural varieties. The advancement in crop breeding has come up with some climate-smart crop varieties which were initially susceptible to changing climate. Different submergence and drought-tolerant rice varieties have been developed to withstand drought and submergence stress. Different drought-tolerant genes have been incorporated in rice and drought-tolerant varieties such as Sahabhagi Dhan, DRR Dhan 42, 43, and 44 are cultivated in India. Similarly, Sambha Mahsuri, CR1009, etc. are known for submergence tolerance in rice. Whereas Sukkha Dhan 6, CR Dhan 801, and CR Dhan 802 are tolerant to both submergence and drought conditions. Similarly, drought is responsible for yield reduction in maize and wheat. Drought causes yield loss of up to 15–20% in maize under South Asian conditions (Zaidi et al., 2004). Indian varieties such as Pusa Hybrid Makka 1, HM 4, and Pusa Hybrid Makka 5 are tolerant to drought. Similarly, RCRMH2 and Lall-454 maize varieties are heat-tolerant maize genotypes, and HQPM-1, HHM-1, and HM-1 are cold- and frost-tolerant genotypes. These tolerant genotypes aid in yield enhancement and climate change adaptation.

For the benefit of humanity, several biotechnological approaches for genetic manipulation and genome engineering are in use. Currently, a number of technologies and strategies are being used to assist and improve features associated with biotic and abiotic stress tolerance.

11.3 CONVENTIONAL TECHNIQUES FOR DEVELOPING CLIMATE RESILIENT CROP

High-yielding crop varieties are developed by using conventional crop improvement techniques. Farmers and breeders have used some of these strategies for thousands of years. These include artificial selection based on morphological characters, introgression of traits from wild relatives, conventional breeding where offspring with superior desirable characteristics were selected, and hybridization-based methods. The time it takes to generate a new crop variety with desirable features is a substantial constraint. These techniques only exploit the variation already available in nature. Crop improvement was based only on pyramiding or recombining the superior desirable characters already present in nature (Ahmar et al., 2020; Khusg, 2013). However, with the growing population of the world, it is nearly impossible to feed the world with only the use of conventional crop improvement techniques. In a continuously changing climate, the current demand is to develop climate-smart crops which can withstand biotic and abiotic stresses. Due to extensive cultivation of some common improved varieties, the genetic variation has been decreased. Variation breeding can be used to induce new non-existent characters in crop varieties that could be created using some chemicals and radiation also called mutagens (Acquaah, 2016; Steinwand and Ronald, 2020). In molecular breeding, conventional breeding methods are assisted by molecular techniques. These include next-generation sequencing techniques that provide a comprehensive understanding of the genes, Allele mining tells regarding the dissection of allelic differences in candidate genes, genome-wide association study (GWAS), molecular marker assisted selection such as marker-assisted backcrossing (MABC) and marker-assisted recurrent selection (MARS), SNP detection, and quantitative trait loci association analysis (Oladosu et al., 2019). The expanded knowledge of genomics and bioinformatics when associated with traditional crop improvement methods can solve the major problems of food scarcity. Improvement in major world-feeding cereal crops takes place by producing better biotic and abiotic stress resistance. These techniques helped to reach the food demand of the globe (Singh et al., 2016). A vast quantity of data on the genetic composition and phenotype of important crops has been gathered to date. Better knowledge of gene function has been available due to advances in functional genomics. The advanced techniques involve plant transformation with novel genes across species, genetic engineering to modify the genetic background of crops, plant tissue culture for transformation and regeneration, somaclonal variation, and somatic hybridization are the modern techniques that can modify the genetic makeup of the plants to improve their usefulness. The most recent advancements are the development of specific nucleases such as zinc finger nucleases (ZFN), transcription activator-like effector nucleases (TALENs), and clustered regularly interspaced short palindromic repeats-Cas (CRISPR-Cas) systems that can modify or edit the genomes at specific predefined positions by using their molecular machinery.

11.4 GENETIC ENGINEERING TECHNIQUES AND THEIR USE IN DEVELOPING CLIMATE-SMART CROPS

11.4.1 RNAI

Genetic engineering (GE) is a useful technique that can directly introduce agronomically important traits such as insect pest resistance, herbicide tolerance, and efficient nutrient use. It modifies the genetic material of any organism in very less time as compared to conventional and molecular breeding techniques. Genetic material can be transferred across species in a few generations even in vegetatively propagated plants (Dong and Ronald, 2019; Kumar et al., 2020). Genes could be overexpressed to amplify the gene effect or their effect could be minimized by gene silencing. RNA interference RNAi utilizes small noncoding RNAs such as small interfering RNA and microRNA which have the power of precise gene regulation without affecting the expression of other genes. MiRNA and SiRNA comprise 20–24 nucleotides but they have differences in their biogenesis, precursors, and their mechanism of action (Kamthan et al., 2015). In reverse genetics, RNAi has been employed to uncover gene functions (Gaj et al., 2013). Nutritional qualities have been improved related to bio-fortification and elimination of anti-nutritional components in crops by tuning gene expression patterns with this technology for better quality attributes (Saurabh et al., 2014). In crop improvement programs, RNAi has proven its importance by combating the biotic (viruses, fungi, bacteria, nematodes, etc.) and abiotic stresses (Draught, salinity, cold, etc.), and improving the shelf life of different vegetables and fruits (Younis et al., 2014). However, RNAi is incomplete for its practical use because it targets at the mRNA level so provides a temporary solution to gene repression.

11.4.2 GENE EDITING WITH NUCLEASES: ZFNS, TALENS, CRISPR

Gene editing is a multipurpose technique that has become a major part of functional genomics and crop improvement programs that transform genetic engineering into more advanced biotechnology (Hillary and Ceasar, 2019). The availability of sequencing data of some major crops along with advancement of editing technologies, it is now possible to raise nearly any important desirable agronomic trait in plants (Jaganathan et al., 2018). In this frequently changing climate, genome editing technologies have potential to produce climate resilient crops with improved yield and nutritional content (Steinwand and Ronald, 2020). This editing technology is programmable depending upon engineered endonucleases which specifically cleave and edit the genomic DNA at predetermined target sites. These endonucleases cause double-stranded breaks in genomic target DNA, which is subsequently repaired by the cell repair system utilizing error-prone non-homologous end joining (NHEJ), which causes frameshift mutations in a gene sequence, and homology directed Repair (HDR), by providing the insert with homology at flanking regions which results in gene insertion or could be used for insertional mutagenesis (Zhang et al., 2019). The frame of mRNA codons now shifted causing the incorrect amino acid incorporation during elongation of protein. A non-functional protein will be produced due to frameshift mutation which often results in gene knockouts, that create genetic knockouts. These mutant lines can be used for functional and reverse genetic research, as well as agricultural purposes (Malzahn et al., 2017; Zhang et al., 2017). Site directed nucleases are more advanced tools in modern circumstances preferred by researchers. These are categorized in three parts as SDN1, SDN2, and SDN3 based upon the availability of template DNA during repair of cleaved DNA by specific endonucleases.

ZFNs, TALENs, and CRISPR associated with Cas protein variants are the most recent and advanced programmable nucleases that can efficiently and precisely target the gene of interest. The first-generation editing tools are ZFNs and TALENs, which can easily be programmable and with sequence-specific DNA-binding domains fused to non-specific DNA-cleaving nuclease.

11.4.2.1 ZINC FINGER NUCLEASES

ZFNs are composed of a dimer in which each monomer contains a zinc finger DNA-binding domain and a DNA-cleaving endonuclease FokI restriction enzyme. (Cai et al., 2009). The zinc finger domain is composed of repeated groups of cysteine and histidine residues that recognizes 3 nts, three to four zinc fingers making one zinc finger monomer that is able to recognize around 12 nucleotides. Multiple zinc fingers could be put together to recognize a long nucleotide stretch (Davies et al., 2017; Gaj et al., 2013; Petolino, 2015). Herbicide-tolerant Zea mays have been developed by expressing ZFNs supplemented with a herbicide donar heterologous delivery system at anticipated position (Shukla et al., 2009). Imidazolinone herbicide-resistant bread wheat has been developed by precisely modifying a single amino acid sequence in a native gene acetohydroxyacid synthase (AHAS) using ZFNs when supplied with DNA repair template (Ran et al., 2018). An assay for determining cleavage and mutation specificity of ZFNs driven by an inducible heat shock promoter in Arabidopsis has been conducted. ZFNs expression has been induced by heat shock at the seedling stage resulting in the deletion and addition mutations at the ZFNs recognition site which are even transferred to subsequent generations (Lloyd et al., 2005; Richter et al., 2010). DNA has been integrated at specific transgenic and native loci using ZFNs. A GFP reporter gene was reconstructed and pat herbicide resistance gene was successfully integrated at ZFNs induced cleavage site using HDR in tobacco (Cai et al., 2009). Mutagenesis efficiency of ZFNs has been determined by targeting Dicer-like and RNA silencing genes cloned in vectors driven by the inducible estrogen promoter. Seven out of nine genes were found to be mutated when efficiency was determined by the hairy root transformation system. Plant transformation by ZFNs in soybean shows mutation in Dicer such as DCL4a and DCL4b genes that are subsequently inherited to the next generation (Curtin et al., 2011).

11.4.2.2 TALENS

Plant pathogenic bacteria of the genus Xanthomonas emit transcriptional activator-like effectors (TALEs), which impact the transcription machinery of plants and imitate transcription factors. They recognize a single base pair in the promoter region with the help of 33–35 amino acid residue DNA-binding tandem repeat domains. 12th and 13th amino acid residues are repeat variable diresidue (RVD), which are hypervariable and reprogrammable. These RVDs identify the different nucleotides—NI identifies A, NG identifies T, HD identifies C, and NN identifies G or A (Boch and Bonas, 2010). After deciphering DNA recognition code and DNA-binding mechanism of TALEs, they become part of genome editing technologies. FokI nuclease was fused with TALEs to create TALENs so developed new technology having a site-specific DNA-binding domain with a nuclease that nicks the DNA. Dimer form of TALENs is programmed so that two adjacent nicks create a DNA break (Cermak et al., 2011; Li and Yang, 2013). An easy protocol, reagents, software, and constructs with custom repeats are designed to carry out TALEN-related experiments. ADH1 gene in Arabidopsis thaliana protoplasts and the HPRT1 gene in human cells were successfully mutated using TALENs by the same group (Cermak et al., 2011). A new and more efficient Hax3-based hybrid TALE nuclease has been developed which can be programmed to any specific genomic target. A transient expression study in tobacco leaves shows Mutations arising by NHEJ repair after DSB created by hybrid TALENs (Mahfouz et al., 2011). Another variant of regular TALENs called designer TALEs (dTALEs) has been established which can regulate the expression of genes. TALENs have been used to target the acetolactate synthase gene in tobacco protoplasts and got TALENs induced base substitution and gene replacement in ALS gene in protoplast cells as well as protoplast-raised tobacco Calli (Yong Zhang et al., 2013). In general, they have great power to modify plant genomes at specific targets. Both ZFNs and TALENs depend upon DNA–protein interaction for target specificity but CRISPR-Cas systems interact with RNA–DNA interaction. DNA breaks induced by ZFNs, TALENs, and the CRISPR-Cas system activate the cell's endogenous repair process, which is repaired through error-prone NHEJ and homology guided repair mechanisms. NHEJ causes single nucleotide reshuffling, deletion, and addition of nucleotides that results in disruption of gene function by frameshift mutation (Bonawitz et al., 2019; Gaj et al., 2013). ZFNs and TALENs are very cumbersome, time consuming, and expensive processes along with many unpredictable off-targets due to their monomer repetition nature. Designing and programming of ZFNs and TALENs required difficult protocols involving protein engineering but the more advanced technology CRISPR is easier to use and cost effective. The construct designing and programming in CRISPR is not too tedious as compared to previous nucleases.

11.4.2.3 CRISPR

During bacteriophage challenges on Bacteria and archaea, the invading genetic material of bacteriophages is silenced by a special immune response that is CRISPR/CRISPR-associated Cas systems that confer the adaptive immunity to bacteria and archaea. At each new viral attack, new spacers produced from viral genomic sequences were incorporated into bacterial genomic regions. When these spacers are transcribed into mature CRISPR RNA (CrRNA) along with trans-activating crRNA (tracrRNA) make an RNAfold structure. This dual RNA provides direction to CRISPR associated protein Cas protein toward the CrRNA similarity sequence target in the viral genome as a result the active domains of Cas protein produce double-stranded breaks in invading genome (Barrangou et al., 2007; Bhaya et al., 2011; Sampson et al., 2013). When created as a single RNA hybrid, dual-tracrRNA:crRNA also instructs sequence-specific Cas9 dsDNA destruction (Jinek et al., 2012). The error-prone NHEJ and HDR DNA repair mechanisms in cells operate to repair the DSBs, resulting in mutations and sequence replacement (Zhang et al., 2017). The CRISPR-Cas systems are difficult to characterize because of differences in protein composition, genomic locus location and architecture, effector molecule structure, and techniques of pre-CRISPR (cr) RNA processing and silencing. Still, CRISPR/Cas bacterial defense systems are divided into two major classes, namely, classes 1 and 2 (Chaudhuri et al., 2022). Class 1 is composed of multi-subunit effectors but class 2 contains single unit effector molecules. Class 1 is further subdivided into three subtypes, namely, type I, type III, and type IV, even Class 2 is also subdivided into type II, type V, type VI. Type I has eight distinct Cas operons; type II has four Cas operons and trans-activating CRISPR RNA (tracRNA: crRNA); type III has eight Cas operons and Csm/Cmr complexes; and type IV has two Cas operons and four DinG/Csf proteins, Four Cas operons and four Cpf2 proteins are found in type V, while three Cas operons and three C2c2 proteins are found in type VI. Despite CRISPR/Cas systems having a large number of Cas operons the Cas9 protein, which is part of the type II CRISPR/Cas defense system, has been used extensively as a simple programmable genome editing tool. (Butiuc-Keul et al., 2021; Liu et al., 2020; Makarova et al., 2015). This bacterial adaptive immunity derived CRISPR-Cas9 technique is more Robust, precise in editing, cost effective, easy to program, less time consuming, and generates heritable changes.

11.5 CRISPR-CAS9

Natural mature crRNA and tracrRNA are generated and combined to form a single guide RNA (sgRNA) capable of guiding Cas9 nuclease to the target protospacer DNA and cutting around three base pairs upstream of a PAM protospacer adjacent motif (PAM), resulting in double-stranded breaks. Broken DNA ends are repaired by NHEJ, which results in minor indels, or HDR, which allows for exact gene or nucleotide substitution (Zhang et al., 2017). The engineered sgRNA has two components one is an 18–20 nucleotides long target sequence that is a substitute for crRNA and another is a GuideRNA scaffold that replaced tracrRNA, So, it is an RNA-guided DNA endonuclease (RGEN). The PAM sequence helps in the recognition of the target sequence and enhances the efficiency of Cas9 endonuclease. This PAM sequence is different for various Cas nucleases (Mojica et al., 2009). The catalytic activity of Cas9 depends upon its two active domains, one is Ruv-C which cleaves at GuideRNA similar target strand and another is the HNH nuclease domain which makes a cut in target sequence complementary to SgRNA (Jinek et al., 2012; Shan et al., 2013). CRISPR-Cas9 nuclease is a versatile tool of genome editing technologies. CRISPR/Cas systems employ different techniques such as single gene knockout, multiple gene knockouts, and large deletions, to create desired point mutations, correction of a point mutation in the gene (precise base editing), transcriptional regulation (overexpression and repression) (Zhang et al., 2017). This bacterial adaptive immunity-derived CRISPR-Cas9 technique is more Robust, precise in editing, cost effective, easy to program, less time consuming, and generates heritable changes. CRISPR-Cas9 has largely supplanted earlier genome editing methods like as ZFNs and TALENs since its discovery.

11.5.1 CRISPR-CAS9 AND ITS MODIFIED VERSIONS IN GENOME MODIFICATIONS

A large globular recognition (REC)-specific functional domain and a smaller nuclease (NUC)-specific functional domain make up the natural Cas9 endonuclease. There are two nuclease sites in the NUC domain, RuvC, and HNH, as well as a PAM-interacting domain (Doudna and Charpentier, 2014). The DSB created by nuclease domains is repaired by NHEJ or HDR which results in Indels causing gene knockout or gene replacement (Wyman and Kanaar, 2006). The naturally occurring spcas9 from *Streptococcus pyrogenes* is changed to nickase Cas9 SpCas9n by substituting the aspartate to alanine at the 10th position in the RuvC domain. CRISPR-mediated targeted genomic insertions are more successfully carried out with this SpCas9n. Off-targets are mitigated by paired Cas9n, which create nicks in two genomic targets (Cong, 2013).

When both of the cas9 nuclease domains are deactivated by creating mutants of native Cas9 they result in deactivated dCas9 having no nuclease activity. When dCas9 is combined with other effectors, such as gene repressors or activators, it may modify gene expression at the transcriptional level, resulting in novel Cas9 variations known as CRISPR interference (CRISPRi) and CRISPR activator (CRISPRa), which are extremely selective and efficient (Gilbert et al., 2013). dCas9 has been coupled with transcriptional activator VP64 or transcriptional repressor domain SRDX, resulting in gene expression regulation In arabidopsis and tobacco (Lowder et al., 2015). When dCas9 is combined with RNA-guided FokI nucleases (RFNs), they can detect and edit endogenous genes without off-targets by employing two guide RNAs at a time, with proper spacing and orientations. (Tsai et al., 2014). The fusion of CRISPR-Cas9 with the cytidine deaminase enzyme resulted in a novel technique for base editing, which changed C to T or G to A and could be used to fix point mutations (Komor et al., 2016). Mutant Cas9 mediates the conversion of A•T to G•C in genomic DNA when coupled with RNA adenosine deaminase (Gaudelli et al., 2017).

Likewise, Cas9 variants and different methods of guide expression systems have been developed which finally results in mature 17–20 nts sgRNAs. These methods include truncated guide RNA truRNA of 17 bp working with paired Cas9n to reduce off-targets (Fu et al., 2014), Ribozyme-gRNA-Ribozyme (RGR) is an artificial gene that can be expressed with any type of promoter and uses the self-catalyzing action of ribozymes to make desired guide RNA molecules (Gao and Zhao, 2014). A polycistronic tRNA-gRNA synthetic gene was employed to express a polycistronic mRNA with tRNA and gRNA alternately spaced, and numerous mature desired guide RNA were generated during tRNA processing, which can direct Cas9 to distinct targets (Xie et al., 2015). Different CRISPR systems, variants of Cas, cleavage site, the target of cleavage, their PAM sequence along with the mutational site have been enlisted in Table 11.1.

11.5.2 CAS9 VARIANTS IN GENETIC ENGINEERING TECHNOLOGY

CRISPR-cpf1 is a newly identified CRISPR-Cas system with RNA-guided DNA endonuclease activity that requires smaller guideRNA, requires T-rich PAM, and produces staggered ends more appropriate for knock-in DNA fragments (Kim et al., 2017). Two rice genes OsPDS and OsBEL are specifically targeted and mutated in rice using CRISPR-cpf1 system (Xu et al., 2017). Recombinant Cpf1 protein and in vitro transcribed CrRNA complexes are used for DNA-free genome editing in soybean and wild tobacco protoplast (Kim et al., 2017). CRISPR-Cas12b/C2c1 system recognizes T-rich PAM and creates staggered DSB. It has been applied in Rice for multiplex targeted mutagenesis and also developed targeted transcriptional activation and repression (Ming et al., 2020). Another CRISPR effector variant Cas13/ C2c2 has been applied in plants which targets single-stranded RNA in eukaryotes to help in combating plant viruses. Its HEPN (higher eukaryotes and prokaryotes nucleotide-binding) domain is responsible for its RNase activity. The fusion of Cas13 with other Cas nucleases provides new prospects for crop improvement by targeting both DNA and RNA in the same plant (Wolter and Puchta, 2018). Cas14a/ Cas14f, an RNA-guided DNA nuclease that only cleaves single-stranded DNA, is another CRISPR. This unique property makes it special and useful in combating plant-infecting viruses with single-stranded DNA (Harrington et al., 2018).

TABLE 11.1 CRISPR Systems, Cas Variant, Type of Cleavage, Target, PAM Sequence, and Mutation.

System name	Cas nuclease	Cleavage type	Target	PAM sequence	Mutation type	References
CRISPR-Cas9	Cas9	DSB	dsDNA	NGG	Knockout, gene replacement	Jinek et al. (2012)
CRISPR-Cas9n	Cas9nickase	Nick	dsDNA	NGG	Knockout, gene replacement	Cong (2013)
CRISPR-Cas9d	Deactivated Cas9	Non cleavage activity	dsDNA	NGG	Transcriptional regulation	Gilbert et al. (2013)
CRISPR interference CRISPRi	dCas9 fused with transcriptional repressor	Repression	dsDNA	NGG	Transcriptional repression of gene	Gilbert et al. (2013); Lowder et al. (2015)
CRISPR activator CRISPRa	dCas9 fused with transcriptional Activator	overexpression	dsDNA	NGG	Transcriptional expression of gene	Gilbert et al. (2013); Lowder et al. (2015)
Base editor Cytidine base editors	dCas9 fused to Cytidine Deaminase	C to T, G to A base editing	dsDNA	NGG	Correction and Creation point mutations	Komor et al. (2016)
Base editor Adenine base editors	dCas9 fused with adenosine deaminase	A to T, G to C Base editing	dsDNA	NGG	Correction and Creation point mutations	Gaudelli et al. (2017)
CRISPR/cpf1	Cpf1	Cohesive DSB	dsDNA	TTTN	Gene knock-in	Kim et al. (2017).
CRISPR-Cas12b/C2c1	Cas12b/C2c1	Cohesive DSB	dsDNA	TTN, ATTN,	Transcriptional repression and activation	Ming et al. (2020)
CRISPR-Cas13/ C2c2	Cas13/ C2c2	Cleave single-stranded RNA	ssRNA	PFS (Protospacer Flanking Sequence)	Combat against ssrna viruses	Wolter and Puchta (2018).
CRISPR/Dead Cas13/ C2c2	Dead Cas13/C2c2	No cleavage	ssRNA	PFS	Specifically edit ribonucleotide in ssrna	Wolter and Puchta (2018)
CRISPR- Cas14a/ Cas14f	Cas14a/ Cas14f	Cleave ssDNA	ssDNA	No PAM required	Cas14-detectr snp genotyping	Harrington et al. (2018)

11.6 APPLICATION OF CRISPR-CAS IN CEREALS

Several crop-enhancement programs have used CRISPR-Cas9 with tremendous results. Biotechnologists created the modified Cas9, which is codon optimized for each of the major crops. Currently, CRISPR-Cas systems have been used in crop enhancement projects considering yield, quality, biotic and abiotic stress tolerance, and herbicide resistance traits. Cereal crops like wheat, rice, and maize have already been developed with sufficient tolerance to Drought and Salinity using CRISPR-Cas9 technique to cope with a constantly changing climate (Hillary and Ceasar, 2019). Different outcomes of CRISPR-Cas genome editing on different cereals have been summarized in Table 11.2.

11.6.1 APPLICATION OF CRISPR-CAS9 IN RICE

Rice provides more than half of the world's food, and its small genome size, whole genome annotation, and availability of entire genomic sequences have made it a model plant for cereals, with the majority of research focused on it. Rice disease resistance has been improved by using the CRISPR-Cas9 system to target and create mutations in the OsMPK5 gene, which codes a stress-responsive rice mitogen-activated protein kinase (Xie and Yang, 2013). Rice that is resistant to salinity stress has been produced by targeting the OsRR22 gene with guideRNA and the Cas9 protein expressed by Cas9-OsRR22-gRNA expression vector. When compared to wild-type genotype, the regenerated altered plants demonstrate extraordinary resilience to salt stress (Zhang et al., 2019). Gn1a, DEP1, GS3, and IPA1 are four rice regulatory genes that are concurrently targeted and altered utilizing multiplex CRISPR-Cas9 editing, resulting in rice with increased grain size and number with dense erect panicles (Li et al., 2016). Knockout of an early developmental gene, epidermal patterning factor-like 9 (EPFL9), a positive regulator of stomatal development in Arabidopsis) orthologue in rice, CRISPR-Cas9 and CRISPR-Cpf1 technology is being utilized to generate drought-tolerant rice (Yin et al., 2017). CRISPR-cas9 modifies the SRL1 and SRL2 genes, resulting in mutant rice plants with rolled leaves that have improved agronomic traits and are drought tolerant (Liao et al., 2019). CRISPR-Cas9 is used to knock out the OsEBP89 gene in rice, resulting in enhanced drought tolerance (Zhang et al., 2020). CRISPR-cas9-generated dst gene mutant alleles increased drought and salt tolerance, as well as more grain production, in

TABLE 11.2 CRISPR-Cas Gene Editing in Major Cereals Against Abiotic Stresses.

SN	Crop	Gene	Function	Genome editing tool	Outcome	Reference
1	Rice	OsMPK5	Negative regulator of rice defense response	CRISPR-Cas9	Enhanced resistance to disease	Xie and Yang (2013)
2	Rice	OsEBP89	Involved in a novel signal transduction associated with the tolerance to drought and submergence stress	Knock out	Improved seed germination under submerged conditions and also enhanced the tolerance to drought stress	Zhang et al. (2020)
3	Rice	OsERF109	Negatively regulates drought tolerance by regulating ethylene biosynthesis in rice	RNAi	Improved drought stress tolerance	Yu et al. (2017)
4	Rice	OsMYB30	Cold Tolerance	CRISPR-Cas9	Increased cold tolerance	Lv et al. (2017)
5	Rice	TIFY1a/TIFY1b	Cold Tolerance	CRISPR-Cas9	Increased cold tolerance	Huang et al. (2017)
6	Rice	OsDST	Drought and salt-tolerant gene	CRISPR-Cas9	Moderate level tolerance to osmotic stress and high level of salt stress in seedlings	Santosh Kumar et al. (2020)
7	Rice	OsSAPK2	Key regulator of hyperosmotic stress signaling and abscisic acid (ABA)-dependent development in various plants	CRISPR-Cas9	Elucidated functional properties of SAPK2 using loss-of-function mutants produced with the CRISPR-Cas9 system	Lou et al. (2017)
8	Rice	OsRR22	Negative regulator of salinity tolerance	CRISPR-Cas9	Salinity tolerance	Zhang et al. (2019)
9	Rice	qSOR1	qSOR1 is negatively regulated by auxin, predominantly expressed in root columella cells, and involved in the gravitropic responses of roots	CRISPR-Cas9	Better root structure in saline soil	Kitomi et al. (2020)
10	Rice	OsNramp5	Higher Cd Accumulation	CRISPR-Cas9	Low Cd-accumulating indica rice without compromising yield	Tang et al. (2017)
11	Rice	OsLCT1	Low-affinity cation transporter OsLCT1 regulates cadmium transport into rice grains	CRISPR-Cas9	Less Cd accumulation in grains	Songmei et al. (2019)

TABLE 11.2 *(Continued)*

SN	Crop	Gene	Function	Genome editing tool	Outcome	Reference
12	Maize	ARGOS8	Causes sensitivity	CRISPR-Cas9	Improved grain yield under drought stress conditions	Shi et al. (2017)
13	Wheat	TaDREB2, TaERF3	Drought susceptible	CRISPR-Cas9	Improved survival under severe drought conditions	Kim et al. (2018)
14	Wheat	TaHAG1	Positive regulator to salt tolerance in wheat	CRISPR-Cas9	Increased salt tolerance	Zheng et al. (2021)
15	Wheat	TaARE1	Low NUE	CRISPR-Cas9	Enhanced NUE and improved grain yield	Guo et al. (2021)

indica rice varieties (Santosh Kumar et al., 2020). Rice with early blooming is created by employing CRISPR/Cas technology to target three heading date genes at once (Li et al., 2017). OsmiR535 is a stress-response gene-negative regulator. Rice plants with the OsmiR535 gene knocked out are more resistant to NaCl, ABA, dehydration, and PEG-induced stressors (Yue et al., 2020). Rice mutant knockout lines of the annexin gene OsAnn3 were produced using CRISPR-Cas9 genome editing and resulted in a cold tolerance phenotype, indicating that OsAnn3 is involved in rice cold tolerance (Shen et al., 2017). By using the CRISPR-Cas9 technology to concurrently alter three genes, OsPIN5b, GS3, and OsMYB30, novel rice mutants with high yield and great cold tolerance are created (Zeng et al., 2020). CRISPR-Cas system knockdown of the Cs+ -permeable K+ transporter OsHAK1 drastically decreased Cs+ absorption by rice plants (Nieves-Cordones et al., 2017). In rice, knockout mutants of the OsNramp5 metal transporter gene result in reduced cadmium buildup without impacting yield (Tang et al., 2022). Rice with improved abiotic stress tolerance like salinity tolerance, drought tolerance, cold resistance, heavy metals tolerance, and salinity tolerance has been generated utilizing the CRISPR/Cas method.

11.6.2 APPLICATION OF CRISPR-CAS9 IN WHEAT

Two stress-responsive transcription factor genes wheat dehydration-responsive element binding protein 2 (TaDREB2) and wheat ethylene-responsive factor 3 (TaERF3) are edited in wheat protoplast (Kim et al., 2018). Histone acetyltransferase TaHAG1 overexpression, silencing, and CRISPR-mediated knockout provide details on the role of TaHAG1 in wheat salinity tolerance (Zheng et al., 2021). By targeting four-grain quality-related genes, pinb, waxy, ppo, and psy, novel wheat germplasms with increased abiotic stress resistance and improved grain quality in hardness, starch content, and dough color were developed (Hu et al., 2015). Male sterile lines in wheat are obtained by targeting all three homeologs of male fertility factor Ms45. Ms45-based NGMS might be used to create a hybrid seed production system in wheat similar to Seed Production Technology (SPT) (Singh et al., 2018). In wheat, CRISPR-Cas9-mediated targeted alteration of TaARE1 homeologs produces a variety of transgene-free mutant lines that are more efficient in Nitrogen usage, have delayed senescence, and have higher grain production (Zhang et al., 2021). In wheat, CRISPR/Cas technology has been utilized to enhance salinity tolerance, abiotic stress resistance, efficient nitrogen usage, and delayed senescence.

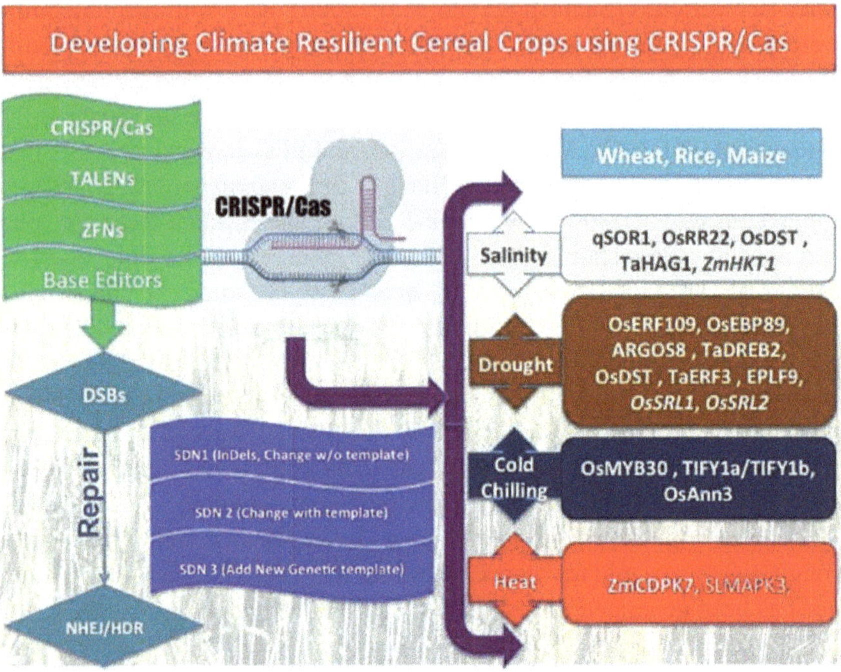

FIGURE 11.1 Use of genome editing techniques to develop climate-resilient crops along with their target genes. TALENs, Transcriptional activator-like effectors nucleases; ZFNs, Zinc finger nucleases; CRISPR/Cas, Clustered regularly interspaced short palindromic rerpeats and associated proteins; DSBs, Double-stranded Breaks; SDN, Site directed nucleases; qSOR1, *quantitative trait locus for SOIL SURFACE ROOTING 1*); OsRR22, Oryza sativa response regulators 22; OsDST, Oryza sativa drought and salt tolerance; TaHAG1, Histone acetyltransferase, a putative ortholog of AtHAG1/GCN5 in arabidopsis; ARGOS8, Auxin-regulated gene involved in organ size; ZmHKT1, Zea mays high-affinity potassium transporter; OsERF109, *Oryza sativa* ethylene-responsive factor 109; TaERF3, Triticum aestivum ethylene-responsive factor 3; OsEBP89, *Oryza sativa Ethylene-responsive element binding protein, clone 89*; TaDREB, Triticum aestivum dehydration responsive element binding protein; EPLF9, Oryza sativa epidermal patterning factor-like 9; OsSRL1, Oryza sativa semi-rolled leaves; OsMYB30, Oryza sativa; TIFY1a/b, Transcription factors in TIFY domain; OsAnn3, Oryza sativa annexin gene; ZmCDPK, Zea mays calcium-dependent protein kinase; SLMAPK3, Solanum lycopersicum mitogen-activated protein kinases

11.6.3 APPLICATION OF CRISPR-CAS9 IN MAIZE

The native promoter of the maize ARGOS8 locus is replaced with the maize GOS2 promoter using CRISPR-Cas, and the transgenic maize plants demonstrate enhanced grain production in water deficit conditions (Shi et al., 2017). CRISPR-Cas9 system with Maize codon optimized Cas9 nuclease is used to develop a male sterile line for hybrid seed production in maize by targeting the MS8 gene (Chen et al., 2018). The novel pZ1WS prime editor was used to create W542L and S621I double mutations in the two ALS genes, which provide resistance to a variety of herbicides that act as ALS inhibitors in Maize (Jiang et al., 2020). So CRISPR-Cas9 successfully has been used in maize to improve drought tolerance, herbicide resistance, and other agronomic traits.

11.7 CONCLUSIONS

CRISPR-Cas genome editing for targeted trait improvement is a robust technique for crop improvement. Although relatively few achievements have been made in CRA in comparison to other aspects of crop improvement, we can expect significant outputs in developing climate-smart cultivars. CRISPR-Cas9 system of genome editing has transformed plant genome editing as it is simple and versatile. Since this system can be rapidly improved in flexibility and accuracy, it guarantees broader and more effective future applications. Different traits like stress tolerance, nutritional improvement, yield, pest resistance, and so on are significantly improved and an ample number of traits will soon be adapted to changing climate by this technique. Also, CRISPR-Cas tool can be used for improving traits of wind species/orphan crops so they can be domesticated. This will also aid in global food security.

However, despite of versatility and flexibility of genome editing tools, all the need for genome manipulation is yet to be achieved. Since some traits are governed by QTLs, editing a single gene will not produce the desired phenotypic change. Thus, it would be better to develop CRISPR-Cas-mediated targeted insertion and chromosome rearrangement technology to combine mutated alleles. In addition to applications of CRISPR-Cas, new avenues can be unlocked by editing mitochondrial and chloroplast genomes, studying patterns of development via cell lineages, signal integration and transduction by genetic circuits, plant biosensor development to detect internal and external signals, and other applications in plant biotechnology (Zhu et al., 2020).

KEYWORDS

- **abiotic stress**
- **CRISPR**
- **cereals**
- **genome editing**

REFERENCES

Acquaah, G. Conventional Plant Breeding Principles and Techniques. In *Advances in Plant Breeding Strategies: Breeding, Biotechnology and Molecular Tools*, Vol. 1; 2016; pp 115–158. https://doi.org/10.1007/978-3-319-22521-0_5

Ahmar, S.; Gill, R. A.; Jung, K. H.; Faheem, A.; Qasim, M. U.; Mubeen, M.; Zhou, W. Conventional and Molecular Techniques from Simple Breeding to Speed Breeding in Crop Plants: Recent Advances and Future Outlook. *Int. J. Mol. Sci.* **2020,** *21* (7). https://doi.org/10.3390/ijms21072590

Altieri, M. A.; Koohafkan, P. *Enduring Farms: Climate Change, Smallholders and Traditional Farming Communities*, Vol. 6; Third World Network (TWN) Penang, 2008.

Asseng, S.; Ewert, F.; Martre, P.; Rötter, R. P.; Lobell, D. B.; Cammarano, D.; Kimball, B. A.; Ottman, M. J.; Wall, G. W.; White, J. W. Rising Temperatures Reduce Global Wheat Production. *Nat. Clim. Change* **2015,** *5* (2), 143–147.

Barrangou, R.; Fremaux, C.; Deveau, H.; Richards, M.; Boyaval, P.; Moineau, S.; Romero, D. A.; Horvath, P. CRISPR Provides Acquired Resistance Against Viruses in Prokaryotes. *Science* **2007,** *315* (5819), 1709–1712. https://doi.org/10.1126/science.1138140

Bhaya, D.; Davison, M.; Barrangou, R. CRISPR-cas Systems in Bacteria and Archaea: Versatile Small RNAs for Adaptive Defense and Regulation. *Annu. Rev. Genet.* **2011,** *45*, 273–297. https://doi.org/10.1146/annurev-genet-110410-132430

Boch, J.; Bonas, U. Xanthomonas AvrBs3 family-type III effectors: Discovery and function. *Annu. Rev. Phytopathol.* **2010,** *48*, 419–436. https://doi.org/10.1146/annurev-phyto-080508-081936

Bonawitz, N. D.; Ainley, W. M.; Itaya, A.; Chennareddy, S. R.; Cicak, T.; Effinger, K.; Jiang, K.; Mall, T. K.; Marri, P. R.; Samuel, J. P.; Sardesai, N.; Simpson, M.; Folkerts, O.; Sarria, R.; Webb, S. R.; Gonzalez, D. O.; Simmonds, D. H.; Pareddy, D. R. Zinc Finger Nuclease-Mediated Targeting of Multiple Transgenes to an Endogenous Soybean Genomic Locus Via Non-Homologous End Joining. *Plant Biotechnol. J.* **2019,** *17* (4), 750–761. https://doi.org/10.1111/pbi.13012

Boyer, J. S. Plant Productivity and Environment. *Science* **1982,** *218* (4571), 443–448.

Butiuc-Keul, A.; Farkas, A.; Carpa, R.; Iordache, D. CRISPR-Cas System: The Powerful Modulator of Accessory Genomes in Prokaryotes. *Microb. Physiol.* **2021**. https://doi.org/10.1159/000516643

Cai, C. Q.; Doyon, Y.; Ainley, W. M.; Miller, J. C.; DeKelver, R. C.; Moehle, E. A.; Rock, J. M.; Lee, Y. L.; Garrison, R.; Schulenberg, L.; Blue, R.; Worden, A.; Baker, L.; Faraji, F.; Zhang, L.; Holmes, M. C.; Rebar, E. J.; Collingwood, T. N.; Rubin-Wilson, B. et al. Targeted Transgene Integration in Plant Cells Using Designed Zinc Finger Nucleases. *Plant Mol. Biol.* **2009,** *69* (6), 699–709. https://doi.org/10.1007/s11103-008-9449-7

Cermak, T.; Doyle, E. L.; Christian, M.; Wang, L.; Zhang, Y.; Schmidt, C.; Baller, J. A.; Somia, N. V.; Bogdanove, A. J.; Voytas, D. F. Efficient Design and Assembly of Custom TALEN and Other TAL Effector-Based Constructs for DNA targeting. *Nucl. Acids Res.* **2011,** *39* (12). https://doi.org/10.1093/nar/gkr218

Ch, S.; Prasad, R. S.; Mohapatra, T. Climate change and Indian Agriculture: Impacts, Coping Strategies, Programmes and Policy. *Technical Bulletin/Policy Document*, 2019.

Chaudhuri, A.; Halder, K.; Datta, A. Classification of CRISPR/Cas System and Its Application in Tomato Breeding. *Theor. Appl. Genet.* **2022,** *135* (2), 367–387. https://doi.org/10.1007/s00122-021-03984-y

Chen, R.; Xu, Q.; Liu, Y.; Zhang, J.; Ren, D.; Wang, G.; Liu, Y. Generation of Transgene-Free Maize Male Sterile Lines Using the CRISPR/Cas9 System. *Front. Plant Sci.* **2018,** *9*. https://doi.org/10.3389/fpls.2018.01180

Chivenge, P.; Mabhaudhi, T.; Modi, A. T.; Mafongoya, P. The Potential Role of Neglected and Underutilised Crop Species As Future Crops Under Water Scarce Conditions in Sub-Saharan Africa. *Int. J. Environ. Res. Public Health* **2015,** *12* (6), 5685–5711.

Cong, L. Multiplex Genome Engineering Using. *Science* **2013,** *25* (7), 778–785. http://www.nature.com/doifinder/10.1038/nbt1319

Curtin, S. J.; Zhang, F.; Sander, J. D.; Haun, W. J.; Starker, C.; Baltes, N. J.; Reyon, D.; Dahlborg, E. J.; Goodwin, M. J.; Coffman, A. P.; Dobbs, D.; Joung, J. K.; Voytas, D. F.; Stupar, R. M. Targeted Mutagenesis of Duplicated Genes in Soybean with Zinc-Finger Nucleases. *Plant Physiol.* **2011,** *156* (2), 466–473. https://doi.org/10.1104/pp.111.172981

Davies, J. P.; Kumar, S.; Sastry-Dent, L. Use of Zinc-Finger Nucleases for Crop Improvement. *Progress Mol. Biol. Transl. Sci.* **2017,** *149*, 47–63. https://doi.org/10.1016/bs.pmbts.2017.03.006

de Oliveira, E. D.; Bramley, H.; Siddique, K. H. M.; Henty, S.; Berger, J.; Palta, J. A. Can Elevated CO2 Combined with High Temperature Ameliorate the Effect of Terminal Drought in Wheat? *Funct. Plant Biology* **2012,** *40* 2, 160–171.

Dong, O. X.; Ronald, P. C. Genetic Engineering for Disease Resistance in Plants: Recent Progress and Future Perspectives. *Plant Physiol.* **2019,** *180* (1), 26–38. https://doi.org/10.1104/pp.18.01224

Doudna, J. A.; Charpentier, E. The New Frontier of Genome Engineering with CRISPR-Cas9. *Science* **2014,** *346* 6213, 1258096.

Fu, Y.; Sander, J. D.; Reyon, D.; Cascio, V. M.; Joung, J. K. Improving CRISPR-Cas Nuclease Specificity Using Truncated Guide RNAs. *Nat. Biotechnol.* **2014,** *32* (3), 279–284. https://doi.org/10.1038/nbt.2808

Gaj, T.; Gersbach, C. A.; Barbas, C. F. ZFN, TALEN, and CRISPR/Cas-Based Methods for Genome Engineering. *Trends Biotechnol.* **2013,** *31* (7), 397–405. https://doi.org/10.1016/j.tibtech.2013.04.004

Gao, Y.; Zhao, Y. Self-Processing of Ribozyme-Flanked RNAs into Guide RNAs In Vitro and In Vivo for CRISPR-Mediated Genome Editing. *J. Integr. Plant Biol.* **2014,** *56* (4), 343–349. https://doi.org/10.1111/jipb.12152

Gaudelli, N. M.; Komor, A. C.; Rees, H. A.; Packer, M. S.; Badran, A. H.; Bryson, D. I.; Liu, D. R. Programmable Base Editing of A• T to G• C in Genomic DNA Without DNA Cleavage. *Nature* **2017,** *551* 7681, 464–471.

Gilbert, L. A.; Larson, M. H.; Morsut, L.; Liu, Z.; Brar, G. A.; Torres, S. E.; Stern-Ginossar, N.; Brandman, O.; Whitehead, E. H.; Doudna, J. A. CRISPR-Mediated Modular RNA-Guided Regulation of Transcription in Eukaryotes. *Cell* **2013,** *154* 2, 442–451.

Godfray, H. C. J.; Garnett, T. Food Security and Sustainable Intensification. *Phil. Trans. R. Soc. B: Biol. Sci.* **2014,** *369* 1639, 20120273.

Griffin, J. J.; Ranney, T. G.; Pharr, D. M. Heat and Drought Influence Photosynthesis, Water Relations, and Soluble Carbohydrates of Two Ecotypes of Redbud *Cercis canadensis*. *J. Am. Soc. Horticult. Sci.* **2004,** *129* 4, 497–502.

Guo, M.; Wang, Q.; Zong, Y.; Nian, J.; Li, H.; Li, J.; Wang, T.; Gao, C.; Zuo, J. Genetic Manipulations of TaARE1 Boost Nitrogen Utilization and Grain Yield in Wheat. *J. Genet. Genom.= Yi Chuan Xue Bao* **2021,** *48* 10, 950–953.

Harrington, L. B.; Burstein, D.; Chen, J. S.; Paez-Espino, D.; Ma, E.; Witte, I. P.; Cofsky, J. C.; Kyrpides, N. C.; Banfield, J. F.; Doudna, J. A. Programmed DNA Destruction by Miniature CRISPR-Cas14 Enzymes. *Science* **2018,** *362,* (6416), 839–842. https://doi.org/10.1126/science.aav4294

Hillary, V. E.; Ceasar, S. A. Application of CRISPR/Cas9 Genome Editing System in Cereal Crops. *Open Biotechnol. J.* **2019,** *13* (1), 173–179. https://doi.org/10.2174/1874070701913010173

Hu, B.; Wang, W.; Ou, S.; Tang, J.; Li, H.; Che, R.; Zhang, Z.; Chai, X.; Wang, H.; Wang, Y. Variation in NRT1: 1B Contributes to Nitrate-Use Divergence Between Rice Subspecies. *Nat. Genet.* **2015,** *47* 7, 834.

Huang, X.; Zeng, X.; Li, J.; Zhao, D. Construction and Analysis of Tify1a and Tify1b Mutants in Rice *Oryza sativa* Based on CRISPR/Cas9 Technology. *J. Agric. Biotechnol.* **2017,** *25* (6), 1003–1012.

IPCC. *Global Warming of 1.5° C: an IPCC Special Report on the Impacts of Global Warming of 1.5°C Above Pre-Industrial Levels and Related Global Greenhouse Gas Emission Pathways, in the Context of Strengthening the Global Response to the Threat of Climate Change*; Intergovernmental Panel on Climate Change, 2018.

Jaganathan, D.; Ramasamy, K.; Sellamuthu, G.; Jayabalan, S.; Venkataraman, G. CRISPR for Crop Improvement: An Update Review. *Front. Plant Sci.* **2018**. https://doi.org/10.3389/fpls.2018.00985

Jiang, Y. Y.; Chai, Y. P.; Lu, M. H.; Han, X. L.; Lin, Q.; Zhang, Y.; Zhang, Q.; Zhou, Y.; Wang, X. C.; Gao, C.; Chen, Q. J. Prime Editing Efficiently Generates W542L and S621I Double Mutations in Two ALS Genes in Maize. *Genome Biol.* **2020,** *21* (1). https://doi.org/10.1186/s13059-020-02170-5

Jinek, M.; Chylinski, K.; Fonfara, I.; Hauer, M.; Doudna, J. A.; Charpentier, E. A Programmable Dual-RNA-Guided DNA Endonuclease in Adaptive Bacterial Immunity. *Science* **2012**. https://doi.org/10.1126/science.1225829

Kamthan, A.; Chaudhuri, A.; Kamthan, M.; Datta, A. Small RNAs in plants: Recent development and application for crop improvement. *Front. Plant Sci.* **2015,** *6* (APR). https://doi.org/10.3389/fpls.2015.00208

Khusg, G. Strategies for Increasing the Yield Potential of Cereals: Case of Rice as an Example. *Plant Breed.* **2013,** *132*, 433–436.

Kim, D.; Alptekin, B.; Budak, H. CRISPR/Cas9 Genome Editing in Wheat. *Funct. Integr. Genom.* **2018,** *18* 1, 31–41.

Kim, H.; Kim, S.-T.; Ryu, J.; Kang, B.-C.; Kim, J.-S.; Kim, S.-G. CRISPR/Cpf1-Mediated DNA-Free Plant Genome Editing. *Nat. Commun.* **2017,** *8* 1, 1–7.

Kitomi, Y.; Hanzawa, E.; Kuya, N.; Inoue, H.; Hara, N.; Kawai, S.; Kanno, N.; Endo, M.; Sugimoto, K.; Yamazaki, T. Root Angle Modifications by the DRO1 Homolog Improve Rice Yields in Saline Paddy Fields. *Proc. Natl. Acad. Sci.* **2020,** *117* 35, 21242–21250.

Komor, A. C.; Kim, Y. B.; Packer, M. S.; Zuris, J. A.; Liu, D. R. Programmable Editing of a Target Base in Genomic DNA Without Double-Stranded DNA Cleavage. *Nature* **2016,** *533*, 420–424. https://doi.org/10.1038/nature17946

Kumar, K.; Gambhir, G.; Dass, A.; Tripathi, A. K.; Singh, A.; Jha, A. K.; Yadava, P.; Choudhary, M.; Rakshit, S. Genetically Modified Crops: Current Status and Future Prospects. *Planta* **2020,** *251* (4). https://doi.org/10.1007/s00425-020-03372-8

Li, M.; Li, X.; Zhou, Z.; Wu, P.; Fang, M.; Pan, X.; Lin, Q.; Luo, W.; Wu, G.; Li, H. Reassessment of the Four Yield-Related Genes Gn1a, DEP1, GS3, and IPA1 in Rice Using a CRISPR/Cas9 System. *Front. Plant Sci.* **2016,** *7* (MAR2016). https://doi.org/10.3389/fpls.2016.00377

Li, T.; Yang, B. TAL Effector Nuclease TALEN Engineering. *Methods Mol. Biol.* **2013,** *978*, 63–72. https://doi.org/10.1007/978-1-62703-293-3_5

Li, X.; Zhou, W.; Ren, Y.; Tian, X.; Lv, T.; Wang, Z.; Fang, J.; Chu, C.; Yang, J.; Bu, Q. High-Efficiency Breeding of Early-Maturing Rice Cultivars via CRISPR/Cas9-Mediated Genome Editing. *J. Genet. Genom.* = *Yi Chuan Xue Bao* **2017,** *44* 3, 175–178.

Liao, S.; Qin, X.; Luo, L.; Han, Y.; Wang, X.; Usman, B.; Nawaz, G.; Zhao, N.; Liu, Y.; Li, R. CRISPR/Cas9-Induced Mutagenesis of Semi-Rolled Leaf1, 2 Confers Curled Leaf Phenotype and Drought Tolerance by Influencing Protein Expression Patterns and ROS Scavenging in Rice *Oryza sativa* L.. *Agronomy* **2019,** *9* 11, 728.

Liu, Z.; Dong, H.; Cui, Y.; Cong, L.; Zhang, D. Application of Different Types of CRISPR/Cas-Based Systems in Bacteria. *Microb. Cell Factories* **2020,** *19* (1). https://doi.org/10.1186/s12934-020-01431-z

Lloyd, A.; Plaisier, C. L.; Carroll, D.; Drews, G. N. Targeted Mutagenesis Using Zinc-Finger Nucleases in Arabidopsis. *Proc. Natl. Acad. Sci.* **2005,** *102* 6, 2232–2237.

Lobell, D. B.; Field, C. B. Global Scale Climate–Crop Yield Relationships and the Impacts of Recent Warming. *Environ. Res. Lett.* **2007,** *2* 1, 14002.

Lou, D.; Wang, H.; Liang, G.; Yu, D. OsSAPK2 Confers Abscisic Acid Sensitivity and Tolerance to Drought Stress in Rice. *Front. Plant Sci.* **2017,** *8*, 993.

Lowder, L. G.; Zhang, D.; Baltes, N. J.; Paul, J. W.; Tang, X.; Zheng, X.; Voytas, D. F.; Hsieh, T. F.; Zhang, Y.; Qi, Y. A CRISPR/Cas9 Toolbox for Multiplexed Plant Genome Editing and Transcriptional Regulation. *Plant Physiol.* **2015,** *169* (2), 971–985. https://doi.org/10.1104/pp.15.00636

Lv, Y.; Yang, M.; Hu, D.; Yang, Z.; Ma, S.; Li, X.; Xiong, L. The OsMYB30 Transcription Factor Suppresses Cold Tolerance by Interacting with a JAZ Protein and Suppressing β-Amylase Expression. *Plant Physiol.* **2017,** *173* 2, 1475–1491.

Mahfouz, M. M.; Li, L.; Shamimuzzaman, M.; Wibowo, A.; Fang, X.; Zhu, J.-K. De novo-Engineered Transcription Activator-Like Effector TALE Hybrid Nuclease with Novel DNA Binding Specificity Creates Double-Strand Breaks. *Proc. Natl. Acad. Sci.* **2011,** *108* 6, 2623–2628.

Makarova, K. S.; Wolf, Y. I.; Alkhnbashi, O. S.; Costa, F.; Shah, S. A.; Saunders, S. J.; Barrangou, R.; Brouns, S. J. J.; Charpentier, E.; Haft, D. H.; Horvath, P.; Moineau, S.; Mojica, F. J. M.; Terns, R. M.; Terns, M. P.; White, M. F.; Yakunin, A. F.; Garrett, R. A.; Van Der Oost, J. et al. An Updated Evolutionary Classification of CRISPR-Cas Systems. *Nat. Rev. Microbiol.* **2015,** *13* (11), 722–736. https://doi.org/10.1038/nrmicro3569

Malzahn, A.; Lowder, L.; Qi, Y. Plant Genome Editing with TALEN and CRISPR. *Cell Biosci.* **2017,** *7* (1). https://doi.org/10.1186/s13578-017-0148-4

Ming, M.; Ren, Q.; Pan, C.; He, Y.; Zhang, Y.; Liu, S.; Zhong, Z.; Wang, J.; Malzahn, A. A.; Wu, J. CRISPR–Cas12b Enables Efficient Plant Genome Engineering. *Nat. Plants* **2020,** *6* 3, 202–208.

Mojica, F. J. M.; Díez-Villaseñor, C.; García-Martínez, J.; Almendros, C. Short Motif Sequences Determine the Targets of the Prokaryotic CRISPR Defence System. *Microbiology* **2009,** *155* (3), 733–740. https://doi.org/10.1099/mic.0.023960-0

Murray, V.; Ebi, K. L. IPCC Special Report on Managing the Risks of Extreme Events and Disasters to Advance Climate Change Adaptation SREX. *J. Epidemiol. Community Health* **2012,** *66* (9), 759–760. BMJ Publishing Group Ltd.

Nieves-Cordones, M.; Mohamed, S.; Tanoi, K.; Kobayashi, N. I.; Takagi, K.; Vernet, A.; Guiderdoni, E.; Périn, C.; Sentenac, H.; Véry, A. Production of Low-Cs+ Rice Plants by Inactivation of the K+ Transporter Os HAK 1 with the CRISPR-Cas System. *Plant J.* **2017,** *92* 1, 43–56.

Oladosu, Y.; Rafii, M. Y.; Samuel, C.; Fatai, A.; Magaji, U.; Kareem, I.; Kamarudin, Z. S.; Muhammad, I.; Kolapo, K. Drought Resistance in Rice from Conventional to Molecular Breeding: A Review. *Int. J. Mol. Sci.* **2019,** *20* (14). https://doi.org/10.3390/ijms20143519

Parry, M. L.; Canziani, O.; Palutikof, J.; Van der Linden, P.; Hanson, C. *Climate Change 2007-Impacts, Adaptation and Vulnerability: Working Group II Contribution to the Fourth Assessment Report of the IPCC*, Vol. 4; Cambridge University Press, 2007.

Petolino, J. F. Genome Editing in Plants via Designed Zinc Finger Nucleases. *In Vitro Cell. Dev. Biol.—Plant* **2015,** *51* (1). https://doi.org/10.1007/s11627-015-9663-3

Porter, J. R.; Xie, L.; Challinor, A. J.; Cochrane, K.; Howden, S. M.; Iqbal, M. M.; Lobell, D. B.; Travasso, M. I. *Food Security and Food Production Systems*, 2014.

Ran, Y.; Patron, N.; Kay, P.; Wong, D.; Buchanan, M.; Cao, Y.; Sawbridge, T.; Davies, J. P.; Mason, J.; Webb, S. R. Zinc Finger Nuclease-Mediated Precision Genome Editing of an Endogenous Gene in Hexaploid Bread Wheat *Triticum aestivum* Using a DNA Repair Template. *Plant Biotechnol. J.* **2018,** *16* 12, 2088–2101.

Raza, A.; Razzaq, A.; Mehmood, S. S.; Zou, X.; Zhang, X.; Lv, Y.; Xu, J. Impact of Climate Change on Crops Adaptation and Strategies to Tackle Its Outcome: A Review. *Plants* **2019,** *8* 2, 34.

Richter, R.; Behringer, C.; Müller, I. K.; Schwechheimer, C. The GATA-Type Transcription Factors GNC and GNL/CGA1 Repress Gibberellin Signaling Downstream from DELLA Proteins and Phytochrome-Interacting Factors. *Genes Dev.* **2010,** *24* 18, 2093–2104.

Sampson, T. R.; Saroj, S. D.; Llewellyn, A. C.; Tzeng, Y. L.; Weiss, D. S. A CRISPR/Cas System Mediates Bacterial Innate Immune Evasion and Virulence. *Nature* **2013,** *497* (7448), 254–257. https://doi.org/10.1038/nature12048

Santosh Kumar, V. V.; Verma, R. K.; Yadav, S. K.; Yadav, P.; Watts, A.; Rao, M. V.; Chinnusamy, V. CRISPR-Cas9 Mediated Genome Editing of Drought and Salt Tolerance OsDST Gene in Indica Mega Rice Cultivar MTU1010. *Physiol. Mol. Biol. Plants* **2020,** *26* (6), 1099–1110. https://doi.org/10.1007/s12298-020-00819-w

Sapkota, T. B.; Jat, R. K.; Singh, R. G.; Jat, M. L.; Stirling, C. M.; Jat, M. K.; Bijarniya, D.; Kumar, M.; Saharawat, Y. S.; Gupta, R. K. Soil Organic Carbon Changes After Seven Years of Conservation Agriculture in a Rice–Wheat System of the Eastern Indo-Gangetic Plains. *Soil Use Manage.* **2017,** *33* 1, 81–89.

Saurabh, S.; Vidyarthi, A. S.; Prasad, D. RNA Interference: Concept to Reality in Crop Improvement. *Planta* **2014,** *239* (3), 543–564. https://doi.org/10.1007/s00425-013-2019-5

Schlenker, W.; Roberts, M. J. Nonlinear Temperature Effects Indicate Severe Damages to US Crop Yields Under Climate Change. *Proc. Natl. Acad. Sci.* **2009,** *106* 37, 15594–15598.

Shan, Q.; Wang, Y.; Li, J.; Zhang, Y.; Chen, K.; Liang, Z.; Zhang, K.; Liu, J.; Xi, J. J.; Qiu, J.-L. Targeted Genome Modification of Crop Plants Using a CRISPR-Cas System. *Nat. Biotechnol.* **2013,** *31* 8, 686–688.

Shen, C.; Que, Z.; Xia, Y.; Tang, N.; Li, D.; He, R.; Cao, M. Knock Out of the Annexin Gene OsAnn3 via CRISPR/Cas9-Mediated Genome Editing Decreased Cold Tolerance in Rice. *J. Plant Biol.* **2017,** *60* (6), 539–547. https://doi.org/10.1007/s12374-016-0400-1

Shi, J.; Gao, H.; Wang, H.; Lafitte, H. R.; Archibald, R. L.; Yang, M.; Hakimi, S. M.; Mo, H.; Habben, J. E. ARGOS8 Variants Generated by CRISPR-Cas9 Improve Maize Grain Yield

Under Field Drought Stress Conditions. *Plant Biotechnol. J.* **2017,** *15* (2), 207–216. https://doi.org/10.1111/pbi.12603

Shukla, P. R.; Skeg, J.; Buendia, E. C.; Masson-Delmotte, V.; Pörtner, H.-O.; Roberts, D. C.; Zhai, P.; Slade, R.; Connors, S.; van Diemen, S. *Climate Change and Land: an IPCC Special Report on Climate Change, Desertification, Land Degradation, Sustainable Land Management, Food Security, and Greenhouse Gas Fluxes in Terrestrial Ecosystems*, 2019.

Shukla, V. K.; Doyon, Y.; Miller, J. C.; Dekelver, R. C.; Moehle, E. A.; Worden, S. E.; Mitchell, J. C.; Arnold, N. L.; Gopalan, S.; Meng, X.; Choi, V. M.; Rock, J. M.; Wu, Y. Y.; Katibah, G. E.; Zhifang, G.; McCaskill, D.; Simpson, M. A.; Blakeslee, B.; Greenwalt, S. A. et al. Precise Genome Modification in the Crop Species Zea Mays Using Zinc-Finger Nucleases. *Nature* **2009,** *459* 7245, 437–441. https://doi.org/10.1038/nature07992

Singh, J.; Kaur, S.; Majithia, H. Emerging Genetic Technologies for Improving the Security of Food Crops. In *Emerging Technologies for Promoting Food Security: Overcoming the World Food Crisis*, 2016; pp 23–41. https://doi.org/10.1016/B978-1-78242-335-5.00002-0

Singh, M.; Kumar, M.; Albertsen, M. C.; Young, J. K.; Cigan, A. M. Concurrent Modifications in the Three Homeologs of Ms45 Gene with CRISPR-Cas9 Lead to Rapid Generation of Male Sterile Bread Wheat *Triticum aestivum* L.. *Plant Mol. Biol.* **2018,** *97* (4–5), 371–383. https://doi.org/10.1007/s11103-018-0749-2

Songmei, L.; Jie, J.; Yang, L.; Jun, M.; Shouling, X.; Yuanyuan, T.; Youfa, L.; Qingyao, S.; Jianzhong, H. Characterization and Evaluation of OsLCT1 and OsNramp5 Mutants Generated Through CRISPR/Cas9-Mediated Mutagenesis for Breeding Low Cd Rice. *Rice Sci.* **2019,** *26* 2, 88–97.

Stamm, P.; Ramamoorthy, R.; Kumar, P. P. Feeding the Extra Billions: Strategies to Improve Crops and Enhance Future Food Security. *Plant Biotechnol. Rep.* **2011,** *5* 2, 107–120.

Steinwand, M. A.; Ronald, P. C. Crop Biotechnology and the Future of Food. *Nat. Food* **2020,** *1* (5), 273–283. https://doi.org/10.1038/s43016-020-0072-3

Tang, L.; Dong, J.; Qu, M.; Lv, Q.; Zhang, L.; Peng, C.; Hu, Y.; Li, Y.; Ji, Z.; Mao, B. Knockout of OsNRAMP5 Enhances Rice Tolerance to Cadmium Toxicity in Response to Varying External Cadmium Concentrations via Distinct Mechanisms. *Sci. Total Environ.* **2022,** 155006.

Tang, L.; Mao, B.; Li, Y.; Lv, Q.; Zhang, L.; Chen, C.; He, H.; Wang, W.; Zeng, X.; Shao, Y.; Pan, Y.; Hu, Y.; Peng, Y.; Fu, X.; Li, H.; Xia, S.; Zhao, B. Knockout of OsNramp5 Using the CRISPR/Cas9 System Produces Low Cd-Accumulating Indica Rice Without Compromising Yield. *Sci. Rep.* **2017,** *7* (1). https://doi.org/10.1038/s41598-017-14832-9

Tsai, S. Q.; Wyvekens, N.; Khayter, C.; Foden, J. A.; Thapar, V.; Reyon, D.; Goodwin, M. J.; Aryee, M. J.; Joung, J. K. Dimeric CRISPR RNA-Guided FokI Nucleases for Highly Specific Genome Editing. *Nat. Biotechnol.* **2014,** *32* 6, 569–576.

UN. World Population Prospects: the 2017 Revision, Key Findings and Advance Tables; Department of Economics and Social Affairs PD, Ed.; United Nations: New York, **2017**; p 46.

Vaughan, M. M.; Block, A.; Christensen, S. A.; Allen, L. H.; Schmelz, E. A. The Effects of Climate Change Associated Abiotic Stresses on Maize Phytochemical Defenses. *Phytochem. Rev.* **2018,** *17* 1, 37–49.

Watts, N.; Scales, I. R. Seeds, Agricultural Systems and Socio-Natures: Towards an Actor–Network Theory Informed Political Ecology of Agriculture. *Geog. Compass* **2015,** *9* 5, 225–236.

Wolter, F.; Puchta, H. The CRISPR/Cas Revolution Reaches the RNA World: Cas13, a New Swiss Army Knife for Plant Biologists. *Plant J.* **2018,** *94* (5), 767–775. https://doi.org/10.1111/tpj.13899

Wyman, C.; Kanaar, R. DNA Double-Strand Break Repair: All's Well That Ends Well. *Annu. Rev. Genet.* **2006,** *40*, 363–383.

Xie, K.; Minkenberg, B.; Yang, Y. Boosting CRISPR/Cas9 Multiplex Editing Capability with the Endogenous tRNA-Processing System. *Proc. Natl. Acad. Sci. USA* **2015,** *112* (11), 3570–3575. https://doi.org/10.1073/pnas.1420294112

Xie, K.; Yang, Y. RNA-Guided Genome Editing in Plants Using a CRISPR–Cas System. *Mol. Plant* **2013,** *6* 6, 1975–1983.

Xu, R.; Qin, R.; Li, H.; Li, D.; Li, L.; Wei, P.; Yang, J. Generation of Targeted Mutant Rice Using a CRISPR-Cpf1 System. *Plant Biotechnol. J.* **2017,** *15* (6), 713–717. https://doi.org/10.1111/pbi.12669

Xu, Z. Z.; Zhou, G. S. Combined Effects of Water Stress and High Temperature on Photosynthesis, Nitrogen Metabolism and Lipid Peroxidation of a Perennial Grass *Leymus chinensis*. Planta **2006,** *224* 5, 1080–1090.

Yin, X.; Biswal, A. K.; Dionora, J.; Perdigon, K. M.; Balahadia, C. P.; Mazumdar, S.; Chater, C.; Lin, H. C.; Coe, R. A.; Kretzschmar, T.; Gray, J. E.; Quick, P. W.; Bandyopadhyay, A. CRISPR-Cas9 and CRISPR-Cpf1 Mediated Targeting of a Stomatal Developmental Gene EPFL9 in Rice. *Plant Cell Rep.* **2017,** *36* (5), 745–757. https://doi.org/10.1007/s00299-017-2118-z

Younis, A.; Siddique, M. I.; Kim, C. K.; Lim, K. B. RNA Interference RNAi Induced Gene Silencing: A Promising Approach of Hi-Tech Plant Breeding. *Int. J. Biol. Sci.* **2014,** *10* (10), pp. 1150–1158. https://doi.org/10.7150/ijbs.10452

Yu, Y.; Yang, D.; Zhou, S.; Gu, J.; Wang, F.; Dong, J.; Huang, R. The Ethylene Response Factor OsERF109 Negatively Affects Ethylene Biosynthesis and Drought Tolerance in Rice. *Protoplasma* **2017,** *254* 1, 401–408.

Yue, E.; Cao, H.; Liu, B. Osmir535, a Potential Genetic Editing Target for Drought and Salinity Stress Tolerance in Oryza sativa. *Plants* **2020,** *9* (10), 1–10. https://doi.org/10.3390/plants9101337

Zaidi, P. H.; Srinivasan, G.; Cordova, H. S.; Sanchez, C. Gains from Improvement for Mid-Season Drought Tolerance in Tropical Maize *Zea mays* L.. *Field Crops Res.* **2004,** *89* 1, 135–152.

Zeng, Y.; Wen, J.; Zhao, W.; Wang, Q.; Huang, W. Rational Improvement of Rice Yield and Cold Tolerance by Editing the Three Genes OsPIN5b, GS3, and OsMYB30 With the CRISPR–Cas9 System. *Front. Plant Sci.* **2020,** *10*. https://doi.org/10.3389/fpls.2019.01663

Zhang, A.; Liu, Y.; Wang, F.; Li, T.; Chen, Z.; Kong, D.; Bi, J.; Zhang, F.; Luo, X.; Wang, J. Enhanced Rice Salinity Tolerance via CRISPR/Cas9-Targeted Mutagenesis of the OsRR22 Gene. *Mol. Breed.* **2019,** *39* 3, 1–10.

Zhang, H. X.; Zhang, Y.; Yin, H. Genome Editing with mRNA Encoding ZFN, TALEN, and Cas9. *Mol. Therap.* **2019,** *27* 4, 735–746. https://doi.org/10.1016/j.ymthe.2019.01.014

Zhang, H.; Zhang, J.; Lang, Z.; Botella, J. R.; Zhu, J. K. Genome Editing—Principles and Applications for Functional Genomics Research and Crop Improvement. *Crit. Rev. Plant Sci.* **2017,** *36* (4), 291–309. https://doi.org/10.1080/07352689.2017.1402989

Zhang, J.; Zhang, H.; Li, S.; Li, J.; Yan, L.; Xia, L. Increasing Yield Potential Through Manipulating of an ARE1 Ortholog Related to Nitrogen Use Efficiency in Wheat by

CRISPR/Cas9. *J. Integr. Plant Biol.* **2021,** *63* (9), 1649–1663. https://doi.org/10.1111/jipb.13151

Zhang, Yong, Zhang, F.; Li, X.; Baller, J. A.; Qi, Y.; Starker, C. G.; Bogdanove, A. J.; Voytas, D. F. Transcription Activator-Like Effector Nucleases Enable Efficient Plant Genome Engineering. *Plant Physiol.* **2013,** *161* (1), pp. 20–27. https://doi.org/10.1104/pp.112.205179

Zhang, Yu, Li, J.; Chen, S.; Ma, X.; Wei, H.; Chen, C.; Gao, N.; Zou, Y.; Kong, D.; Li, T.; Liu, Z.; Yu, S.; Luo, L. An APETALA2/Ethylene Responsive Factor, OsEBP89 Knockout Enhances Adaptation to Direct-Seeding on Wet Land and Tolerance to Drought Stress in Rice. *Mol. Genet. Genom.* **2020,** *295* (4), 941–956. https://doi.org/10.1007/s00438-020-01669-7

Zheng, M.; Lin, J.; Liu, X.; Chu, W.; Li, J.; Gao, Y.; An, K.; Song, W.; Xin, M.; Yao, Y. Histone Acetyltransferase TaHAG1 Acts as a Crucial Regulator to Strengthen Salt Tolerance of Hexaploid Wheat. *Plant Physiol.* **2021,** *186* 4, 1951–1969.

Zhu, H.; Li, C.; Gao, C. Applications of CRISPR–Cas in Agriculture and Plant Biotechnology. *Nat. Rev. Mol. Cell Biol.* **2020,** *21* (11), 661–677. https://doi.org/10.1038/s41580-020-00288-9

CHAPTER 12

Recent Advances in CRISPR/Cas9 for Climate-Resilient Agriculture in Vegetable Crops

VISHAL DINKAR[1], AMAR KANT KUSHWAHA[2], ASHISH KUMAR SINGH[3], KULESHWAR PRASAD SAHU[4], SAYANTAN SARKAR[5], ALKA BHARATI[6], AMARJEET KUMAR[7], and BADAL SINGH[8]

[1]*ICAR–Central Institute of Temperate Horticulture, Srinagar, J&K, India*

[2]*ICAR–Central Institute of Sub-tropical Horticulture, Lucknow, India*

[3]*ICAR–Vivekananda Parvatiya Krishi Anusandhan Sansthan, Almora, India*

[4]*School of Agricultural Science, GD Goenka University, Gurugram, Haryana, India*

[5]*Blackland Research and Extension Center, Texas A&M Agrilife Research, Temple, TX, USA*

[6]*ICAR–Central Agroforestry Research Institute, Jhansi, UP, India*

[7]*Department of Genetics and Plant Breeding, MTTC & VTC, Selesih, CAU, Imphal, India*

[8]*Division of Conservation, ICAR-National Bureau of Plant Genetic Resources, New Delhi, New Delhi, India*

ABSTRACT

Climate change is now a well-established fact that inevitably affects every aspect of our lives and, most importantly, agriculture. Climate change directly inflicts crop losses due to unprecedented weather extremes, increase in mean

Climate-Resilient Agriculture: A Molecular Perspective. Ashutosh Singh, Saurabh Pandey (Eds.)

temperature, precipitation and CO_2 level, and the emergence of new diseases and pests. The indirect impact of climate change on crop production can be attributed to the enormous use of irrigation, fertilizers, and chemicals that led to soil erosion and degradation, depletion of soil nutrients and metal toxicity. Vegetables are an important source of nutrition in the human diet, and vegetable cultivation fetches high profit for the farmers. However, vegetable cultivation has suffered losses due to erratic climatic changes over the years. Modern genomic tools like genome editing techniques have revolutionized the pace of crop improvement. Recently, the CRISPR/Cas9 technique has been applied to generate several loss-of-function and gain-of-function mutants for many traits/ genes. This chapter provides a brief overview of the role of CRISPR/Cas9-based genome editing in improving vegetable crops to meet their biological demand for combating the potential impact of climate change. The chapter discusses the various experimental studies that have explored the power of the CRISPR/Cas9-based genome in producing an array of genomic stocks tolerant to new biotic and abiotic stresses and their usefulness in conventional plant breeding experiments. The chapter has highlighted potential traits and genomic targets for CRISPR/Cas9 application in the development of cultivars and genetic mutants contributing to resilience toward unprecedented flood, drought, thermal stress, and degraded soil health, etc.

12.1 INTRODUCTION

Vegetables are an important source of nutrition all over the world, providing food security and amelioration of nutrient deficiencies (Acharya et al., 2014; Prasad et al., 2014). Vegetable production is an important source of farm income and provides livelihood to several farming households (Prasad and Chakravorty 2015). Within the last two decades (2000–2020), global vegetable production has gone up, and its current trade value is higher than that of cereals. During this period, vegetable yields in East Asia, having temperate and sub-temperate climatic conditions, have increased 2.5 times, especially in China and India (Huang 2004; Ayyogari et al., 2014). However, vegetable crops are sensitive to biotic and abiotic stresses such as drought, high temperature, salinity, flooding, fungal and viral diseases, and insect damage (Bharadwaj, 2012).

In recent years, these stress conditions have been enhanced due to changes in normal weather and climatic conditions (IPCC, 2013). These climatic changes include abnormal fluctuations in atmospheric temperature, relative humidity, gas composition, and precipitation. These variations in climatic conditions that persist for a longer period and can be attributed directly to human activity or natural variability are known as climate change (UNFCC, 2011). The global temperature (combined land and ocean) has increased at an average rate of 0.18°C since 1981, which is twice the rate of increase before 1980 (NOAA, 2021). Climate simulations have suggested that the earth could be 1.1 to 5.4°C warmer in the year 2100 than today (Herring, 2012; Tellefson, 2020). It has been predicted that heatwaves will be more intense, frequent, and long-lasting in a future warmer climate (Meehl et al., 2007). India is among the top 10 most vulnerable countries to climate change (Eckstein et al., 2021). This is also supported by the variability data, which show the increasing trend of frequency, total duration, and maximum heat waves over central and northwestern parts of India (Rohini et al., 2016). Twelve of India's 15 warmest years have occurred since 2006, and the year 2016 was the hottest year, with a mean increase of temperature by 0.71°C (NOAA, 2021). Vegetable cultivation is generally sensitive to environmental extremes (Ahmed et al., 2011; Ayyogari et al., 2014). Higher temperatures, due to increasing greenhouse gases like CO_2 and CH_4, lower vegetable yield by reducing carbon assimilation and photosystem II in leaves. Higher temperatures hasten the pathogen's life cycle, causing higher infection pressure, thereby causing more damage to crops (Boonekamp, 2012). Globally, higher temperatures also cause changes in wind patterns causing erratic precipitation leading to droughts, floods, and storms. Drought stress causes a wide array of symptoms in crops, such as poor germination, stunted shoots, smaller leaves, less flowering, aborted pollens and ovules, poor seed set, reduced pod or tuber formation, and aflatoxin contamination, causing lower yields (Nuruddin, 2001; Rakshit et al., 2009; Minaxi et al., 2011; Sarkar, 2020; Bennett et al., 2021). For instance, crop modeling studies have shown that with the rise in global temperature and drought incidences, potato yield might decline by 13.72% by 2050 in temperate and sub-tropical regions (Singh et al., 2010). Hot and dry environments result in high soil evapotranspiration, causing a higher concentration of residual solutes, thereby increasing soil salinity (Prasad and Chakravorty, 2015). Salinity is a global problem affecting over 932 mha area worldwide (Kumar and Sharma, 2020). Salt stress affects crops during growth and development via modulating cellular processes and ultimately reducing economic crop quantity and quality attributes. Salinity stress has been shown to adversely affect economically important parts of vegetable crops by causing cellular and anatomical modifications (Maksimovic and Ilin, 2012). Rapid and inadvertent industrialization, urbanization, sewer waste, mining activities, fossil fuel burning, use of agrochemicals, etc., are the major source of heavy metals such as arsenic (As), chromium (Cr), cadmium (Cd), copper (Cu), cobalt (Co), nickel (Ni), zinc (Zn), mercury (Hg), and lead (Pb), etc., contamination of agricultural soil and water bodies (Ahmad and Goni, 2010; Ratul et al., 2018; Yang et al., 2018; Sawut et al., 2018; Hindarwati et al., 2018). The toxic levels of heavy metals in soils eventually affect human health via biomagnification. The concentration of toxic metals in soil system, water, and the atmosphere is increasing tremendously and has been reported to be reached beyond tolerance level at several geographical locations around the globe. In India, groundwater of >700 districts is affected by toxic heavy metals (Kumar et al., 2019). Vegetable production under such conditions for safe human consumption is a difficult task from a health perspective. Thus, we see that minute changes and variations in climate can disturb the whole ecology of a region, which in turn affects the cropping system of vegetables in the region. Attributes of climate change significantly diminish the yield potential of the majority of vegetable crops. Appropriate adaptation and mitigation measures are required for sustainable vegetable production. Development of stress-tolerant cultivars and alteration of crop calendar have been suggested; however, these might increase the production cost of cultivation, which needs to be evaluated (Parajuli et al., 2019).

The advancement of gene/genome editing technology has made it possible to precisely and efficiently modify genomes in a variety of species, including vegetable crops, to boost production and quality (Chen and Gao 2014). Advanced technologies of molecular biology-based on SSNs such as transcription activator-like effector nucleases (TALENs), zinc-finger nucleases (ZFNs), as well as the CRISPR/Cas system have allowed plant scientists to undertake accurate and efficient targeted genome editing (Kim et al., 1996; Christian et al., 2010; Jinek et al., 2012; Chen and Gao, 2014; Gao, 2014). Genome editing and CRISPR/Cas9 systems, in particular, have been widely used on a variety of vegetable crops. In most cases, it has been employed for the total knockout by precise indel mutations, either for functional analysis of the target genes or to generate breeding lines with desired features (Corte et al., 2019).

12.2 CRISPR/CAS9 FOR ENGINEERING BIOTIC STRESS TOLERANCE IN VEGETABLES

Diseases caused by fungi, bacteria, or viruses cause significant losses to the yield of the crop. Apart from direct production loss, it also reduces the quality of produce and ultimately leads to economic losses to the grower. Vegetable crops are also attacked by a variety of insect pests and nematodes, which causes yield loss and reduces the quality of produce, leading to lesser acceptance by the consumer. The agricultural production system is mainly dependent on chemical pesticides to manage diseases and pests, which are normally not compatible with the environment and trade (Sivasakthi et al., 2014; Bisht et al., 2019). Losses caused by diseases and pests can be minimized by developing tolerant cultivars. The local climate is an important factor in the spatial and temporal distribution and population dynamics of pathogens and pests. Climate change has emerged as a potential threat to global agroecosystem productivity (IPCC, 2001). The impact of climate change is being predicted to seriously affect insect pests and nematodes' abundance, distribution, behavior, and physiology (Woiwod, 1997; Parmesan, 2007; Merrill et al., 2008). Insects are poikilothermic and expected to respond with each degree change in temperature of the surrounding environment. Hence, it is predicted that with the increase in temperature, life cycle period of insects will be reduced, and the number of generations (voltinmism) per crop cycle will be increased. Therefore, to reduce dependency on chemicals and conservation of nature, it is imperative to stay ahead of notorious pathogen and pest adaptation, emergence, and need to consider research planning and decision making in devising pest management tools in response to climate change.

Plant defense response to pathogen attack is a multi-layered defense strategy. The first step begins with the pathogen-associated molecular patterns (PAMPs) recognition by plant-based recognition receptors (PRRs) and activation of PAMP-triggered immunity (PTI). A successful pest utilizes an effector to suppress PTI to enable compatible interaction with the host to derive nutrition. Initiation of step two is triggered with recognition of effector molecule by R-protein encoded by R-gene and results in effector-triggered immunity (ETI) in host plants. So, in the crux, the pathogen may evade host resistance by evolving and developing diverse effector molecules and parallelly host may evolve diverse R genes to suppress effectors by ETI (Hogenhout and Bos, 2011). Therefore, a thorough understanding of pathogen/pest effectors, underlying R-genes and their stacking into the host is of prime importance. But this is a daunting task through the classical breeding approach. With new developments in genetic engineering technologies, diseases and pests management has gained a new horizon. *Bt* gene obtained from the bacterium *Bacillus thuringiensis* is among the most successful examples that significantly impacted crop productivity (Wang et al., 2016). However, due to governmental regulations, this technology is limited to a few crops and regions. CRISPR/Cas9-mediated genome engineering can bring desirable changes without making the plant transgenic. Plant changes are being carried out to boost their effectiveness in discouraging insects, while creative research is being carried out to modify insects to reduce their efficacy to attack using genome editing tools (Tyagi et al., 2020). However, the progress in the use of genome editing technologies for the production of insect-resistant plants has been limited. In the current section, we will discuss the current status of CRISPR/Cas9 application for the development of cultivar's resistance to different biotic stresses.

12.2.1 ENGINEERING RESISTANCE AGAINST DISEASES

CRISPR/Cas9 has been proved as a potent technology that can assist in the development of varieties tolerant against phytopathogens that are difficult to control (Karkute et al., 2017; Khatodia et al., 2017). As we are aware that plants harbor disease resistance genes (R-genes); in the same way, they also harbor genes responsible for susceptibility (S-genes) to a particular disease (Zaidi et al., 2018). Knockout of that S-gene makes it resistant to specific diseases. Applying this method, Nekrasov et al., (2017) developed powdery mildew-resistant tomato by knockout of the *Mildew resistant locus o1* (*SlMlo1*) gene responsible for susceptibility to powdery mildew fungus, *Oidium neolycopersici*. Here, transgene-free tomato resistant to powdery mildew was created by targeting two positions of the gene *SlMlo1* by double sgRNA. Employing the CRISPR/Cas9 system, watermelon *Clpsk1* gene (S-gene), encoding the phytosulfokine precursor (essential for pathogenicity), was silenced, which further conferred resistance to *Fusarium oxysporum* f.sp. *niveum* causing wilt disease (Zhang et al., 2020). Most plants possess S-genes which play a key role in infection caused by pathogens. In tomato, *Downy Mildew Resistance 6-1* (*SlDMR6-1*) is a gene which mediates pathogenesis events by three important pathogens, namely *Phytophthora capsici, Pseudomonas syringae,* and *Xanthomonas spp.* (Schwartz et al., 2015). This *SlDMR6-1* lowers the salicylic acid level in plants leading to increased susceptibility to disease. CRISPR/Cas9-mediated *SlDMR6-1* disruption enabled the plants to show resistance against these three pathogens without having any ill effect on the growth and physiology of the plant (De Toledo Thomazella et al., 2016). Thus, there is a possibility to achieve broad-spectrum resistance by targeting a single gene through CRISPR/Cas9 technology. Coronatine produced by *P. syringae* pv. *tomato* induces stomatal reopening, supporting the virulence of bacteria, and thus allowing bacterial colonization (Melotto et al., 2006). In tomatoes, *Jasmonate ZIM-Domain-2* (*SlJAZ2*) is responsible for stomatal response to coronatine. To prevent the coronatine-mediated stomatal opening, *SlJAZ2* was targeted by CRISPR/Cas9, and the mutant plants displayed resistance to the disease (Ortigosa et al., 2019).

The plant cell machinery is essential for RNA viruses to sustain their life cycle, and the eukaryotic translation initiation factor *eIF4E* is one of the most important host factors for viral growth. Melon (*Cucumber melon*) plants were generated with silenced *eIF4E* gene and identified as resistant upon challenge with a range of viruses, including *Zucchini yellow mosaic virus* (ZYMV), *Melon necrotic spot virus* (MNSV), *Cucumber vein yellowing virus* (CVYV), and *Moroccan watermelon mosaic virus* (MWMV) (Rodríguez-Hernández et al., 2012). A similar experiment was conducted in tomatoes where two genes of the *eIF4E* family, that is, *eIF4E1* and *eIF4E2,* were silenced using genome editing techniques and the mutant lines showed increased tolerance against potyviruses (Mazier et al., 2011). Simultaneous editing of *eIF4E* isoforms, viz., *novel Cap-Binding Protein* (*nCBP-1* and *nCBP-2*) in cassava, reduces cassava brown streak disease (CBSD) symptom severity and incidence (Gomez et al. 2019). The first application of CRISPR/Cas9 in cucumber was made by Chandrasekaran et al. (2016), where they generated broad-spectrum resistance against *Cucumber vein yellowing virus* (*Ipomovirus*), *Zucchini yellow mosaic virus* (Potyvirus), and *Papaya ringspot mosaic virus-W* (Potyvirus) through CRISPR/Cas9-mediated knockout of two different sites of *eIF4E* gene using two gRNA. In contrast to plants where gRNA targeting viral DNA are inserted in plants resulting in transgenic virus-resistant plants, this CRISPR/Cas9 technique requires sufficient knowledge about susceptibility genes and creates non-transgenic plants (Ali et al., 2015a).

12.2.2 ENGINEERING RESISTANCE AGAINST INSECT PESTS

Understanding host–pest interaction is critical for the development of a novel pest management strategy. Host and pest (insect/nematode) interactions are conserved and universal phenomena among diverse pests of the crop. During the course of co-evolution, insect and plant-herbivore insects have adapted diverse modes of feeding strategy and behaviors. Various warfare arsenals assist the battle between two systems. For example, suppose insects possess a chewing/sucking type mouth part to bite or penetrate the host plant surface. In that case, the plant will revert back with its defense system of physical barrier (hard cuticle surface) and chemical defense (secondary metabolites, etc.). Insect pest usually releases elicitor compound to initiate host–plant interactions. In the compatible reaction, the insect wins the battle, and the host manifests the visible expression on the cell/tissue level (symptoms). While in incompatible reaction, host cell commits suicide (hypersensitive response) to prevent insect damage (Harris et al., 2003; Wu and Baldwin, 2010). Hemipteran insect pest (Bugs and Aphids) poses stylet to pierce and suck the nutrition from host cells. These pests actively suppress host cells by careful penetration to host cells, saliva injection, and ingestion of digested cell contents (Powell et al., 1995). The use of RNAi offered a ray of hope, but use is limited due to off-target effects and low reproducibility of results. Recently, CRISPR/Cas9-based genome editing tool offered a novel insect resistance strategy (Hall et al., 2018). CRISPR/Cas9 application can precisely modify any target gene, but any mismatch between gRNA and targeted DNA may result in poor specificity and reduced effectiveness in disruption. Tools like CRISPR *flydesign (flyCRISPR)* help in designing enhanced matching specificity and off-target activity (Gokcezade et al., 2014; Prykhozhij et al., 2015). Once the target is determined, the Cas9 and sgRNA construct can be delivered to the target through direct (transfection, virus-mediated delivery, and microinjection) and indirect methods (transformation).

Since the inception of CRISPR systems, various loci of model insect Drosophila have been mutated for functional gene studies, proving their potential to manage insect pests efficiently. Diamondback moth (DBM) gene *Pxabd-A* targeted through CRISPR demonstrated abnormal gonad development (Huang et al. 2016). Among locusts, antennae are an important organ to perceive signals and orient insects to their cues (chemoreception) for mating, feeding, foraging, etc. The *Orco* gene (odorant receptor) of Locusta migratoria targeted through CRISPR/Cas9 resulted in the disruption of the olfaction mechanism (Li and Scott, 2016). Similarly, the abdominal-related gene, *Slabd-A* in *Spodoptera litura* targeted through CRISPR, resulted in irregular segmentation and pigmentation (Bi et al., 2016). In this line, various insect pests have been successfully targeted and utilized through CRISPR/Cas9 approach, have been mentioned in Table 1.

TABLE 12.1 Summary of CRISPR/Cas9-based insect mutations.

Target insect	Targeted gene	Type of mutation	Obtained phenotype	References
Spodoptera litura	*Slabd-A*	Mutation	Abnormal body development and segmentation observed	Bi et al., (2016)
Spodoptera litura	*SlitPBP3*	Mutation	Mutant male showed decreased response to pheromone	Zhu et al., (2015)
Spodoptera littoralis	*Orco*	Mutation	Chemoreception against plant odors and pheromone was blocked	Koutroumpa et al., (2016)
Locusta migratoria	*Orco*	Mutation	Disruption in olfaction	Li and Scott, (2016)
Helicoverpa armigera	*HaCad*	Mutation	Enhanced resistance by 549-fold than *cry1ac* gene	Wang et al., (2016)
Plutella xylostella	*Pxabd-A*	Mutation	Abnormal gonad development	Huang et al., (2016)

The area that many researchers have exploited is changing the insect's genome so that it no longer has the ability to attack the crop. This strategy can be employed successfully in two steps involving the modification of insects followed by their release in the target area (McFarlane et al., 2018). *Bt* gene is one of the most successful transgene technologies, but there is a great chance of developing resistance to the *Bt* insecticidal proteins in the target pest. Thus, receptors of insects can be targeted and engineered to manage the resistance. The silencing of cadherin receptors associated with "Cry1Ac toxin" tolerance in *Helicoverpa armigera* is evidence of effective genome editing (Wang et al., 2016). Insects also contain several enzymes responsible for detoxifying the plant's chemical defense response. Targeting those genes coding for detoxifying enzymes in insect pests can be an effective option. Insect vulnerability was increased when insecticidal detoxifying genes such as gossypol-inducing *cytochrome P450* were knocked out (Hafeez et al., 2019). Similar results were obtained when *CYP6AE* gene *H. armigera* was knocked out using CRISPR/Cas9, resulting in higher susceptibility to plant's toxic chemicals (Wang et al., 2018). Targeting mating behavior by disrupting the chemical communication through the genome is another method of insect management. CRISPR/Cas9-mediated disruption of *Orco* (olfactory receptor coreceptor) has resulted in the loss of mating partner recognition in *Spodoptera litura* (Koutroumpa et al., 2018). Similar results were observed in the case of *H. armigera* males, where *Odorant Receptor 16* (*OR16*) was knocked out using CRISPR/Cas9 leading to a loss in pheromone signal recognition (Sun et al., 2017). Thus, knocking down the *OR16* receptor may be a unique and successful technique for managing lepidopteran pests in agricultural crops. CRISPR/Cas9 could be utilized to develop new Bt toxin to break resistance developed by insects against Bt crops through the CRSIPR-mediated direct evolution approach (Badran et al., 2016). The mechanism of insecticide resistance development is another important area where CRISPR/Cas9 system has demonstrated its value in identifying the mutant gene responsible for insecticide resistance upregulation (Somers et al., 2015). Thus, the list of applications for insect pest management through CRISPR/Cas9 is still in the evolution phase, and many more could be incorporated in future.

12.2.3 CRISPR/CAS9 IN NEMATODE MANAGEMENT

Plant-parasitic nematodes (PPNs) are the most destructive hidden enemies of the farmers. More than 4100 PPNs have been reported to parasitize a wide range of crops and attributing to 173 billion dollars globally (Decraemer and Hunt, 2006; Elling, 2013). Jones et al. (2013) described the top 10 most obnoxious PPNs threatening the global food security from the worldwide survey, viz., (1) Root-knot nematodes (*Meloidogyne spp.*); (2) Cyst nematodes (*Heterodera* and *Globodera* spp.); (3) Root lesion nematodes (*Pratylenchus spp.*); (4) The burrowing nematode *Radopholus similis*; (5) Onion and bulb nematodes *(Ditylenchus dipsaci)*; (6) Pine wilt nematode (*Bursaphelenchus xylophilus)*; (7) Reniform nematode *Rotylenchulus reniformis*; (8) Dagger nematodes (*Xiphinema index,* a virus vector); (9) *Nacobbus aberrans*; and (10) *Aphelenchoides besseyi*. Therefore, it is necessary to develop and validate an effective management strategy. So far, various management strategies have been utilized but provided limited success.

Regarding the health hazards and environmental concerns, most of the chemical nematicides are under phased withdrawal provisioned in the UN Montreal protocol. The use of bioagents for nematode management is still in the development process and not viable for large acreage. Another major approach which is farmer 12 and environment friendly is the use of natural resistance (Cook and Evans, 1987; Atkinson, 1995). Robert (1992) has reviewed both the success and limitations of this approach; mostly, problems are centered on the availability of vertical resistance, long breeding period, and durability under different genetic variabilities of the pathogen.

A major hope for the future is engineering plant resistance, a novel, time-efficient, cost-effective, eco-friendly, durable, and sustainable strategy for limiting crop damage due to PPNs. Recombinant DNA technology offers us avenues to transfer putative genes across species barriers for engineering resistant crop varieties. In planta, RNAi strategies targeted at crucial genes of nematodes offer good hope for engineering resistance against the nematodes. However, their robustness under field conditions is yet to be tested. The inception of CRISPR technique-based genome editing has been reported as a revolutionary tool for induction of mutation in target-specific sequences and induction of disease resistance (Wang et al., 2019). The application of CRISPR in nematode management is in the beginning phase and has not been used to develop resistance in the crop against nematodes.

At present, the CRISPR technique has been deployed to understand the functional role of R-gene in plants against nematodes. In soybean cv., forest is containing *GmSHMT08* gene at *Rhg4* locus attributes for resistance against soybean cyst nematode (SCN). Constructs (*SHMT8*-1 and *SHMT8*-4) using two single guiding RNA were found to be effective in bringing disruption in gene function of *GmSHMT08* and significantly increased the nematode infection in the plant (Kang et al., 2016). Likewise, CRISPR/Cas9 mutation in two syntaxin genes and their two homeologs in SCN resistance plants reduced the resistance level to control (Dong et al., 2020). CRISPR systems can be utilized to identify gene favoring host susceptibility (S-genes) in response to nematode attack by developing targeted mutant/deletion lines (Siddique et al., 2022). These S-genes will be the key target in vegetable crops to enhance durable resistance against nematodes.

12.2.4 ENGINEERING PARASITIC WEED TOLERANCE

Noxious plant pests, including genera *Striga*, *Orobanche*, and *Cuscuta,* have become a threat to global food security. Some problematic parasitic weeds in vegetable crops are *O. crenata* in leguminous vegetables, *Apiaceae* in lettuce; *O. ramosa* in *Solanaceous,* and Cucurbitaceous vegetables; *Cuscuta* spp. in leguminous, tomato, carrots, onion, asparagus) (Zaragoza, 2003). Studies have shown that specific interactions exist between plant pests and crop plants. Recently, Hegenauer et al. (2016) identified a cell surface receptor-like protein kinase (RLK), *Cuscuta Receptor 1 (CuRe1),* in tomato, which recognizes the parasite-associated molecular pattern of parasitic *Cuscuta reflexa* and brings the resistance response. Likewise, sunflower RLK *HaOr7* confers resistance to *Orobanche cumana* race F (Duriez et al., 2019). Thus, CRISPR/Cas9 can be employed to identify such specific genes in plant pests and crop plants, and their expression can be manipulated to develop plant pest tolerant vegetable cultivars. Manipulating hormone signaling can be a viable idea to engineer vegetable varieties' tolerance against infestation by parasitic weeds. Carotenoids derived strigolactones (SLs) are phytohormones that have a proven role as shoot branching inhibition hormones and act as signaling molecules for the rhizosphere microflora such as *Orobranche spp., Phelipanche spp.,* and *Striga spp.* (Walter et al., 2017). Recently, CRISPR/Cas9-mediated targeted editing of a key enzyme in SL production, *Carotenoid Cleavage Dioxygenases 8* (*CCD8*), and *More Axillary Growth 1* (*MAX1*) gene in tomato has led to developed lines resistance against root parasitic weed *Phelipanche aegyptiaca* (Bari et al. 2019; 2021).

12.3 CRISPR/CAS9 FOR ENGINEERING ABIOTIC STRESS TOLERANCE

Plants have several physiological and biochemical mechanisms to adapt to an abiotic stress environment. These genetic mechanisms have been broadly classified as avoidance mechanisms or tolerance mechanisms. Various genetically regulated coping mechanisms for abiotic stress include plasma membrane integrity, maintaining ROS homeostasis, ROS scavengers, accumulation of compatible solutes, kinases and phosphatases signaling cascades, and chaperone signaling transcriptional regulation (Wahid et al., 2007). Plants may avoid heat stress by transpirational cooling, dense trichomes, thick cuticle, wax coat, changing leaf orientation (paraheliotropism), leaf rolling, and alteration of membrane lipid compositions, etc. On the other hand, inherent plant features may impart tolerance to prevailing heat stress via modulation of molecular mechanisms of ion transporters, stress-related proteins (heat shock proteins, ubiquitin, Pir proteins, dehydrins, late embryogenesis abundant [LEA] proteins), osmoprotectants, antioxidant defense, and transcription factors (TFs) and ultimately confer plant's ability to produce economic yield under heat stress (Hasanuzzaman et al., 2013). CRISPR/Cas9 applications in vegetable crops for engineering abiotic stress tolerance are listed in Table 2.

TABLE 12.2 List of CRISPR/Cas9 applications in vegetable crops for climate-resilient agriculture.

Crop	Targeted gene	Type of mutation induced	Results of targeted mutagenesis	Reference
Lettuce	*LsNCED4* (9-cis-epoxycarotenoid dioxygenase4) – regulatory enzyme in ABA biosynthesis	Loss-of-function	Seed germination at a high temperature	Bertier et al., (2018)
Tomato	MADS-box gene *SlAGL6* (*Sl Agamous-like 6*)	Loss-of-function	Mutants produced improved fruit set via facultative parthenocarpy under heat stress	Klap et al., (2017)
Tomato	TF – *BZR1* (*Brassinazole Resistant 1*)	Loss-of-function	Impaired accumulation of H_2O_2 in apoplast and decreased H_2O_2 directed heat stress tolerance	Yin et al., (2018)
Tomato	Kinase – *SlMAPK3*	Loss-of-function	Mutants exhibit enhanced heat tolerance, reduce ROS accumulation and upregulated HSPs/HSFs genes expressions	Yu et al., (2019)
Tomato	Kinase – *SlMAPK3*	Loss-of-function	Reduced chilling tolerance in tomato fruits via weakening the inhibition effect of ferulic acid on H_2O_2 accumulation	Shu et al., (2022)
Tomato	TF – *PIF4* (*Phytochrome-Interacting Factor4*)	Loss-of-function	Enhanced cold tolerance in pollens due to reduced temperature sensitivity of the tapetum	Pan et al., (2021)
Tomato	TF – S*lCBF1*	Loss-of-function	Membrane damage, decreased proline content, toxic ROS levels, altered hormone signaling, and diminished induction of CBF-regulated genes	Li et al., (2019)
Tomato	*SlSBPase* (Sedoheptulose-1,7-bis phosphatase)	Loss-of-function	Mutants were more sensitive to chilling-induced oxidative stress due to suppressed glutathione biosynthesis and ascorbate-glutathione recycling	Wang et al., (2020)
Tomato	CGFS-type glutaredoxin (*GRX*) genes	Loss-of-function	Elevated sensitivity to multiple abiotic stresses, including heat, cold and cadmium stress, due to disturbed ROS homeostasis	Kakeshpour et al., (2021)
Tomato	RING-type E3 ubiquitin ligase – *Sl1*	Loss-of-function	Mutants show a higher accumulation of Cadmium in root and shoot	Liu et al., (2022)
Tomato	*SlALMT9* (Al-Activated Malate Transporter9)	Loss-of-function	Vacuole membrane-localized *SlALMT9* increases malate transport and enhances Al resistance.	Ye et al., (2017)

12.3.1 ENGINEERING TOLERANCE AGAINST ELEVATED TEMPERATURE

Elevated temperature in the environment can be sensed by plant macromolecules and show altered behavior, consequently leading to a cascade of heat adaptive responses by plants. Such heat sensors can be changed in plasma membranes, unfolded proteins in plant organelles and cytosol, histone modifications, and accumulation of free radicles. The plasma membrane is the primary heat sensor, and changes in membrane fluidity occur first during high heat stress (Sajid et al., 2018). This triggers Ca^{2+} ion channels and an inward flux of Ca^{2+} ions, which triggers multiple signaling pathways. The Ca^{2+} ions bind with calmodulins and activate kinases (MAPKs, CDPKs, etc.) and heat stress-responsive TFs such as HSF, MBFs, Myb, WRKY, CBFs, DREB, etc. (Saidi et al., 2011; Kai and Iba, 2014). The plasma membrane also harbors ROS producing enzymes [NADPH oxidases (NOXs), also called respiratory burst oxidase homologues (RBOHs) in plants], which get activated either directly via an elevated level of cytosolic Ca^{2+} ions or via phosphorylation activity of Ca^{2+}-regulated proteins, such as calcium-dependent protein kinases (CDPKs) and calcineurin B-like protein-interacting protein kinases (CIPKs) (Drerup et al., 2013; Baxter et al., 2014). NOXs-mediated apoplastic ROS burst triggers the oxidative stress signaling network, activating MBFs and HSFs and affecting the transcriptional regulation of antioxidant encoding genes (Mittler et al., 2004, 2012). Altered membrane stability/fluidity may act as substrates for the production of different signaling lipid molecules such as phosphatidic acid, phosphoinositides, sphingolipids, lysophospholipids, oxylipins, etc., which trigger lipid signaling via lipid-metabolic enzymes such as phospholipase C and D (PLC/PLD), diacylglycerol kinase (DGK), and phosphatidylinositol-4-phosphate 5-kinase (PIP5K) (Huo et al., 2015; Hayes et al., 2021). Unfolded protein response (UPR) pathways, viz., cytosolic (Cyt-UPR) and endoplasmic reticulum (ER-UPR), functions through sense molecules and TFs like HSFs, bZIPs, etc. (Walter and Ron, 2011; Kim et al., 2018). Enzymatic machinery involved in the degree of lipid saturation in thylakoid membranes of chloroplast has a role in plants' thermal stress tolerance and photosynthetic activity. During transgene-based experiments, it is noted that increasing unsaturation via upregulation of glycerol 3-phosphate acyltransferase synthesis led to enhanced cold tolerance (Murata et al., 1992), whereas increasing the degree of saturation via disruption of gene encoding chloroplastic omega-3 desaturase (synthesizes trienoic fatty acids) enhances the plant acclimatization capacity under elevated heat (Murakami et al., 2000). Hormones signaling can play a significant role in heat stress tolerance. For instance, ethylene mediates the induction of genes including 1-aminocyclopropane-1-carboxylic acid (ACC) synthase, ethylene-responsive factors (ERFs), and the transcriptional coactivator multiprotein bridging factor 1 (MBF1) play a specific role in heat stress response in microspores of tomato under high heat (Frank et al., 2009).

Crop establishment is a critical phase of crop production in regions with high temperatures, which can interfere with the rapid seed germination in winter annual or biennial species (Huo et al., 2013). Bertier et al. 2018 performed CRISPR/Cas9-mediated knockout experiments and mutants for *LsNCED4* (9-cis-epoxycarotenoid dioxygenase4), a gene conditioning thermo-inhibition of seed germination in lettuce, could germinate on high temperatures and demonstrate up to 57% editing efficiency. Elevated heat disturbs the functions of chloroplasts and mitochondria, which can lead to toxic levels of ROS. ROS-mediated oxidative damage is manifested by protein oxidation, lipid peroxidation, and nucleic acid damage and may cause activation of programmed cell death (PCD) (Sharma et al., 2012). Hormones play an important role in the regulation of abiotic stress tolerance in crop plants. Brassinosteroids (BRs) are a group of steroidal hormones that regulate a number of cellular processes, including abiotic stress tolerance (Bajguz and Hayat, 2009). BRs trigger apoplastic H_2O_2 accumulation by inducing NADPH oxidase, which in turn activates MAPKs leading to increased stress tolerance (Nie et al., 2013). Under stress, binding of BR to the receptor-like kinase (RLK), *Brassinosteroid Insensitive 1* (*BRI1*) at the cell surface gets activated and initiates the downstream signaling of BR-responsive genes. *Brassinosteroid Insensitive 2* (*BIN2*) negatively regulates the TF *Brassinazole Resistant 1* (*BZR1*), which gets inactivated by elevated BR (Clouse, 2011). Yin et al. (2018) have unraveled the role of TF *BZR1* in heat stress tolerance in tomato under BR signaling. They showed that CRISPR/Cas9-derived null mutants of the *BZR1* gene in tomato impair the activation of *Respiratory Burst Oxidase Homolog1* (*RBOH1*), which encodes NADPH oxidase. This caused the accumulation of low levels of apoplastic H_2O_2 and led to impaired oxidative stress signaling, and ultimately led to reduced heat stress tolerance in *BZR1* gene mutant plants. In addition to redox signaling, BR-induced stress responses also involve mitogen-activated protein kinases (MAPKs) cascades (Xia et al., 2009). BRs triggered induction of NADPH oxidase, and H_2O_2 accumulation in apoplast activates MAPKs. Different MAPKs serve either as positive or negative regulations of abiotic stress tolerance in tomato. Silencing of MAPKs (*SlMAPK1, SlMAPK2,* and *SlMAPK3*) reduced

RBOH1 expression in tomato and reduced H_2O_2 accumulation, consequently enhancing positive heat stress response as well as tolerance to other abiotic stresses (Nie et al., 2013; Ding et al., 2018; Yu et al., 2019). CRISPR/Cas9-derived tomato *SlMAPK3* knock-out mutant exhibited enhanced heat tolerance, reduced ROS accumulation, and showed upregulated HSPs/HSFs gene expressions (Yu et al., 2019). Thus, *RBOH1 and* MAPKs encoding genes can be targeted for engineering high heat stress responses using the CRISPR approach in vegetable crops.

HSPs act as molecular chaperones, involved in the regulation of protein folding, accumulation, cellular localization as well as degradation in plants. HSPs play a pivotal role in refolding proteins during heat stress conditions (Tripp et al., 2009). In plants, HSPs have been classified into five major families based on their molecular weight, viz., HSP100, HSP90, HSP70, HSP60, and <40 kDa small heat stress proteins (sHSPs). A detailed review of HSPs regarding their structure, function, classification, and regulation is presented by Al-Whaibi (2011); Andrási et al., (2021). HSP70 positively regulates spinach's heat and cold stress tolerance by lessening the cell membrane damage and ROS accumulation and improving antioxidant enzyme activity (Qi et al., 2019). Moreover, sHSPs have the ability to bind with partially folded or denatured proteins and prevent their irreversible unfolding or wrong protein aggregation and thus, enhance plant-acquired thermotolerance (Sun et al., 2002). Zhao et al., 2018 identified 48 HSP20 genes in the potato genome and analyzed their relative expression levels. They found 14 *StHsp20* genes were up-regulated by >100 times under heat stress. Similarly, the HSP20 gene family has been characterized in several vegetable crops, including 33 in pumpkin (Hu et al., 2021), 35 in pepper (Guo et al., 2015), 22 in lettuce (Kim et al., 2021), which can be targeted for the development of heat stress-tolerant varieties.

HSP transcription level is regulated by HSFs. Wang et al. (2020) performed genome-wide identification, functional characterization, and classification of HSF encoding genes in brinjal. They identified 20 HSF genes differentially expressing under cold, heat, salinity, and drought treatments. Additionally, they found *SmeHsfA6b* gene was upregulated under cold, heat, and salinity stress in the leaves but not under drought stress and concluded that homologous HSFs could have different regulatory functions in different crop plants. Several HSFs have been identified in vegetable crops, including *Brassica spp.* (Song et al., 2014; Lohani et al., 2019), brinjal (Gong et al., 2020), and lettuce (Kim et al., 2021). For a detailed review of structural and functional diversity in plant HSFs, their regulation and interactions, see Waters and Vierling, 2020; Yadav et al., 2020; Andrási et al., 2021.

12.3.2 ENGINEERING CHILLING STRESS TOLERANCE

In the consensus model, cold/chilling stress causes decreased membrane fluidity and cytoskeletal rearrangement, which leads to cold stress perception by the membrane and membrane-bound proteins such as Ca^{2+} permeable channels (e.g., *Chilling Tolerance Divergence 1, COLD1* in rice), histidine kinases (*Hik33* in cyanobacteria), receptor-like kinases (RLKs) (*CRLK1/2* in *Arabidopsis*), phospholipases (phospholipase Dδ in *Arabidopsis*), and chloroplast (*Sensitive to Freezing 2, SFR2* gene), etc. (Li et al., 2004; Muarata and Los, 2006; Fourrier et al., 2008; Crosatti et al., 2013; Guo et al., 2018; Song et al., 2021). Rapid activation of Ca^{2+} channels followed by the influx of Ca^{2+} which is sensed by Ca^{2+} decoding proteins calmodulin (CaM), CaM-like proteins (CML), Ca^{2+}-dependent protein kinases (CDPKs), and calcineurin B-like proteins (CBLs). Ca^{2+} as a secondary messenger, activates ROS production, ABA accumulation, and MAPK cascade activation (Yuan et al., 2018; Guo et al., 2018). An elevated concentration of cytosolic Ca^{2+} induces the expression of early cold-responsive transcription factors (early TFs). In turn, the early TFs regulate the expression of downstream *cold-responsive* (*COR*) genes that confer cold stress tolerance. Concomitantly, several early TFs can co-regulate many *COR* genes (Park et al., 2015; Zhao et al., 2015). Concurrently, CBFs and early TFs can regulate other early TFs. In *Arabidopsis*, CBF genes are mainly regulated by bHLH TFs, *Inducer of CBF Expression 1 and 2* (*ICE1* and *ICE2*) and three closely related calmodulin-binding transcription activator (CAMTA) TFs, *CAMTA1-3* (Chinnusamy et al., 2010). *ICE1* is a key regulator (regulates negatively) in the ICE1-CBF-COR pathway itself regulated post-translationally. *A* Ser/Thr protein kinase *Open Stomata 1* (OST1) phosphorylates *ICE1* and activates it. *ICE1* can be ubiquitinated by E3 ubiquitin ligase, *High Expression of Osmotically Responsive Gene 1 (HOS1)*, leading to its degradation (Ding et al., 2015). However, ICE1 SUMOylation by SUMO E3 ligase *(SIZ1* and *SIZ2)* enhances its stability and, thereby, induces the expression of CBF genes (Miura et al., 2007). But *ICE1* phosphorylation by *OST1* can inhibit its degradation (Ding et al.,2015). Moreover, *MAPK3* and *MAPK6* can phosphorylate *ICE1* in a cold stress-dependent manner and lead to the degradation of *ICE1* by E3 ligases, thus negatively regulating the plant's cold stress tolerance (Li et al., 2017). In *Arabidopsis*, *CaM-regulated Receptor-like Kinase* (*CRLK1*) regulates CBF regulons by modulation of MAPK kinase activity (Yang et al., 2010). Hormonal may have a regulatory role in cold stress response in crop plants. In the ABA signaling pathway, TF ABF1/2 interacts with ABRE-containing genes and regulates their expression (Sah et al., 2016). Cytokinin receptor kinases (*Arabidopsis Histidine Kinases—AHK2 and AHK3*) function as cold stress sensors and are involved in inhibiting ABA signaling under freezing conditions (Kim and Jeon, 2013).

Most of the studies regarding the functional characterization of genes involved in low-temperature acclimatization are done either in model plants such as *Arabidopsis thaliana* (Jia et al., 2016; Cho et al., 2017; Shkryl et al., 2020) or in field crops (Shen et al., 2017; Nawaz et al., 2019; Hillmann, 2019; Zeng et al., 2020). Reports regarding CRISPR/Cas9 for cold stress response manipulation in horticultural crops are scarce. To our knowledge, tomato is the only vegetable crop where CRISPR/Cas9-mediated genome editing is applied for targeting molecular pathways involved in the low-temperature response. Such studies with targeted mutagenesis have advanced our current knowledge regarding the mechanism of cold stress responses by vegetable crops. Such as integration of the role of secondary metabolites such as ferulic acid as an inducer of the cold stress signaling cascade, unraveling the role of regulatory genes for ROS homeostasis and key TFs in the cold-response system and their regulators (Li et al., 2019; Kakeshpour et al., 2021; Wang et al., 2020; Pan et al., 2021; Shu et al., 2022).

12.3.3 ENGINEERING DROUGHT TOLERANCE

Vegetable yield loss under drought is a manifestation of perturbed plant developmental, biochemical, and physiological processes and tolerance to drought is under genetic control (Yordanov et al., 2000; Barnabás et al., 2008). Identification and utilization of drought-tolerant genes in germplasm collections are a continuous process. Crop, where genetic variability is limited in breeding programs, suffers drastically. In such a scenario, the scope needs to extend up to CWRs or even unrelated genera. CRISPR/Cas9 can serve better under such circumstances in inducing mutants tolerant to drought, and if any gene is already identified, it can be functionally validated. A master regulator, that is, *SlMAPK3*, its role and regulatory mechanism in drought tolerance has been deciphered in tomato utilizing CRISPR/cas9 (Wang et al. 2017). The knockout mutant of *SlMAPK3* has shown drought susceptibility and modified regulation of drought-responsive genes (*SlLOX*, *SlGST*, *SlDREB*, etc.) when compared with the wild type under drought stress conditions. Similarly, a target that had earlier been reported to be involved in pathogen defense-related pathways was modified using CRISPR/cas9 and reported to have a role in drought response. This speculation was marked by an enhanced aperture of stomata, electrolyte leakage, ROS level, and low level of antioxidant activities and lower expression level of drought-responsive genes viz., *DREB, GST,* and *DHN* in the slnpr1 mutant when compared with wild type in drought condition (Li et al. 2019). Lateral organs, for example, fruits and roots, are kinds of organs which are highly sensitive to drought conditions. *Lateral Boundary Organ* (*LBD*) genes are very important in the development of such organs. In tomato, one of such genes, that is, *SlLBD40,* is inducible in the PEG and salt environment. Liu et al. (2020) reported that this *SlBD40* negatively regulates the drought tolerance by comparing over-expressed and knock-down mutants of the gene. The over-expressed individuals showed severe wilting while the knock-down was slightly wilted under PEG-induced drought. Though very limited stances are available for CRISPR/cas9 system utilization for inducing drought tolerance, the targets viz., *CBP 80, CBP20, IPT,* etc., identified earlier via RNAi and other approaches can be utilized to have the full-scale advantage of CRISPR technology.

12.3.4 ENGINEERING SALINITY TOLERANCE

Genetic improvement is a sustainable and cost-effective technique to combat the global salinity issue. Identification and modification of salt stress-related targets can lead to the development of a salt stress-tolerant genotype. The mechanisms providing salt tolerance range from molecular processes to morphological features (Behra et al., 2022). Many genes imparting tolerance to salt stress have been identified and characterized to the level of molecular functions. Most of these genes have been classified as transcriptions factors (viz., *NACs, WRKY, bZIP, ERF,* and *MYB*), genes encoding osmoticum (*Osmotins, P5CS, mtl1D,* etc.), antiporter genes (*NHX1, AtHKT1, LeNHX4,* etc.) and genes encoding scavengers of reactive oxygen species (*PaSOD, CaCKR1*). Genes from other sources have also been utilized to transfer crop species using genetic engineering methods to validate their functions and development of salt-tolerant genotypes, but they suffer from the limitations of unintended effects on the environment and biodiversity, etc.

CRISPR/Cas9 has been employed in a wide range of crop species to crack the key pathways and players responsible for the salinity stress response of plants. Though numerous potential targets have been identified for salt stress tolerance in vegetable crops, there are limited reports available from tomato, pumpkin, potato, etc. For instance, Tran et al. (2020) precisely excluded proline rich domain and eight cysteine domains of tomato SlHyPRP1 gene and induced salt tolerance under controlled conditions for germination, and vegetative growth showed *SlHyPRP1* gene negative regulator of multiple abiotic stress. *Auxin response factors (ARF)* have been reported to regulate the transcription of many genes. Bouzroud et al. (2019) have reported their role in salinity stress tolerance by regulating stomatal conductance, leaf water retention, and ABA content. They induced targeted mutation using CRISPR/cas9 in one such ARF gene, that is, *SlARF4* and deciphered its role in salt tolerance as the mutant had better withstanding potential under salt and osmotic stress. Ion toxicity is one of the detrimental effects imposed by salinity stress in plants. Such an effect has been known to be alleviated by cellular homeostasis. *HKT1* and *SOS1* have been involved in such regulation. Wang et al. (2020) reported tomato variation for salt tolerance is contributed by genic variation in *SlSOS1* using association analysis and validated using knock-out mutants generated using CRISPR/cas9. In potato, the *coilin* gene was targeted using CRISPR/Cas9 to induce viral resistance. It was found from the phenotyping of mutants generated that the gene also has a role in imparting resistance against salt tolerance (Makhotenko et al., 2018). Through transcriptome analysis, Huang et al. (2019) identified DEGs under salinity stress in pumpkin, notably *RBOHD (respiratory burst oxidase homolog D)* encoding *NADPH oxidase*. They further targeted to knock out the gene using CRISPR/cas9 to validate the results and show that in mutant was salt sensitive with marked reduction in H_2O_2 and K+ content in root apex and in the expression of salinity-related genes (*GRF12, AHA1,* and *HAK5*). Geng et al. (2022) have shown the role of *CmoHKT1;1* in salinity stress tolerance by communication of Na^+ from root-to-shoot by generating mutants using the CRISPR/cas9 system. The mutants so generated were more sensitive to salt stress, and more of the Na^+ was reported from the shoot (in xylem sap) than in root when compared with control.

12.3.5 ENGINEERING TOLERANCE AGAINST NUTRIENT DEFICIENCY AND METAL TOXICITY

Plants have evolved numerous physiological, biochemical, and molecular mechanisms to adapt to nutrient deficiency or toxicity stress conditions. One such universal adaptation under mineral deficiency is high root surface area and root-to-shoot ratio by alterations in root architecture (Nath and Tuteja, 2016). Expression of nutrient transporters in plant roots is enhanced in response to mineral nutrient starvation. The nutrient signaling network has the ability for two-way long-distance communication in plants. Plants have developed essential systemic signaling to relay signals about the capacity of root development, nutrient uptake, and transport from root-to-shoot and nutrient partitioning/remobilization as per the nutritional demand (i.e., shoot-to-root signals) (Ruffel, 2018). Major signaling factors include phytohormones, ROS, sugars, nutrient responsive genes/transcription factors, etc., for maintaining ion homeostasis under nutrient starvation or excess conditions.

Thus, a better understanding of plant mechanism of mineral nutrient and metal absorption and assimilation in plants is important to engineer plants with better nutrient use efficiency and tolerance to toxic levels in the soil as well in the plant system. Several studies with model plants such as *Arabidopsis*, tomato and rice have provided a better insight into plant responses to nutrients and toxic heavy metals (Yang et al., 2020; El-Sappah et al., 2021; Lhamo et al., 2021); and this knowledge is growing continuously. AMT family ammonium (NH_4^+) transporters (e.g., *SlAMT1* in tomato) involved in direct uptake of NH_4^+ from soil (Filiz and Akbudak 2020). Similarly, several root-specific transporters (e.g., *BraNRT2*.1 in Chinese cabbage and *CsNRT2*.1 in cucumber) coordinate nitrate (NO_3^-) uptake (Liu et al., 2014a; Li et al., 2018). Inorganic phosphorus (Pi) is directly taken through phosphate transporters (PHTs) (e.g., *SlyPHO1;1* in tomato), and their activity is regulated by the *phosphate starvation response (PHR)* gene (e.g., *SlPHL1* in tomato) to maintain Pi homeostasis; and *PHR* expression, in turn, can be negatively regulated by *SPX* proteins in Pi-dependent manner (Wang et al., 2014; Zhao et al., 2019; Zhang et al., 2021). A number of gene families such as *KUP/HAK/KT, HKT, NHX*, and *CHX* constitute K^+ transporters in the plant, such as *LKT1* in tomato, and *SKT1* in potato (Ragel et al., 2019). Likewise, sulfate transporters (SULTRs) in root cells mediate sulfate (NO_4^{2-}) as well as selenate uptake from soil (Hu et al., 2022).

Plants have the inherent capacity to sequester excess metals in various above-ground parts, such as trichomes and within the cellular structure, such as vacuoles of leaf epidermal and mesophyll cells. Plants have adaptive mechanisms such as chelation, excretion, and subcellular compartmentalization to combat metal toxicity (Mehes-Smith et al., 2013). Baker 1981 classified plants are colonizing metalliferous soil into three categories based on their physiological strategies for minimizing the toxic effects of metals—excluders, indicators, and accumulators. Excluder restricts the metal uptake and their translocation into shoots. Indicators have the ability to regulate the uptake and transport of metal ions into the shoot so that internal concentration reflects soil metal levels. Metal accumulators/hyperaccumulators can transport and accumulate metals in their above-ground parts at concentration levels higher than the soil. Heavy metals freely enter plant roots and passively diffuse in the apoplast through the cell wall. Thus, roots act as the first physical barrier for heavy metals. Negatively charged sites in the cell wall bind to positively charged heavy metal ions which change cell wall structure and damage cell membrane, particularly plasma membrane, endoplasmic reticulum, and cell organelles (Liu et al., 2014b; Gutsch et al., 2019; Berni et al., 2019). Cell-wall structural alterations generate a damage-associated molecular pattern (DAMP) type response which is perceived by receptors involved in sensing cell wall integrity. DAMPs initiate the downstream signal transduction via Ca^{2+} ions and may trigger genes involved in jasmonic acid synthesis (Berni et al., 2019). Few heavy metal ions can be trapped in the cell wall, and the remaining metallic ions move through the apoplast or symplast pathway. Plant cell to cell perception and downstream signaling are carried out by membrane-bound receptors/transporters system (also called transceptors) which is involved in metal homeostasis (Mendoza-Cózatl et al., 2019). Due to heavy metals' different structural and chemical characteristics, plants could not develop heavy metal-specific receptors. Thus, little is known about such heavy metal sensors and receptors in plasma membrane compared to existing knowledge of nutrient transceptors. In fact, toxic heavy metals show structural similarity with essential plant nutrients and thus, plant nutrient receptors/transporters could be potentially involved in sensing the heavy metals at the plasma membrane (Keyster et al., 2020).

Metal ion influx into the cytoplasm is performed either by plasma membrane-bound transporters or by transporters located on intracellular organellar membranes. Such transporter proteins have been identified in *Arabidopsis*, rice, maize, etc., and classified into proteins families viz., *Zinc-regulated transporter (ZIP)*, *Yellow Stripe 1-Like family (YSL)*, *Ctr/Copper Transporter (COPT)* family, and *Natural Resistance-Associated Macrophage Protein (NRAMP)* (Guerinot, 2000; Sancenón et al., 2003; Schaaf et al., 2004; Nevo and Nelson, 2006). Likewise, the efflux of metal ions from the cytoplasm to extracellular space or into organelles can be performed by metal transporters belonging to either *Heavy Metal ATPase* (*HMAs*) or *Cation Diffusion Facilitator (CDF)* family (Ueno et al., 2010; Gustin et al., 2011). Comprehensive information about nutrient specific and non-specific transceptors is now stated to be developed. *Arabidopsis* IRT1 transports Fe, Mn, and Zn and mediates Cd accumulation in the iron-deficient plant but cannot uptake Cu (Connolly et al., 2002; Lee et al., 2009). Oryza spp. employ as a combined strategy for iron uptake via precisely regulation of transporters like *NRAMP1, YSL15,* and *IRO2* as major strategies and iron transporters *OsIRT1* and *OsIRT2* as minor strategies under different situations (Wairich et al., 2019). The ZIP and IRT transporters carry Zn^{2+} ions into the cytoplasm (Krishna et al., 2017). *OsZIP1, OsZIP3, OsZIP4*, and *OsZIP5* are Zn transporters in rice induced under Zn deficiency. Different transporters can be upregulated at different organs under nutrient deficiency conditions. Under Zn deficiency, *OsIRT1, OsZIP5,* and *OsZIP4* are upregulated in shoots and *OsZIP4, OsZIP5,* and *OsZIP7* are upregulated in rice shoots (Mani and Sankaranarayanan 2018). Rice *OsZIP6* is reported to transport Mn^{2+}, Co^{2+}, and Cd^{2+} (Kavitha et al., 2015).

Heavy metals can interfere with the nutrient metal uptake by the plants. Küpper and Kochian 2010 reported that cadmium inhibited the uptake of Zn, Fe, and Mn either by competing for transporters or by interfering with transporter gene expression. In fact, transporters/transceptors may show an affinity for an array of metallic ions. NRAMP transporters can carry Fe, Mn, Co, Cd, Al, and Zn (Qin et al., 2017). HMA transporters have been grouped based on their metal specificity—Cu/Ag group and Zn/Co/Cd/Pb group (Mani and Sankaranarayanan, 2018). *OsMTP1* mainly transports Zn but can also transport Co, Fe, Cd, as well as Ni revealed by transgenic assays (Menguer et al., 2013). Conversely, deprivation of one nutrient ion can lead to lower expression of metal transporters and possibly decrease the absorption of other metal ions in the root, although present in abundance. Kabir et al. (2021) showed that Zn-deprivation caused a significant decline in Fe concentration along with low Zn in root/shoot of tomato plants due to low expression *of iron-regulated transporter 1, zinc transporter-like,* and *zinc transporter.* The upward movement of heavy metals in plants occurs in chelated form with ligands such as organic acids, amino acids, and thiols (Singh et al., 2016). For long-distance transport of metals, several small molecules can serve candidate ligands such as any organic acids (carboxylates)—citrate and malate, any amino acids (nicotinamine, histidine, cysteine, and high-affinity Fe^{3+} chelating compounds derived from nicotinamide called as phytosiderophores—mugineic and 2′-deoxymugineic acids) or any proteins/peptides (metallothioneins) (Álvarez-Fernández et al., 2014). Nicotianamine (NA) is the only known phloem metal transporter, and it facilitates the chelation, transport, and localization of Fe, Zn, and Mn cations from phloem to sink organs and also has a role in tolerance against Ni toxicity (Schuler et al., 2012; Nozoye, 2018). In turn, NA synthesis is dependent on *Nicotianamine synthase* (*NAS*) gene (Nozoye, 2018). Specific plasma membrane transporter proteins like Y*SL* proteins mainly expressed in the phloem cells transport Fe^{2+}–NA complex and Mn^{2+}–NA complex (Morrissey et al., 2009). Phytochelatins peptides chelate free toxic heavy metal ions such as Cd^{2+}, Hg^{2+}, Zn^{2+}, Pb^{2+}, and As^{3+} and control the toxic levels in plant fluids (Elizabeth et al., 2003; Tennstedt et al., 2009; Chia, 2021).

In addition to the storage role, vacuoles are also involved in metal transport. Vacuolar sequestration capacity (VSC) is modulated by cytosolic metal chelators and tonoplast-localized transporters. During mineral deficiency, nutrients released from vacuoles into the cytosol by reducing the VSC mediated via upregulation of NRAMPs. On the other hand, strong VSC is needed during excessive metal concentration, which is mediated via upregulation of *Metal Transporter (MTP), Zinc-Induced Facilitator (ZIF), Vacuolar Iron Transporter (VIT),* and *Cation Exchanger (CAX)* genes (Peng and Gong, 2014). For instance, a vacuolar membrane localized *OsHMA3* transports Cd into vacuoles in rice roots and controls Cd accumulation in the rice shoots and grains (Miyadate et al., 2011; Sasaki et al., 2014). This enhances Cd tolerance in rice. Likewise, rice *OsVIT1* and *OsVIT2* involve in Fe and Zn long-distance translocation between flag leaves and seeds. Rice *NRAT1 (NRAMP Aluminium Transporter 1)* performs efflux of Al into vacuoles of root cells (Li et al., 2014). CDF proteins involve in the sequestration of Zn, Fe, Co, Cd, and Mn into vacuole or efflux them outside of the cell (Gustin et al., 2011).

Melatonin-mediated upregulation of the *metallothionein 1* (*RsMT1)* gene expression promotes cadmium tolerance in radish plants (Xu et al., 2020). Metallothionein is a small protein that that can potentially chelate numerous heavy metals and those ions can be upregulatethed by the CRISPR/dCas9 system in vegetable crops (Roca Paixão et al., 2019). Zhang et al. 2020 identified five *CsNRAMPs* and four *ClNRAMPs* encoding genes during a genome-wide survey of cucumber (*Cucumis sativus*) and watermelon (*Citrullus lanatus*) genomes, respectively. Das et al. 2021 showed that Fe-toxicity tolerant cv. Ratan limits Fe-accumulation through the downregulation of iron-transporter genes (*SlIRT1, SlNRAMP1,* and *SlFRO1*). Based on combined transcriptome and proteome analyses by Li et al. (2021) concluded that ROS scavengers, HSPs, acetylases, and metal transporters are key players in Cd detoxification in a Cd-hyperaccumulating turnip landrace. Transporter function is genetic controlled by transcription factors. For instance, *Sulfur Limitation 1* (*SLIM1*) controls the transcription of *SULTR* genes in *Arabidopsis* (Maruyama-Nakashita et al., 2006). Likewise, tomato TF *SlNAC063* has a key role in aluminum toxicity tolerance (Jin et al., 2022). *SlNAC063* functions as the master regulator of Acyl Activating Enzyme encoding gene (*SlAAE3-2*) expression under Al stress in both a time-dependent and dose-dependent manner. Thus, key transporter genes and their modulators can be identified in vegetable crops. The CRISPR/Cas9 approach can prove to be a valuable tool for engineer crops tolerant of deficiency and toxicity stresses.

12.3.6 ENGINEERING HERBICIDE TOLERANCE

Weeds can inflict crop losses up to 20–60% (Elena et al., 2018). To assure optimum availability of nutrients and waterweed containment is necessary. Herbicides are effective in weed management, but several recommended herbicides can have a dangerous impact on the terrestrial and aquatic ecosystem. Among GM crop varieties approved for cultivation across the world, herbicide tolerance is the dominating trait (about 84%) and covers around 146.4 million ha (Schütte et al., 2017). There are several mechanisms of herbicide resistance, that is, changes at the site of enzyme molecule interacting with herbicides; chemical changes by conjugation or degradation in herbicide molecules; exclusion of interaction between herbicide molecule and target site target, either structural barriers (e.g., cuticular properties) or physiologically (e.g., transporters) and biochemical properties to avoid toxic metabolites produced upon chemical application such ROS scavengers (Sammons and Gaines, 2014). Crops plants employ one or a combination of these mechanisms for coping against herbicides. Thus, specific changes in the plant's enzymatic machinery can alter crop response to herbicides. In GM glyphosate-resistant crops, *Agrobacterium*-derived glyphosate insensitive *5-enolpyruvylshikimate-3-phosphate synthase (EPSPS)* enzyme, glyphosate degrading *glyphosate oxidoreductase (GOX)* enzyme, or glyphosate degrading *glyphosate acetyltransferase (GAT)* enzymes have been targeted (Schütte et al., 2017). These enzymes are potential targets for modifications via CRISPR/Cas9 system to achieve one to several herbicides (Wei et al., 2022). *Acetolactate synthase (ALS)* gene is the key enzyme for branched-chain amino acids biosynthesis has been base-edited via CRISPR/Cas9 to confer resistance against herbicide in tomato, potato, watermelon, and rice (Butler et al., 2015; Tian et al., 2018; Veillet et al., 2019; Wang et al., 2021).

12.4 OPPORTUNITIES AND CHALLENGES IN CRISPR/CAS9 APPLICATIONS

Potential applications of CRISPR/Cas9 in vegetable improvement are vast. Initial successes are now realized, including breeding for biotic and abiotic stress management, higher yield and productivity, improved nutritional quality, etc. The elevated mean temperature will affect the flowering and fruiting time and can shift the geographical trend of vegetable cultivation (Ayyogari et al., 2014). But complex molecular pathways underlying the control of flowering in response to elevated temperature are now uncovered (Strasser et al., 2009; Capovilla et al., 2015). A better understanding of flowering time regulation help in engineering signaling pathways for better flowering, fruiting, and yield under abiotic stress conditions. Crop wild relatives (CWRs) are valuable reservoirs for genetic diversity and are paramount in breeding vegetables against changing climate. Various traits/genes for biotic and abiotic stress tolerance have been transferred in cultivated species (Kaur et al., 2018; Dinkar et al., 2020). Breeding for only a few marketable traits has led to a narrow genetic base in the modern vegetable and other economically important crops. This unwilling ignorance of landraces and wild relatives, along with other factors such as unwanted traits, loss of habitat, deforestation, invasive species, global warming, etc., ultimately results in the loss of genetic diversity. CRISPR/Cas9-based genome manipulations can induce domestication of wild species or combine agronomically desirable traits from them. For example, Zsögön et al. (2018) performed *de novo* domestication of wild tomato, *Solanum pimpinellifolium,* by editing six loci related to fruit number, fruit size, fruit shape, lycopene content, and plant architecture simultaneously. Mutant lines were higher in fruit size by 3-fold, fruit number by 10-fold and fruit lycopene content by 5-fold compared with the widely cultivated *S. lycopersicum*. Likewise, Lemmon et al. 2018 modified domestication related traits in an orphan Solanaceae crop *Physalis pruinosa* (Groundcherry). Such forced domestication of crops will definitely widen crop diversity which could be useful for breeding climate-resilient varieties. A successful commercial achievement can be realized in potato, that is, reduced enzymatic browning of potato tubers; browning lead to loss of organoleptic properties and nutritional value. CRISPR/Cas9-induced mutations in four alleles of the *StPPO2* gene of tetraploid *Solanum tubersoum* cultivar Desiree led to reduced *polyphenol oxidases* (PPOs) activity by 69% in tubers and 73% reduction in enzymatic browning (González et al., 2020). Likewise, CRISPR-mediated disruption of cytochrome P450 genes, *CYP79D1,* and *CYP79D2,* involved in the first step in cyanogenic glucoside biosynthesis, eliminated cyanide in the leaves and roots of cassava (Gomez et al., 2021). Available genome sequence database and other genomic tools can be used to identify stress-responsive homologous genes in vegetable crops. Several genome-wide studies have identified numerous differentially expressed genes (DEGs) under biotic and abiotic stresses. These genes can serve as potential targets for gene editing to fine-tune the signaling pathways for the optimum response of vegetable crops for the realization of sustainable yield under unprecedented climatic conditions.

Several gaps need to be filled to use the full potential of CRISPR/Cas9-based genome editing in vegetables (Gan and Ling, 2022). *Firstly*, for precise editing, highly accurate gene sequence data are a prerequisite. For several vegetable crops, the lack of robust genome sequence and gene annotation data limits the CRISPR-target identification in a gene and limits genome editing efforts in those species. *Secondly*, there is a lack of standard transformation protocols for delivering CRISPR/Cas9 components in many vegetable crops. Thorough research is needed for an effective *Agrobacterium*-mediated delivery process and regeneration to enhance transformation efficiency in plants. *Thirdly,* there are lack of proper governmental laws and regulations regarding the CRISPR-edited crop plants' research and use. Most countries have a common set of rules for transgenics and genome-edited crops, which need to be amended to keep pace with modern demands. *Fourthly*, CRISPR can have off-target effects, and CRISPR nuclease activity may persist in subsequent generations, which can induce unanticipated mutations in an otherwise genetically stable line (Ahmad et al., 2020). Moreover, not all the time, CRISPR-edited crops fetch desired results, and we must be aware of the associated threats. Mehta et al. (2019) cautioned on the evolution of CRISPR-resistant pathogens while CRISPR/Cas9-mediated editing with a single sgRNA in cassava and how long CRISPR-derived products should be monitored prior to release.

KEYWORDS

- **genome editing**
- **vegetable crops**
- **CRISPR-Cas9**
- **biotechnological interventions**

REFERENCES

Acharya, T.; Fanzo, J.; Gustafson, D.; Ingram, J.; Schneeman, B.; Allen, L.; Boote, K.; Drewnowski, A.; Ewert, F.; Hall, S.; Hendley, P. *Assessing Sustainable Nutrition Security: The Role of Food Systems*; The International Life Sciences Institute, Research Foundation, Center for Integrated Modeling of Sustainable Agriculture and Nutrition: Washington, DC, 2014.

Ahmad, J.; Alam, D.; Haseen, M. S. Impact of Climate Change on Agriculture and Food Security in India. *Int. J. Agric. Environ. Biotechnol.* **2011,** *4*, 129–137.

Ahmad, J. U.; Goni, M. Heavy Metal Contamination in Water, Soil, and Vegetables of the Industrial Areas in Dhaka, Bangladesh. *Environ. Monitor. Assess.* **2010,** *166*, 347–357.

Ahmad, N.; Rahman, M. U.; Mukhtar, Z.; Zafar, Y.; Zhang, B. A Critical Look on CRISPR-Based Genome Editing in Plants. *J. Cell. Physiol.* **2020,** *235*, 666–682.

Ajeesh Krishna, T. P.; Ceasar, S. A.; Maharajan, T.; Ramakrishnan, M.; Duraipandiyan, V.; Al-Dhabi, N. A.; Ignacimuthu, S. Improving the Zinc-Use Efficiency in Plants: A Review. *SABRAO J. Breed. Genet.* **2017,** *49*, 211–230.

Ali, Z.; Abulfaraj, A.; Idris, A.; Ali, S.; Tashkandi, M.; Mahfouz, M. M. CRISPR/Cas9-Mediated Viral Interference in Plants. *Genome Biol.* **2015,** *16*, 1–11.

Álvarez-Fernández, A.; Díaz-Benito, P.; Abadía, A.; López-Millán, A. F.; Abadía, J. Metal Species Involved in Long Distance Metal Transport in Plants. *Front. Plant Sci.* **2014,** *5*, 105.

Al-Whaibi, M. H. Plant Heat-Shock Proteins: A Mini Review. *J. King Saud Univ.-Sci.* **2011,** *23*, 139–150.

Andrási, N.; Pettkó-Szandtner, A.; Szabados, L. Diversity of Plant Heat Shock Factors: Regulation, Interactions, and Functions. *J. Exp. Bot.* **2021,** *72*, 1558–1575.

Atkinson, H. J.; Urwin, P. E.; McPherson, M. J. Engineering Plants for Nematode Resistance. *Annu. Rev. Phytopathol.* **2003,** *41*, 615–639.

Ayyogari, K.; Sidhya, P.; Pandit, M. K. Impact of Climate Change on Vegetable Cultivation—A Review. *Int. J. Agric. Environ. Biotechnol.* **2014,** *7*, 145.

Badran, A. H.; Guzov, V. M.; Huai, Q.; Kemp, M. M.; Vishwanath, P.; Kain, W.; Nance, A. M.; Evdokimov, A.; Moshiri, F.; Turner, K. H.; Wang, P. Continuous Evolution of *Bacillus thuringiensis* Toxins Overcomes Insect Resistance. *Nature* **2016,** *533*, 58–63.

Bajguz, A.; Hayat, S. Effects of Brassinosteroids on the Plant Responses to Environmental Stresses. *Plant Physiol. Biochem.* **2009,** *47*, 1–8.

Bari, V. K.; Nassar, J. A.; Aly, R. CRISPR/Cas9 Mediated Mutagenesis of More Axillary Growth 1 in Tomato Confers Resistance to Root Parasitic Weed *Phelipanche aegyptiaca*. *Sci. Rep.* **2021,** *11*, 1–11.

Bari, V. K.; Nassar, J. A.; Kheredin, S. M.; Gal-On, A.; Ron, M.; Britt, A.; Steele, D.; Yoder, J.; Aly, R. CRISPR/Cas9-Mediated Mutagenesis of Carotenoid Cleavage Dioxygenase 8 in Tomato Provides Resistance Against the Parasitic Weed *Phelipanche aegyptiaca*. *Sci. Rep.* **2019,** *9*, 1–12.

Barnabás, B.; Jäger, K.; Fehér, A. The Effect of Drought and Heat Stress on Reproductive Processes in Cereals. *Plant Cell Environ.* **2008,** *31*, 11–38.

Baxter, A.; Mittler, R.; Suzuki, N. ROS as Key Players in Plant Stress Signalling. *J. Exp. Bot.* **2014,** *65*, 1229–1240.

Behera, T. K.; Krishna, R.; Ansari, W. A.; Aamir, M.; Kumar, P.; Kashyap, S. P.; Pandey, S.; Kole, C. Approaches Involved in the Vegetable Crops Salt Stress Tolerance Improvement: Present Status and Way Ahead. *Front. Plant Sci.* **2021,** *12*, 787292.

Bennett, RS, Chamberlin, K.; Morningweg, D.; Wang, N.; Sarkar, S.; Balota, M.; Burow, M.; Chagoya, J.; Pham, H. Response to Drought Stress in a Subset of the U. S. Peanut Mini-Core Evaluated in Three States. *Peanut Sci.* **2021**.

Berni, R.; Luyckx, M.; Xu, X.; Legay, S.; Sergeant, K.; Hausman, J. F.; Lutts, S.; Cai, G.; Guerriero, G. Reactive Oxygen Species and Heavy Metal Stress in Plants: Impact on the Cell Wall and Secondary Metabolism. *Environ. Exp. Bot.* 161: 98–106.

Bertier, L. D.; Ron, M.; Huo, H.; Bradford, K. J.; Britt, A. B.; Michelmore, R. W. High-Resolution Analysis of the Efficiency, Heritability, and Editing Outcomes of CRISPR/Cas9-Induced Modifications of *NCED4* in Lettuce (*Lactuca sativa*). G3: Genes Genomes Genet. **2018,** *8*, 1513–1521.

Bi, H. L.; Xu, J.; Tan, A. J.; Huang, Y. P. CRISPR/Cas9-Mediated Targeted Gene Mutagenesis in *Spodoptera litura. Insect Sci.* **2016,** *23*, 469–477.

Bisht, D. S.; Bhatia, V.; Bhattacharya, R. Improving Plant-Resistance to Insect-Pests and Pathogens: The New Opportunities Through Targeted Genome Editing. *Semin. Cell Dev. Biol.* **2019,** *96*, 65–76.

Boonekamp, P. M. Are Plant Diseases Too Much Ignored in the Climate Change Debate? *Eur. J. Plant Pathol.* **2012,** *133*, 291–294.

Bouzroud, S.; Gasparini, K.; Hu, G.; Barbosa, M. A. M.; Rosa, B. L.; Fahr, M.; Bendaou, N.; Bouzayen, M.; Zsögön, A.; Smouni, A.; Zouine, M. Down Regulation and Loss of *Auxin Response Factor 4* Function Using CRISPR/Cas9 Alters Plant Growth, Stomatal Function and Improves Tomato Tolerance to Salinity and Osmotic Stress. *Genes* **2020,** *11*, 272.

Butler, N. M.; Atkins, P. A.; Voytas, D. F.; Douches, D. S. Generation and Inheritance of Targeted Mutations in Potato (*Solanum tuberosum* L.) Using the CRISPR/Cas System. *PloS One*. **2015,** *10*: e0144591.

Capovilla, G.; Schmid, M.; Posé, D. Control of Flowering by Ambient Temperature. *J. Exp. Bot*. 66: 59–69.

Chandrasekaran, J.; Brumin, M.; Wolf, D.; Leibman, D.; Klap, C.; Pearlsman, M.; Sherman, A.; Arazi, T.; Gal-On, A. Development of Broad Virus Resistance in Non-Transgenic Cucumber Using CRISPR/Cas9 Technology. *Mol. Plant Pathol.* **2016,** *17*, 1140–1153.

Chen, K.; Gao, C. Targeted Genome Modification Technologies and Their Applications in Crop Improvements. *Plant Cell Rep.*. **2014,** *33*, 575–583.

Chia, J. Phytochelatin Synthase in Heavy Metal Detoxification and Xenobiotic Metabolism. In: *Biodegradation Technology of Organic and Inorganic Pollutants*; Mendes, P. K. F., de Sousa, R N., Mielke, K. C., Eds.; IntechOpen: London, 2021.

Chinnusamy, V.; Zhu, J. K.; Sunkar, R. Gene Regulation During Cold Stress Acclimation in Plants. In: *Plant Stress Tolerance: Methods in Molecular Biology (Methods and Protocols)*, Vol. 639; Humana Press, Springer, 2010; pp 39–55.

Cho, S.; Yu, S. I.; Park, J.; Mao, Y.; Zhu, J. K.; Yun, D. J.; Lee, B. H. Accession-Dependent *CBF* Gene Deletion by CRISPR/Cas System in *Arabidopsis*. *Front. Plant Sci.* **2017,** *8*, 1910.

Christian, M.; Cermak, T.; Doyle, E. L.; Schmidt, C.; Zhang, F.; Hummel, A.; Bogdanove, A. J.; Voytas, D. F. Targeting DNA Double-Strand Breaks with TAL Effector Nucleases. *Genetics* **2010,** *186*, 757–761.

Clouse, S. D. Brassinosteroid Signal Transduction: From Receptor Kinase Activation to Transcriptional Networks Regulating Plant Development. *Plant Cell.* **2011,** *23*, 1219–1230.

Cook, R.; K. Evans. Resistance and Tolerance. In: *Principles and Practice of Nematode Control in Crops*; Brown, R. H., Kerry, B. R., Eds.; Academic Press: Marrickville, 1987; pp 179–231.

Corte, E. D.; Mahmoud, L.; Moraes, T.; Mou, Z.; Grosser, J.; Dutt, M. Development of Improved Fruit, Vegetable, and Ornamental Crops Using the CRISPR/Cas9 Genome Editing Technique. *Plants* **2019,** *8*, 601.

Crosatti, C.; Rizza, F.; Badeck, F. W.; Mazzucotelli, E.; Cattivelli, L. Harden the Chloroplast to Protect the Plant. *Physiol. Plantarum* **2013,** *147*, 55–63.

Das, U.; Islam, M. R.; Akther, M. S.; Prity, S. A.; Parvez, M. S.; Kabir, A. H. The Downregulation of Fe-Acquisition Genes in the Plasma Membrane Along with Antioxidant Defence and Nitric Oxide Signalling Confers Fe Toxicity Tolerance in Tomato. *Sci. Horticult.* **2021,** *279*, 109897.

de Toledo Thomazella, D. P.; Brail, Q.; Dahlbeck, D.; Staskawicz, B. CRISPR-Cas9 Mediated Mutagenesis of a *DMR6* Ortholog in Tomato Confers Broad-Spectrum Disease Resistance. *bioRxiv* **2016,** 064824.

Decraemer, W.; Hunt, D. J. Structure and Classification. In: *Plant Nematology*; Perry, R. N., Moens, N., Eds.; CABI Publishing: Wallingford, 2006; pp 3–32.

Ding, H.; He, J.; Wu, Y.; Wu, X.; Ge, C.; Wang, Y.; Zhong, S.; Peiter, E.; Liang, J.; Xu, W. The Tomato Mitogen-Activated Protein Kinase *SlMPK1* Is as a Negative Regulator of the High-Temperature Stress Response. *Plant Physiol.* **2018,** *177*, 633–651.

Ding, Y.; Li, H.; Zhang, X.; Xie, Q.; Gong, Z.; Yang, S.; *OST1* Kinase Modulates Freezing Tolerance by Enhancing *ICE1* Stability in *Arabidopsis*. *Dev. Cell.* **2015,** *32*, 278–289.

Dinkar, V.; Jha, S. K.; Mallick, N.; Niranjana, M.; Agarwal, P.; Sharma, J. B. Molecular Mapping of a New Recessive Wheat Leaf Rust Resistance Gene Originating from *Triticum spelta*. *Sci. Rep.* **2020,** *10*, 1–9.

Dong, J.; Zielinski, R. E.; Hudson, M. E. t-SNAREs Bind the *Rhg1 alpha-SNAP* and Mediate Soybean Cyst Nematode Resistance. *Plant J.* **2020,** *104*, 318–331.

Drerup, M. M.; Schlücking, K.; Hashimoto, K.; Manishankar, P.; Steinhorst, L.; Kuchitsu, K.; Kudla, J. The Calcineurin B-Like Calcium Sensors *CBL1* and *CBL9* Together with Their Interacting Protein Kinase *CIPK26* Regulate the *Arabidopsis* NADPH Oxidase *RBOHF*. *Mol. Plant.* **2013,** *6*, 559–569.

Eckstein, D.; Künzel, V.; Schäfer, L. *Global Climate Risk Index 2021*; Germanwatch e.V.: Bonn, 2021. https://www.germanwatch.org/en/19777

Elena, G. M.; Ramona, B. E.; Holban, A. M. Approved Genetically Engineered Foods: Types, Properties, and Economic Concerns. In: *Genetically Engineered Foods*; Holban, A. M., Grumezescu, A. M., Eds.; Academic Press: Elsevier, 2018; pp 85–107.

Elling, A. A. Major Emerging Problems with Minor *Meloidogyne* Species. *Phytopathology* **2013,** *103*, 1092–1102.

El-Sappah, A. H.; Elrys, A. S.; Desoky, E. S. M.; Zhao, X.; Bingwen, W.; El-Sappah, H. H.; Zhu, Y.; Zhou, W.; Zhao, X.; Li, J. Comprehensive Genome Wide Identification and Expression Analysis of MTP Gene Family in Tomato (*Solanum lycopersicum*) Under Multiple Heavy Metal Stress. *Saudi J. Biol. Sci.* **2021,** *28*, 6946–6956.

Filiz, E.; Akbudak, M. A. *Ammonium transporter 1 (AMT1)* Gene Family in Tomato (*Solanum lycopersicum* L.): Bioinformatics, Physiological and Expression Analyses Under Drought and Salt Stresses. *Genomics* **2020,** *112*, 3773–3782.

Fourrier, N.; Bédard, J.; Lopez-Juez, E.; Barbrook, A.; Bowyer, J.; Jarvis, P.; Warren, G.; Thorlby, G. A Role for *Sensitive to Freezing2* in Protecting Chloroplasts Against Freeze-Induced Damage in *Arabidopsis*. *Plant J.* **2008,** *55*, 734–745.

Frank, G.; Pressman, E.; Ophir, R.; Althan, L.; Shaked, R.; Freedman, M.; Shen, S.; Firon, N. Transcriptional Profiling of Maturing Tomato (*Solanum lycopersicum* L.) Microspores

Reveals the Involvement of Heat Shock Proteins, ROS Scavengers, Hormones, and Sugars in the Heat Stress Response. *J. Exp. Bot.* **2009,** *60*, 3891–3908.

Gan, W. C.; Ling, A. P. CRISPR/Cas9 in Plant Biotechnology: Applications and Challenges. BioTechnologia. *J. Biotechnol. Computat. Biol. Bionanotechnol.* **2022,** *103*.

Gao, C. Genome Editing in Crops: From Bench to Field. *Natl. Sci. Rev.* **2015,** *2*, 13–15.

Geng, S.; Sohail, H.; Cao, H.; Sun, J.; Chen, Z.; Zhou, L.; Wang, W.; Ye, R.; Yang, L.; Bie, Z. An Efficient Root Transformation System for CRISPR/Cas9-Based Analyses of Shoot-Root Communication in Cucurbit Crops. *Horticult. Res.* **2022,** uhab082.

Gokcezade, J.; Sienski, G.; Duchek, P. Efficient CRISPR/Cas9 Plasmids for Rapid and Versatile Genome Editing in Drosophila. *G3: Genes, Genomes, Genet.* **2014,** *4*, 2279–2282.

Gomez, M. A.; Berkoff, K. C.; Gill, B. K.; Iavarone, A. T.; Lieberman, S. E.; Ma, J. M.; Schultink, A.; Wyman, S. K.; Chauhan, R. D.; Taylor, N. J.; Staskawicz, B. J. CRISPR-Cas9-Mediated Knockout of *CYP79D1* and *CYP79D2* in Cassava Attenuates Toxic Cyanogen Production. *bioRxiv* **2021,** 462827.

Gomez, M. A.; Lin, Z. D.; Moll, T.; Chauhan, R. D.; Hayden, L.; Renninger, K.; Beyene, G.; Taylor, N. J.; Carrington, J. C.; Staskawicz, B. J.; Bart, R. S. Simultaneous CRISPR/Cas9-Mediated Editing of Cassava *eIF4E* Isoforms *nCBP*-1 and *nCBP*-2 Reduces Cassava Brown Streak Disease Symptom Severity and Incidence. *Plant Biotechnol. J.* **2019,** *17*, 421–434.

Gong, C.; Pang, Q.; Li, Z.; Li, Z.; Chen, R.; Sun, G.; Sun, B. Genome-Wide Identification and Characterization of *HSF* and *HSP* Gene Families and Gene Expression Analysis Under Heat Stress in Eggplant (*Solanum melongema* L.). *Horticulturae* **2021,** *7*, 149.

González, M. N.; Massa, G. A.; Andersson, M.; Turesson, H.; Olsson, N.; Fält, A. S.; Storani, L.; Décima Oneto, C. A.; Hofvander, P.; Feingold, S. E. Reduced Enzymatic Browning in Potato Tubers by Specific Editing of a *Polyphenol Oxidase* Gene via Ribonucleoprotein Complexes Delivery of the CRISPR/Cas9 System. *Front. Plant Sci.* 10: 1649.

Guerinot, M. L. The *ZIP* Family of Metal Transporters. *Biochim et Biophys. Acta (BBA)-Biomembr.* **2000,** *1465*, 190–198.

Guo, M.; Liu, J. H.; Lu, J. P.; Zhai, Y. F.; Wang, H.; Gong, Z. H.; Wang, S. B.; Lu, M. H. Genome-Wide Analysis of the CaHsp20 Gene Family in Pepper: Comprehensive Sequence and Expression Profile Analysis Under Heat Stress. *Front. Plant Sci.* **2015,** 6: 806.

Guo, X.; Liu, D.; Chong, K. Cold Signalling in Plants: Insights into Mechanisms and regulation. *J. Integr. Plant Biol.* **2018,** *60*, 745–756.

Gustin, J. L.; Zanis, M. J.; Salt, D. E. Structure and Evolution of the Plant Cation Diffusion Facilitator Family of Ion Transporters. *BMC Evol. Biol.* **2011,** *11*, 1–13.

Gutsch, A.; Sergeant, K.; Keunen, E.; Prinsen, E.; Guerriero, G.; Renaut, J.; Hausman, J. F.; Cuypers, A. Does Long-Term Cadmium Exposure Influence the Composition of Pectic Polysaccharides in the Cell Wall of *Medicago sativa* Stems? *BMC Plant Biol.* **2019,** *19*, 1–17.

Hafeez, M.; Liu, S.; Jan, S.; Shi, L.; Fernández-Grandon, G. M.; Gulzar, A.; Ali, B.; Rehman, M.; Wang, M. Knock-Down of Gossypol-Inducing Cytochrome P450 Genes Reduced Deltamethrin Sensitivity in *Spodoptera exigua* (Hübner). *Int. J. Mol. Sci.* **2019,** *20*, 2248.

Hall, J. C.; Donnelly-Vanderloo, M. J.; Hume, D. J. Triazine-Resistant Crops: The Agronomic Impact and Physiological Consequences of Chloroplast Mutation. In: *Herbicide-Resistant Crops*; Duke, S. O., Ed.; CRC Press: Boca Raton, 2018; pp 107–126.

Harris, M. O.; Stuart, J. J.; Mohan, M.; Nair, S.; Lamb, R. J.; Rohfritsch, O. Grasses and Gall Midges: Plant Defence and Insect Adaptation. *Annu. Rev. Entomol.* **2003,** *48*, 549–577.

Hasanuzzaman, M.; Nahar, K.; Alam, M.; Roychowdhury, R.; Fujita, M. Physiological, Biochemical, and Molecular Mechanisms of Heat Stress Tolerance in Plants. *Int. J. Mol. Sci.* **2013,** *14*, 9643–9684.

Hayes, S.; Schachtschabel, J.; Mishkind, M.; Munnik, T.; Arisz, S. A. Hot Topic: Thermosensing in Plants. *Plant Cell Environ.* **2021,** *44*, 2018–2033.

Hegenauer, V.; Fürst, U.; Kaiser, B.; Smoker, M.; Zipfel, C.; Felix, G.; Stahl, M.; Albert, M.; Detection of the Plant Parasite *Cuscuta reflexa* by a Tomato Cell Surface Receptor. *Science* **2016,** *353* (6298), 478–481.

Herring, D. Climate Change: Global Temperature Projections. Climate.Gov, Oct 2012. https://www.climate.gov/news-features/understanding-climate/climate-change-global-temperature-projections

Hillmann, K. Looking for Maize Genes Involved in Cold Response: Producing Knockouts for *Arabidopsis* Homologs of Maize Candidate Genes Using a CRISPR/Cas9 Approach. Departmental Honors Projects. Minnesota, 2019. https://digitalcommons.hamline.edu/dhp/82

Hindarwati, Y.; Soeprobowati, T. R. Heavy Metal Content in Terraced Rice Fields at Sruwen Tengaran Semarang-Indonesia. *E3S Web Conf., EDP Sci.* **2018,** 31. https://doi.org/10.1051/e3sconf/20183103009

Hogenhout, S. A.; Bos, J. I. Effector Proteins That Modulate Plant-Insect Interactions. *Curr. Opin. Plant Biol.* **2011,** *14*, 422–428.

Hou, Q.; Ufer, G.; Bartels, D. Lipid Signalling in Plant Responses to Abiotic Stress. *Plant Cell Environ.* **2016,** *39*, 1029–1048.

Hu, J.; Wang, Z.; Zhang, L.; Peng, J.; Huang, T.; Yang, X.; Jeong, B. R.; Yang, Q. Seleno-Amino Acids in Vegetables: A review of Their Forms and Metabolism. *Front. Plant Sci.* **2022,** *13*, 804368.

Hu, Y.; Zhang, T.; Liu, Y.; Li, Y.; Wang, M.; Zhu, B.; Liao, D.; Yun, T.; Huang, W.; Zhang, W.; Zhou, Y. Pumpkin (*Cucurbita moschata*) *HSP20* Gene Family Identification and Expression Under Heat Stress. *Front. Genet.* **2021,** *12*, 753953.

Huang, S. Global Trade Patterns in Fruits and Vegetables; USDA-ERS Agriculture and Trade Report No. WRS-04–06, 2004.

Huang, Y.; Cao, H.; Yang, L.; Chen, C.; Shabala, L.; Xiong, M.; Niu, M.; Liu, J.; Zheng, Z.; Zhou, L.; Peng, Z. Tissue-Specific Respiratory Burst Oxidase Homolog-Dependent H2O2 Signaling to the Plasma Membrane H+-ATPase Confers Potassium Uptake and Salinity Tolerance in Cucurbitaceae. *J. Exp. Bot.* **2019,** *70*, 5879–5893.

Huang, Y.; Chen, Y.; Zeng, B.; Wang, Y.; James, A. A.; Gurr, G. M.; Yang, G.; Lin, X.; Huang, Y.; You, M. CRISPR/Cas9 Mediated Knockout of the Abdominal—A homeotic gene in the Global Pest, Diamondback Moth (*Plutella xylostella*). *Insect Biochem. Mol. Biol.* **2016,** *75*, 98–106.

Huo, H.; Dahal, P.; Kunusoth, K.; McCallum, C. M.; Bradford, K. J. Expression of *9-cis-EPOXYCAROTENOID DIOXYGENASE4* Is Essential for Thermoinhibition of Lettuce Seed Germination But Not for Seed Development or Stress Tolerance. *Plant Cell.* **2013,** *25*, 884–900.

IPCC Climate Change *Impacts, Adaptation, and Vulnerability*; Cambridge University Press: Cambridge, 2001; 1032 pp.

IPCC Summary for Policymakers. In: Climate Change 2013: The Physical Science Basis. Contribution of Working Group I to the Fifth Assessment Report of the Intergovernmental

Panel on Climate Change; Cambridge University Press, Cambridge and New York, 2013; 222 pp.

Jia, Y.; Ding, Y.; Shi, Y.; Zhang, X.; Gong, Z.; Yang, S. The *cbfs* Triple Mutants Reveal the Essential Functions of *CBFs* in Cold Acclimation and Allow the Definition of *CBF* Regulons in *Arabidopsis*. *New Phytol.* **2016,** *212*, 345–353.

Jin, J. F.; Zhu, H. H.; He, Q. Y.; Li, P. F.; Fan, W.; Xu, J. M.; Yang, J. L.; Chen, W. W.; The Tomato Transcription Factor *SLNAC063* Is Required for Aluminum Tolerance by Regulating *SlAAE3–1* Expression. *Front. Plant Sci.* **2022,** *13*, 826954.

Jinek, M.; Chylinski, K.; Fonfara, I.; Hauer, M.; Doudna, J. A.; Charpentier, E. A Programmable Dual-RNA–Guided DNA Endonuclease in Adaptive Bacterial Immunity. *Science* **2012,** *337*, 816–821.

Jones, J. T.; Haegeman, A.; Danchin, E. G.; Gaur, H. S.; Helder, J.; Jones, M. G.; Kikuchi, T.; Manzanilla-López, R.; Palomares-Rius, J. E.; Wesemael, W. M.; Perry, R. N. Top 10 Plant-Parasitic Nematodes in Molecular Plant Pathology. *Mol. Plant Pathol.* **2013,** *14*, 946–961.

Kabir, A. H.; Akther, M. S.; Skalicky, M.; Das, U.; Gohari, G.; Brestic, M.; Hossain, M. Downregulation of Zn-Transporters Along with Fe and Redox Imbalance Causes Growth and Photosynthetic Disturbance in Zn-Deficient Tomato. *Sci. Rep.* **2021,** *11*, 1–12.

Kai, H.; Iba, K. Temperature Stress in Plants. In: *eLS*; John Wiley & Sons, Ltd: Chichester, 2014. https://doi.org/10.1002/9780470015902.a0001320.pub2

Kakeshpour, T.; Tamang, T. M.; Motolai, G.; Fleming, Z. W.; Park, J. E.; Wu, Q.; Park, S. CGFS-Type Glutaredoxin Mutations Reduce Tolerance to Multiple Abiotic Stresses in Tomato. *Physiol. Plantarum.* **2021,** *173*, 1263–1279.

Kang, J. Application of CRISPR/Cas9-Mediated Genome Editing for Studying Soybean Resistance to Soybean Cyst Nematode; Doctoral dissertation, University of Missouri, Columbia, USA, 2016.

Karkute, S. G.; Singh, A. K.; Gupta, O. P.; Singh, P. M.; Singh, B. CRISPR/Cas9 Mediated Genome Engineering for Improvement of Horticultural Crops. *Front. Plant Sci.* **2017,** *8*, 1635.

Kaur, V.; Shubha, K.; Kumar, P.; Manju. Role of Crop Wild Relatives in Crop Improvement Under Changing Climatic Conditions. In: *Crop Improvement for Sustainability*; Yadav, P. K.; Kumar, S.; Kumar, S.; Yadav, R. C., Eds.; Daya Publishing House: New Delhi, India, 2018; pp 13–35.

Kavitha, P. G.; Kuruvilla, S.; Mathew, M. K. Functional Characterization of a Transition Metal Ion Transporter, *OsZIP6* from Rice (*Oryza sativa* L.). *Plant Physiol. Biochem.* **2015,** *97*, 165–174.

Keyster, M.; Niekerk, L. A.; Basson, G.; Carelse, M.; Bakare, O.; Ludidi, N.; Klein, A.; Mekuto, L.; Gokul, A. Decoding Heavy Metal Stress Signalling in Plants: Towards Improved Food Security and Safety. *Plants* **2020,** *9*, 1781.

Khatodia, S.; Bhatotia, K.; Tuteja, N. Development of CRISPR/Cas9 Mediated Virus Resistance in Agriculturally Important Crops. *Bioengineered.* **2017,** *8*, 274–279.

Kim, J.; Jeon, J. Cold Signaling Via the Two-Component Signaling System. *Mol. Plant* **2013,** *6*, 15–17.

Kim, J. S.; Yamaguchi-Shinozaki, K.; Shinozaki, K. ER-Anchored Transcription Factors *bZIP17* and *bZIP28* Regulate Root Elongation. *Plant Physiol.* **2018,** *176*, 2221–2230.

Kim, T.; Samraj, S.; Jiménez, J.; Gómez, C.; Liu, T.; Begcy, K. Genome-Wide Identification of Heat Shock Factors and Heat Shock Proteins in Response to UV and High Intensity Light Stress in Lettuce. *BMC Plant Biology* **2021,** *21*, 1–20.

Kim, Y. G.; Cha, J.; Chandrasegaran, S. Hybrid Restriction Enzymes: Zinc Finger Fusions to *Fok I* Cleavage Domain. *Proc. Natl. Acad. Sci.* **1996,** *93*, 1156–1160.

Klap, C.; Yeshayahou, E.; Bolger, A. M.; Arazi, T.; Gupta, S. K.; Shabtai, S.; Usadel, B.; Salts, Y.; Barg, R. Tomato Facultative Parthenocarpy Results from *Sl AGAMOUS-LIKE 6* Loss of Function. *Plant Biotechnol. J.* **2017,** *15*, 634–647.

Koutroumpa, F. A.; Monsempes, C.; François, M. C.; de Cian, A.; Royer, C.; Concordet, J. P.; Jacquin-Joly, E. Heritable Genome Editing with CRISPR/Cas9 Induces Anosmia in a Crop Pest Moth. *Sci. Rep.* **2016,** *6*, 1–9.

Kumar, P.; Sharma, P. K. Soil Salinity and Food Security in India. *Front. Sustain. Food Syst.* **2020,** *4*, 533781.

Kumar, S.; Prasad, S.; Yadav, K. K.; Shrivastava, M.; Gupta, N.; Nagar, S.; Bach, Q. V.; Kamyab, H.; Khan, S. A.; Yadav, S.; Malav, L. C. Hazardous Heavy Metals Contamination of Vegetables and Food Chain: Role of Sustainable Remediation Approaches—A Review. *Environ. Res.* **2019,** *179*, 108792.

Küpper, H.; Kochian, L. V. Transcriptional Regulation of Metal Transport Genes and Mineral Nutrition During Acclimatization to Cadmium and Zinc in the Cd/Zn Hyperaccumulator, *Thlaspi caerulescens* (Ganges Population). *New Phytol.* **2010,** *185*, 114–129.

Lee, S.; An, G. Over-Expression of *OsIRT1* Leads to Increased Iron and Zinc Accumulations in Rice. *Plant Cell Environ.* **2009,** *32*, 408–416.

Lemmon, Z. H.; Reem, N. T.; Dalrymple, J.; Soyk, S.; Swartwood, K. E.; Rodriguez-Leal, D.; Van Eck, J.; Lippman, Z. B. Rapid Improvement of Domestication Traits in an Orphan Crop by Genome Editing. *Nat. Plants* **2018,** *4*, 766–770.

Lhamo, D.; Luan, S. Potential Networks of Nitrogen-Phosphorus-Potassium Channels and Transporters in *Arabidopsis* Roots at a Single Cell Resolution. *Front. Plant Sci.* **2021,** *12*, 1189.

Li, F.; Scott, M. J. CRISPR/Cas9-Mediated Mutagenesis of the White and Sex Lethal Loci in the Invasive Pest, *Drosophila suzukii. Biochem. Biophys. Res. Commun.* **2016,** *469*, 911–916.

Li, H.; Ding, Y.; Shi, Y.; Zhang, X.; Zhang, S.; Gong, Z.; Yang, S. *MPK3*- and *MPK6*-Mediated *ICE1* Phosphorylation Negatively Regulates *ICE1* Stability and Freezing Tolerance in *Arabidopsis*. *Dev. Cell* **2017,** *43*, 630–642.

Li, J-Y.; Liu, J.; Dong, D.; Jia, X.; McCouch, S. R.; Kochian, L. V. Natural Variation Underlies Alterations in *Nramp Aluminum Transporter* (*NRAT1*) Expression and Function That Play a Key Role in Rice Aluminum Tolerance. *Proc. Natl. Acad. Sci.* **2014,** *111*, 6503–6508.

Li, R.; Liu, C.; Zhao, R.; Wang, L.; Chen, L.; Yu, W.; Zhang, S.; Sheng, J.; Shen, L. CRISPR/Cas9-Mediated *SlNPR1* Mutagenesis Reduces Tomato Plant Drought Tolerance. *BMC Plant Biology*. **2019,** *19*, 1–13.

Li, W.; Li, M.; Zhang, W.; Welti, R.; Wang, X. The Plasma Membrane–Bound *phospholipase Dδ* Enhances Freezing Tolerance in *Arabidopsis thaliana. Nat. Biotechnol.* **2004,** *22*, 427–433.

Li, X.; Chen, D.; Li, B.; Yang, Y.; Yang, Y. Combined Transcriptomic, Proteomic and Biochemical Approaches to Identify the Cadmium Hyper-Tolerance Mechanism of Turnip Seedling Leaves. *Environ. Sci. Pollut. Res.* **2021,** *28*, 22458–22473.

Li, Y.; Li, J.; Yan, Y.; Liu, W.; Zhang, W.; Gao, L.; Tian, Y. Knock-Down of *CsNRT2.1,* a Cucumber Nitrate Transporter, Reduces Nitrate Uptake, Root Length, and Lateral Root Number at Low External Nitrate Concentration. *Front. Plant Sci.* **2018,** *9*, 722.

Liu, C. X.; Yang, T.; Zhou, H.; Ahammed, G. J.; Qi, Z. Y.; Zhou, J. The E3 Ubiquitin Ligase Gene *Sl1* Is Critical for Cadmium Tolerance in *Solanum lycopersicum* L. *Antioxidants* **2022,** *11*, 456.

Liu, L.; Zhang, J.; Xu, J.; Li, Y.; Guo, L.; Wang, Z.; Zhang, X.; Zhao, B.; Guo, Y. D.; Zhang, N. CRISPR/Cas9 Targeted Mutagenesis of *SlLBD40*, a Lateral Organ Boundaries Domain Transcription Factor, Enhances Drought Tolerance in Tomato. *Plant Sci.* **2020,** *301*, 110683.

Liu, T.; Dai, W.; Sun, F.; Yang, X.; Xiong, A.; Hou, X. Cloning and Characterization of the Nitrate Transporter Gene *BraNRT2.1* in Non-Heading Chinese Cabbage. *Acta Physiol. Plantarum* **2014a,** *36*, 815–823.

Liu, T.; Shen, C.; Wang, Y.; Huang, C.; Shi, J. New Insights into Regulation of Proteome and Polysaccharide in Cell Wall of *Elsholtzia splendens* in Response to Copper Stress. *PLoS One*. **2014b,** *9*, e109573.

Lohani, N.; Golicz, A. A.; Singh, M. B.; Bhalla, P. L. Genome-Wide Analysis of the *Hsf* Gene Family in *Brassica oleracea* and a Comparative Analysis of the *Hsf* Gene Family in *B. oleracea, B. rapa* and *B. napus*. *Funct. Integr. Genom.* **2019,** *19*, 515–531.

Maier, E. A.; Matthews, R. D.; McDowell, J. A.; Walden, R. R.; Ahner, B. A. Environmental Cadmium Levels Increase Phytochelatin and Glutathione in Lettuce Grown in a Chelator-Buffered Nutrient Solution. *J. Environ. Qual.* **2003,** *32*, 1356–1364.

Makhotenko, A. V.; Khromov, A. V.; Snigir, E. A.; Makarova, S. S.; Makarov, V. V.; Suprunova, T. P.; Kalinina, N. O.; Taliansky, M. E. Functional Analysis of Coilin in Virus Resistance and Stress Tolerance of Potato *Solanum tuberosum* Using CRISPR-Cas9 Editing. *Doklady Biochem. Biophys.* **2019,** *484*, 88–91.

Maksimovic, I.; Ilin, Z. Effects of Salinity on Vegetable Growth and Nutrients Uptake. In: *Irrigation Systems and Practices in Challenging Environments*; T. S. Lee, T. S., Ed.; IntechOpen: London, 2012; pp 169–170.

Mani, A.; Sankaranarayanan, K. pp. 149–179. Heavy Metal and Mineral Element-Induced Abiotic Stress in Rice Plant. In: *Rice Crop— Current Developments*; Shah, F., Khan, Z. H., Iqbal, A., Eds.; IntechOpen: London, 2018; pp 149–179.

Maruyama-Nakashita, A.; Nakamura, Y.; Tohge, T.; Saito, K.; Takahashi, H. *Arabidopsis SLIM1* Is a Central Transcriptional Regulator of Plant Sulphur Response and Metabolism. *Plant Cell.* **2006,** *18*, 3235–3251.

Mazier, M.; Flamain, F.; Nicolaï, M.; Sarnette, V.; Caranta, C. Knock-Down of Both *eIF4E1* and *eIF4E2* Genes Confers Broad-Spectrum Resistance Against Potyviruses in Tomato. *PLoS One* **2011,** *6*, e29595.

McFarlane, G. R.; Whitelaw, C. B. A.; Lillico, S. G. CRISPR-Based Gene Drives for Pest Control. *Trends Biotechnol.* **2018,** *36*, 130–133.

Meehl, G. A.; Stocker, T. F.; Collins, W. D.; Friedlingstein, P.; Gaye, A. T.; Gregory, J. M.; Kitoh, A.; Knutti, R.; Murphy, J. M.; Noda, A.; Raper, S. C. B.; Watterson, I. G.; Weaver, A. J.; Zhao, Z. C. Global Climate Projections. In: *AR4 Climate Change 2007: The Physical Science Basis. Contribution of Working Group I to the Fourth Assessment Report of the Intergovernmental Panel on Climate Change*; Solomon, S.; Qin, D.; Manning, M.; Marquis, M.; Averyt, K.; Tignor, M. M. B.; Miller Jr., H. L.; Chen, Z.; Eds.; Cambridge University Press: Cambridge and New York, 2007; pp 747–845.

Mehes-Smith, M.; Nkongolo, K.; Cholewa, E. Coping Mechanisms of Plants to Metal Contaminated Soil. In: *Environmental Change and Sustainability*; Silvern, S.; Young, S., Eds.; IntechOpen: London, 2013; pp 53–90. https://doi.org/10.5772/55124

Mehta, D.; Stürchler, A.; Anjanappa, R. B.; Zaidi, S. S. E. A.; Hirsch-Hoffmann, M.; Gruissem, W.; Vanderschuren, H. Linking CRISPR-Cas9 Interference in Cassava to the Evolution of Editing-Resistant Geminiviruses. *Genome Biol.* **2019,** 20, pp 1–10.

Melotto, M.; Underwood, W.; Koczan, J.; Nomura, K.; He, S. Y. Plant Stomata Function in Innate Immunity Against Bacterial Invasion. *Cell.* **2006,** *126*, 969–980.

Mendoza-Cózatl, D. G.; Gokul, A.; Carelse, M. F.; Jobe, T. O.; Long, T. A.; Keyster, M. Keep Talking: Crosstalk Between Iron and Sulphur Networks Fine-Tunes Growth and Development to Promote Survival Under Iron Limitation. *J. Exp. Bot.* **2019,** *70*, 4197–4210.

Menguer, P. K.; Farthing, E.; Peaston, K. A.; Ricachenevsky, F. K.; Fett, J. P.; Williams, L. E. Functional Analysis of the Rice Vacuolar Zinc Transporter *OsMTP1*. *J. Exp. Bot.* **2013,** *64*, 2871–2883.

Merrill, R.; Gutie¥rrez, D.; Lewis, O.; Gutie¥rrez. J.; Diez, S.; Wilson, R. Combined Effects of Climate and Biotic Interactions on the Elevational Range of a Phytophagous Insect. *J. Anim. Ecol.* **2008,** *77*, 145–155.

Minaxi, R. P.; Acharya, K. O.; Nawale, S. Impact of Climate Change on Food Security. *Int. J. Agric. Environ. Biotechnol.* **2011,** *4*, 125–127.

Mittler, R.; Finka, A.; Goloubinoff, P. How Do Plants Feel the Heat? *Trends Biochem. Sci.* **2012,** *37*, 118–125.

Mittler, R.; Vanderauwera, S.; Gollery, M.; Van Breusegem, F. Reactive Oxygen Gene Network of Plants. *Trends Plant Sci.* **2004,** *9*, 490–498.

Miura, K.; Jin, J. B.; Lee, J.; Yoo, C. Y.; Stirm, V.; Miura, T.; Ashworth, E. N.; Bressan, R. A.; Yun, D. J.; Hasegawa, P. M. *SIZ1*-Mediated Sumoylation of *ICE1* Controls *CBF3/DREB1A* Expression and Freezing Tolerance in *Arabidopsis*. *Plant Cell.* **2007,** *19*, 1403–1414.

Miyadate, H.; Adachi, S.; Hiraizumi, A.; Tezuka, K.; Nakazawa, N.; Kawamoto, T.; Katou, K.; Kodama, I.; Sakurai, K.; Takahashi, H.; Satoh-Nagasawa, N. *OsHMA3*, a P1B-Type of ATPase Affects Root-to-Shoot Cadmium Translocation in Rice by Mediating Efflux into Vacuoles. *New Phytol.* **2011,** *189*, 190–199.

Morrissey, J.; Guerinot, M. L. Iron Uptake and Transport in Plants: The Good, the Bad, and the Ionome. *Chem. Rev.* **2009,** *109*, 4553–4567.

Murakami, Y.; Tsuyama, M.; Kobayashi, Y.; Kodama, H.; Iba, K. Trienoic Fatty Acids and Plant Tolerance of High Temperature. *Science* **2000,** *287*, 476–479.

Murata, N.; Los, D. A. Histidine Kinase *Hik33* Is an Important Participant in Cold-Signal Transduction in Cyanobacteria. *Physiol. Plantarum.* **2006,** *126*, 17–27.

Murata, N.; Ishizaki-Nishizawa, O.; Higashi, S.; Hayashi, H.; Tasaka, Y.; Nishida, I. Genetically Engineered Alteration in the Chilling Sensitivity of Plants. Nature **1992,** *356*, 710–713.

Nath, M.; Tuteja, N. NPKS Uptake, Sensing, and Signaling and miRNAs in Plant Nutrient Stress. *Protoplasma* **2016,** *253*, 767–786.

Nawaz, G.; Han, Y.; Usman, B.; Liu, F.; Qin, B.; Li, R.; Knockout of *OsPRP1*, a Gene Encoding Proline-Rich Protein, Confers Enhanced Cold Sensitivity in Rice (*Oryza sativa* L.) at the Seedling Stage. *3 Biotech* **2019,** *9*, 1–18.

Nekrasov, V.; Wang, C.; Win, J.; Lanz, C.; Weigel, D.; Kamoun, S. Rapid Generation of a Transgene-Free Powdery Mildew Resistant Tomato by Genome Deletion. *Sci. Rep.* **2017,** *7*, 1–6.

Nevo, Y.; Nelson, N.; The NRAMP Family of Metal-Ion Transporters. *Biochim. et Biophys Acta (BBA)—Mol. Cell Res.* **2006,** *1763*, 609–620.

Nie, W. F.; Wang, M. M.; Xia, X. J.; Zhou, Y. H.; Shi, K.; Chen, Z.; Yu, J. Q. Silencing of Tomato *RBOH1* and *MPK2* Abolishes Brassinosteroid-Induced H2O2 Generation and Stress Tolerance. *Plant Cell Environ.* **2013,** *36*, 789–803.

NOAA. National Centers for Environmental Information, State of the Climate: Global Climate Report for Annual **2020,** Jan 2021. https://www.ncdc.noaa.gov/sotc/global/202013

Nozoye, T. The *Nicotianamine Synthase* Gene Is a Useful Candidate for Improving the Nutritional Qualities and Fe-Deficiency Tolerance of Various Crops. *Front. Plant Sci.* **2018,** *9*, 340.

Nuruddin, M. M.; Madramootoo, C. A.; Dodds, G. T. Effects of Water Stress at Different Growth Stages on Greenhouse Tomato Yield and Quality. *HortScience* **2003,** *38*, 1389–1393.

Ortigosa, A.; Gimenez-Ibanez, S.; Leonhardt, N.; Solano, R. Design of a Bacterial Speck Resistant Tomato by CRISPR/Cas9-Mediated Editing of *SlJAZ 2. Plant Biotechnol. J.* **2019,** *17*, 665–673.

Pan, C.; Yang, D.; Zhao, X.; Liu, Y.; Li, M.; Ye, L.; Ali, M.; Yu, F.; Lamin-Samu, A. T.; Fei, Z.; Lu, G. *PIF4* Negatively Modulates Cold Tolerance in Tomato Anthers via Temperature-Dependent Regulation of Tapetal Cell Death. *Plant Cell.* **2021,** *33*, 2320–2339.

Parajuli, R.; Thoma, G.; Matlock, M. D. Environmental Sustainability of Fruit and Vegetable Production Supply Chains in the Face of Climate Change: A Review. *Sci. Total Environ.* **2019,** *650*, 2863–2879.

Park, S.; Lee, C. M.; Doherty, C. J.; Gilmour, S. J.; Kim, Y.; Thomashow, M. F. Regulation of the Arabidopsis CBF Regulon by a Complex Low-Temperature Regulatory Network. *Plant J.* **2015,** *82*, 193–207.

Parmesan, C. Influences of Species, Latitudes and Methodologies on Estimates of Phonological Response to Global Warming. *Glob. Change Biol.* **2007,** *13*, 1860–1872.

Peng, J.; Gong, J. Vacuolar Sequestration Capacity and Long-Distance Metal Transport in Plants. *Front. Plant Sci.* **2014,** *5*, 19.

Powell, G.; Pirone, T.; Hardie, J. Aphid Stylet Activities During Potyvirus Acquisition from Plants and an In Vitro System That Correlate with Subsequent Transmission. *Eur. J. Plant Pathol.* **1995,** *101*, 411–420.

Prasad, B. V. G.; Chakravorty, S. Effects of Climate Change on Vegetable Cultivation–A Review. *Nat. Environ. Pollut. Technol.* **2015,** *14*, 923–929.

Prasad, B. V. G.; Chakravorty, S.; Deb, P. Food-Nutritional and Aesthetic Security for Future India by Horticulture. *Int. J. Life Sci.* **2014,** *1*, 328–331.

Prykhozhij, S. V.; Rajan, V.; Gaston, D.; Berman, J. N. CRISPR Multitargeter: A Web Tool to Find Common and Unique CRISPR Single Guide RNA Targets in a Set of Similar Sequences. *PloS One* **2015,** *10*, e0119372.

Qi, C.; Lin, X.; Li, S.; Liu, L.; Wang, Z.; Li, Y.; Bai, R.; Xie, Q.; Zhang, N.; Ren, S.; Zhao, B.; *SoHSC70* Positively Regulates Thermotolerance by Alleviating Cell Membrane Damage, Reducing ROS Accumulation, and Improving Activities of Antioxidant Enzymes. *Plant Sci.* **2019,** *283*, 385–395.

Qin, L.; Han, P.; Chen, L.; Walk, T. C.; Li, Y.; Hu, X.; Xie, L.; Liao, H.; Liao, X. Genome-Wide Identification and Expression Analysis of NRAMP Family Genes in Soybean (*Glycine max* L.). *Front. Plant Sci.* **2017,** *8*, 1436.

Ragel, P.; Raddatz, N.; Leidi, E. O.; Quintero, F. J.; Pardo, J. M.; Regulation of K+ Nutrition in Plants. *Front. Plant Sci.* **2019,** *10*, 281.

Rakshit, A.; Sarkar, N. C.; Pathak, H.; Maiti, R. K.; Makar, A. K.; Singh, P. L. Agriculture: A Potential Source of Greenhouse Gases and Their Mitigation Strategies. *IOP Conf. Ser.: Earth Environ. Sci.* **2009,** *6*, 24.

Ratul, A. K.; Hassan, M.; Uddin, M. K.; Sultana, M. S.; Akbor, M. A.; Ahsan, M. A. Potential Health Risk of Heavy Metals Accumulation in Vegetables Irrigated with Polluted River Water. *Int. Food Res. J.* **2018,** *25*, 329–338.

Roberts, P. A. Current Status of the Availability, Development and Use of Host Plant Resistance to Nematodes. *J. Nematol.* **1992,** *24*, 213–227.

Roca Paixão, J. F.; Gillet, F. X.; Ribeiro, T. P.; Bournaud, C.; Lourenco-Tessutti, T.; Noriega, D. D.; Melo, B.; Almeida-Engler, J.; Grossi-de-Sa, M. F. Improved Drought Stress Tolerance in Arabidopsis by CRISPR/dCas9 Fusion with a Histone AcetylTransferase. *Sci. Rep.* **2019,** *9*, 8080.

Rodríguez-Hernández, A. M.; Gosalvez, B.; Sempere, R. N.; Burgos, L.; Aranda, M. A.; Truniger, V. Melon RNA Interference (RNAi) Lines Silenced for *Cm-eIF4E* Show Broad Virus Resistance. *Mol. Plant Pathol.* **2012,** *13*, 755–763.

Rohini, P.; Rajeevan, M.; Srivastava, A. K. On the Variability and Increasing Trends of Heat Waves Over India. *Sci. Rep.* **2016,** *6*, 1–9.

Ruffel, S. Nutrient-Related Long-Distance Signals: Common Players and Possible Cross-Talk. *Plant Cell Physiol.* **2018,** *59*, 1723–1732.

Sah, S. K.; Reddy, K. R.; Li, J. Abscisic Acid and Abiotic Stress Tolerance in Crop Plants. *Front. Plant Sci.* **2016,** *7*, 571.

Saidi, Y.; Finka, A.; Goloubinoff, P. Heat Perception and Signalling in Plants: A Tortuous Path to Thermotolerance. *New Phytol.* **2011,** *190*, 556–565.

Sajid, M.; Rashid, B.; Ali, Q.; Husnain, T. Mechanisms of Heat Sensing and Responses in Plants. It Is Not All About Ca2+ Ions. *Biol. Plantarum.* **2018,** *62*, 409–420.

Sammons, R. D.; Gaines, T. A. Glyphosate Resistance: State of Knowledge. *Pest Manag. Sci.* **2014,** *70*, 1367–1377.

Sancenón, V.; Puig, S.; Mira, H.; Thiele, D. J.; Peñarrubia, L. Identification of a Copper Transporter Family in *Arabidopsis thaliana*. *Plant Mol. Biol.* **2003,** *51*, 577–587.

Sarkar, S. Development of High-Throughput Phenotyping Methods and Evaluation of Morphological and Physiological Characteristics of Peanut in a Sub-Humid Environment. Doctoral Dissertation, Virginia Polytechnic Institute and State University, USA, 2020.

Sasaki, A.; Yamaji, N.; Ma, J. F. Overexpression of *OsHMA3* Enhances Cd Tolerance and Expression of Zn Transporter Genes in Rice. *J. Exp. Bot.* **2014,** *65*, 6013–6021.

Sawut, R.; Kasim, N.; Maihemuti, B.; Hu, L.; Abliz, A.; Abdujappar, A.; Kurban, M. Pollution Characteristics and Health Risk Assessment of Heavy Metals in the Vegetable Bases of Northwest China. *Sci. Total Environ.* **2018,** *642*, 864–878.

Schaaf, G.; Ludewig, U.; Erenoglu, B. E.; Mori, S.; Kitahara, T.; Von Wirén, N. *ZmYS1* Functions as a Proton-Coupled Symporter for Phytosiderophore- and Nicotianamine-Chelated Metals. *J. Biol. Chem.* **2004,** *279*, 9091–9096.

Schachtman, D. P.; Shin, R. Nutrient Sensing and Signaling: NPKS. *Annu. Rev. Plant Biol.* **2007,** *58*, 47–69.

Schuler, M.; Rellán-Álvarez, R.; Fink-Straube, C.; Abadía, J.; Bauer, P. Nicotianamine Functions in the Phloem-Based Transport of Iron to Sink Organs, in Pollen Development and Pollen Tube Growth in *Arabidopsis*. *Plant Cell.* **2012,** *24*, 2380–2400.

Schütte, G.; Eckerstorfer, M.; Rastelli, V.; Reichenbecher, W.; Restrepo-Vassalli, S.; Ruohonen-Lehto, M.; Saucy, A. G. W.; Mertens, M. Herbicide Resistance and Biodiversity: Agronomic and Environmental Aspects of Genetically Modified Herbicide-Resistant Plants. *Environ. Sci. Eur.* **2017,** *29*, 1–12.

Schwartz, A. R.; Potnis, N.; Timilsina, S.; Wilson, M.; Patané, J.; Martins Jr, J.; Minsavage, G. V.; Dahlbeck, D.; Akhunova, A.; Almeida, N.; Vallad, G. E. Phylogenomics of *Xanthomonas* Field Strains Infecting Pepper and Tomato Reveals Diversity in Effector Repertoires and Identifies Determinants of Host Specificity. *Frontiers in Microbiol.* **2015,** *6*, 535.

Sharma, P.; Jha, A. B.; Dubey, R. S.; Pessarakli, M. Reactive Oxygen Species, Oxidative Damage, and Antioxidative Defence Mechanism in Plants Under Stressful Conditions. *J. Bot.* **2012,** 2012. https://doi.org/10.1155/2012/217037

Shen, C.; Que, Z.; Xia, Y.; Tang, N.; Li, D.; He, R.; Cao, M. Knock Out of the Annexin Gene *OsAnn3* via CRISPR/Cas9-Mediated Genome Editing Decreased Cold Tolerance in Rice. *J. Plant Biol.* **2017,** *60*, 539–547.

Shkryl, Y.; Yugay, Y.; Avramenko, T.; Grigorchuk, V.; Gorpenchenko, T.; Grischenko, O.; Bulgakov, V. CRISPR/Cas9-Mediated Knockout of *HOS1* Reveals Its Role in the Regulation of Secondary Metabolism in *Arabidopsis thaliana. Plants* **2021,** *10*, 104.

Shu, P.; Li, Y.; Li, Z.; Xiang, L.; Sheng, J.; Shen, L. Ferulic Acid Enhances Chilling Tolerance in Tomato Fruit by Up-Regulating the Gene Expression of *CBF* Transcriptional Pathway in MAPK3-Dependent Manner. *Postharv. Biol. Technol.* **2022,** *185*, 111775.

Sikora, R. A.; Desaeger, J.; Molendijk, L. *Integrated Nematode Management: State-of-the-Art and Visions for the Future*; CABI, 2021.

Singh, J. P.; Lal, S. S.; Govindakrishnan, P. M.; Dua, V. K.; Pandey, S. K. Climate Change and Potato Production in India. In: *Poster presented in Global Potato Conference* **2008**; New Delhi, 2008; pp 09–12.

Singh, S.; Parihar, P.; Singh, R.; Singh, V. P.; Prasad, S. M. Heavy Metal Tolerance in Plants: Role of Transcriptomics, Proteomics, Metabolomics, and Ionomics. *Front. Plant Sci.* **2016,** *6*, 1143.

Sivasakthi, S.; Usharani, G.; Saranraj, P. Biocontrol Potentiality of Plant Growth Promoting Bacteria (PGPR)-*Pseudomonas fluorescens* and *Bacillus subtilis*: A Review. *Afr. J. Agric. Res.* **2014,** *9*, 1265–1277.

Somers, J.; Nguyen, J.; Lumb, C.; Batterham, P.; Perry, T. In Vivo Functional Analysis of the *Drosophila melanogaster* Nicotinic Acetylcholine Receptor Dα6 Using the Insecticide Spinosad. *Insect Biochem. Mol. Biol.* **2015,** *64*, 116–127.

Song, X.; Liu, G.; Duan, W.; Liu, T.; Huang, Z.; Ren, J.; Li, Y.; Hou, X. Genome-Wide Identification, Classification and Expression Analysis of the Heat Shock Transcription Factor Family in Chinese Cabbage. *Mol. Genet. Genom.* **2014,** *289*, 541–551.

Song, Y.; Feng, L.; Alyafei, M. A. M.; Jaleel, A.; Ren, M. Function of Chloroplasts in Plant Stress Responses. *Int. J. Mol. Sci.* **2021,** *22*, 13464.

Strasser, B.; Alvarez, M. J.; Califano, A.; Cerdán, P. D. A Complementary Role for *ELF3* and *TFL1* in the Regulation of Flowering Time by Ambient Temperature. *Plant J.* **2009,** *58*, 629–640.

Sun, D.; Guo, Z.; Liu, Y.; Zhang, Y. Progress and Prospects of CRISPR/Cas Systems in Insects and Other Arthropods. *Front. Physiol.* **2017,** *8*, 608.

Sun, W.; Van Montagu, M.; Verbruggen, N. Small Heat Shock Proteins and Stress Tolerance in Plants. Biochimica et Biophysica Acta (BBA)-Gene Structure and Expression. 1577: 1–9.

Tennstedt, P.; Peisker, D.; Bottcher, C.; Trampczynska, A.; Clemens, S. Phytochelatin Synthesis Is Essential for the Detoxification of Excess Zinc and Contributes Significantly to the Accumulation of Zinc. *Plant Physiol.* **2009,** *149*, 938–948.

Tian, S.; Jiang, L.; Cui, X.; Zhang, J.; Guo, S.; Li, M.; Zhang, H.; Ren, Y.; Gong, G.; Zong, M.; Liu, F. Engineering Herbicide-Resistant Watermelon Variety Through CRISPR/Cas9-Mediated Base-Editing. *Plant Cell Rep.* **2018,** *37*, 1353–1356.

Tollefson, J. How Hot Will Earth Get by 2100? *Nature* **2020,** *580*, 443–446.

Tran, M. T.; Doan, D. T. H.; Kim, J.; Song, Y. J.; Sung, Y. W.; Das, S.; Kim, E. J.; Son, G. H.; Kim, S. H.; Van Vu, T.; Kim, J. Y. CRISPR/Cas9-Based Precise Excision of *SlHyPRP1* Domain(s) to Obtain Salt Stress-Tolerant Tomato. *Plant Cell Rep.* **2021,** *40*, 999–1011.

Tripp, J.; Mishra, S. K.; Scharf, K. D. Functional Dissection of the Cytosolic Chaperone Network in Tomato Mesophyll Protoplasts. *Plant Cell Environ.* **2009,** *32*, 123–133.

Tyagi, S.; Kesiraju, K.; Saakre, M.; Rathinam, M.; Raman, V.; Pattanayak, D.; Sreevathsa, R. Genome Editing for Resistance to Insect Pests: An Emerging Tool for Crop Improvement. *ACS Omega* **2020,** *5*, 20674–20683.

Ueno, D.; Yamaji, N.; Kono, I.; Huang, C. F.; Ando, T.; Yano, M.; Ma, J. F. Gene Limiting Cadmium Accumulation in Rice. *Proc. Natl. Acad. Sci.* **2010,** *107*, 16500–16505.

UNFCC. Fact Sheet: Climate Change Science-the Status of Climate Change Science Today, 2011. https://unfccc.int/

Veillet, F.; Perrot, L.; Chauvin, L.; Kermarrec, M. P.; Guyon-Debast, A.; Chauvin, J. E.; Nogué, F.; Mazier, M. Transgene-Free Genome Editing in Tomato and Potato Plants Using Agrobacterium-Mediated Delivery of a CRISPR/Cas9 Cytidine Base Editor. *Int. J. Mol. Sci.*. **2019,** *20*, 402.

Wahid, A.; Gelani, S.; Ashraf, M.; Foolad, M. R. Heat Tolerance in Plants: An Overview. *Environ. Exp. Bot.* **2007,** *61*, 199–223.

Wairich, A.; de Oliveira, B. H. N.; Arend, E. B.; Duarte, G. L.; Ponte, L. R.; Sperotto, R. A.; Ricachenevsky, F. K.; Fett, J. P. The Combined Strategy for Iron Uptake Is Not Exclusive to Domesticated Rice (*Oryza sativa*). *Sci. Rep.* **2019,** *9*, 1–17.

Walter, P.; Ron, D. The Unfolded Protein Response: From Stress Pathway to Homeostatic Regulation. *Science* **2011,** *334*, 1081–1086.

Wang, F.; Xu, Y.; Li, W.; Chen, Z.; Wang, J.; Fan, F.; Tao, Y.; Jiang, Y.; Zhu, Q. H.; Yang, J. Creating a Novel Herbicide-Tolerance *OsALS* Allele Using CRISPR/Cas9-Mediated Gene Editing. *Crop J.* **2021,** *9*, 305–312.

Wang, H.; Shi, Y.; Wang, L.; Liu, S.; Wu, S.; Yang, Y.; Feyereisen, R.; Wu, Y. CYP6AE Gene Cluster Knockout in *Helicoverpa armigera* Reveals Role in Detoxification of Phytochemicals and Insecticides. *Nat. Commun.* **2018,** *9*, 1–8.

Wang, J.; Hu, H.; Wang, W.; Wei, Q.; Hu, T.; Bao, C. Genome-Wide Identification and Functional Characterization of the Heat Shock Factor Family in Eggplant (*Solanum melongena* L.) Under Abiotic Stress Conditions. Plants. **2020,** *9*, 915.

Wang, J.; Zhang, H.; Wang, H.; Zhao, S.; Zuo, Y.; Yang, Y.; Wu, Y. Functional Validation of Cadherin as a Receptor of Bt Toxin *Cry1Ac* in *Helicoverpa armigera* Utilizing the CRISPR/Cas9 System. *Insect Biochem. Mol. Biol.* **2016,** *76*, 11–17.

Wang, L.; Chen, L.; Li, R.; Zhao, R.; Yang, M.; Sheng, J.; Shen, L. Reduced Drought Tolerance by CRISPR/Cas9-Mediated *SlMAPK3* Mutagenesis in Tomato Plants. *J. Agric. Food Chem.* **2017,** *65*, 8674–8682.

Wang, M.; Ding, F.; Zhang, S. Mutation of *SlSBPase* Aggravates Chilling-Induced Oxidative Stress by Impairing Glutathione Biosynthesis and Suppressing Ascorbate-Glutathione Recycling in Tomato Plants. *Front. Plant Sci.* **2020,** *11*, 2135.

Wang, T.; Zhang, H. Y.; Zhu, H. L. CRISPR Technology Is Revolutionizing the Improvement of Tomato and Other Fruit Crops. *Horticult. Res.* **2019,** *6*, 77.

Wang, Z.; Hong, Y.; Li, Y.; Shi, H.; Yao, J.; Liu, X.; Wang, F.; Huang, S.; Zhu, G.; Zhu, J. K. Natural Variations in *SlSOS1* Contribute to the Loss of Salt Tolerance During Tomato Domestication. *Plant Biotechnol. J.* **2021,** *19*, 20.

Wang, Z.; Ruan, W.; Shi, J.; Zhang, L.; Xiang, D.; Yang, C.; Li, C.; Wu, Z.; Liu, Y.; Yu, Y.; Shou, H. Rice *SPX1* and *SPX2* Inhibit Phosphate Starvation Responses Through Interacting with *PHR2* in a Phosphate-Dependent Manner. *Proc. Natl. Acad. Sci.* **2014,** *111*, 14953–14958.

Waters, E. R.; Vierling, E. Plant Small Heat Shock Proteins–Evolutionary and Functional Diversity. *New Phytol.* **2020,** *227*, 24–37.

Waters, M. T.; Gutjahr, C.; Bennett, T.; Nelson, D. C. Strigolactone Signalling and Evolution. *Annu. Rev. Plant Biol.* **2017,** *68*, 291–322.

Wei, T.; Wen, X.; Niu, C.; An, S.; Wang, D.; Xi, Z.; Wang, N. N. Design of Acetohydroxyacid Synthase Herbicide-Resistant Germplasm Through MB-QSAR and CRISPR/Cas9-Mediated Base-Editing Approaches. *J. Agric. Food Chem.* **2022,** *70*, 2817–2824.

Woiwod, I. Detecting the Effects of Climate Change on Lepidoptera. *J. Insect Conserv.* **1997,** *1*, 149–158.

Wu, J.; Baldwin, I. T. New Insights Into Plant Responses to the Attack from Insect Herbivores. *Annu. Rev. Genet.* **2010,** *44*, 1–24.

Xia, X. J.; Wang, Y. J.; Zhou, Y. H.; Tao, Y.; Mao, W. H.; Shi, K.; Asami, T.; Chen, Z.; Yu, J. Q. Reactive Oxygen Species Are Involved in Brassinosteroid-Induced Stress Tolerance in Cucumber. *Plant Physiol.* **2009,** *150*, 801–814.

Xu, L.; Zhang, F.; Tang, M.; Wang, Y.; Dong, J.; Ying, J.; Chen, Y.; Hu, B.; Li, C.; Liu, L. Melatonin Confers Cadmium Tolerance by Modulating Critical Heavy Metal Chelators and Transporters in Radish Plants. *J. Pineal Res.* **2020,** *69*, e12659.

Yadav, A.; Singh, J.; Ranjan, K.; Kumar, P.; Khanna, S.; Gupta, M.; Kumar, V.; Wani, S. H.; Sirohi, A. Heat Shock Proteins. In: *Heat Stress Tolerance in Plants—Physiological, Molecular and Genetic Perspectives*; Wani, S. H., Kumar, V., Eds.; John Wiley & Sons Ltd: Oxford, 2020; pp 189–211.

Yang, J.; Ma, S.; Zhou, J.; Song, Y.; Li, F. Heavy Metal Contamination in Soils and Vegetables and Health Risk Assessment of Inhabitants in Daye, China. *J. Int. Med. Res.* **2018,** *46*, 3374–3387.

Yang, T.; Shad Ali, G.; Yang, L.; Du, L.; Reddy, A. S. N.; Poovaiah, B. W. Calcium/Calmodulin-Regulated Receptor-Like Kinase *CRLK1* Interacts with *MEKK1* in Plants. *Plant Signal. Behav.* **2010,** *5*, 991–994.

Yang, X.; Xia, X.; Zeng, Y.; Nong, B.; Zhang, Z.; Wu, Y.; Tian, Q.; Zeng, W.; Gao, J.; Zhou, W.; Liang, H. Genome-Wide Identification of the Peptide Transporter Family in Rice and Analysis of the PTR Expression Modulation in Two Near-Isogenic Lines with Different Nitrogen Use Efficiency. *BMC Plant Biol.* **2020,** *20*, 1–15.

Ye, J.; Wang, X.; Hu, T.; Zhang, F.; Wang, B.; Li, C.; Yang, T.; Li, H.; Lu, Y.; Giovannoni, J. J.; Zhang, Y. An InDel in the Promoter of *Al-Activated Malate Transporter9* Selected During Tomato Domestication Determines Fruit Malate Contents and Aluminium Tolerance. *Plant Cell* **2017,** *29*, 2249–2268.

Yin, Y.; Qin, K.; Song, X.; Zhang, Q.; Zhou, Y.; Xia, X.; Yu, J. *BZR1* Transcription Factor Regulates Heat Stress Tolerance Through FERONIA Receptor-Like Kinase-Mediated Reactive Oxygen Species Signalling in Tomato. *Plant Cell Physiol.* **2018,** *59*, 2239–2254.

Yordanov, I.; Velikova, V.; Tsonev, T. Plant Responses to Drought, Acclimation, and Stress Tolerance. *Photosynthetica* **2000,** *38*, 171–186.

Yu, W.; Wang, L.; Zhao, R.; Sheng, J.; Zhang, S.; Li, R.; Shen, L. Knockout of *SlMAPK3* Enhances Tolerance to Heat Stress Involving ROS Homeostasis in Tomato Plants. *BMC Plant Biol.* **2019,** *19*, 1–13.

Zahoor, M. K.; Ahmad, A.; Zahoor, M. A.; Majeed, H. N.; Zulhussnain, M.; Ranian, K. CRISPR/Cas-Based Insect Resistance in Crops. In: *CRISPR Crops*; Ahmad, A.; Khan, S. H.; Khan, Z.; Eds.; Springer: Singapore, 2021; pp 117–149.

Zaidi, S. S. E. A.; Mukhtar, M. S.; Mansoor, S. Genome Editing: Targeting Susceptibility Genes for Plant Disease Resistance. *Trends Biotechnol.* **2018,** *36*, 898–906.

Zaragoza, C. Weed Management in Vegetables. FAO Plant Production and Protection Paper (FAO), 2003. https://www.fao.org/3/y5031e/y5031e0b.htm

Zeng, Y.; Wen, J.; Zhao, W.; Wang, Q.; Huang, W. Rational Improvement of Rice Yield and Cold Tolerance by Editing the Three Genes *OsPIN5b*, *GS3*, and *OsMYB30* with the CRISPR–Cas9 System. *Front. Plant Sci.* **2020,** *10*, 1663.

Zhang, H.; Li, G.; Cao, N.; Yang, H.; Zhu, F. Genome-Wide Identification and Expression Analysis of *NRAMP* Transporter Genes in *Cucumis sativus* and *Citrullus lanatus*. *Can. J. Plant Sci.* **2020,** *101*, 377–392.

Zhang, M.; Liu, Q.; Yang, X.; Xu, J.; Liu, G.; Yao, X.; Ren, R.; Xu, J.; Lou, L. CRISPR/Cas9-Mediated Mutagenesis of *Clpsk1* in Watermelon to Confer Resistance to *Fusarium oxysporum* f. sp. *niveum*. *Plant Cell Rep.* **2020,** *39*, 589–595.

Zhang, Y.; Wang, Y.; Wang, E.; Wu, X.; Zheng, Q.; Han, Y.; Lin, W.; Liu, Z.; Lin, W. *SlPHL1*, a MYB-CC Transcription Factor Identified from Tomato, Positively Regulates the Phosphate Starvation Response. *Physiol. Plantarum.* **2021,** *173*, 1063–1077.

Zhao, C.; Lang, Z.; Zhu, J. K. Cold Responsive Gene Transcription Becomes More Complex. *Trends Plant Sci.* **2015,** *20*, 466–468.

Zhao, P.; Wang, D.; Wang, R.; Kong, N.; Zhang, C.; Yang, C.; Wu, W.; Ma, H.; Chen, Q. Genome-Wide Analysis of the Potato *Hsp20* Gene Family: Identification, Genomic Organization and Expression Profiles in Response to Heat Stress. *BMC Genom.* **2018,** *19*, 1–13.

Zhao, P.; You, Q.; Lei, M. A CRISPR/Cas9 Deletion into the Phosphate Transporter *SlPHO1;1* Reveals Its Role in Phosphate Nutrition of Tomato Seedlings. *Physiol. Plantarum* **2019,** *167*, 556–563.

Zhu, L.; Mon, H.; Xu, J.; Lee, J. M.; Kusakabe, T. CRISPR/Cas9-Mediated Knockout of Factors in Non-Homologous End Joining Pathway Enhances Gene Targeting in Silkworm Cells. *Sci. Rep.* **2015,** *5*, 1–13.

Zsögön, A.; Čermák, T.; Naves, E. R.; Notini, M. M.; Edel, K. H.; Weinl, S.; Freschi, L.; Voytas, D. F.; Kudla, J.; Peres, L. E. P. *De novo* Domestication of Wild Tomato Using Genome Editing. *Nat. Biotechnol.* **2018,** *36*, 1211–1216.

CHAPTER 13

Recent Advances in CRISPR-Cas for Climate-Resilient Horticulture in Fruits Crops

GEETA PRASAD[1], DIVYA CHAUHAN[2], HIMANSHU PANDEY[3], DEVENDRA SINGH[4*], VIVEK KUMAR DHIMAN[5], and VINAY KUMAR DHIMAN[6]

[1]*ICAR-National Bureau of Plant Genetic Resources, New Delhi, India*

[2]*Banasthali Vidyapeeth, Jaipur, Rajasthan, India*

[3]*Dr. YSP University of Horticulture and Forestry, Solan, Himachal Pradesh, India*

[4]*B.N. College of Engineering and Technology, Lucknow, Uttar Pradesh, India*

[5]*Himachal Pradesh University Shimla, Himachal Pradesh, India*

[6]*Dr. YSP University of Horticulture and Forestry, Solan, Himachal Pradesh, India*

ABSTRACT

The changing climatic conditions in today's era have caused adverse impacts on fruit production. Climate change is significantly affecting the yield of fruits resulting in low fruits production. These adverse conditions have pushed plant researchers to expand climate-resilient agricultural technology to resist the effects of climatic change in the field of agriculture to provide nutritional security at the global level. Recent advancements in genetic engineering-based methodology or molecular biology and breeding techniques were

Climate-Resilient Agriculture: A Molecular Perspective. Ashutosh Singh, Saurabh Pandey (Eds.)

quite useful in providing stress resistance to plants. Currently, advancement in the field of genome editing has shown a ray of new hope for functional genomics as well as crop improvement. The CRISPR/Cas9 system is often used for genome editing. CRISPR/Cas9 is a cost-effective technique as it entails highly efficient, simple cloning techniques, and targeted alteration at particular loci-of-interest. Previous studies on different crops, fruits, and vegetables have shown that genome editing through CRISPR/Cas9 has great potential for achieving global food security. Since these modified crops can resist a broad spectrum of abiotic and biotic stresses, thus the CRISPR/Cas9 system can decrease breeding time in fruit crops as compared to conventional breeding. This chapter reviews the research efforts aimed at improving effective and sustainable agricultural production through climate-resilient strategies that can evolve a need-based technology to fight the climatic challenges in agriculture.

13.1 INTRODUCTION

Climate change has become a major factor that affects the productivity of crops since it is enormously interconnected with global processes. Therefore, the horticulture sector can be affected by even a slight change in the climatic situation, leading to adversely decreasing the production rate. Previous studies using crop yield simulation and statistical models have also shown that climatic condition has affected the productivity of important fruit crops because crop growth potential is directly proportional to the amount of sunlight, temperature, and water (Ewert et al., 2015). Currently, the increased temperature has led to an alarming concentration of greenhouse gases for instance N_2O, CH_4, CO_2, and chlorofluorocarbons in the atmosphere due to rapid-scale industrialization. Hence, Climatic variations are now considered a major environmental threat and an economic challenge in today's world (Moretti et al., 2010).

Evidence reported that adverse effects of climatic conditions on horticultural production could be reduced via climate mitigation to overcome the increasing demands for food on a global scale. A comprehensive measurement of climate mitigation practice helps to better understand that horticultural productivity can be enhanced through changes in proper management practices. Further, the mitigation and adaptation process impart sustainable production of fruit crops and ensure food security. Therefore, different adaptation techniques to alleviate the harmful effects of changing climate integrated the crop rotation and the pattern of cropping to retain the fertility of soil in a sustainable way and also via develop practices in addition to feasible alterations in sowing time and germplasm selection. Furthermore, some of the mitigation strategies that have been incorporated include.

13.1.1 CARBON SEQUESTRATION

Mitigation of greenhouse gases (GHG) can be achieved through sequestration of atmospheric carbon in agricultural soils, which consequently increases soil organic carbon. This strategy helps in declining greenhouse gas emissions to the atmosphere (Lal et al., 1999).

13.1.2 RESTORATION OF DEGRADED SOIL AND CHANGING ARABLE LAND TO GRASSLAND OR FOREST

Restoring soil through grassland allows for co-development with renewable energy, such as solar power, and contributes to reducing GHG. The combination of plant diversity and soil restoration interacts with abiotic and biotic factors and helps maximize their positive impacts and minimize the negative impacts on GHG emissions of other GHGs, such as CH_4 and N_2O.

13.1.3 CROPPING SYSTEM

Rotation of crops also facilitates improvement in biomass production. Such as rotation of legumes after non-leguminous crops provides natural fertilization to soils and helps to significantly reduce the incorporation of chemical fertilizers. Further, it also assists in the reduction of fossil fuel consumption in manufacturing fertilizers (Zentner et al., 2004).

13.1.4 WATER MANAGEMENT

Due to climate change, there will be an increase in the demand for water supply for agricultural purposes in the future. So, improving irrigation access and dwindling water supplies have become important. This could be achieved by implying water conservation and water harvesting techniques by farmers.

13.2 TECHNOLOGY ADVANCEMENTS

Climatic variations are considered to be the major issue of the present time and require joint effort across the globe to find an effective solution. Thus, the application of technological innovations can help reduce climate change effects. For example, improved technology enables the use of energy more efficiently through equipment such as vehicles, machinery, and electric instruments in order to reduce the energy used and GHG emissions._Additional energy sources and substances that emit less GHG can be used, such as renewable energy sources or new low-nitrogen fertilizers;_using alternative products that have low GHG emission rates or facilitating larger system-wide alterations such as replacing automotive and air travel with teleconferencing and telecommuting. Therefore, some of the above advanced technological methods can provide measures for the prevention of ill effects of climate change in our environment.

Besides all these earlier, utilizing conventional approaches has been the only option for genetic improvement in various fruit crops. These crops were modified for different morphological and biochemical characteristics such as color, flavor, juiciness, gain resistance toward stress environment, and improved nutritional and production value. Unlike breeding in field crops, limitations to breeding in fruit crops include pro-longed juvenility, outcross reproduction, and huge genome landscapes. Hence, it requires more research to improve on such traits. With the advancement in genetic engineering, new breeding techniques in plants are being implemented. Studies have revealed that targeting DNA DSBs using homologous recombination can lead to transgene integration or genome editing (Gascuel et al., 2017). The most applied technologies for targeting DSBs include transcription activator and ZFNs (Kim et al., 1996, Christian et al., 2010). At the time of infection, arrays (foreign DNA and CRISPR repeats) transcribe themselves and process into siCRISPR RNAs (known as crRNAs) along with a trans-activating CRISPR RNA (also called as tracrRNA) to lead Cas9 nuclease to homologous sequences in the DNA named as protospacers. A PAM sequence of three nucleotides is necessary for cleavage such that upstream of PAM is responsible for targeting specific DNA. Lately, CRISPR-associated protein 9 nucleases (Cas9) have gained a wide platform playing a crucial role in bacterial and Achaea immunity systems (Bortesi and Fischer, 2015). CRISPR technology works by incorporating foreign DNA between repeats of the host bacterial genome. Therefore, this technology firstly requires a two-component system that includes a gRNA and a Cas9 nuclease, expressed in the organism target cell; secondly, 20bps of sgRNA (single guide RNA) that guides cas9 to the target DNA by DNA–RNA complementarity. This chapter deals with the CRISPR-Cas9 technique, which has displayed encouraging results in fruit crop development; one such example is the grapevine (Malnoy et al., 2016). In this recent study, the group of researchers targeted the gene MLO-7, whereas, in apple DIPM-1, DIPM-2, and DIPM-4, gene were found resistance against fire blight and powdery mildew disease. There are different applications of CRISPR in crops some are mentioned in Figure 13.1.

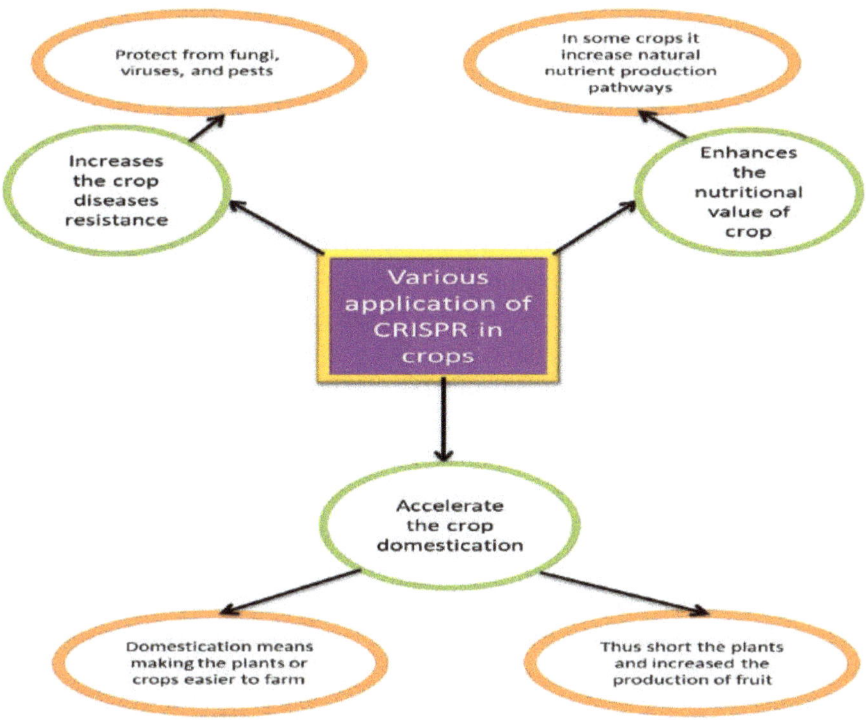

FIGURE 13.1 This chapter deals with studies on the advancement of CRISPR technology in enhancing crop quality and stress mitigation in fruit crops.

13.3 BIOTIC STRESS RESISTANCE USING CRISPR-CAS9 GENOME EDITING

There are two different methods that have been applied to provide resistance to plants against viruses: (1) creating sgRNAs which are useful for hampering the virus genome and (2) altering the genes in fruit crops for antiviral activity. The ability of viral DNA or RNA sequences associated with VPg protein in relation to plant protein eIF4E is an important factor in determining virus infection against the host plant. Any alteration or changes at the important position of eIF4E can have an impact on virus–host plant interaction and is useful in providing tolerance to plants against viral infection (Bastet et al., 2019). In the case of cucumbers, using CRISPR-Cas 9 N′ and C′ terminus of eIF4E were targeted, which generated transgenic free homozygous plants of cucumber in the T3 generation; these were able to exhibit resistance against cucumber vein yellow virus and pumpkin mosaic virus, and PRSV-W (Bastet et al., 2019; Chandrasekaran et al., 2016). CRISPR-Cas9 can alter the non-coding and coding sequences of Gemini virus, which is effective in decreasing its pathogenicity. Further in *Nicotiana sp*, sgRNA-Cas9 cassettes alter beet severe curly top virus, thereby hampering its growth in leaves (Ji et al., 2016). In the case of Gemini virus, sequence alteration in the non-coding region reduces its ability to multiply.

In comparison with mutations present in the coding region, non-coding-region variations produce less viral mutants (Ali et al., 2016). Fungi serve as a major disease-causing organism and threaten fruit yield and quality. For example, powdery and downy mildews cause great loss to tomato growers as it reduces the yield and quality of fruit (Borrelliet al., 2018). *A. thaliana* down mildew resistant 6 is an integral component of the 2-oxoglutarate oxygenase Iron (II)-dependent superfamily and is a part of salicylic acid (SA) homeostasis. Regulation of DMR6 in crops cans decreases vulnerability to downy mildew (Zeilmaker et al., 2015). The DMR6 generated by genome editing involving the CRISPR-Cas9 technique, which was useful in decreasing the expression of homologous genes in tomatoes, provided resistance to *P. syringae*, *Xanthomonas,* and *Phytophthora* (Sharma and Bhattarai 2019). Mildew-resistant locus 1 is responsible for the expression of membrane-linked protein and is sensitive loci for powdery mildew. In the case tomato plant, the Mlo1 variant resulted via gene alteration and showed resistance against *Oidium neolycopers*.

Furthermore, a variant free of mlo1 is generated by T-DNA selfingat T0 generation (Nekrasov et al., 2017). The fungi *Fusarium oxysporum* causes wilt disease in maximum fruit crops (Chaudhary and Atamian 2017). In the case of watermelons, the Clpsk1 gene was altered, which encodes for Phytosulfokine, providing resistance to *Fusarium oxysporum* (Wan et al., 2021). *B. cinerea* is a highly destructive pathogen that infects a variety of fruit crops and is responsible for causing maximum economic losses. Initially, its symptoms are not clear, and due to insufficient spray of fungicides make it difficult to control. The growth of pathogenic organisms can be suppressed by utilizing germplasm that receives heritable resistance from parents' generation. In the case of tomatoes, variation in MAPK3 generated by applying the CRISPR-Cas9 technique provided resistance against *B. cinerea* (Zhang et al., 2018). Leaf spot caused by microbial pathogen *P. syringae* severely affects the yield and deteriorates fruits' parameters. In the case of *A. thaliana*, CRISPR-Cas9 was applied to manipulate the C-terminus of the jasmonate domain of JAZ2, causing regulation of JAZ2 repressors; these repressors help in providing resistance against *P. syringae* infection (Ortigosa et al., 2019).

13.4 ABIOTIC STRESS RESISTANCE BY USING THE TECHNOLOGY CRISPR-CAS9 GENOME EDITING

Changes in climatic conditions resulted in the damaging impact of abiotic stress on crop production. Although conventional breeding approaches can provide stable crop production up to some extent, the utilization of new techniques to increase novel crop genotypes that are able to respond against abiotic stress is important for enhanced cultivation of modern crop varieties (Haque et al., 2018). Further, the development of CRISPR-Cas9 editing technology has decreased the time span for creating novel varieties. BZR1 participates in different BR-mediated developmental pathways. The CRISPR-based genome editing in the BZR1 gene allowed the generation of RBOH1 and the help in the synthesis of H_2O_2. Exogenous H_2O_2 restored the heat resistance mechanism in mutated tomato plants for the gene BZR1 (Yin et al., 2018). Furthermore, cold and drought-resistant genotypes can be produced by applying gene-editing techniques for that purpose, of CBF1, which is responsible for cold tolerance in fruit crops, and MAPK3, which is useful in providing protection to the cell membrane of the tomato plant against peroxidative damage in drought conditions (Li et al., 2018).

13.5 HERBICIDE RESISTANCE IMPROVEMENT USING CRISPR-CAS9 GENOME EDITING

Off-type plants known as weeds are major competitors for main fruit crops, which affect their quality and yield, and selective herbicides are generally applied to check the growth of weeds during the time of cultivation of fruit crops. For the production of herbicide-resistant fruits, CRISPR-Cas9 genome editing was applied for site-specific mutagenesis against herbicide-targeted gene ALS in watermelon, yielding a watermelon variety having resistance to herbicide- (Tianm et al., 2018). CBE was applied for cytidine modification of important ALS gene in tomato, resulting in alteration in amino acid sequence. About 71% of modified tomato plants showed tolerance to the pesticide chlorsulfuron (Butler et al., 2015). *Phelipanche aegyptiaca*, a noxious weedy plant parasite, thrives in the roots of the host, thereby releasing the plant SL for its germinating seeds; CRISPR-Cas9 technology was utilized to alter the CCD8 gene, a significantly important enzyme involved in the production of the carotenoid pathway that helps in accumulation of SLs in tomato, and MAX1 gene, which required for genesis of SLs, thereby considerably decreasing SL quantity & producing *P. aegyptiaca*-resistant plants of tomato (Bari et al., 2021).

13.6 FRUIT QUALITY IMPROVEMENT USING CRISPR-CAS9 GENOME EDITING

The most significant objective in the field of fruit breeding is to improve the shelf life of collected fruit. Quality assessment of fruits is based on external and internal features. External evaluation of fruit quality is based on trait-like fruit shape, texture, size, and color, which can be perceived by simply visualization. Internal attributes for good quality are recorded by using instruments that include contents of nutrients, basically vitamins, protein, fats, secondary metabolites, sugars, and bioactive chemicals synthesized by the plant-like lycopene, anthocyanins, and malate. In tomato, a maximum number of locules present in the ovary are responsible for approximately 50% of the genetic variations in fruit size and is governed by a large number of QTLs (Li et al., 2017). Scientists at CSH Laboratory have designed eight sgRNAs and applied CRISPR-Cas9 to alter the promoter sequence of the tomato stem cell gene CLV3 to produce fruits that are more vigorous and increased in number as compared to wild-tomato fruits (Li et al., 2018). Fruit texture and color are significant attributes for consumers. For instance, people of the European and American subcontinents choose red tomatoes, whereas in general Asian people like to consume pink tomatoes (Kang et al., 2017). CRISPR-Cas was utilized to modify PSY1, MYB12, and ANT2 to produce pink and purple tomatoes (Cermak et al., 2015).

13.7 FRUIT CROPS THAT ARE IMPROVED BY APPLYING CRISPR-CAS9

Major fruits crops that are considered for genome alteration using CRISPR-Cas9 for biotic and abiotic stress include apple, banana, citrus, and papaya, etc. Some of them are enlisted in Table 13.1.

13.7.1 APPLES

In 2020, Apple (Malus × domestic) had a worldwide production of 87 million tons and was the most produced temperate fruit crop. CRISPR-Cas9 methodology was effectively executed by introducing the PDS gene into the apple rootstocks, thereby increasing its efficiency by 31%. In the near future, genes will be delivered through CRISPR-Cas9 RNP plasmid or via complex. Hence, a study conducted by Chevreau et al. (1987) developed an adventitious and agro-bacterium-based CRISPR system in "Gala" apples. As a result, transformation efficiency was increased by 30%. Another study displayed targeting of two scorable genes viz.—PDS and TFL1, where early flowering in nearly 93% of the apples transgenic lines was observed.

The formation of chimeras is one of the major problems related to genetic engineering in fruit crops. In this study, the regeneration step was followed, which resulted in a maximum percentage of truncated alleles at RT0, where 67% of heterogenous editing was observed compared to T0 (Malabarba et al., 2021). One of the major threats to apples is *E. amylovora, which causes fire blight disease leading to a* decline in apple production. It works by binding of DspA/E effector to MdDIPM4. Hence, the knockdown via CRISPR/CAS 9 resulted in 75% efficiency in 57 genetically engineered lines. Also, seven lines on exposure to heat showed complete elimination of transgene and reduced susceptibility (Peil et al., 2021).

13.7.2 BANANA

Banana is considered to be an important food crop and has been commercialized in many countries. This fruit harbors breeding challenges traditionally, but due to advancements in gene editing (GE), CRISPR-Cas holds the solution to this problem. A study in bananas reported targeting two conserved domains (RAS-PDS) by sgRNA present in the embryonic cell suspension culture of cv. Rasathali. A range of variegated to albino phenotypes with a 59% mutation rate was observed in regenerated plantlets (Awasthi et al., 2022). Another study in banana cv. Cavendish displayed targeting of exon 1 and PDS genes by utilizing a polycistronic tRNA-gRNA system. The study further showed 19 regenerated plants with albino phenotypes with 100% mutation efficiency (Zhou et al., 2020). Hence, these studies depicted the possibility of GE through CRISPR-based targeting of the genome.

Disruption of the MaACO1 gene, which is responsible for ethylene generation and fruit ripening by CRISPR-Cas9 in the Brazilian cultivar of banana, has resulted in prolonged ripening in both fields and post-harvest storage (Liu et al., 2021). Also, the Cavendish cultivar, which is rich in β-carotene, was produced by targeting the exon of the lycopene epsilon-cyclase gene. Sequence investigation reported enhanced production of β-carotene using the system (CRISPR/Cas9) such that the production reached up to six times without displaying stunted agro-morphological traits (Kaur et al., 2020)

13.7.3 CITRUS

Citrus is rich in vitamin C and many other nutrients, and therefore, is one of the most important fruit crops across the world. However,, it is susceptible to both biotic as well as abiotic stresses. Conventional breeding techniques have proved to be tedious and time-consuming due to their characteristic of having prolonged juvenility and crossing life cycle, high heterozygosity, and juvenility. CRISPR-Cas9 was deployed in this crop to target the PDS gene in sweet orange leaves, which resulted in up to 4% of mutation efficiency (Huang et al., 2021). Moreover, a similar group of researchers edited the PDS gene by successfully implanting CRISPR-LbCas12a in citrus. Hence, modification of the CsLOB1 canker susceptibility gene present in grapefruit through CRISPR-Cas9 can be achieved at the mutation efficiency of up to 89%. The results showed six regenerated edited lines when inoculated with pathogen Xanthomonascitri ssp. Citri. Among them, two lines showed canker symptoms in wild grapefruit, whereas the other lines were asymptomatic but later showed minor symptoms at a later stage (Jia et al., 2016).

13.7.4 PAPAYA

Papaya belongs to the tropical fruit class and has commercial importance because of its high nutritional and medicinal value. The production of papaya almost reached 14 million metric tons in 2020 (Yeh 2020). The papaya protects itself from the invasion of *P. palmivora*, an oocyte, by secreting cysteine protease called papain. The homozygous mutant PpalEPIC8 generated through CRISPR/Cas9 played a key role in displaying virulence by combating papain. A study was conducted to demonstrate the role of pathogenicity and the progression of infection caused by P. palmivora. The results from the study depicted the presence of six Ppal15kDa mutants having homozygous mutation using CRISPR/Cas9, whereas heterozygous mutants showed different infection levels (Sattar et al., 2021).

13.7.5 PEAR

The annual production of both *Pyrus communis* and *Pyrus pyrifolia* is 13,919,075 tons under 1,379,075 ha of area (Sattar et al., 2021). Research conducted by Charrier et al. (2019) resulted in early flowering after knocking off the TFL1 gene in pears. The sequencing of nine transgenic clines revealed the positions of mutation at different frequencies. Also, the results showed the dominance of insertional mutations over substitutions and deletions. However, chimerism remains a problem in this process which can be further resolved by regenerating the leaves of transgenic plants at the T0 stage. Pang et al. (2019) reported enhanced yield in dwarf pear (*P. bretschneideri*) due to loss in function because of mutation in the PbPAT14 gene by the technology (CRISPR/Cas9).

TABLE 13.1 Fruit Crops That Are Edited Using the CRISPR-Cas9 Technique for a Particular Trait.

Fruits crops	Trait	Targeted gene	References
Grapevine	Resistance to pathogen	*MLO7*	Malnoy et al. (2016)
	Biosynthesis of Carotenoid	*IdnDH*	
	Biosynthesis of Carotenoid	*VvPDS*	
Watermelon	biosynthesis of Carotenoid	*ClPDS*	Kaur et al. (2021)
Cucumber	Resistance to viral infection	*eIF4E*	Shahriar et al. (2021)
Banana	Biosynthesis of Carotenoid	*PDS*	Tripathi et al. (2022)
Kiwifruit	Biosynthesis of Carotenoid	*AcPDS*	Savadi et al. (2021)
Sweet orange	Biosynthesis of Carotenoid	*CsPDS*	Arbatli et al. (2021); Huang et al. (2021);
	Resistance to pathogen	*CsLOB1*	Kaur et al. (2021)
Duncan grapefruit	Resistance to pathogen	*CsLOB1*	Jia et al. (2022)
Woodland strawberry	Auxin synthesis and signaling pathway	TAA1 ARF8	Kaur et al. (2020)
Cultivated strawberry	Floral development	*FaTM6*	Martín et al. (2019)
Apple	Biosynthesis of Carotenoid	*PDS* *DIPM1,2*and *4*	Sandhya et al. (2020)

13.8 CONCLUSIONS

The CRISPR/Cas9 technology is now being utilized for engineering fruit crops for different attributes. However, it is quite difficult to transform large genomes, which have slowed developmental cycles, and in the absence of genomic information, it becomes a more tedious task to utilize CRISPR/Cas9 for most fruit trees. The advancement in the field of CRISPR/Cas9 technique and its proper execution may be useful in developing novel horticultural crops with the desired trait.

KEYWORDS

- **CRISPR**
- **climate resilience**
- **genome editing**
- **crop improvement**
- **fruits**

REFERENCES

Ali, Z.; Ali, S.; Tashkandi, M.; Zaidi, S. S.; Mahfouz, M. M. CRISPR/Cas9-Mediated Immunity to Gemini Viruses: Differential Interference and Evasion. *Sci. Rep.* **2016,** *6*, 26912. https://doi.org/10.1038/srep26912

Arbatli, S.; Weiss, J.; Egea-Cortines, M. Gene and Genome Editing with CRISPR/Cas systems for Fruit and Vegetable Improvement. In *Concepts and Strategies in Plant Sciences*; Springer, 2021; pp 227–245. https://doi.org/10.1007/978-3-030-64994-4_11

Awasthi, P.; Khan, S.; Lakhani, H.; Chaturvedi, S.; Shivani, S.; Kaur, N.; Singh, J.; Kesarwani, A. K.; Tiwari, S. Transgene-Free Genome Editing Supports CCD4 Role as a Negative Regulator of β-Carotene in Banana. *J. Exp. Bot.* **2022.** https://doi.org/10.1093/jxb/erac042

Bari, V. K.; Nassar, J. A.; Aly, R. CRISPR/Cas9 Mediated Mutagenesis of More Axillary Growth 1 in Tomato Confers Resistance to Root Parasitic Weed Phelipancheaegyptiaca. *Sci. Rep.* **2021,** *11* (1), 3905. https://doi.org/10.1038/s41598-021-82897-8

Bastet, A.; Zafirov, D.; Giovinazzo, N.; Guyon-Debast, A.; Nogué, F.; Robaglia, C.; Gallois, J. L. Mimicking Natural Polymorphism in eIF 4E by CRISPR-Cas9 Base Editing Is Associated with Resistance to Potyviruses. *Plant Biotechnol. J.* **2019,** *17* (9), 1736–1750. https://doi.org/10.1111/pbi.13096

Bill, M.; Korsten, L.; Remize, F.; Glowacz, M.; Sivakumar, D. Effect of Thyme Oil Vapours on Phenylalanine Ammonia-Lyase (PAL) and Lipoxygenase (LOX) Genes Expression, and Control of Anthracnose in "Hass" and "Ryan" Avocado Fruit. *Sci. Hortic.* **2017,** *224*, 232–237.

Borrelli, V. M. G.; Brambilla, V.; Rogowsky, P.; Marocco, A.; Lanubile, A. The Enhancement of Plant Disease Resistance Using CRISPR/Cas9 Technology. *Front. Plant Sci.* **2018,** *9*, 1245. https://doi.org/10.3389/fpls.2018.01245

Bortesi, L.; Fischer, R. The CRISPR/Cas9 System for Plant Genome Editing and Beyond. *Biotechnol. Adv.* **2015,** *33* (1), 41–52. https://doi.org/10.1016/j.biotechadv.2014.12.006

Butler, N. M.; Atkins, P. A.; Voytas, D. F.; Douches, D. S. Generation and Inheritance of Targeted Mutations in Potato (*Solanumtuberosum* L.) Using the CRISPR/Cas System. *PLOS ONE* **2015,** *10* (12), e0144591. https://doi.org/10.1371/journal.pone.0144591

Cadle-Davidson, L. Monitoring Pathogenesis of Natural *Botrytis cinerea* Infections in Developing Grape Berries. *Am. J. Enol. Vitic.* **2008,** *59*, 387–395.

Čermák, T.; Baltes, N. J.; Čegan, R.; Zhang, Y.; Voytas, D. F. High-Frequency, Precise Modification of the Tomato Genome. *Genome Biol.* **2015,** *16* (1), 232. https://doi.org/10.1186/s13059-015-0796-9

Chaâbene, Z.; Rorat, A.; Rekik Hakim, I.; Bernard, F.; Douglas, G. C.; Elleuch, A.; Vandenbulcke, F.; Mejdoub, H. Insight into the Expression Variation of Metal-Responsive Genes in the Seedling of Date Palm (*Phoenix dactylifera*). *Chemosphere* **2018,** *197*, 123–134. https://doi.org/10.1016/j.chemosphere.2017.12.146

Chandrasekaran, J.; Brumin, M.; Wolf, D.; Leibman, D.; Klap, C.; Pearlsman, M.; Sherman, A.; Arazi, T.; Gal-On, A. Development of Broad Virus Resistance in Non-Transgenic Cucumber Using CRISPR/Cas9 Technology. *Mol. Plant Pathol.* **2016,** *17* (7), 1140–1153. https://doi.org/10.1111/mpp.12375

Charrier, A.; Vergne, E.; Dousset, N.; Richer, A.; Petiteau, A.; Chevreau, E. Efficient Targeted Mutagenesis in Apple and First Time Edition of Pear Using the CRISPR-Cas9 System. *Front. Plant Sci.* **2019,** *10*, 40. https://doi.org/10.3389/fpls.2019.00040

Chaudhary, R.; Atamian, H. S. Resistance-Gene-Mediated Defense Responses Against Biotic Stresses in the Crop Model Plant Tomato. *J. Plant Pathol. Microbiol.* **2017,** *8* (2).

Chen, Y.; Wilde, H. D. Mutation Scanning of Peach Floral Genes. *BMC Plant Biol.* **2011,** *11*, 96. https://doi.org/10.1186/1471-2229-11-96

Chevreau, E.; Laurens, F. The Pattern of Inheritance in Apple (Malus × domesticaBorkh.): Further Results from Leaf Isozyme Analysis. *Theor. Appl. Genet.* **1987,** *75* (1), 90–95. https://doi.org/10.1007/BF00249147

Christian, M.; Cermak, T.; Doyle, E. L.; Schmidt, C.; Zhang, F.; Hummel, A.; Bogdanove, A. J.; Voytas, D. F. Targeting DNA Double-Strand Breaks with TAL Effector Nucleases. *Genetics* **2010,** *186* (2), 757–761. https://doi.org/10.1534/genetics.110.120717

Ewert, F.; Rötter, R. P.; Bindi, M.; Webber, H.; Trnka, M.; Kersebaum, K. C.; Olesen, J. E.; van Ittersum, M. K.; Janssen, S.; Rivington, M.; Semenov, M. A.; Wallach, D.; Porter, J. R.; Stewart, D.; Verhagen, J.; Gaiser, T.; Palosuo, T.; Tao, F.; Nendel, C. et al. Crop modelling for Integrated Assessment of Risk to Food Production from Climate Change. *Environ. Modell. Softw.* **2015,** *72*, 287–303. https://doi.org/10.1016/j.envsoft.2014.12.003

Fister, A. S.; Landherr, L.; Maximova, S. N.; Guiltinan, M. J. Transient Expression of CRISPR/Cas9 Machinery Targeting TcNPR3 Enhances Defense Response in Theobroma Cacao. *Front. Plant Sci.* **2018,** *9*, 268. https://doi.org/10.3389/fpls.2018.00268

Gascuel, Q.; Diretto, G.; Monforte, A. J.; Fortes, A. M.; Granell, A. Use of Natural Diversity and Biotechnology to Increase the Quality and Nutritional Content of Tomato and Grape. *Front. Plant Sci.* **2017,** *8*, 652. https://doi.org/10.3389/fpls.2017.00652

Gomez, M. A.; Lin, Z. D.; Moll, T.; Luebbert, C.; Chauhan, R. D.; Vijayaraghavan, A. et al. Simultaneous CRISPR/Cas9-Mediated Editing of Cassava eIF4E Isoforms nCBP-1 and nCBP-2 Confers Elevated Resistance to Cassava Brown Streak Disease. *bioRxiv*. **2017,** PubMed: 209874.

Gumtow, R.; Wu, D.; Uchida, J.; Tian, M. A Phytophthorapalmivora Extracellular Cystatin-Like Protease Inhibitor Targets Papain to Contribute to Virulence on Papaya. *Mol. Plant-Microbe Interact.* **2018,** *31* (3), 363–373. https://doi.org/10.1094/MPMI-06-17-0131-FI

Haque, E.; Taniguchi, H.; Hassan, M. M.; Bhowmik, P.; Karim, M. R.; Śmiech, M.; Zhao, K.; Rahman, M.; Islam, T. Application of CRISPR/Cas9 Genome Editing Technology for the Improvement of Crops Cultivated in Tropical Climates: Recent Progress, Prospects, and Challenges. *Front. Plant Sci.* **2018,** *9*, 617. https://doi.org/10.3389/fpls.2018.00617

Huang, X.; Wang, Y.; Wang, N. Highly Efficient Generation of Canker-Resistant Sweet Orange Enabled by an Improved CRISPR/Cas9 System. *Front. Plant Sci.* **2021,** *12*, 769907. https://doi.org/10.3389/fpls.2021.769907

Ji, X.; Zhang, H.; Zhang, Y.; Wang, Y.; Gao, C. Establishing a CRISPR-Cas-Like Immune System Conferring DNA Virus Resistance in Plants. *Nat. Plants* **2015,** *1*, 15144. https://doi.org/10.1038/nplants.2015.144

Jia, H.; Omar, A. A.; Orbović, V.; Wang, N. Biallelic Editing of the LOB1 Promoter via CRISPR/Cas9 Creates Canker-Resistant 'Duncan' Grapefruit. *Phytopathology* **2022,** *112* (2), 308–314. https://doi.org/10.1094/PHYTO-04-21-0144-R

Jia, H.; Orbovic, V.; Jones, J. B.; Wang, N. Modification of the PthA4 Effector Binding Elements in Type I Cs LOB 1 Promoter Using Cas9/sgRNA to Produce Transgenic Duncan Grapefruit Alleviating XccΔpthA4: dCs LOB 1.3 Infection. *Plant Biotechnol. J.* **2016,** *14* (5), 1291–1301. https://doi.org/10.1111/pbi.12495

Kang, S. I.; Hwang, I.; Goswami, G.; Jung, H. J.; Nath, U. K.; Yoo, H. J.; Lee, J. M.; Nou, I. S. Molecular Insights Reveal Psy1, SGR, and SlMYB12 Genes Are Associated with Diverse Fruit Color Pigments in Tomato (*Solanum Lycopersicum* L.). *Molecules* **2017,** *22* (12), 2180. https://doi.org/10.3390/molecules22122180

Kaur, H.; Pandey, D. K.; Goutam, U.; Kumar, V. CRISPR/Cas9-Mediated Genome Editing Is Revolutionizing the Improvement of Horticultural Crops: Recent Advances and Future Prospects. *ScientiaHorticulturae* **2021,** *289*. https://doi.org/10.1016/j.scienta.2021.110476, PubMed: 110476.

Kaur, N.; Alok, A.; Shivani, Kumar, P.; Kaur, N.; Awasthi, P.; Chaturvedi, S.; Pandey, P.; Pandey, A.; Pandey, A. K.; Tiwari, S. CRISPR/Cas9 Directed Editing of Lycopene Epsilon-Cyclase Modulates Metabolic Flux for β-Carotene Biosynthesis in Banana Fruit. *Metab. Eng.* **2020,** *59*, 76–86. https://doi.org/10.1016/j.ymben.2020.01.008

Kaur, N.; Awasthi, P.; Tiwari, S. Fruit Crops Improvement Using CRISPR/Cas9 System. In *Genome Engineering via CRISPR-Cas9 System*; Academic Press, 2020; pp 131–145.

Kim, T. H.; Lee, Y. S.; Cho, C. K.; Park, S.; Choi, S. Y.; Yool, S. Y. Protective Effect of Ginseng on Radiation-Induced DNA Double Strand Breaks and Repair in Murine Lymphocytes. *Cancer Biother. Radiopharm* **1996,** *11* (4), 267–272. https://doi.org/10.1089/cbr.1996.11.267

Lal, R.; Follett, R. F.; Kimble, J.; Cole, C. V. Managing US Cropland to Sequester Carbon in Soil. *J. Soil Water Conserv.* **1999,** *54*, 374–381.

Lemmon, Z. H.; Reem, N. T.; Dalrymple, J.; Soyk, S.; Swartwood, K. E.; Rodriguez-Leal, D.; Van Eck, J.; Lippman, Z. B. Rapid Improvement of Domestication Traits in an Orphan Crop by Genome Editing Nature. *Plants* **2018,** *4*, 766–770.

Li, H.; Qi, M.; Sun, M.; Liu, Y.; Liu, Y.; Xu, T.; Li, Y.; Li, T. Tomato Transcription Factor SlWUS Plays an Important Role in Tomato Flower and Locule Development. *Front. Plant Sci.* **2017,** *8*, 457. https://doi.org/10.3389/fpls.2017.00457

Li, R.; Zhang, L.; Wang, L.; Chen, L.; Zhao, R.; Sheng, J.; Shen, L. Reduction of Tomato-Plant Chilling Tolerance by CRISPR–Cas9-Mediated SlCBF1 Mutagenesis. *J. Agric. Food Chem.* **2018,** *66* (34), 9042–9051. https://doi.org/10.1021/acs.jafc.8b02177

Li, T.; Yang, X.; Yu, Y.; Si, X.; Zhai, X.; Zhang, H.; Dong, W.; Gao, C.; Xu, C. Domestication of wild tomato is accelerated by Genome Editing. Nat. Biotechnol. **2018,** *36* (12), 1160–1163. https://doi.org/10.1038/nbt.4273

Liu, Q.; Yang, F.; Zhang, J.; Liu, H.; Rahman, S.; Islam, S.; Ma, W.; She, M. Application of CRISPR/Cas9 in Crop Quality Improvement. *Int. J. Mol. Sci.* **2021,** *22* (8), 4206. https://doi.org/10.3390/ijms22084206

Malabarba, J.; Chevreau, E.; Dousset, N.; Veillet, F.; Moizan, J.; Vergne, E. New Strategies to Overcome Present CRISPR/Cas9 Limitations in Apple and Pear: Efficient Dechimerization and Base Editing. *Int. J. Mol. Sci.* **2020,** *22* (1), 319. https://doi.org/10.3390/ijms22010319

Malnoy, M.; Viola, R.; Jung, M. H.; Koo, O. J.; Kim, S.; Kim, J. S.; Velasco, R.; Nagamangala Kanchiswamy, C. DNA-Free Genetically Edited Grapevine and Apple Protoplast Using CRISPR/Cas9 Ribonucleoproteins. *Front. Plant Sci.* **2016,** *7*, 1904. https://doi.org/10.3389/fpls.2016.01904

Martín-Pizarro, C.; Triviño, J. C.; Posé, D. Functional Analysis of the TM6 MADS-Box Gene in the Octoploid Strawberry by CRISPR/Cas9-Directed Mutagenesis. *J. Exp. Bot.* **2019,** *70* (3), 885–895. https://doi.org/10.1093/jxb/ery400

Moretti, C. L.; Mattos, L. M.; Calbo, A. G.; Sargent, S. A. Climate Changes and Potential Impacts on Post-Harvest Quality of Fruit and Vegetable Crops: A Review. *Food Res. Int.* **2010,** *43* (7), 1824–1832. https://doi.org/10.1016/j.foodres.2009.10.013

Nekrasov, V.; Wang, C.; Win, J.; Lanz, C.; Weigel, D.; Kamoun, S. Rapid Generation of a Transgene-Free Powdery Mildew Resistant Tomato by Genome Deletion. *Sci. Rep.* **2017,** *7* (1), 482. https://doi.org/10.1038/s41598–017–00578-x

Ortigosa, A.; Gimenez-Ibanez, S.; Leonhardt, N.; Solano, R. Design of a Bacterial Speck Resistant Tomato by CRISPR/Cas9-Mediated Editing of sl JAZ 2. *Plant Biotechnol. J.* **2019,** *17* (3), 665–673. https://doi.org/10.1111/pbi.13006

Pang, H.; Yan, Q.; Zhao, S.; He, F.; Xu, J.; Qi, B.; Zhang, Y. Knockout of the S-Acyltransferase Gene, PbPAT14, Confers the Dwarf Yellowing Phenotype in First Generation Pear by ABA Accumulation. *Int. J. Mol. Sci.* **2019,** *20* (24), 6347. https://doi.org/10.3390/ijms20246347

Peil, A.; Emeriewen, O. F.; Khan, A.; Kostick, S.; Malnoy, M. Status of Fire Blight Resistance Breeding in Malus. *J. Plant Pathol.* **2021,** *103* (S1), 3–12. https://doi.org/10.1007/s42161–020–00581–8

Peng, A.; Chen, S.; Lei, T.; Xu, L.; He, Y.; Wu, L.; Yao, L.; Zou, X. Engineering Canker-Resistant Plants Through CRISPR/Cas9-Targeted Editing of the Susceptibility Gene CsLOB1 Promoter in Citrus. *Plant Biotechnol. J.* **2017,** *15* (12), 1509–1519. https://doi.org/10.1111/pbi.12733

Rajesh, M. K.; Rachana, K. E.; Naganeeswaran, S. A.; Shafeeq, R.; Thomas, R. J.; Shareefa, M.; Merin, B.; Anitha, K. Identification of Expressed Resistance Gene Analog Sequences in Coconut Leaf Transcriptome and Their Evolutionary Analysis. *Turk. J. Agric. Forest.* **2015,** *39*, 489–502. https://doi.org/10.3906/tar-1409–75

Sandhya, D.; Jogam, P.; Allini, V. R.; Abbagani, S.; Alok, A. The Present and Potential Future Methods for Delivering CRISPR/Cas9 Components in Plants. *J. Genet. Eng. Biotechnol.* **2020,** *18* (1), 25. https://doi.org/10.1186/s43141–020–00036–8

Sattar, M. N.; Iqbal, Z.; Al-Khayri, J. M.; Jain, S. M. Induced Genetic Variations in Fruit Trees Using New Breeding Tools: Food Security and Climate Resilience. *Plants* **2021,** *10* (7), 1347. https://doi.org/10.3390/plants10071347

Savadi, S.; Mangalassery, S.; Sandesh, M. S. Advances in Genomics and Genome Editing for Breeding Next Generation of Fruit and Nut Crops. *Genomics* **2021,** *113* (6), 3718–3734. https://doi.org/10.1016/j.ygeno.2021.09.001

Shahriar, S. A.; Islam, M. N.; Chun, C. N. W.; Rahim, M. A.; Paul, N. C.; Uddain, J.; Siddiquee, S. Control of Plant Viral Diseases by CRISPR/Cas9: Resistance Mechanisms, Strategies and Challenges in Food Crops. *Plants* **2021,** *10* (7), 1264. https://doi.org/10.3390/plants10071264

Sharma, S.; Bhattarai, K. Progress in Developing Bacterial Spot Resistance in Tomato. *Agronomy* **2019,** *9* (1), 26. https://doi.org/10.3390/agronomy9010026

Su, Y.; Guo, J.; Ling, H.; Chen, S.; Wang, S.; Xu, L.; Allan, A. C.; Que, Y. Isolation of a Novel Peroxisomal Catalase Gene from Sugarcane, Which Is Responsive to Biotic and Abiotic Stresses. *PLOS ONE* **2014,** *9* (1), e84426. https://doi.org/10.1371/journal.pone.0084426

Sun, B.; Zheng, A.; Jiang, M.; Xue, S.; Yuan, Q.; Jiang, L.; Chen, Q.; Li, M.; Wang, Y.; Zhang, Y.; Luo, Y.; Wang, X.; Zhang, F.; Tang, H. CRISPR/Cas9-Mediated Mutagenesis of Homologous Genes in Chinese Kale. *Sci. Rep.* **2018,** *8* (1), article no.: 16786. https://doi.org/10.1038/s41598-018-34884-9

Tian, S.; Jiang, L.; Cui, X.; Zhang, J.; Guo, S.; Li, M.; Zhang, H.; Ren, Y.; Gong, G.; Zong, M.; Liu, F.; Chen, Q.; Xu, Y. Engineering Herbicide-Resistant Watermelon Variety Through CRISPR/Cas9-Mediated Base-Editing. *Plant Cell Rep. Sep.* **2018,** *37* (9), 1353–1356. https://doi.org/10.1007/s00299-018-2299-0

Tripathi, J. N.; Ntui, V. O.; Ron, M.; Muiruri, S. K.; Britt, A.; Tripathi, L. CRISPR/Cas9 Editing of Endogenous Banana Streak Virus in the B Genome of Musa spp. Overcomes a Major Challenge in Banana Breeding. *Commun. Biol.* **2019,** *2*, 46. https://doi.org/10.1038/s42003-019-0288-7

Tripathi, L.; Ntui, V. O.; Tripathi, J. N. Control of Bacterial Diseases of Banana Using CRISPR/Cas-Based Gene Editing. *Int. J. Mol. Sci.* **2022,** *23* (7), 3619. https://doi.org/10.3390/ijms23073619

Varkonyi-Gasic, E.; Wang, T.; Voogd, C.; Jeon, S.; Drummond, R. S. M.; Gleave, A. P.; Allan, A. C. Mutagenesis of Kiwifruit CENTRORADIALIS-Like Genes Transforms a Climbing Woody Perennial with Long Juvenility and Axillary Flowering into a Compact Plant with Rapid Terminal Flowering. *Plant Biotechnol. J.* **2019,** *17* (5), 869–880. https://doi.org/10.1111/pbi.13021

Wan, L.; Wang, Z.; Tang, M.; Hong, D.; Sun, Y.; Ren, J.; Zhang, N.; Zeng, H. CRISPR-Cas9 Gene Editing for Fruit and Vegetable Crops: Strategies and Prospects. *Horticulturae* **2021,** *7* (7), 193. https://doi.org/10.3390/horticulturae7070193

Wang, Y.; Liu, X.; Ren, C.; Zhong, G.-Y.; Yang, L.; Li, S.; Liang, Zhenchang. Identification of Genomic Sites for CRISPR/Cas9-Based Genome Editing in the *Vitis vinifera* Genome. *BMC Plant Biol.* **2016,** 16, article no.: 96.

Yeh, S. D. Industry Note of Technological Innovation for Promoting Global Papaya Production. *Int. J. Agric. Innov. Technol. Global.* **2020,** *1* (3).

Yin, Y.; Qin, K.; Song, X.; Zhang, Q.; Zhou, Y.; Xia, X.; Yu, J. BZR1 Transcription Factor Regulates Heat Stress Tolerance Through FERONIA Receptor-Like Kinase-Mediated Reactive Oxygen Species Signaling in Tomato. *Plant Cell Physiol.* **2018,** *59* (11), 2239–2254. https://doi.org/10.1093/pcp/pcy146

Zeilmaker, T.; Ludwig, N. R.; Elberse, J.; Seidl, M. F.; Berke, L.; Van Doorn, A.; Schuurink, R. C.; Snel, B.; Van den Ackerveken, G. Downy Mildew Resistant 6 and DMR 6-Like Oxygenase 1 Are Partially Redundant But Distinct Suppressors of Immunity in *Arabidopsis*. *Plant J.: Cell Mol. Biol.* **2015,** *81* (2), 210–222. https://doi.org/10.1111/tpj.12719

Zentner, R. P.; Campbell, C. A.; Biederbeck, V. O.; Selles, F.; Lemke, R.; Jefferson, P. G.; Gan, Y. Long-Term Assessment of Management of an Annual Legume Green Manure Crop for Fallow Replacement in the Brown Soil Zone. *Can. J. Plant Sci.* **2004,** *84* (1), 11–22. https://doi.org/10.4141/P02–188

Zhang, S.; Wang, L.; Zhao, R.; Yu, W.; Li, R.; Li, Y.; Sheng, J.; Shen, L. Knockout of SlMAPK3 Reduced Disease Resistance to *Botrytis cinerea* in Tomato Plants. *J. Agric. Food Chem.* **2018,** *66* (34), 8949–8956. https://doi.org/10.1021/acs.jafc.8b02191

Zhou, J.; Li, D.; Wang, G.; Wang, F.; Kunjal, M.; Joldersma, D.; Liu, Z. Application and Future Perspective of CRISPR/Cas9 Genome Editing in Fruit Crops. *J. Integr. Plant Biol.* **2020,** *62* (3), 269–286. https://doi.org/10.1111/jipb.12793

PART IV

Climate-Resilient Agriculture: Success Stories

CHAPTER 14

Designing the Modern Crop Genome Architecture by Harnessing the Genes from Crop Wild Relatives (CWRs): A Population Genetics and Genomics Perspective

PRASANTA KUMAR MAJHI[1], BISWARANJAN PATRA[2], PARTHA PRATIM BEHERA[3], TANMAYA KUMAR BHOI[4], AALOK SHIV[5], DEVRAJ LENKA[6], ASHUTOSH PUROHIT[6], and AYAZ AHAMAD[7]

[1]*Department of Plant Breeding and Genetics, Regional Research and Technology Transfer Station (RRTTS), Odisha University of Agriculture and Technology (OUAT), Keonjhar, Odisha, India*

[2]*Center for Biotechnology, Siksha 'O' Anusandhan Deemed to be University, Bhubaneswar, Odisha, India*

[3]*Department of Plant Breeding and Genetics, Assam Agricultural University, Jorhat, Assam, India*

[4]*Forest Protection Division, ICFRE-Arid Forest Research Institute (AFRI), Jodhpur, Rajasthan, India*

[5]*Division of Crop Improvement, ICAR-Indian Institute of Sugarcane Research (ICAR-IISR), Lucknow, Uttar Pradesh, India*

[6]*Department of Plant Breeding and Genetics, College of Agriculture, Odisha University of Agriculture and Technology (OUAT), Bhubaneswar, Odisha, India*

[7]*Biosciences Group, Homi Bhabha National Institute (HBNI), Bhabha Atomic Research Centre, Mumbai, Maharashtra, India*

Climate-Resilient Agriculture: A Molecular Perspective. Ashutosh Singh, Saurabh Pandey (Eds.)

ABSTRACT

The increasing global population, erratic climatic changes, and the emergence of new races of pathogens and insect pests in many cultivated crop plants are serious threats to crop productivity and food security. To meet this challenge, we have to develop climate-resilient varieties for multiple environments and will have to fulfill our aim to produce a sufficient quantity of food to nourish the growing population of the globe. To overcome this vulnerability, we have to search for the best alternatives, and the crop wild relatives (CWRs) are reservoirs of several resistant genes and are capable enough to attenuate the impact of climate change as their genetic constitution provides greater resistance/tolerance against biotic and abiotic stresses. The CWRs are a valuable resource with desirable characteristics such as agronomic, nutritional excellence, and tolerance to abiotic and biotic stressors. Because low-cost and high-throughput DNA sequencing and resequencing methods are now available, CWR genomes can be surveyed for genetic diversity and novel genes and alleles. The genome editing (GE) tools help to edit the undesirable genes transferred to the cultivated species from the wild cousins. Although little effort has gone into this beyond a handful of the basic crops, a lack of diversity in modern germplasm may further limit our ability to breed for better nutritional levels. However, in the case of alien gene introgression, the inherent problems of crossability, sexual incompatibility, linkage drag, fertility, and gene recombination must be dealt with and resolved by using appropriate procedures in conjunction with molecular biology and bioengineering tools such as quantitative trait locus (QTL) mapping, marker-assisted selection, and genome-wide association (GWA study. Therefore, this chapter will shed light on (i) crop Synteny and Comparative Genomics and sequencing study on crop wild-related species; (ii) designing modern crop genome architecture through advanced molecular breeding techniques such as the development of multiparent advanced generation intercross (MAGIC) and MAGIC-like populations for QTL mapping, marker-assisted backcrossing (MAB), using GE (CRISPR-Cas) tools; and (iii) the challenges of transferring genes from wild sources to modern varieties and the solute ions to overcome.

14.1 INTRODUCTION

Crop productivity and food security are impacted by the growing world population (9.8 billion by 2050), irregular changes in climatic factors (drought, floods, and high temperatures), and the advent of pathogens and pests that are immune to many domesticated crop plants (Hunter et al., 2017; Khan et al., 2020). In the vast majority of countries, the effects have been seen in per-person food availability. To deal with this problem, we must design climate-resilient cultivars for various locations to achieve our goal of producing 70% extra food by 2050 than we do now to feed the world's rising population (Hunter et al., 2017). To address this insecurity, we must search for better alternative solutions. Crop wild relatives (CWR) are resources of many resistance genes and thus are effective in minimizing the effects of climate change due to their genetic constitution, which gives better immunity to biotic (pathogens and insects) as well as abiotic (drought, frost, and flood) stresses (Li et al., 2014; Luo et al., 2013; Zhao et al., 2018). CWRs are a valuable source of characteristics such as agronomical and nutritional properties, and resistance to abiotic and biotic stressors. These are identified as an essential factor for ensuring food security, as well as sustainability for the environment, in the current century (Fig. 14.1). CWRs are the wild species that are significantly interrelated to crop species, such as wild ancestors and species that are commonly associated with agricultural plants (Dempewolf et al., 2017). We have seen some outstanding cases of bringing desirable features or traits of CWR to our cultivated crops in the past few decades. Unique pest and disease tolerance genes have been transferred from annual crops (e.g., rye) or perennial crops of the tribe Triticeae (e.g., *Agropyron*, *Thinopyrum*, and *Lophopyrum*) into crops such as wheat using engineering of chromosomes such as alteration of pairing control mechanisms and introduction of chromosomal translocations (Vasil, 2007). For protection against blast in rice cultivars, the gene *Pi9* (Amante–Bordeos et al., 1992; Liu et al., 2002), *Pi40(t)* on the short arm of chromosome 6 of *Oryza australiensis* (Jeung et al., 2007), and *Pi-ta+* (Geng et al., 2008) were transplanted from wild crop species. Utilizing a positional cloning method, a commonly utilized resistance gene in rice, known as *Xa21* for BLB resistance, was identified and discovered to contain a receptor kinase-like polypeptide. *Oryza glaberrima*, despite its low-yielding wild ancestors, has been discovered to be an incredible resource of drought resistance in rice (Menguer et al., 2017). Drought endurance, salinity tolerance, high-temperature tolerance, resistance to pests, and lint color are only a few of the distinctive agronomic properties found in the wild species of cotton (*Gossypium tomentosum*). The moisture stress tolerance, Verticillium wilt, and Fusarium wilt resistance are among the many important features of *Gossypium darwinii*, a wild allotetraploid plant with an AD5 genome (Liu et al., 2015). *Vigna unguiculata* group *sesquipedalis* (*Vigna* luteola, *Vigna marina*, *Vigna riukiuensis*, *Vigna nakashimae*, and *Vigna vexillata* var. macrosperma) grows in seasonally warm climates and tolerates high saline levels (van Zonneveld et al., 2020). Except for rice beans (*Vigna umbellate*), neither of the legume crops is immune to bruchid attack. Furthermore, a wild mung bean cultivar of Madagascar (TC1966: *Vigna radiata* var. *sublobata*) exhibited full resistance to important bruchid species such as *Callosobruchus chinensis*, *Callosobruchus phaseoli*, *Callosobruchus maculatus*, and *Clinocottus analis* (Kashiwaba et al., 2003; Majhi and Mogali, 2020). *Brassica cretica* has indeed been postulated as a promising source of numerous resistant genes to a range of brassica crops, namely, broccoli, cabbage, Brussels sprouts, cauliflower, swede, kale, oilseed rape, and turnip (Kioukis et al., 2019). Most of these characteristics are related to biotic stress tolerance, such as protection against late blight caused by *Phytophthora infestans* in the wild potato *Solanum demissum* Lindl (Witek et al., 2016). Cucurbita wild species are more resistant to viral infections than cultivated taxa and have also been utilized to improve resistance in several cultivars. *Cucurbita lundelliana* L. has already been particularly useful as a genetic link for introducing resistance to disease from wild species to modern cultivars (Khoury et al., 2020). Aside from abiotic or biotic tolerances, several quantitative trait locus (QTL) analyses have been well-known and incorporated, including grain hardiness from wild barley (*Hordeum spontaneum*) to cultivated rice (*Oryza sativa*) (Septiningsih et al., 2003) and grain quality from *Oryza rufipogon* (Li et al., 2010). Crop wild species have a lot of genetic variation and variability, making them a significant resource toward crop improvement. Many of their features can improve crops to become more robust and durable, as well as to adapt to the new requirements that climate change will bring. An enormous global effort is underway to conserve numerous wild crop ancestors and to make their use in breeding programs for food security (Kioukis et al., 2019). In this chapter, we have discussed the wild relatives of several cultivated field crops (cereals, pulses, and oilseeds) and horticultural crops (Table 14.1), as well as their applicability in crop improvement programs.

CWRs complete genome sequencing can be utilized to determine wild populations and support conservation strategies. Examining the sequencing information can reveal genetic diversity and important genetic traits that have been lost due to cultivation and could be transferred to cultivated species to aid in climate adaptation, nutrient enhancement, and food security (Brozynska et al., 2016). Because cost-effective and

TABLE 14.1 CWRs of Important Cultivated Crops and Their Desirable Features

Sl. No.	CWRs	Desirable features	Remarks	References
Rice (O. sativa L.)				
	O. australiensis	Contributes the blast tolerance	The genes such as *Pi9, Pi40(t), and Pi-ta+* contribute blast resistance to the cultivated rice genotypes. These genes are present on the short arm of chromosome 6.	Liu et al. (2002); Jeung et al. (2007); Geng et al. (2008)
	O. rufipogon	Bacterial blight resistance	*Xa23*, a solitary completely dominant gene outlined from *O. rufipogon*, has been recognized as the most beneficial gene which is highly resistant to 20 strains of *Xoo*, with effectiveness at all phases of rice growth with broad-spectrum tolerance.	Zhang et al. (2002)
	Oryza officinalis and *O. rufipogon*	Brown plant hopper (BPH) resistance	*Bph14* and *bph29*, from *O. officinalis* and *O. rufipogon*, correspondingly, are two of the four resistance genes that have been identified and transferred so far. Introgressions of BPH resistance into domesticated/elite lines have also been undertaken in parallel with gene discovery and cloning activities.	Sarao et al. (2016)
	O. glaberrima	Tolerance to drought	*O. glaberrima*, for instance, has been discovered to be an effective source of drought tolerance despite its low yield.	Ndjiondjop et al. (2010); Menguer et al. (2017)
Wheat (Triticum spp.)				
	Annual (e.g., rye) and perennial (e.g., *Thinopyrum*, *Agropyron*, and *Lophopyron*)	Resistance to disease and pests	Wheat-grass hybridization has benefited in the genetic evolution of the crop, as has cytogenetic manipulation of the hybrid composite.	Chaudhary et al. (2014); Vasil (2007)

TABLE 14.1 *(Continued)*

Sl. No.	CWRs	Desirable features	Remarks	References
	Aegilops speltoides	Broaden the gene pool	Constitute the secondary gene pool of wheat.	Riley and Chapman (1958)
Maize (Zea mays L.)				
	Zea luxurians, *Zea diploperennis*, *Zea perennis*	Resistance to biotic and abiotic stresses	Annual (*Z. luxurians*), diploid perennial (*Z. diploperennis*), tetraploid perennial (*Z. perennis*), and polytypic annual *Z. mays*, including four subspecies: ssp. mays (maize), ssp. *parviglumis*, ssp. *mexicana*, and ssp. huehuetanangensis; Tripsacum (*Z. mays* ssp. *parviglumis*) and Teosinte are two CWRs which have been thoroughly characterized as contributors of economically relevant features/traits that could be employed for maize improvement.	Fukunaga et al. (2005)
	Z. mays ssp. *parviglumis*	Maize spotted stalk borer (MSSB)	Teosinte germplasms of *Z. mays* ssp. *mexicana* which showed resistance to the MSSB generated more benzoxazinoids (BXs) than sensitive maize lines.	Niazi et al. (2014)
	Z. diploperennis	Resistant to southern corn leaf blight, northern corn leaf blight, and corn leaf spot diseases	The chromosomal positions of *Z. diploperennis*-derived portions in maize and *Z. diploperennis* crossing that carried loci resistant to fungal infections such as southern corn leaf blight, northern corn leaf blight, and corn leaf spot diseases utilizing genomic in situ hybridization methods.	Wei et al. (2001)
	Z. mays ssp. mexicana	Fusarium wilt and downy mildew resistance	These teosinte species found to be a source of Fusarium wilt and downy mildew resistance.	Maazou et al. (2017)
Cotton (Gossypium spp.)				

TABLE 14.1 *(Continued)*

Sl. No.	CWRs	Desirable features	Remarks	References
	Gossypium gossipioides; *Gossypium raimondii*	Progenitor of cultivated species	The diploid (2*n*) species *G. gossipioides* L. and *G. raimondii* Ulbrich ($2n = 2x = 26$) resulted in the emergence of new global tetraploid (4n) species.	Wendel and Cronn (2003)
	G. tomentosum; *G. darwinii*	Drought adaptation, salinity tolerance, Fusarium wilt, fiber fineness, and Verticillium wilt resistance	Drought adaptation, salinity tolerance, tolerance to high temperature, nectarilessness, resistance to pests, and lint color are only a few of the distinctive agronomic properties found in the wild cotton species (*G. tomentosum*). Drought tolerance, Fusarium wilt, fiber fineness, and Verticillium wilt resistance are among the many important features of *G. darwinii*, which is a wild allotetraploid plant having an AD5 genome.	Liu et al. (2015)
	Gossypium anomalum (B1), *G. raimondii* (D5), *Gossypium armourianum* (D2), *Gossypium davidsoni* (D3), and *Gossypium thurberi* (D1)	Resistance to Jassids	Jassids were found in low numbers on cotton plants having low tannin and high phenolic contents.	Shindel et al. (2014)
	G. anomalum	Resistance to cotton rust	Interspecific hybridization with continual resistance screening in back-cross populations, and polyploidy induction, cotton rust tolerance genes were managed to transfer to *Gossypium hirsutum* from *G. anomalum*.	Blank and Leathers (1963)
	G. thurberi	Resistance to Verticillium wilt	*G. thurberi* is another wild relative with great Verticillium wilt resistance, as evidenced by research that identified 57 distinct proteins in *G. thurberi* affected with *Verticillium dahliae* using tandem time-of-flight mass spectrometry (MALDI-TOF-MS), and 2D electrophoresis (2-DE).	Zhao et al. (2012)

TABLE 14.1 *(Continued)*

Sl. No.	CWRs	Desirable features	Remarks	References
	Gossypium arboretum; *Gossypium stocksii*	Resistance to cotton leaf curl virus (CLCuV)	Using standard hybridization and backcrossing, *Gossypium arboreum* was employed as a source of CLCuV tolerance to transmit it to *G. hirsutum*. Using interspecific hybridization, tolerance to CLCuV from *G. stocksii* was effectively transferred into "MNH-786," which are *G. hirsutum* cultivars.	Nazeer et al. (2014)
Grain Legumes (Vigna spp.)				
	Vigna trilobata; *V. vexillata* var. *ovata; Vigna monantha*, and *Vigna aconitifolia*	Resistance to drought and salinity	*V. trilobata* and *V. vexillata* var. *ovata* thrive in hot, humid climates and can withstand high degrees of drought and salinity. *V. monantha* and *V. aconitifolia* grow in arid, seasonally warm climates with little rainfall during the rainy quarter.	van Zonneveld et al. (2020)
	Vigna aridicola; *Vigna exilis*; *Vigna laurentii*	Resistance to heat and water stress	*V. aridicola* thrives in perpetually hot climates and can withstand severe dehydration. *V. exilis* thrives in hot climates, both year-round and seasonally. *V. laurentii* grows in warm and dry climates.	van Zonneveld et al. (2020)
	V. unguiculata group *sesquipedalis*; *V. unguiculata*, and *V. vexillata*	Resistance to drought, heat, and salinity	The *V. unguiculata* group *sesquipedalis* grows in hot, humid climates and can endure highly saline levels. The cultivated *V. unguiculata* and *V. vexillata*, as well as the wild relatives *V. luteola*, *V. marina*, *V. riukiuensis*, *V. nakashimae*, and *V. vexillata* var. *macrosperma*, showed high levels of salt tolerance.	van Zonneveld et al. (2020)
	V. adiate; *V. unguiculata* group *unguiculata*; *V. unguiculata* group *sesquipedalis*	Resistance to Yellow mosaic disease (YMD)	YMD tolerance was found (corresponds to 100%) and belonging to gene pools A and C of V. adiate and certain other Asian Vigna crops, as well as *V. unguiculata* group *unguiculata* with *V. unguiculata* group *sesquipedalis*.	van Zonneveld et al. (2020)

TABLE 14.1 *(Continued)*

Sl. No.	CWRs	Desirable features	Remarks	References
	V. unguiculata group *sesquipedalis*, and *V. unguiculata* group unguiculata	Resistance to bacterial blight and thrips	Germplasms from gene pool C (*V. unguiculata* group *sesquipedalis* and *V. unguiculata* group *unguiculata*) was found to be resistant to bacterial blight, Thrips spp., as well as legume pod borer.	van Zonneveld et al. (2020)
Soybean (Glycine max)				
	Glysine soja	Resistance to biotic and abiotic stresses	*G. soja* has a large geographic distribution and significant genetic variability due to its capacity to adapt to a variety of environmental stresses.	Qiu et al. (2013); Sherman-Broyles et al. (2014)
	G. soja	Resistance to soybean cyst nematode	Several studies on classical genetic linkage mappings or linkage disequilibrium (LD)-based association mappings have been reported on the identification of novel SCN-resistance QTL(s) in *G. soja*	Winter et al. (2007); Yu and Diers (2017)
	Glycine tabacina, *Glycine clandestine*, *Glycine argyrea*, *Glycine microphylla*, *Glycine latifolia*, and *Glycine tomentella*	Resistance to Asian soybean rust (ASR)	The fungal infection of ASR is a severe problem; however, these perennial *Glycine* species have shown resilience to it.	Hartman et al. (1992)
	Glycine canescens, and *G. latifolia*	Resistance to drought	Because of their reduced epidermal conductivity, higher osmotic adjustment, and substantially higher water content, accessions of *G. canescens*, and *G. latifolia* are more certain to be drought resistant than *G. soja* and *G. max*.	James et al. (2008)
Brassica species				

TABLE 14.1 *(Continued)*

Sl. No.	CWRs	Desirable features	Remarks	References
	Brassica oleracea	Rich nutritional value	*B. oleracea*, sometimes known as wild cabbage, and for its higher proportion of minerals including vitamin and anticancer capabilities.	Kioukis et al. (2019); Higdon et al. (2007)
	B. cretica	Resistance to diseases and wild progenitor of cultivated brassica species	*B. cretica* Lam is a wild species of *B. oleracea* L., the domesticated Brassica group, and is thought to be the origin of broccoli, cabbage, Brussel sprouts, cauliflower, turnip, swede, kale, and oilseed rape.	Kioukis et al. (2019)
Sugarcane (Saccarum spp.)				
	Saccharum robustum	Progenitor of cultivated sugarcane	*Saccharum offcinarum* types were developed from wild *S. robustum* some 8,000 years back in New Guinea, according to legend.	Ming et al. (2010)
	Saccharum spontaneum	Resistance to drought salinity and flood	*S. spontaneum*, on the other hand, has a larger ratooning capability, a higher level of fibers, and an overall improved adaptation capability, driven by the ability to operate better in adverse environmental conditions like drought, high salt, or flood and these characteristics are transferred to *S. offcinarum* after crossing to develop improved genotype.	Mohan Naidu and Sreenivasan (1987)
Tomato (Solanum lycopersicum L.)				
	Solanum pimpinellifolium *Solanum pennellii*	Fusarium wilt	These genes, *I* (on chromosome 7, 11), *I-2 and*, *I-3* are responsible for resistance against *Fusarium oxyporum f.sp. lucopersici*	Scott and Gardner (2007)
	S. pimpinellifolium, *Solanum habrochaites*, *S. pennellii*, *S. habrochaites*	Resistance to late blight	The genes *Ph-1*, *Ph-2*, and *Ph-3* have played a role against *P. infestans* pathogens.	Merk et al. (2012); Foolad et al. (2008)

TABLE 14.1 *(Continued)*

Sl. No.	CWRs	Desirable features	Remarks	References
	S. pimpinellifolium, S. peruvianum, S. chilense, S. habrochaites	Resistance to tomato (yellow) leaf curl virus	The genes *Ty1* (on chromosome 6, 3), *Ty2* (on chromosome 11) *Ty3*, and *Ty4* show resistance against various Begomovirus species.	Ji et al. (2007); Ji et al. (2008)
	S. peruvianum, S. peruvianum	Resistance to Root knot nematode	The genes *Mi*, *Mi-2* to *Mi-9* show resistance against the nematode species *Melidogyne incognita*, *Melidogyne javanica*, *Melidogyne arenaria*.	Pereira-Carvalho et al. (2010); Wang et al. (2013a, 2013b)
Brinjal (Solanum melongena L.)				
	Solanum incanum	Resistant to biotic and abiotic stress	*S. incanum*, a member of the typical eggplant's secondary gene pool has been noted as a rich source of phenolic contents, with amounts multiple times greater than *S. melongena*, and is resistant to a variety of abiotic and biotic stress, most notably drought.	Syfert et al. (2016), Prohens et al. (2013); Knapp et al. (2013)
Cucurbits (Cucurbita spp.)				
	C. lundelliana	Disease resistance	*C. lundelliana* has been particularly useful as a genetic link for introducing resistance to diseases from wild populations into domesticated crops.	Khoury et al. (2020)
Sweet potato (Ipomoea batatas L.)				
	Ipomoea trifida	Progenitor of hexaploid cultivated sweet potato.	The wild diploid (2*n*) species *I. trifida* (Kunth.) ($2n = 2x = 30$) has been widely documented as the closest ancestor of the farmed sweet potato.	Munoz–Rodrguez et al. (2018); Roullier et al. (2013)

high-throughput DNA sequencing and resequencing methods are now available, CWR genomes can be examined for genetic diversity, unique genes, and alleles. Genome editing (GE) technologies help in the editing of unwanted genes transmitted from wild relatives to domesticated species. New ways to analyze the genetic alterations that have been accompanied by plant domestication have been spawned by recent developments in high throughput sequencing with computational methods (Kantar et al., 2017). We still do not know how genetic variability in crops corresponds to nutritional features, even though the population-wide genetic implications of crop domestication provide several hypotheses about the extent of genetic variation in crops. The levels of critical minerals and vitamins in modern cultivars have reduced. While little research has been done on this, we believe that choosing edibility and higher yield during cultivation and postcultivation divergence exacerbated several crop's low nutritional levels. Although some efforts have been taken into this above a handful of basic crops, a lack of variability in the present germplasm would further hinder our ability to breed for better nutritional levels. In a fast-changing world, a knowledge of domestication over a wide range of crop taxa may give the crucial understanding for breeding superior nutrient-rich crops (Smykal et al., 2018; Cortes et al., 2022). The current review will focus on (i) crop handling and genetic analysis, as well as sequencing studies on CWRs; (ii) modeling modern crop genetic framework using advanced molecular breeding methods such as the construction of multiparent advanced generation intercross (MAGIC) and MAGIC-like populations for QTL mapping, marker-assisted backcrossing (MAB), and genome editing (GE) tools; and (iii) difficulties in gene delivery from wild relatives to modern cultivated varieties, as well as the potential solutions.

14.2 CROP DOMESTICATION AND LOSS OF ALLELIC DIVERSITY: IN A POPULATION GENETICS PERSPECTIVE

Domestication is a well-studied phenomenon that dates back to Darwin's studies (Darwin, 1860) on how selection affects crop species (Gepts, 2010; Meyer and Purugganan, 2013). The process involves numerous regularly seen phases that transform a wild entity into a domesticate which is more suited to human usage. The phases are essential as follows: (a) harvesting wild relatives; (b) optimization of plant characteristics by conscious and unconscious selection (e.g., framework, the timing of flowering, size, dormancy, and reduced defensive structures); and (c) critical selection of plants for particular locations and uses with the plant usually throwing away its capability to sustain without human supervision and care (Harlan, 1992) (Fig. 13.2). Following Darwin's pioneering study of this selection-driven strategy as a model for how natural selection can work, a large and prolific area of research in evolutionary biology has evolved that emphasizes evaluating the strengths and weaknesses and information of directional selection-driven evolution (Meyer and Purugganan, 2013). In the last 12,000 years, humans have cultivated hundreds of species of plants and animals for food, fiber, fodder, and tools, with numerous consequences for both human culture and the genetic makeup of domesticated species (Smykal et al., 2018). There are approximately 7000 plant species used only for nourishment by preagricultural human communities, of which only a small percentage were domesticated (Khoury et al., 2014). As a result, domesticated crops comprise just a small part of the global plant variety (Smykal et al., 2018).

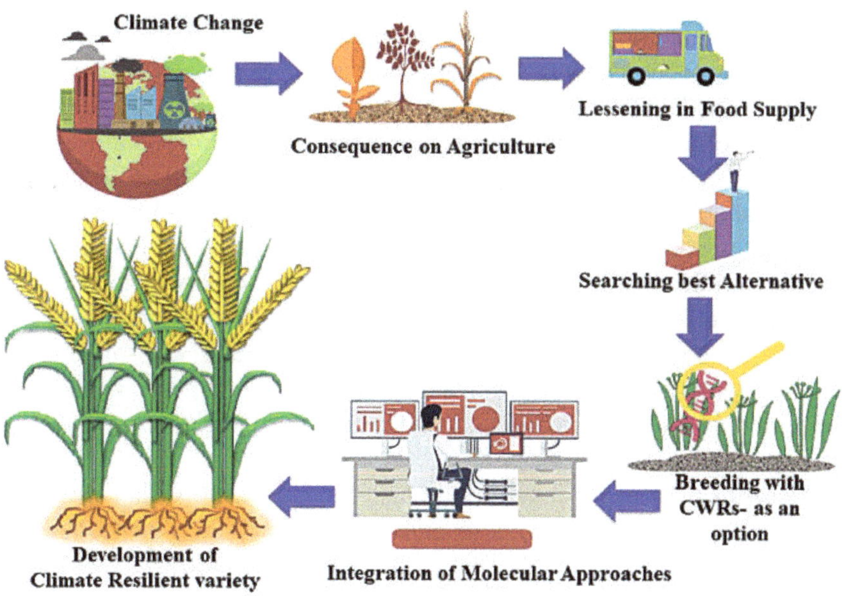

FIGURE 14.1 A conceptual framework for development of climate resilient genotype with the crop wild genes considering the effect of climate change.

Over 160 taxonomic groups contain domesticated species of plants (Meyer et al., 2012). Poaceae, Brassicaceae, and Fabaceae are the most notable. Around 2500 species have been domesticated to some extent, but still, only 250 are considered completely domesticated (Meyer et al., 2012). Many are only used in limited circumstances or certain places. Conversely, the major part of people's daily diet is based on a small number of crop plants. Indeed, roughly 95% of the world's calories come from just 20 plant species. The crops such as; bananas/plantains, cassava, sweet potatoes, beans, maize, sugarcane, potatoes, millets, rice, soybean, sorghum, wheat, and legumes like lentils, peas, and chickpeas are just a few examples (FAO, 2015). Although many plant species are used in agricultural and food markets, merely 30 crops account for 95% of global food security (Brozynska et al., 2016). The global food supply vulnerability to the consequences of climate change, the appearance of important new insect pests, and illnesses is due to this reliance on a small number of food species. Because there are very few species planted as staples, pests and disease outbreaks are more likely (Smykal et al., 2018). Domesticated crops being used for food production have less genetic variation than their wild species as time goes on. Due to (i) uneven breeding practices concentrating only on a few traits (like yield and its component traits) (ii) regular use of a few germplasm accessions as parents in the development of varieties, and (iii) introduction of a few excellent lines to several nations, natural diversity has been depleted, resulting in greater genetic similarity among modern crop varieties. Domestication's genetic "bottleneck" (Tanksley and McCouch, 1997) led to a loss of important genes. Introgression from wild relatives during the domestication methods, on the contrary, may provide more genetic variety in domesticated cultivars (Hufford et al., 2013; Sawler et al., 2013). There is a risk of helpful and undesired genes being incorporated due to the natural crossover of wild relatives with their counter-cultivated relatives. Agricultural workers are concerned about decreased genetic differences and variation among crop species (Bhanu, 2017).

Selections brought about by human choices, cultivating practices, agricultural surroundings, and several other population genetic developments resulting from the reduction in effective population size affected the consequences of crop domestication. Any choice minimizes variation by default by promoting preferred variations, such as nonshattering seeds or higher palatability. Agricultural techniques also reduced the number of the effective population size of the crop, allowing genetic drift to shift genotypic frequencies (Smykal et al., 2018). The transition from wild to farmed species is a difficult process. These activities resulted in the so-called domestication phenomenon, a set of characteristics that might emerge as a result of human choices for ease of harvesting and growth benefits under human cultivation. Diminished seed and fruit dissemination ability, alterations in plant structure, and modifications in plant defense properties and palatability were among these features. Smykal et al., 2018 defined domestication as the influence of selective pressures on existing genetic diversity as well as new genetic differences introduced via mutations or introgression. These "selected sweeps" may allow slightly harmful alleles to stabilize, potentially resulting in a genetic load within the cultivated gene pool. Traditional genetic and linkage mapping have revealed a few hundred genes and loci as drivers of domestication and postdomestication variability to date. Only a few of them have been identified, and even lesser have the function of the wild-type gene in natural populations figured out. Only advantageous haplotypes are kept around targeted genes during domestication, resulting in a genetic valley with very little genetic variability (Smykal et al., 2018). Human selection acted on several loci during as well as after crop domestication, according to recent developments in molecular techniques, specifically genome sequencing. Population-level molecular analysis can also help us understand the demographic backgrounds of the domestication methods and shed light on crop lineage.

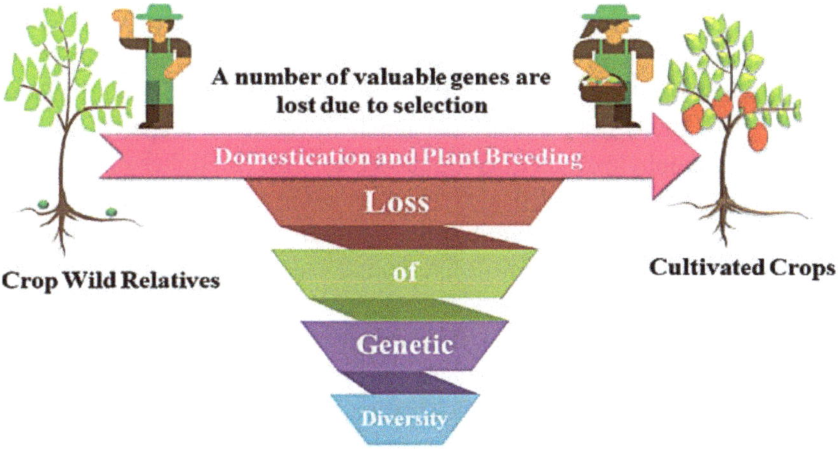

FIGURE 14.2 Loss of genetic diversity through the process of domestication by conscious and unconscious selection.

14.3 GENE POOL TO PREBREEDING: BRIDGING THE GAP BETWEEN CULTIVATED AND WILD SPECIES

Whether geologically or biologically, biodiversity is not equally distributed. Alfonse de Candolle (Candolle, 2012) introduced the theory that crops were domesticated in the centers of corresponding species diversity, which Nikolai I. Vavilov polished and broadened (Vavilov, 1926). Genetic diversity is dispersed biologically among primary, secondary, and tertiary gene pools (Harlan and Wet, 1971), which are characterized by their hybridization compliance and hence not equally available to domesticated species. The idea of gene pools (Harlan and Wet, 1971) was formulated based upon reproductive isolation, in which the nearest crop wild species that are interfertile with crop species depict a readily available or primary gene pool for crop improvement, while more distant species from which genes can be obtained with intensifying difficulty levels create secondary and tertiary gene pools. In certain cases, the wild relatives, from which a crop plant was evolved, are well-known and easily recognizable in present wild plant groups, while in others, polyploidization and hybridization have resulted (Kantar et al., 2017). Domestication as a centric process at the beginning of agriculture (Abbo and Gopher, 2017; Lev-Yadun et al., 2000) against many origins across long time scales (Civán et al., 2013; Fuller et al., 2011) has now been the subject of intense discussion. Domesticated crops may have a single or several sources, a linear or septate ancestry from an ancestor population(s), and gene flow within and between domesticated and wild populations across the crop's evolution. All of these theories appear to be supported by evidence in our domesticated crop species. Given the limited territory of its wild species (Berger et al., 2003) and the restricted genetic diversity, chickpea is assumed to have a monophyletic ancestry (Abbo et al., 2003). Higher polyphyletic patterns are found in domesticated barley (Pankin et al., 2018) and emmer wheat (Civán et al., 2013). Both had ancestors from the eastern and western margins of the Near Eastern Fertile Crescent. Domesticated rice also provides the most evidence for polyphyletic ancestors, with separate, spatially diverse domestication throughout China (japonica gene pool), Indochina (indica gene pool), and central India to Bangladesh (aus gene pool) (Civán et al., 2015). Following that, there was gene flow between each of these gene pools, which may or may not have been crucial for the implementation of domestication syndrome features from japonica to the indica gene pool (Civá and Brown, 2017; Choi et al., 2017).

14.4 GENE-GENOME-PANGENOME TO SUPERPANGENOME: TRANSLATING THE CONCEPTS FOR CWRS

Tettelin et al. (2005) coined the term "pangenome" to describe the genome study of numerous pathogenic strains of *Streptococcus agalactiae*. The superpangenome framework was proposed by Khan et al. (2020), and it relates to the gene pool of the crops comprising many species, notably wild relatives possessing stock that is genetically diverse. They proposed that for the advancement of a more complete and detailed pangenome, accessions from all available species of a given genus be used. In their opinion column, Khan et al. (2020) stated that the superpangenome must encompass wild species and provide a whole-genetic variation portfolio of a genus, as well as unique future crop development prospects. The pangenome can be categorized into two components: core genome and dispensable genome. Several research on plants has found that the core genome is larger and contributes the most genes (Golicz et al., 2016; Li et al., 2014). There are two sorts of pangenomes available: open and closed pangenomes. The dispensable genome is thought to contain genes involved in survival and adaptation in changing environments. The impact of domestication is revealed by comparing the core genomes of wild species with the dispensable genomes of domesticated crop species (Li et al., 2014). The requirement for a pangenome has been proven in the study of prokaryotes and eukaryotes, as a single reference genome is insufficient to reflect a species' whole-genetic range (Cao et al., 2011). Pangenome analysis can also aid in the discovery of genes not found in reference genomes. Pangenome investigations, which were first used in prokaryotes, are increasingly being used in species of plants as well (Li et al., 2014; Zhao et al., 2018).

Pangenomes have recently been used to investigate the diversity of species based on structural variation (SV) in rice (Schatz et al., 2014), soybean (Li et al., 2014), wheat (Montenegro et al., 2017), brassica (Golicz et al., 2016), sesamum (Jingyin Yu et al., 2019), sunflower (Hübner et al., 2019), and tomato (Gao et al., 2019). In rice pangenome research, several genomic regions with particular desired features, such as submergence resilience, were found. In Dongxiang wild rice (*O. rufipogon* Griff.), the sub-1 locus, leucine-rich repeat receptor-like kinase (LRK) genes, that are linked to a 16%- yield enhancement (Zha et al., 2009), and the *phosphorus uptake1* (*Pup1*) gene cluster, that provides phosphate shortage tolerance in soil (Schatz et al., 2014). The Rice Pan Genome Browser is a web-based pangenome for rice that was created using readings from 3010 germplasms that were aligned on the IRGSP-1.0 genome, resulting in the discovery of 23,914 core genes, 4986 potential core genes, and 22,095 dispersed genes (Sun et al., 2017). Pangenomes were also created for *O. rufipogon* as well as *O. sativa* by employing 67 different rice germplasms, resulting in the discovery of 10,872 genes that were substantially lacking in the Nipponbare reference sequence (Zhao et al., 2018). We must move away from using genus-level pangenomes to construct a complete pangenome, or "Superpangenome". Because numerous species are present under a common genus in a gene pool, genes from distinct species can be transplanted to a specific crop using normal breeding, distant hybridization, or current chromosomal engineering methods. The superpangenome approach entails the following steps (Khan et al., 2020): (1) recognition and recruitment of the most diversified germplasms from a given species (species-I) (2) de novo genome assembly of one of the germplasms, and then the mapping of resequencing information from the remainder germplasms onto this assembly to build a species-level pangenome (species I pangenome) (3) picking the 10 most diversified germplasms from species I, then arrange one of these germplasms and map the sequencing information from the residual access (4) A pangenome for another species (species II) will be created in the same way. As a result, distinct species-level pangenomes for the genera of the specified crop species will be constructed. Such species-level pangenomes will be combined to create the pangenome for a genus. The resulting pangenome will be known as a superpangenome, and it will have the capacity to reflect the genus' whole-genetic diversity. The superpangenome will also be a valuable resource for phylogenetic studies since it will allow for precise identification of species divergence times and true estimation of the various evolutionary processes that have shaped the current genomic architecture of diverse species (Fig. 14.3). The superpangenome would identify new haplotypes that could be useful for agricultural development and conservation in the future.

14.5 CROP SYNTENY, COMPARATIVE GENOMICS, AND SEQUENCING INFORMATION: MADE EASY THE BREEDING PROGRAMME

Rapid advances in sequencing technologies have made it possible to examine gene and genomic evolution using comparative genomics-based on fully sequenced genomes (Alföldi and Lindblad-Toh, 2013). In the last decade, marker techniques have expanded in the agricultural field, allowing for the quick generation of resistant varieties and nutritious rice variations

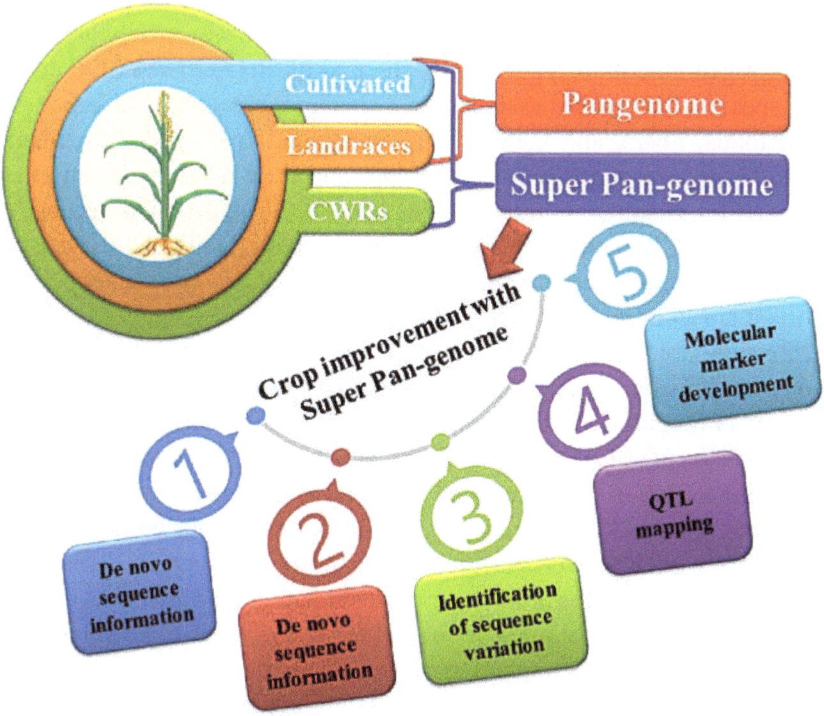

FIGURE 14.3 Development of Super Pan-genome concept with CWRs for crop improvement.

in a variety of food crops. Innovative methods such as genetic engineering, GE, and GAB have recently been developed to enhance modern varieties of crops with better resistance to diseases, insect pests, and nutritional properties (Varshney et al., 2005). With major developments in next-generation sequencing (NGS) technology, sequencing and genetic profiling techniques have become more readily inexpensive and available, boosting genomics for the improvement of crops (Yuan et al., 2017). The presence of genetic variability in the context of single nucleotide polymorphisms (SNPs) and InDels (insertions/deletions) (Saxena et al., 2014a) has been facilitated by the sequence analysis and resequencing information of reference genomes of many cultivated and wild species (Varshney et al., 2017; Wang et al., 2018). These could be used to produce adequate varieties as markers. Other types of changes (deletions, transversions, inversions, translocations, etc.) have recently been researched using the same data as SVs, presence or absence variations (PAVs), and copy numbers variations (CNVs) (Cook et al., 2012; Wang et al., 2015).

Whole-genome resequencing offers valuable information regarding crop genetics, hybridization, and evolutionary history, and a vast variety of species now have genomic data available. By the end of 2017, 37.7% (236 angiosperm species) of recorded vascular plant species were employed largely for human foods, fibers, and fuels, 17.7% constituted crop wild descendants, and 22.3% represented noncrop prototypes and their relatives (Chen et al., 2018). CWRs genomic architecture has been better understood because of high-throughput sequencing techniques, which have facilitated their usage in current crop modification. The accomplishment of genome sequencing for important agricultural species has revealed the need for a larger gene pool, which can be done by concentrating on CWR. Several research utilizing de novo assembling and resequencing techniques utilizing CWR have been published in the last decade. Resequencing of 14 domesticated and 17 wild soybeans (*Glycine max*) germplasms revealed that wild germplasms have more allelic variation than cultivated germplasms, as well as substantial LD (Lam et al., 2010). Also, in rice (*O. sativa*), 40 domesticated and 10 wild germplasms were resequenced, giving rise to the discovery of genes with varying levels of genetic variation across wild and domesticated accessions. Domestication, resistance to diseases, and flowering were all linked to these genes (Xu et al., 2012). Furthermore, a rice diversity map was created from 446 wild accessions of rice crop species, including *O. rufipogon*, which revealed loci connected to domestication-related genes such as hull color, the width of grain, and seed-shattering properties (Huang et al., 2012a). Using a single-molecule real-time (SMRT) sequencing technique paired with fosmid libraries, Wu et al. (2018a) provide the 777 Mb de novo generated genome sequence of *Oryza granulate*. Numerous functional routes involved in DNA integration, protein translation, photosynthesis, and ATP production are enhanced in *O. granulate*, according to this analysis (Wu et al., 2018a). They created a high-quality *O. granulata* standard genome sequence. Comparative genomic analysis of *Oryza* genomes shows the mechanisms causing genome expansion, gene adaptation, and centromere evolution, among others (Wu et al., 2018a). In some other analyses, 75 maize lines were resequenced to analyze the development of modern maize, encompassing wild type, landraces (LRs), and enhanced cultivars. The research found several genes associated with selection and indications of favorable genes being introgressed from wild ancestors (Huang et al., 2012a). Strong selective sweep signs associated with fruit ripening, resistance to diseases, dormancy of seeds, and transcription factors including ethylene-responsive factors and basic helix-loop-helix were found in 2.6% of the genome of pepper (Piper nigrum)

(Qin et al., 2014). Whole-genome replication, reorganization of sequences, and proliferation of transposable elements (TEs) have all intensified the complexity of plant genome organization (Adams and Wendel, 2005). Even closely related species, like those in the genus *Oryza* or *Arabidopsis*, have shown significant differences in gene correlations and genomic size (Chen et al., 2013; Piegu et al., 2006). Comparative examination of orthologous sections or whole-genomic sequences between similar species, particularly within the same genus, may provide more insight into genome evolution mechanisms. Domesticated crops have experienced centuries of recombination, genetic drift, and natural selection, whereas wild species are much more ancient and more diversified. We can differentiate between diversity which has developed as a result of demographic changes and locations as an outcome of isolation by distance (IBD) and diversification which has developed as a result of choice via isolation by environment (IBE) with a better understanding of CWR population genomics (Smykal et al., 2018). This knowledge will allow for the introduction of adaptive diversification rather than diversification for its own sake, giving breeders additional tools for crop improvement against a variety of biotic and abiotic challenges and the generation of climate-resilient cultivars.

14.6 DESIGNING MODERN CROP GENOME ARCHITECTURE BY USING CWRS

Future research should focus on expanding the gene pool by finding and introgression of additional QTLs from various wild relatives of cultivated plants, especially for traits with high importance but low variability (Brar and Khush, 2018). To accurately interpret alien chromosomal segments introgression, wild population germplasm mining, cloning of essential genes, and clustering of alien genes, ideally with various mechanisms of resistance to abiotic and biotic stress, and the quality attributes, molecular markers, and cutting-edge genomics techniques are required. Yield-enhancing loci (QTLs) or "wild species alleles" must be located and introgressed into superior breeding lines to increase existing cultivar yield levels. The development of chromosomal segment substitution lines (CSSL) from diverse wild relatives for pf gene/QTL mapping, functional genomics, and breeding will aid in the development of resistant breeding stock. Transferring genes from secondary and tertiary gene pools can benefit from identifying genes that promote homoeologous coupling. The potential for using wild populations in genetic modification of domesticated crops appears to be more intriguing than ever before, due to advances in plant tissue culture, genomics methods, and molecular markers. Figures 14.4 and 14.5 show the breeding strategy for developing climate-resistant varieties employing advanced GAB techniques, and the application of cutting-edge breeding approaches for improving our modern varieties is explored further below.

14.6.1 MAGIC, MAGIC-PLUS, AND GLOBAL MAGIC POPULATION DEVELOPMENT FOR MAPPING THE NOVEL GENES

Traditionally, biparental populations were employed to determine the position and influence of QTLs influencing the desired trait. Typically, population groups like F2, F2:3, as well as Backcross populations are exploited to map QTLs. These populations' gene mapping drawbacks are that they only looked at two alleles and underlying genetic recombination, which restricts the accuracy of QTL identification. As a result, a newly proposed MAGIC technique to probe various alleles and provide enhanced recombination and mapping precision has been recommended (Cavanagh et al., 2008). The fundamental goal of creating MAGIC populations is to encourage intercrossing and genetic rearrangement. The benefits of utilizing multi-parent populations are: (1) more desired traits from each of the parents could be examined based upon the choice of parents being used to make the multi-parent crossings; and (2) because of the high levels of recombination, QTLs could be spotted with better accuracy and resolution (Cavanagh et al., 2008). Scientists are deeply focused on multi-parent populations as a result of the advent of high-throughput SNP genotyping technologies and developments in statistical approaches for analyzing data from this kind of population. As a result, these eight germplasms are the pioneers of an eight-way MAGIC population that is now being developed and has the potential to have a substantial impact on the breeding of eggplants (Pietro et al., 2018). Even if numerous methods and molecular markers have already been established in these crops (Pascual et al., 2015) highlighted the value of the data acquired by resequencing the pioneer parents of the tomatoes MAGIC population in the mapping accuracy of QTLs. Furthermore, only a few investigations have provided a substantial collection of molecular markers within the eggplant gene pool (Acquadro et al., 2017; Barchi et al., 2011), even though they are insufficient for deep molecular analysis of a multi-parent population and the construction of the respective saturated genetic linkage map from the founding members (Pascual et al., 2015). As a result, a few of the main goals of this study was to create a large genome variance set of polymorphic marker data between several accessions that could be used as a source of markers for the advancement of genomic platforms, reproduction, and future eggplant genetics and genomics analyses, such as genetic analysis of the MAGIC population (Pietro et al., 2019).

14.6.2 MARKER-ASSISTED BACKCROSS BREEDING TO INTROGRESS THE RESISTANT GENES

Marker-assisted selection (MAS) has aided in accurate germplasm characterization, recognition, and tracing of large numbers of genes, pyramiding of QTLs, genomic selection, allele mining, map-based cloning of QTLs/genes, and pushing the boundaries of "Omics" and genomic studies to better understand the purpose of agronomically essential genes. Genomics research has proven to be an effective tool for characterizing alien genetic differences, dissecting hidden genetic differences in wild accessions, identifying favorable wild species alleles to broaden the rice gene pool for resistance to abiotic and biotic stresses, and identifying new QTLs for yield-related attributes. Furthermore, the genotyping of various wild species has provided significant information for interpreting Oryza species evolutionary links and using this knowledge to the discovery of novel alleles for rice breeding. The International *Oryza* Map Alignment Project (I-OMAP) at Arizona Genomics Institute in Tucson, USA, has concentrated study on 3 activities: (i) production of molecular map data, CSSLs, and RILs involving diverse teams to be used in functional as well as breeding studies (ii) employing genomics to new sets of natural *Oryza* stocks for conservation, diversity, and evolutionary inquiries, and (iii) using genomics to study diversity, evolution, and conservation in new sets of naturally existing *Oryza* populations. Many wild species, including *Oryza nivara*, *Oryza barthii*, *O. rufipogon*, *Oryza glumaepatula*, *Oryza punctata*, *Oryza meridionalis*, *Oryza brachyantha*, and *Leersia perrieri*, have had their genomes analyzed. These components can be found on the National Center for Biotechnology Information (NCBI) database. Assembling of *Oryza rhizomatis*, *O. officinalis*, and *O. glaberrima* is also accessible in partnership with partners in Japan, China, Taiwan, the Philippines, and France. The genome sequence information of *Oryza* species becomes a valuable genetic asset for future studies in rice breeding, and genomics, including the knowledge of the molecular processes behind *Oryza* genetic variability and evolution. On comparative genomic of wild relatives and molecular differences during the development of the genus *Oryza*, a wealth of knowledge and information has become public. Stein et al. (2018) used 13 reference genomes from across the Oryza species tree to show that rapid species diversity is reflected by lineage-specific onset and retention of several novel components, such as potential new coding genes, transposons, and noncoding genes, despite few large-scale chromosomal alternations. This work highlights the importance of functionally related disease tolerance genes, as well as a slew of new haplotypes that could be valuable for crop protection in the future. Despite the creation of several kinds with a single resistant gene, resistance usually wears off after 3–5 years due to the emergence of new classes of rice blasts. As a result, it is critical to create varieties with broad-spectrum and long-lasting resistance spectra by combining several R genes with contiguous resistance spectra (Sharma et al., 2012). This implies the demand for a pool of R genes.

14.6.3 DESIGNING NEW CROPS VARIETIES BY CRISPR-CAS GE TECHNOLOGY

Conventional breeding methods continue to produce high-yielding crop types with increased yield, but they rely on random mutagenesis or germplasm mating. This method can take anywhere from 7 to 12 years to create an better-quality variety, so it is improbable to catch up with the expected mandate for enhanced crops (Acquaah, 2012). Though character-linked molecular markers could boost breeding efficacy, the nontargeted feature of LD, random mutagenesis, and recombination make germplasm enhancement ineffective. Specific GE techniques, such as hybrid enzymes or the clustered regularly interspaced short palindromic repeat (CRISPR)/CRISPR-associated protein (Cas) system (Jinek et al., 2012), have recently emerged, allowing for the specific target modifications of around any crop genomic sequences to produce novel variability and speed up breeding attempts. CRISPR-Cas is an RNA-mediated disruption of transgenes such as viral and plasmid DNAs that originated in bacteria or archaea immune systems, but it is now widely used in most field and horticultural crops. CRISPR-Cas has been shown to be surprisingly successful in a range of major crop species in the last three years, including rice (Jiang et al., 2013), wheat (Shan et al., 2013), and corn (Svitashev et al., 2016). Without being confined by limited available diversity, the genomes of elite cultivars can be altered in a precise manner to develop new cultivars with higher stress resistance and nutrient-use efficacy. Furthermore, GE does not add potentially harmful alleles through recombination, or crossing, nor does it necessitate time-consuming recurrent backcrossing to restore the elite genetic makeup. Crop enhancement through GE is not restricted by existing variability or the need to choose alleles across several breeding generations (Scheben et al., 2017). To incorporate specific DNA changes into the genetic makeup, DNA double-strand breaks (DSB) are induced at specified genome regions and homology-directed repair (HDR) or nonhomologous end joining (NHEJ) is stimulated. When used in plant breeding, GE can quickly produce superior varieties that are transgene-free. Several crops, particularly soybean (Curtin et al., 2011), and wheat have had their genomes edited using hybrid enzymes with attached DNA-binding domains and nucleases, like transcription activator-like effector nucleases (TALENs) and zinc finger nucleases (ZFNs). There are more than 60 crop genomes available (of higher quality) (Michael and VanBuren, 2015), and we may expect the lowering expense of sequencing to make the sequencing and assembling of every genome of minor and major crops, as well as their wild species, more accessible in the near future. Since plant genome assembly and annotation are frequently insufficient, transcriptomes may serve a significant function as an initial point for GE. Like, third generation sequencing methods, which produce long reads, and help resolve earlier limitations of de novo genome assembly confined to shorter reads, genome assembly, and annotation are projected to enhance in thoroughness and correctness (Yuan et al., 2017).

Recent articles have highlighted the possibilities for crop cultivation through GE, identifying possible domestication genes that are involved in characteristics like shattering of seeds, the architecture of growth, and blooming period. Soyk et al. (2017), for example, described and manipulated the flowering suppressor SP5G in tomatoes, decreasing the duration of harvesting by two weeks. SP5G was discovered to cause delayed blooming in long days after fine-mapping of a QTL site connected to photoperiodic responses. The researchers used genome GE to knock out combinations of SP5G and the dosage-dependent blooming activator SFT, demonstrating that SP5G is likely a suppressor of SFT. This makes SP5G a viable candidate for assisting the cultivation of tomato cousins and other species by allowing them to grow at higher latitudes. Attempting to target orthologues of domestication genes such as SP5G could hasten the domestication of a variety of wild crops. Shattering of seeds in the remote Australian rice cousin weeping grass (*Microlaena stipoides*) was reduced by chemical mutagenesis of the

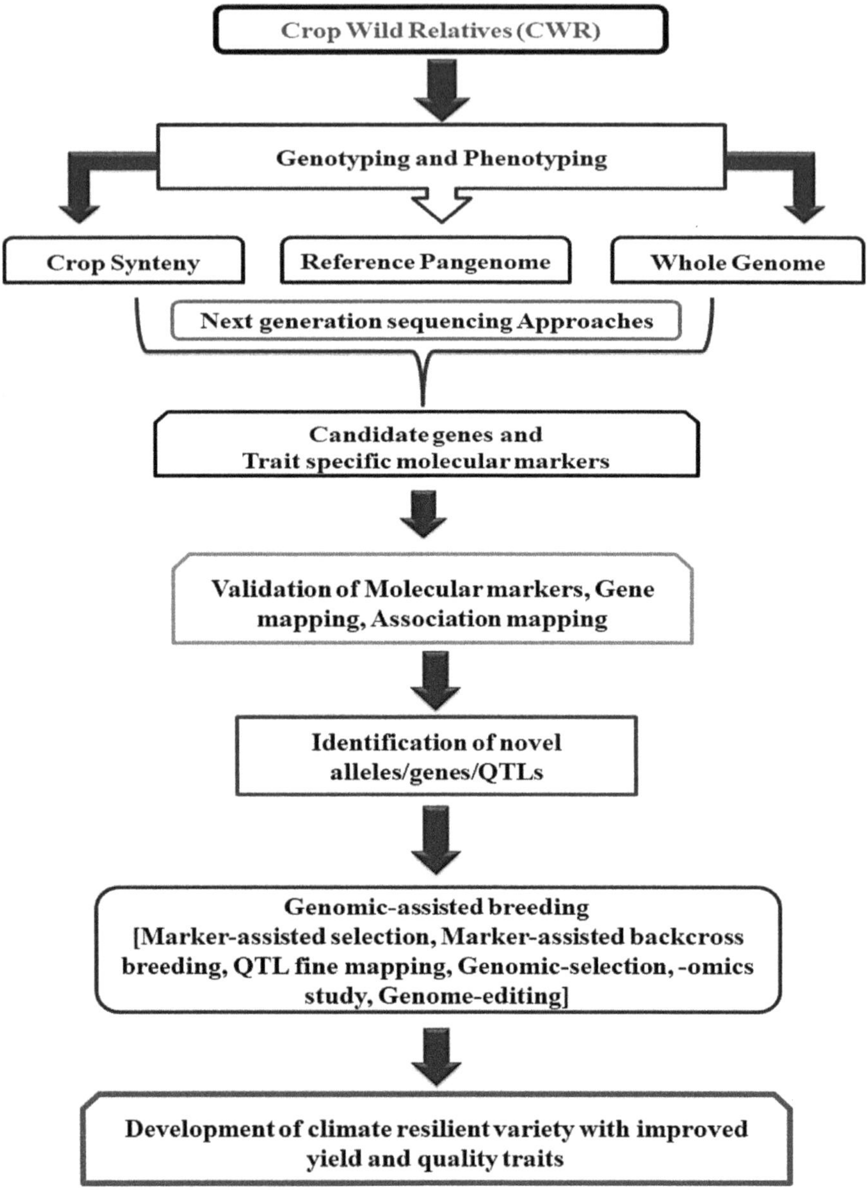

FIGURE 14.4 The procedure for the development of climate-resilient variety with improved yield and quality traits by using the CWRs.

rice orthologues sh4 and *qSH1* (Konishi et al., 2006). Pennycress (*Thlaspi arvense*), a common weed in Eurasia and North America, might be farmed into a viable, cold-resistant oilseed crop if seed pod shatter and oil quality were improved (Sedbrook et al., 2014). Wild emmer wheat (*Triticum dicoccoides*), maize progenitor teosinte (*Z. mays* ssp. *parviglumis*), and typical wild rice (*O. rufipogon*) are among the contenders for domestication, as they may possess valuable adaptive features that are lost due to domestication bottlenecks. In recent years, GE was introduced into the rubber crop *Taraxacum kok-saghyz* to disrupt a probable inhibitor of rubber production (Iaffaldano et al., 2016). As more genomes of wild populations and minor crops are sequenced, it will be easier to find orthologous domestication genes identified through well-studied plants that may be focused on GE. If suitable domestication genes are identified and the challenges of gene dosage and epistasis can be overcome, GE in minor or wild crops will considerably help to extend the crop genetic pool.

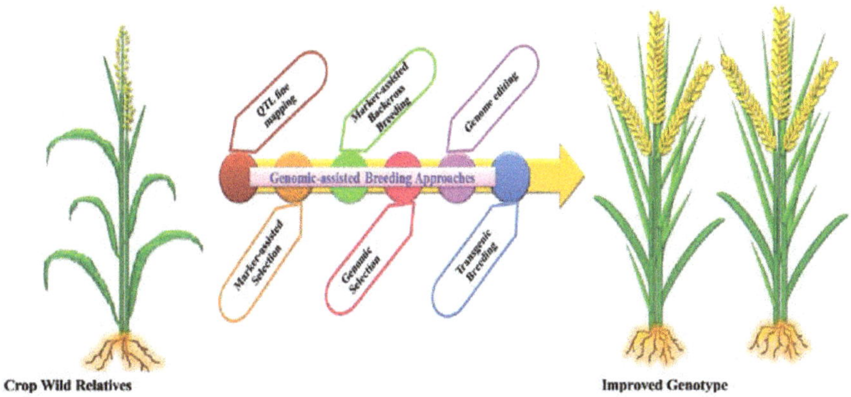

FIGURE 14.5 An illustration for development of improved genotypes by harnessing the genes from CWRs through genomic-assisted breeding approaches.

14.7 CHALLENGES IN GENE TRANSFER FROM THE WILD SOURCES AND THEIR SOLUTIONS

Despite their potential benefits, CWRs are undervalued in global gene banks (Castaeda-lvarez et al., 2016; Maxted and Kell, 2009) and underappreciated in crop breeding programs (Dulloo et al., 2013). The excess use of CWR is limited by a lack of essential data such as taxonomic classification, level of ploidy, gene-trait link, and a limitation in the prebreeding phase (Fig. 14.6). Breeders are typically concerned about introducing wild traits into elite domesticated gene pools. CWR has therefore been used in breeding to introduce a limited number of significant genes, such as pests and diseases resistance (Dulloo et al., 2013; Hajjar and Hodgkin, 2007) or cytoplasmic male sterility (CMS) line creation in crops including maize, pigeon pea, and sunflower. In bi-parental populations, essential genes are quite easy to discover and subsequently implement in mating practices. Numerous agriculturally important qualities, such as yield, adaptation, and quality, are regulated by a large number of genes, necessitating the use of more complex population development tools to detect and incorporate these genes for breeding programs. Rice (McCouch et al., 2007), maize (Yu et al., 2008), wheat (Huang et al., 2012), tomato (Tanksley and Nelson, 1996), or even narrow-leafed lupin are all good examples (Cowling et al., 2009). These methods can be refined even more by carefully selecting pioneer CWR parent germplasms and employing passport data to describe their evolutionary and ecological context, allowing for the introduction of locally relevant traits into the crop (Warschefsky et al., 2014). It is possible that wild parents of various soil conditions or locations were purposefully chosen to give adaptation to certain environments (Berger et al., 2017). When larger populations are generated using ecological and population genetic approaches to maintain as much relevant variation as possible, whole-genome sequencing information can be used to build massive introgression populations (Warschefsky et al., 2014). With a few of the attributes of Nested Association Mapping populations (Lipka et al., 2015), improved backcross introgression populations can be suitable for mapping domestication and agronomic characters as well as transmitting adaptive characters governed by one or several genes into the cultivar genetic makeups (Tanksley and McCouch, 1997). Apart from the need for massively larger collections based on ecological concepts, which we have consistently urged for, there is an increasing need to create methods for harnessing wild resources in genome-enabled crossing programs (Von Wettberg et al., 2018). Despite the fact that genome-wide profiling of most crops is becoming more feasible, even in species with large and complex genomes, such as lentil (Ogutcen et al., 2018), sequencing alone is insufficient to predict relevant phenotypes. Genomic forecasting and selection, which has become more successful in crops ranging from maize and rice to chickpeas, has gained in popularity (Zhang et al., 2017a), is basically focused on elite variety training sets. As additional variety is introduced into breeding programs, forecasting models will have to be rebuilt in every cycle, resulting in increased phenotypic and computing loads and expenses. Despite the fact that genome-wide profiling of most crops is becoming more feasible, even in species with large and complex genomes, such as lentil (Ogutcen et al., 2018), sequencing alone is insufficient to predict relevant phenotypes. Genomic forecasting and selection, which has become more successful in crops ranging from maize and rice to chickpeas, has gained in popularity. Although crop improvement through the utilization of wild relatives of farmed crop species is desirable, linkage drag limits its effectiveness. Using new advanced GE techniques like the CRISPR-Cas9 system, this is now conceivable. Zsögön et al. (2018) used the same strategy to modify tomato (*S. lycopersicum*) varieties by modifying the transferred genome of wild species.

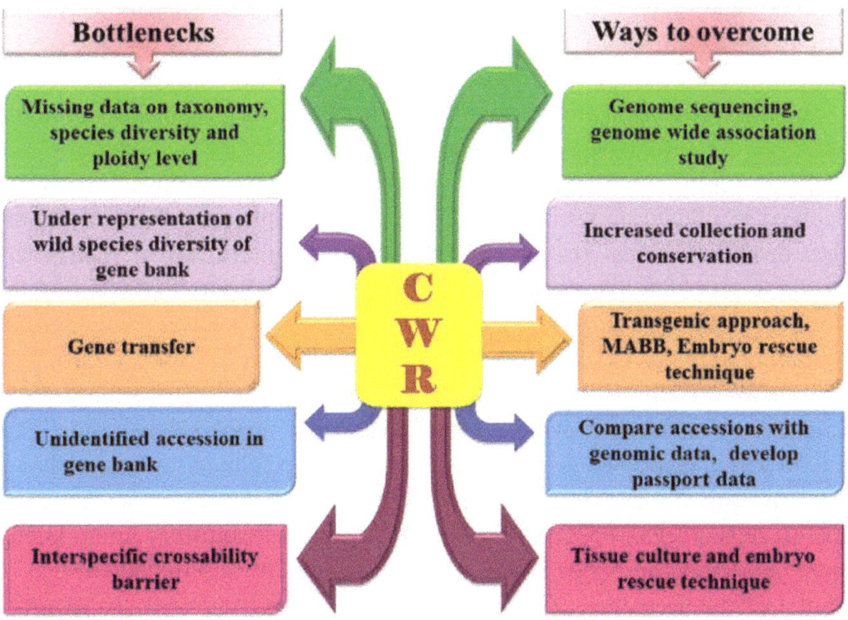

FIGURE 14.6 The bottlenecks in breeding with CWRs and their ways to overcome the problems.

Abbreviation: CWR, Crop wild relative; MABB, Marker-assisted backcross breeding.

14.8 THE THREAT OF CLIMATE CHANGE AND CONSERVATION STRATEGY FOR CWRS

Crop wild species are important genetic assets for improving crops. Our farmed crops are encountering several issues as a result of climate change. Consequently, we must add resistant genes from natural sources, such as wild species of domesticated crop plants. Because of their genetic background, they are prepared to face any form of climatic fluctuation. Their utility in plant breeding is limited by knowledge gaps, which include distributions, taxonomy, and characterization for desired characteristics. The conservation strategies for CWRs looking at the consequence of climate change are illustrated in Figure 14.7. These problems also have an impact on conservation programs, both in terms of the gathering of a new variety for ex situ conservation and in situ habitat protection (Lane, 2006). Twenty-four to 31 out of 51 *Arachis* species, 8–13 out of 107 *Solanum* species, and 1–3 out of 48 *Vigna* species are anticipated to become lost by 2055. The analysis also predicted that ecosystem patches would become more scattered, putting them at greater risk of depletion (Lane, 2006). Crop wild species with close genetic ties to cultivated plants are valuable genetic resources for crop development, but knowledge gaps limit their preservation and utility.

In 2017, 96 organizations ex situ preserved 89,288 germplasms of *Vigna* taxa. Over 53,756 of these germplasms were kept by eight organizations. The Svalbard Global Seed Vault (SGSV, 2018, https://www.nordgen.org/sgsv/) received 31,500 germplasms from 25 taxa as a backup copy in 2018, mostly from the International Institute of Tropical Agriculture (IITA), WorldVeg, and the Centro Internacional de Agricultura Tropical (CIAT). IITA had the greatest stocks of *V. unguiculata* (cowpea) and *Vigna subterranea* (Bambara groundnut) among the cultivated taxa (van Zonneveld et al., 2020). The provenance of the *V. unguiculata* group sesquipedalis (yard-long bean) collections is uncertain because not all gene banks provided taxonomic data at the species level, which is required for this crop. The largest set of *Vigna angularis* and *V. radiata* (mung bean) were found at WorldVeg (azuki bean). The greatest stocks of *Vigna mungo* (urd bean), *V. aconitifolia* (moth bean), and *Vigna umbellata* (rice bean) were maintained by the Indian Bureau of Plant Genetic Resources (NBPGR). The CIAT had the most *V. vexillata* collections (tuber cowpea). The Japanese Agriculture and Food Science Organization's Genetic Resources Center had the biggest collection of *Vigna reflexo-pilosa* (creole bean). IITA and The Australian Grains Genebank housed significant sets of African wild Vigna, whereas NBPGR and the Japanese Agriculture and Food Science Organization housed significant collections of wild Asian *Vigna*. Meise, Belgium's national botanic garden, has the most diversified Vigna stocks, but only a small number of germplasms per taxon (van Zonneveld et al., 2020).

Crop diversity might be necessary for future climatic adaptation, and its preservation is dependent on human activities and choices. Aside from that, we highlight how measures favoring the use of modern cultivars might hasten genetic alterations in indigenous LRs and CWRs/cousins (WRs). Mexico is known for being the birthplace and domestication hub of maize. Concerns about the impacts of gene flow from current maize varieties (MVs) toward maize LRs and their wild cousins, teosintes, were raised when MVs were introduced into Mexico. Rojas–Barrera et al. (2019) evaluated maize WRs and LRs across a wide variety of temporal datasets. For nearly 60 years, modern varieties, LRs, and their WRs have already been in sympatry. The present genomic findings of *Zea mays* L. continuous evolution as a result of MV introgression. These results should encourage monitoring programs and policies that use and protect the genetic diversity of maize and WRs at their source (Rojas-Barrera et al., 2019). Moreover, over 60 years after the first MVs were released, the influence of sympatry on LRs and associated WRs has yet to be investigated using genetic data. We examined alterations in the genomes of 2 WR subspecies, and 7 maize LRs from 70-year-old stocks in this study. We investigated the genomes of WRs and LRs derived by genotyping by sequencing (GBS) prior to and after the adoptions of MVs and demonstrated sympatric MV gene flow into LRs and the WR *Zea mays* ssp. mexicana studied after the year 2000 (Rojas–Barrera et al., 2019). These observations can be utilized to develop agricultural policies and monitoring strategies aimed at achieving biodiversity goals for the protection of gene repositories at the core of diversity (Rojas–Barrera et al., 2019). The following were necessary for a simple conceptual model for crop wild cousins' preservation:

14.8.1 SPREADING AWARENESS OF THE RELEVANCE OF CWR

Recognizing the commercial worth of such a biological resource helps in encouraging conservation efforts. To promote knowledge of the importance of CWR on a global and regional scale, concerted initiatives are required. Botanical gardens can help raise public awareness and educate people about the relevance of CWR and the consequences of climate change.

14.8.2 INCREASE THE USAGE OF CWR

Many crops wild species are actively employed in breeding programs. Several African wild cowpea species contribute toward food security by consuming their tubers, seeds, and fruit (Padulosi and Ng, 1990). In Madagascar, wild yams (*Dioscorea* spp.) constitute a major source of carbs and income, while in the Caucuses, and Central Asia, wild fruits including apple, sea buckthorn, and pistachio are picked for food. Building materials, animal fodder, and pharmaceuticals are all produced by CWR species. On a local level, over-producing of some CWR out from their wild relatives poses a threat to their survival. This threat could be mitigated by increasing local communities' ability to cultivate harvested plants.

14.8.3 IMPROVING CWR PROFILE KNOWLEDGE

While some models currently consider species migration potential and morphological plasticity, it is impossible to account for all events that influence a species actively or passively. Combining modeling with field assessment will lead to more precise projections of the effects of climate change on species distribution, even though it will be more expensive.

14.8.4 STRATEGY FOR CHANGING CLIMATE

When formulating a method for choosing new conservation areas and preserving existing ones, the changing composition, dynamics, and structure, of ecosystems due to climate change must be taken into account. Climate change and habitat fragmentation are severe conservation challenges when they are combined. Fragmentation obstructs migration and recolonization, and as habitat regions shrink, they get more vulnerable.

14.8.5 MIGRATION ROUTES

Areas protected to link landscape pieces can help mobile species migrate ranges and increase the conservation zone.

14.8.6 PRIORITIZING SPECIES AND PLACES FOR CONSERVATION MEASURES

Prioritizing species and areas would include points like economic and social advantages, population viability, conservation status, and economic feasibility for proper management. The genetic diversity of CWR populations is very significant. Disease endurance and tolerance are frequently linked to a small group of plants in a given place, like resistance to the Russian wheat aphid from Iran, resistance to the Ethiopian Highlands barley yellow dwarf virus, and adjacent countries. This emphasizes the value of field-based research and preservation efforts tailored to individual areas and species. Rare or narrowly dispersed species should be considered for conservation since they will face larger consequences than more widespread and broadly distributed species. Restricted climate tolerances and decreasing distributions are linked to a greater risk of extinction. The significance of conservation areas as climatic refugia for vulnerable species, genes for ex situ collections, sources of propagules for recolonization, should all be considered. The number of individuals found in a given location determines the likelihood that the species will sustain. Genetic reserves should comprise target populations big enough to assure persistence. Environmental change is expected to result in the extinction of species in conservation areas. Reservation networks can be developed for species with spanning distributions to reflect both present and future anticipated distributions; for organisms with no overlapping in distribution, different areas must be established. These could include places where climate change is projected to have a minor influence and places where new environments could provide habitat for emigration species. Because plant populations prefer to migrate to higher elevations (Parmesan and Yohe, 2003), creating conservation lands in hilly areas could be a goal.

14.8.7 RELOCATION

Populations could be relocated to locations with a more pleasant climate. Physically inactive species translocation is especially crucial for preventing irreversible extinction of species and gene pools. Botanic gardens can play an important part in this research by disseminating germplasm to different gardens across the world and tracking the progress of species colonization. Preservation can be more beneficial when in situ conservation is combined with strategic restoration, such as connecting conservation areas through corridors.

14.8.8 EX SITU CONSERVATION (COLLECTING VULNERABLE POPULATIONS AND SPECIES)

Despite in situ conservation programs, certain species may be unable to relocate or adapt to changing climates. Ex situ techniques are essential for these species. Maintenance in gene banks and creation as live collections are two ex situ approaches. Botanic gardens could be a workable alternative for species that are vulnerable to climate change and cannot be conserved in their natural habitat.

14.8.9 TRACKING CHANGES IN VEGETATION

Vegetation tracking is often employed to monitor and anticipate alterations in species makeup and migration of plants induced by habitation loss, fires, and pests, as well as change in the climate. Observing species over time sets the foundation for detecting climate-induced changes in vegetation. In addition to CWR groups that frequently occur, indicator species that are not resistant to environmental conditions and inhabit confined habitats within higher altitude and humidity envelopes should be examined.

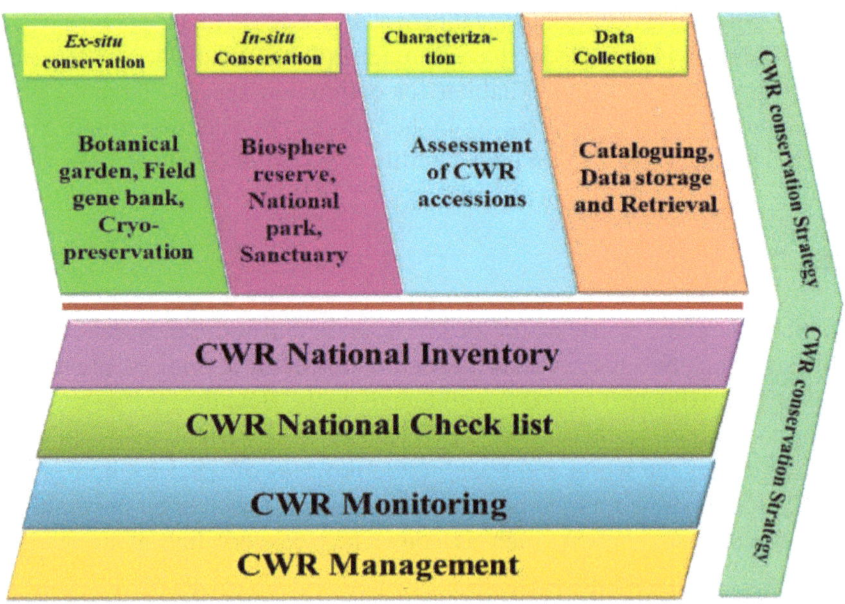

FIGURE 14.7 The conservation strategies for CWRs looking to the consequence of climate change.

14.9 CONCLUSION AND FUTURE OUTLOOK

CWRs are the main basis of genetic variation intended for crop development and genetic variability/diversity, and they are becoming progressively crucial as a result of climate changes and other unpredictable circumstances. CWR could be a repository for a variety of new features that confer resistance to various biotic and abiotic challenges. We will require higher-yielding and more nutritional food crop types that are resilient to a variety of abiotic or biotic challenges in the near future as a consequence of climate change. Consequently, plant breeders are concerned about the long-term genetic enhancement of crops. In the age of global climate change, a continuous infusion of fresh genetic diversity is required for genetic strategy and breeding to be effective. Many different lines can be developed using genetic variability for commercial type defect corrections, production of transgressive segregants, and production of new varieties with higher levels of heterosis. Within the basic gene pool, variability for diverse traits is constrained or even depleted. Gene pools in the secondary, tertiary, and even quaternary stages are a vast and underused resource of appealing alien genes which can be integrated to cultivated genotypes using techniques such as wide hybridization, genetic transformation with transgenic methods, GE technologies, and marker-assisted backcross breeding. Future explorative studies need to be done on the C4-photosynthetic mechanism of wild cereal relatives and apomixis in wild relatives, novel sources for increased water and nutrient acquisition, salt tolerance, and the innovation of a fast-breeding program for developing wild species to synchronize the inflorescence period and improve seed set when hybridized with cultivated crops. For a more in-depth look of nonessential and species-specific genes, pangenome development is critical. It could help with the uncovering of adaptation genes and the development of methods for importing or producing environmentally responsive genotypes. The variations revealed through pangenome research can be used as markers for marker-added selection, allowing for the introduction of desirable CWR traits into cultivated crops. A super pangenome combines the different pangenomes from many species of a genus to reflect the whole-genomic architecture of that genus. The whole-genetic diversity would be realized by superimposing distinct germplasms from different species. Despite the fact that extensive hybridization has been used to introduce alien genes into cultivated types all over the world, its application as a frontline technique is limited because of unavoidable factors such as F_1 sterility, sexual incompatibility, hybrid inviability. Gene transfer via genetic engineering is a well-existing method that can offer high-yielding superior lines with biotic and abiotic stress resilience by introgression of genes from a noncultivable gene pool, while also diversifying the primary gene pool in a long-term manner. Furthermore, in the case of alien gene introgression, the inherent issues of crossability, sexual incompatibility, fertility, gene recombination, and linkage drag must be discussed and resolved using proper processes in combination with modern molecular and biotechnological tools such as marker-added selection, QTL analysis, and genome-wide association studies (GWAS).

KEYWORDS

- **genome**
- **crop wild relatives**
- **climate change**
- **biotic and abiotic stresses**
- **MAGIC population**
- **genome editing**
- **CRISPR-Cas**
- **QTL Mapping**
- **marker-assisted backcrossing**
- **comparative genomics**

REFERENCES

Abbo, S.; Gopher, A. Near Eastern Plant Domestication: A History of Thought. *Trends Plant Sci.* **2017,** *22* (6), 491–511. https://doi.org/10.1016/j.tplants.2017.03.010

Abbo, S.; Berger, J.; Turner, N. C. Evolution of Cultivated Chickpea: Four Bottlenecks Limit Diversity and Constrain Adaptation. *Funct. Plant Biol.* **2003,** *30* (10), 1081–1087. https://doi.org/10.1071/FP03084

Acquaah, G. *Principles of Plant Genetics and Breeding*; John Wiley & Sons Ltd., 2012. ISBN: 9781118313718. |DOI:10.1002/9781118313718

Acquadro, A.; Barchi, L.; Gramazio, P.; Portis, E.; Vilanova, S.; Comino, C.; Plazas, M.; Prohens, J.; Lanteri, S. Coding SNPs Analysis Highlights Genetic Relationships and Evolution Pattern in Eggplant Complexes. *PLoS ONE* **2017,** *12* (7), e0180774. https://doi.org/10.1371/journal.pone.0180774

Adams, K. L.; Wendel, J. F. Polyploidy and Genome Evolution in Plants. *Curr. Opin. Plant Biol.* **2005,** *16* (7), 1667–1678. https://doi.org/10.1016/j.pbi.2005.01.001

Alföldi, J.; Lindblad-Toh, K. Comparative Genomics as a Tool to Understand Evolution and Disease. *Genome Res.* **2013,** *23* (7), 1063–1068.https://doi.org/10.1101/gr.157503.113

Amante-Bordeos, A.; Sitch, L. A.; Nelson, R.; Dalmacio, R. D.; Oliva, N. P.; Aswidinnoor, H.; Leung, H. Transfer of Bacterial Blight and Blast Resistance from the Tetraploid Wild

Rice *Oryza minuta* to Cultivated Rice. *Theor. Appl. Genet.* **1992,** *84* (3), 345–354. https://doi.org/10.1007/BF00229493

Barchi, L.; Lanteri, S.; Portis, E.; Acquadro, A.; Valè, G.; Toppino, L.; & Rotino, G. L. Identification of SNP and SSR Markers in Eggplant Using RAD Tag Sequencing. *BMC Genom.* **2011,** *12* (1), 1–9. https://doi.org/10.1186/1471–2164–12–304

Berger, J.; Abbo, S.; Turner, N. C. Ecogeography of Annual Wild Cicer Species: The Poor State of the World Collection. *Crop Sci.* **2003,** *43* (3), 1076–1090. https://doi.org/10.2135/cropsci2003.1076

Berger, Jens D.; Shrestha, D.; Ludwig, C. Reproductive Strategies in Mediterranean Legumes: Trade-Offs Between Phenology, Seed Size and Vigor Within and Between Wild and Domesticated Lupinus Species Collected Along Aridity Gradients. *Front. Plant Sci.* **2017,** *8*, 548. https://doi.org/10.3389/fpls.2017.00548

Bhanu, A. N. *Adv. Plants Agric. Res.* **2017,** *7* (3), 279–286. https://doi.org/10.15406/apar.2017.07.00255

Blank, L. M.; Leathers, C. R. Environmental and Other Factors Influencing Development of Southwestern Cotton Rust (*Puccinia stakmanii*). Phytopathology **1963,** *53*, 921–928.

Brar, D. S.; Khush, G. S. Wild Relatives of Rice: A Valuable Genetic Resource for Genomics and Breeding Research. **2018,** 1–25. https://doi.org/10.1007/978–3-319–71997–9_1

Brozynska, M.; Furtado, A.; Henry, R. J. Genomics of Crop Wild Relatives: Expanding the Gene Pool for Crop Improvement. *Plant Biotechnol. J.* **2016,** *14* (4), 1070–1085. https://doi.org/10.1111/pbi.12454

Candolle, A. de. Origin of Cultivated Plants. In *Origin of Cultivated Plants*, Vol. 48; 2012. https://doi.org/10.5962/bhl.title.57520

Cao, J.; Schneeberger, K.; Ossowski, S.; Günther, T.; Bender, S.; Fitz, J.; Koenig, D.; Lanz, C.; Stegle, O.; Lippert, C.; Wang, X.; Ott, F.; Müller, J.; Alonso-Blanco, C.; Borgwardt, K.; Schmid, K. J.; & Weigel, D. Whole-Genome Sequencing of Multiple *Arabidopsis thaliana* Populations. *Nat. Genet.* **2011,** *43* (10), 956–963. https://doi.org/10.1038/ng.911

Castañeda-Álvarez, N. P.; Khoury, C. K.; Achicanoy, H. A.; Bernau, V.; Dempewolf, H.; Eastwood, R. J.; Guarino, L.; Harker, R. H.; Jarvis, A.; Maxted, N.; Müller, J. V.; Ramirez-Villegas, J.; Sosa, C. C.; Struik, P. C.; Vincent, H.; Toll, J. Global Conservation Priorities for Crop Wild Relatives. *Nat. Plants.* **2016,** *2* (4), 1–6. https://doi.org/10.1038/NPLANTS.2016.22

Cavanagh, C.; Morell, M.; Mackay, I.; Powell, W. From Mutations to MAGIC: Resources for Gene Discovery, Validation and Delivery in Crop Plants. *Curr. Opin. Plant Biol.* **2008,** *11*, 215–221.

Chaudhary, H. K.; Kaila, V.; Rather, S. A.; Badiyal, A.; Hussain, W.; Jamwal, N. S.; Mahato, A. In *Alien Gene Transfer in Crop Plants. Wheat* **2014,** *2*, 1–26. https://doi.org/10.1007/978-1-4614-9572-7_1

Chen, F.; Dong, W.; Zhang, J.; Guo, X.; Chen, J.; Wang, Z.; Lin, Z.; Tang, H.; Zhang, L. The Sequenced Angiosperm Genomes and Genome Databases. *Front. Plant Sci.* **2018,** *9*, 418. https://doi.org/10.3389/fpls.2018.00418

Chen, J.; Huang, Q.; Gao, D.; Wang, J.; Lang, Y.; Liu, T.; Li, B.; Bai, Z.; Luis Goicoechea, J.; Liang, C.; Chen, C.; Zhang, W.; Sun, S.; Liao, Y.; Zhang, X.; Yang, L.; Song, C.; Wang, M.; Shi, J. et al. Whole-Genome Sequencing of *Oryza brachyantha* Reveals Mechanisms Underlying Oryza Genome Evolution. *Nat. Commun.* **2013,** *4* (1), 1–9. https://doi.org/10.1038/ncomms2596

Choi, J. Y.; Platts, A. E.; Fuller, D. Q.; Hsing, Y. I.; Wing, R. A.; Purugganan, M. D.; Kim, Y. The Rice Paradox: Multiple Origins But Single Domestication in Asian Rice. *Mol. Biol. Evol.* **2017,** *34* (4), 969–979. https://doi.org/10.1093/molbev/msx049

Civáň, P.; Brown, T. A. Origin of Rice (*Oryza sativa* L.) Domestication Genes. *Genet. Resour. Crop Evol.* **2017,** *64* (6), 1125–1132. https://doi.org/10.1007/s10722–017–0518–0

Civán, P.; Craig, H.; Cox, C. J.; Brown, T. A. Three Geographically Separate Domestications of Asian rice. *Nat. Plants.* **2015,** *1* (11), 1–5. https://doi.org/10.1038/nplants.2015.164

Civáň, P.; Ivaničová, Z.; Brown, T. A. Reticulated Origin of Domesticated Emmer Wheat Supports a Dynamic Model for the Emergence of Agriculture in the Fertile Crescent. *PLoS ONE* **2013,** *8* (11), e81955. https://doi.org/10.1371/journal.pone.0081955

Cook, D. E.; Lee, T. G.; Guo, X.; Melito, S.; Wang, K.; Bayless, A. M.; Wang, J.; Hughes, T. J.; Willis, D. K.; Clemente, T. E.; Diers, B. W.; Jiang, J.; Hudson, M. E.; Bent, AF. Copy Number Variation of Multiple Genes at Rhg1 Mediates Nematode Resistance in Soybean. *Science* **2012,** *338* (6111), 1206–1209. https://doi.org/10.1126/science.1228746

Cortés, A. J.; Cornille, A.; Yockteng, R. Evolutionary Genetics of Crop-Wild Complexes. *Genes* **2022,** *13*, 1. https://doi.org/10.3390/ genes13010001

Cowling, W. A.; Buirchell, B. J.; Falk, D. E. A Model for Incorporating Novel Alleles from the Primary Gene Pool into Elite Crop Breeding Programs While Reselecting Major Genes for Domestication or Adaptation. *Crop Pasture Sci.* **2009,** *60* (10), 1009–1015. https://doi.org/10.1071/CP08223

Curtin, S. J.; Zhang, F.; Sander, J. D.; Haun, W. J.; Starker, C.; Baltes, N. J. et al. Targeted Mutagenesis of Duplicated Genes in Soybean with Zinc-Finger Nucleases. *Plant Physiol.* **2011,** *156* (2), 466–473.

Dempewolf, H.; Baute, G.; Anderson, J.; Kilian, B.; Smith, C.; Guarino, L. Past and Future Use of Wild Relatives in Crop Breeding. *Crop Sci.* **2017,** *57* (3), 1070–https://doi.org/10.2135/cropsci2016.10.0885

Dulloo, M. E.; Thormann, I.; Fiorino, E.; De Felice, S.; Rao, V. R.; Snook, L. Trends in Research Using Plant Genetic Resources from Germplasm Collections: From 1996 to 2006. *Crop Sci.* **2013.** https://doi.org/10.2135/cropsci2012.04.0219

FAO and A. O. of the U. N. *Climate Data Explorer*; World Resources Institute, 2015.

Foolad, M. R.; Merk, H. L.; Ashrafi, H. Genetics, Genomics and Breeding of Late Blight and Early Blight Resistance in Tomato. *Crit. Rev. Plant Sci.* **2008,** *27* (2), 75–107.

Fukunaga, K.; Hill, J.; Vigouroux, Y.; Matsuoka, Y.; Sanchez, J.; Liu, K.; Buckler, E. S.; Doebley, J. Genetic Diversity and Population Structure of Teosinte. *Genetics* **2005,** *169*, 2241–2254.

Gao, L.; Gonda, I.; Sun, H.; Ma, Q.; Bao, K.; Tieman, D. M.; Burzynski-Chang, E. A.; Fish, T. L.; Stromberg, K. A.; Sacks, G. L.; Thannhauser, T. W.; Foolad, M. R.; Diez, M. J.; Blanca, J.; Canizares, J.; Xu, Y.; van der Knaap, E.; Huang, S.; Klee, H. J. et al. The Tomato Pan-Genome Uncovers New Genes and a Rare Allele Regulating Fruit Flavor. *Nat. Genet.* **2019,** *51* (6), 1044–1051. https://doi.org/10.1038/s41588–019–0410–2

Geng, X. S.; Yang, M. Z.; Huang, X. Q.; Cheng, Z. Q.; Fu, J.; Sun, T.; Li, J. Cloning and Analyzing of Rice Blast Resistance Gene Pi-ta+ Allele from Jinghong Erect Type of Common Wild Rice (*Oryza rufipogon* Griff) in Yunnan. *Yi Chuan = Hereditas/ Zhongguo Yi Chuan Xue Hui Bian Ji* **2008,** *30* (1), 109–114. https://doi.org/10.3724/SP. J.1005.2008.00109

Gepts, P. Crop Domestication as a Long-Term Selection Experiment. In Plant Breed. *Reviews* **2010,** *24* (2), 1–44. https://doi.org/10.1002/9780470650288.ch1

Golicz, A. A.; Bayer, P. E.; Barker, G. C.; Edger, P. P.; Kim, H. R.; Martinez, P. A.; Chan, C. K.K.; Severn-Ellis, A.; McCombie, W. R.; Parkin, I. A.P.; Paterson, A. H.; Pires, J. C.; Sharpe, A. G.; Tang, H.; Teakle, G. R.; Town, C. D.; Batley, J.; Edwards, D. The Pangenome of an Agronomically Important Crop Plant *Brassica oleracea*. *Nat. Commun.* **2016,** *7* (1), 1–8. https://doi.org/10.1038/ncomms13390

Hajjar, R.; Hodgkin, T. The Use of Wild Relatives in Crop Improvement: A Survey of Developments Over the Last 20 Years. *Euphytica*. 2007. https://doi.org/10.1007/s10681–007–9363–0

Harlan, J. R.; Wet, J. M.J. Toward a Rational Classification of Cultivated Plants. *Taxon* 1971. https://doi.org/10.2307/1218252

Hartman, G. L.; Wang, T. C.; Hymowitz, T. Sources of Resistance to Soybean Rust in Perennial *Glycine* Species. *Plant Dis.* **1992,** *76*, 396–399. DOI: 10.1094/PD-76–0396

Higdon, J. V.; Delage, B.; Williams, D. E.; Dashwood, R. H. Cruciferous Vegetables and Human Cancer Risk: Epidemiologic Evidence and Mechanistic Basis. *Pharmacol. Res.* **2007,** *55* (3), 224–236. https://doi.org/10.1016/j.phrs.2007.01.009

Huang, X.; Kurata, N.; Wei, X.; Wang, Z. X.; Wang, A.; Zhao, Q.; Zhao, Y.; Liu, K.; Lu, H.; Li, W.; Guo, Y.; Lu, Y.; Zhou, C.; Fan, D.; Weng, Q.; Zhu, C.; Huang, T.; Zhang, L.; Wang, Y. et al. A Map of Rice Genome Variation Reveals the Origin of Cultivated Rice. *Nature* **2012,** *490* (7421), 497–501. https://doi.org/10.1038/nature11532

Hübner, S.; Bercovich, N.; Todesco, M.; Mandel, J. R.; Odenheimer, J.; Ziegler, E.; Lee, J. S.; Baute, G. J.; Owens, G. L.; Grassa, C. J.; Ebert, D. P.; Ostevik, K. L.; Moyers, B. T.; Yakimowski, S.; Masalia, R. R.; Gao, L.; Ćalić, I.; Bowers, J. E.; Kane, N. C. et al. Sunflower Pan-Genome Analysis Shows That Hybridization Altered Gene Content and Disease Resistance. *Nat. Plants* **2019,** *5* (1), 54–62. https://doi.org/10.1038/s41477-018-0329-0

Hufford, M. B.; Lubinksy, P.; Pyhäjärvi, T.; Devengenzo, M. T.; Ellstrand, N. C.; Ross-Ibarra, J. The Genomic Signature of Crop-Wild Introgression in Maize. *PLoS Genet.* **2013,** *9* (5), e1003477. https://doi.org/10.1371/journal.pgen.1003477

Hunter, M. C.; Smith, R. G.; Schipanski, M. E.; Atwood, L. W.; Mortensen, D. A. Agriculture in 2050: Recalibrating Targets for Sustainable Intensification. *BioScience* **2017,** *67* (4), 386–391. https://doi.org/10.1093/biosci/bix010

Iaffaldano, B.; Zhang, Y.; Cornish, K. CRISPR/Cas9 Genome Editing of Rubber Producing Dandelion *Taraxacum kok-saghyz* Using *Agrobacterium rhizogenes* Without Selection. *Ind. Crops Prod.* **2016,** *89*, 356–362.

James, R. A.; von Caemmerer, S.; Condon, A. T.; Zwart, A. B.; Munns, R. Genetic Variation in Tolerance to the Osmotic Stress Component of Salinity Stress in Durum Wheat. *Funct. Plant Biol.* **2008,** *35*, 111–123. DOI: 10.1071/FP07234

Jeung, J. U.; Kim, B. R.; Cho, Y. C.; Han, S. S.; Moon, H. P.; Lee, Y. T.; Jena, K. K. A Novel Gene, Pi40 (t), Linked to the DNA Markers Derived from NBS-LRR Motifs Confers Broad Spectrum of Blast Resistance in Rice. *Theor. Appl Genet.* **2007,** *115* (8), 1163–1177. https://doi.org/10.1007/s00122–007–0642-x

Ji, Y.; Scott, J. W.; Hanson, P.; Graham, E. & Maxwell, D. P. Sources of Resistance, Inheritance, and Location of Genetic Loci Conferring Resistance to Members of the Tomato-Infecting Begomoviruses. In: *Tomato Yellow Leaf Curl Virus Disease: Management, Molecular Biology, Breeding for Resistance*; Czosnek, H., Ed.; Kluwer: Dordrecht, 2007; pp 343–362.

Ji, Y.; Scott, J. W.; Maxwell, D. P. & Schuster, D. J. Ty-4, a Tomato Yellow Leaf Curl Virus Resistance Gene on Chromosome 3 of Tomato. *Rep. Tomato Genet. Coop.* **2008,** *58*, 29–31.

Jiang, W.; Bikard, D.; Cox, D.; Zhang, F.; Marraffini, L. A. RNA-Guided Editing of Bacterial Genomes Using CRISPR-Cas Systems. *Nat. Biotechnol.* **2013,** *31* (3), 233–239.

Jinek, M.; Chylinski, K.; Fonfara, I.; Hauer, M.; Doudna, J. A.; Charpentier, E. A Programmable Dual-RNA–Guided DNA Endonuclease in Adaptive Bacterial Immunity. *Science* **2012,** *337* (6096), 816–821.

Jingyin, Y.; Golicz, A. A.; Lu, K.; Dossa, K.; Zhang, Y.; Chen, J.; Wang, L.; You, J.; Fan, D.; Edwards, D.; Zhang, X. Insight into the Evolution and Functional Characteristics of the Pan-Genome Assembly from Sesame Landraces and Modern Cultivars. *Plant Biotechnol. J.* **2019,** *17* (5), 881–892. https://doi.org/10.1111/pbi.13022

Kantar, M. B.; Nashoba, A. R.; Anderson, J. E.; Blackman, B. K.; Rieseberg, L. H. The Genetics and Genomics of Plant Domestication. *BioScience* **2017,** *67* (11), 971–982. https://doi.org/10.1093/biosci/bix114

Kashiwaba, K.; Tomooka, N.; Kaga, A.; Han, O. K.; Vaughan, D. A. Characterization of Resistance to Three Bruchid Species (*Callosobruchus* spp.; *Coleoptera, Bruchidae*) in Cultivated Rice Bean (*Vigna umbellata*). *J. Econ. Entomol.* **2003,** *96* (1), 207–213.

Khan, A. W.; Garg, V.; Roorkiwal, M.; Golicz, A. A.; Edwards, D.; Varshney, R. K. Super-Pangenome by Integrating the Wild Side of a Species for Accelerated Crop Improvement. *Trends Plant Sci.* **2020,** *25* (2), 148–158. https://doi.org/10.1016/j.tplants.2019.10.012

Khoury, C. K.; Bjorkman, A. D.; Dempewolf, H.; Ramirez-Villegas, J.; Guarino, L.; Jarvis, A.; Rieseberg, L. H.; Struik, P. C. Increasing Homogeneity in Global Food Supplies and the Implications for Food Security. *Proc. Natl. Acad. Sci. USA* **2014,** *111* (11), 4001–4006. https://doi.org/10.1073/pnas.1313490111

Khoury, C. K.; Carver, D.; Kates, H. R.; Achicanoy, H. A.; Zonneveld, M.; Thomas, E.; Heinitz, C.; Jarret, R.; Labate, J. A.; Reitsma, K.; Nabhan, G. P.; Greene, S. L. Distributions, Conservation Status, and Abiotic Stress Tolerance Potential of Wild Cucurbits (*Cucurbita* L.). *Plants People Planet* **2020,** *2* (3), 269–283. https://doi.org/10.1002/ppp3.10085

Kioukis, A.; Michalopoulou, V. A.; Briers, L.; Pirintsos, S.; Studholme, D. J.; Pavlidis, P.; Sarris, P. F. Draft Genome Sequence and Intraspecific Diversification of the Wild Crop Relative *Brassica cretica* Lam. Using Demographic Model Selection. *BioRxiv* **2019,** 1–27. https://doi.org/10.1101/521138

Knapp, S.; Vorontsova, M. S.; Prohens, J. Wild Relatives of the Eggplant (*Solanum melongena* L.: Solanaceae), New Understanding of Species Names in a Complex Group. PLoS One **2013**. https://doi.org/10.1371/journal.pone.0057039

Konishi, S.; Izawa, T.; Lin, S. Y.; Ebana, K.; Fukuta, Y.; Sasaki, T.; Yano, M. An SNP Caused Loss of Seed Shattering During Rice Domestication. *Science* **2006,** *312* (5778), 1392–1396.

Labarta, R. In *Crop Improvement, Adoption, and Impact of Improved Varieties in Food Crops in Sub-Saharan Africa*; Walker, T. S., Alwang, J., Eds.; CABI, 2015; pp 164–182.

Lam, H. M.; Xu, X.; Liu, X.; Chen, W.; Yang, G.; Wong, F. L.; Li, M. W.; He, W.; Qin, N.; Wang, B.; Li, J.; Jian, M.; Wang, J.; Shao, G.; Wang, J.; Sun, S. S.M.; Zhang, G. Resequencing of 31 Wild and Cultivated Soybean Genomes Identifies Patterns of Genetic Diversity and Selection. *Nat. Genet.* **2010,** *42* (12), 1053–1059. https://doi.org/10.1038/ng.715

Lane, A.; Jarvis, A.; Hijmans, R. Climate Change Threatens Wild Relatives with Extinction. *Geneflow '06* **2006,** 18.

Lev-Yadun, S.; Gopher, A.; Abbo, S. The Cradle of Agriculture. *Science* **2000,** *288*, 1062–1063. https://doi.org/10.1126/science.288.5471.1602

Li, W. T.; Huang, X.; Wang, J. R.; Chen, G. Y.; Nevo, E.; Zheng, Y. L.; Wei, Y. M. Genetic Analysis and Ecological Association of Hina Genes Based on Single Nucleotide Polymorphisms (SNPs) in Wild Barley, *Hordeum spontaneum. Hereditas*. https://doi.org/10.1111/j.1601–5223.2009.2151.x

Li, Y. H.; Zhou, G.; Ma, J.; Jiang, W.; Jin, L. G.; Zhang, Z.; Guo, Y.; Zhang, J.; Sui, Y.; Zheng, L.; Zhang, S. S.; Zuo, Q.; Shi, X. H.; Li, Y. F.; Zhang, W. K.; Hu, Y.; Kong, G.; Hong, H. L.; Tan, B. et al. De novo Assembly of Soybean Wild Relatives for Pan-Genome Analysis of Diversity and Agronomic Traits. *Nat. Biotechnol.* **2014,** *58* (8), 813–815. https://doi.org/10.1038/nbt.2979

Lipka, A. E.; Kandianis, C. B.; Hudson, M. E.; Yu, J.; Drnevich, J.; Bradbury, P. J.; Gore, M. A. From Association to Prediction: Statistical Methods for the Dissection and Selection of Complex Traits in Plants. *Curr. Opin. Plant Biol.* **2015,** *24*, 110–118. https://doi.org/10.1016/j.pbi.2015.02.010

Liu, G.; Lu, G.; Zeng, L.; Wang, G. L. Two Broad-Spectrum Blast Resistance Genes, Pi9 (t) and Pi2 (t), Are Physically Linked on Rice Chromosome 6. *Mol. Genet. Genom.* **2002,** *267* (4), 472–480. https://doi.org/10.1007/s00438–002–0677–2

Liu, S.; Liu, Y.; Yang, X.; Tong, C.; Edwards, D.; Parkin, I. A.P.; Zhao, M.; Ma, J.; Yu, J.; Huang, S.; Wang, X.; Wang, J.; Lu, K.; Fang, Z.; Bancroft, I.; Yang, T. J.; Hu, Q.; Wang, X.; Yue, Z.; … Paterson, A. H. The *Brassica Oleracea* Genome Reveals the Asymmetrical Evolution of Polyploid Genomes. *Nat. Commun.* **2015,** *5* (1), 1–11. https://doi.org/10.1038/ncomms4930

Luo, X.; Xi Bai, Sun, X.; Zhu, D.; Liu, B.; Ji, W.; Cai, H.; Cao, L.; Wu, J.; Hu, M.; Liu, X.; Tang, L.; Zhu, Y. Expression of Wild Soybean WRKY20 in *Arabidopsis* Enhances Drought Toleranc. *J. Exp. Bot.* **2013,** *64* (8), 2155–https://doi.org/10.1093/jxb/ert073

Maazou, A. R.S.; Qiu, J.; Mu, J.; Liu, Z. Utilization of Wild Relatives for Maize (*Zea mays* L.) Improvement. *Afr. J. Plant Sci.* **2017,** *11*, 105–113. DOI: 10.5897/AJPS2017.1521

Majhi, P. K.; Mogali, S. C. Characterization and Selection of Bruchid [*Callosobruchus maculatus* (F.)] Tolerant Greengram [*Vigna radiata* (L.) Wilczek] Genotypes. *Indian J. Agric. Res.* **2020,** *54* (6), 679–688.

Maxted, N.; Kell, S. P. Establishment of a Global Network for the In Situ Conservation of Crop Wild Relatives: Status and Needs. FAO Commission on Genetic Resources for Food and Agriculture, 2009; pp 1–266. http://www.fao.org/3/i1500e/i1500e18d.pdf

McCouch, S. R.; Sweeney, M.; Li, J.; Jiang, H.; Thomson, M.; Septiningsih, E.; Edwards, J.; Moncada, P.; Xiao, J.; Garris, A.; Tai, T.; Martinez, C.; Tohme, J.; Sugiono, M.; McClung, A.; Yuan, L. P.; Ahn, S. N. Through the Genetic Bottleneck: *O. rufipogon* as a Source of Trait-Enhancing Alleles for *O. sativa*. *Euphytica* **2007,** *154* (3), 317–339. https://doi.org/10.1007/s10681–006–9210–8

Menguer, P. K.; Sperotto, R. A.; Ricachenevsky, F. K. A Walk on the Wild Side: Oryza Species as Source for Rice Abiotic Stress Tolerance. *Genet. Mol. Biol.* **2017,** *40*, 238–252.

Merk, H. L.; Ashrafi, H.; Foolad, M. R. Slective Genotyping to Identify Late Blight Resistance Genes in an Accession of the Tomato Wild Species *Solanum pimpinellifolium. Euphytica* **2012,** *187*, 63–75.

Meyer, R. S.; Purugganan, M. D. Evolution of Crop Species: Genetics of Domestication and Diversification. *Nat. Rev. Genet.* **2013,** *14* (12), 840–852. https://doi.org/10.1038/nrg3605

Michael, T. P.; VanBuren, R. Progress, Challenges and the Future of Crop Genomes. *Curr. Opin. Plant Biol.* **2015,** *24*, 71–81.

Ming, R.; Moore, P. H.; Wu, K. K.; D'Hont, A.; Glaszmann, J. C.; Tew, T. L.; Mirkov, T. E.; da Silva, J.; Jifon, J.; Rai, M.; et al. *Plant Breed Review*, Vol. 27; Wiley Blackwell: Hoboken; pp 15–118. ISBN 978–0-4706–5034–9.

Mohan Naidu, K.; Sreenivasan, T. Conservation of Sugarcane Germoplasm. In *Proceedings of the Copersucar International Sugarcane BreedingWorkshop, São Paulo, Brazil*, 1987; pp 33–53.

Montenegro, J. D.; Golicz, A. A.; Bayer, P. E.; Hurgobin, B.; Lee, H. T.; Chan, C. K.K.; Visendi, P.; Lai, K.; Doležel, J.; Batley, J.; Edwards, D. The Pangenome of Hexaploid Bread Wheat. *Plant J.* 2017. https://doi.org/10.1111/tpj.13515

Muñoz-Rodríguez, P. et al. Reconciling Conflicting Phylogenies in the Origin of Sweet Potato and Dispersal to Polynesia. *Curr. Biol.* **2018,** *28* 1246–1256. e12.

Nazeer, W.; Ahmad, S.; Mahmood, K.; Tipu, A.; Mahmood, A.; Zhou, B. Introgression of Genes for Cotton Leaf Curl Virus Resistance and Increased Fiber Strength from *Gossypium stocksii* into Upland Cotton (*Gossypium hirsutum*). *Genet. Mol. Res.* **2014,** *13*, 1133–1143. DOI: 10.4238/2014.February.21.2

Ndjiondjop, M. N.; Manneh, B.; Cissoko, M.; Drame, N. K.; Kakai, R. G.; Bocco, R. et al. Drought Resistance in an Interspecific Backcross Population of Rice (Oryza spp.) Derived from the Cross WAB56–104 (*O. sativa*) × CG14 (*O. glaberrima*). *Plant Sci.* **2010,** *179* (4), 364–373.

Niazi, I. A.K.; Rafique, A.; Rauf, S.; da Silva, J. A.T.; Afzal, M. Simultaneous Selection for Stem Borer Resistance and Forage Related Traits in Maize (*Zea mays* ssp. *mays* L.) × Teosinte *(Zea mays* ssp. *mexicana* L.*)* Derived Populations. *Crop Prot.* **2014,** *57*, 27–34. DOI: 10.1016/j.cropro.2013.10.026

Ogutcen, E.; Ramsay, L.; von Wettberg, E. B.; Bett, K. E. Capturing Variation in Lens (Fabaceae), Development and Utility of an Exome Capture Array for Lentil. *Appl. Plant Sci.* **2018,** *6* (7), e01165. https://doi.org/10.1002/aps3.1165

Padulosi, S.; Ng, N. Q. Wild Vigna Species in Africa: Their Collection and Potential Utilization. In: *Cowpea Genetic Resources*; Ng, N. Q., Monti, L. M., Eds., IITA, Ibadean: Nigeria, 1990; pp 58–77.

Pankin, A.; Altmüller, J.; Becker, C.; von Korff, M. Targeted Resequencing Reveals Genomic Signatures of Barley Domestication. *New Phytolog.* **2018,** *218* (3), 1247–1259.

Parmesan, C.; Yohe, G. A Globally Coherent Fingerprint of Climate Change Impacts Across Natural Systems. *Nature* **2003,** *421* (6918), 37–42.

Pascual, L.; Desplat, N.; Huang, B. E.; Desgroux, A.; Bruguier, L.; Bouchet, J. P.; Le, Q. H.; Chauchard, B.; Verschave, P.; Causse, M. Potential of a Tomato MAGIC Population to Decipher the Genetic Control of Quantitative Traits and Detect Causal Variants in the Resequencing Era. *Plant Biotechnol J.* **2015.** https://doi.org/10.1111/pbi.12282

Pereira-Carvalho, R. C.; Boiteux, L. S.; Fonseca, M. E.N. et al. Multiple Resistance to Meloidogyne spp.; to Bipartite and Monopartite Begomovirus spp. in Wild Solanum (Lycopersicon) Accessions. *Plant Dis.* **2010,** *94* (2), 179–185.

Piegu, B.; Guyot, R.; Picault, N.; Roulin, A.; Saniyal, A.; Kim, H.; Collura, K.; Brar, D. S.; Jackson, S.; Wing, R. A.; Panaud, O. Doubling Genome Size Without Polyploidization: Dynamics of Retrotransposition-Driven Genomic Expansions in *Oryza australiensis*, a Wild Relative of Rice. *Genome Res.* **2006,** *16* (10), 1262–https://doi.org/10.1101/gr.5290206

Pietro, G.; Prohens, J.; Plazas, M.; Mangino, G.; Herraiz, F. J.; García-Fortea, E.; Vilanova, S. Genomic Tools for the Enhancement of Vegetable Crops: A Case in Eggplant. *Notulae Botanicae Horti Agrobotanici Cluj-Napoca* **2018,** *46* (1), 1–13. https://doi.org/10.15835/nbha46110936

Prohens, J.; Whitaker, B. D.; Plazas, M.; Vilanova, S.; Hurtado, M.; Blasco, M.; Gramazio, P.; Stommel, J. R. Genetic Diversity in Morphological Characters and Phenolic Acids Content Resulting from an Interspecific Cross Between Eggplant, *Solanum melongena*, and Its Wild Ancestor (*S. incanum*). *Ann. Appl. Biol.* **2013,** *162* (2), 242–257. https://doi.org/10.1111/aab.12017

Qin, C.; Yu, C.; Shen, Y.; Fang, X.; Chen, L.; Min, J.; Cheng, J.; Zhao, S.; Xu, M.; Luo, Y.; Yang, Y.; Wu, Z.; Mao, L.; Wu, H.; Changying, L. H.; Zhou, H.; Lin, H.; González-Morales, S.; Trejo-Saavedra, D. L. et al. Whole-Genome Sequencing of Cultivated and Wild Peppers Provides Insights into Capsicum Domestication and Specialization. *Proc. Natl. Acad. Sci. USA* **2014,** *111* (14), 5135–https://doi.org/10.1073/pnas.1400975111

Qiu, L. J.; Xing, L. L.; Guo, Y.; Wang, J.; Jackson, S. A.; Chang, R. Z. A Platform for Soybean Molecular Breeding: The Utilization of Core Collections for Food Security. *Plant Mol. Biol.* **2013,** *83*, 41–50. DOI: 10.1007/s11103–013–0076–6

Riley R, Chapman V. Genetic Control of Cytologically Diploid Behaviour of Hexaploid Wheat. *Nature* **1958,** *182*, 713–715.

Rojas-Barrera, I. C.; Wegier, A.; De Jesús Sánchez González, J.; Owens, G. L.; Rieseberg, L. H.; Piñero, D. Contemporary Evolution of Maize Landraces and Their Wild Relatives Influenced by Gene Flow with Modern Maize Varieties. *Proc. Natl. Acad. Sci. USA* **2019,** *116* (42), 21302–21311. https://doi.org/10.1073/pnas.1817664116

Roullier, C.; Benoit, L.; McKey, D. B.; Lebot, V. Historical Collections Reveal Patterns of Diffusion of Sweet Potato in Oceania Obscured by Modern Plant Movements and Recombination. *Proc. Natl Acad. Sci. USA* **2013,** *110*, 2205–2210.

Sarao, P. S.; Sahi, G. K.; Neelam, K.; Mangat, G. S.; Patra, B. C.; Singh, K. Donors for Resistance to Brown Planthopper *Nilaparvata lugens* (Stål) from Wild Rice Species. *Rice Sci.* **2016,** *23* (4), 219–224.

Sawler, J.; Reisch, B.; Aradhya, M. K.; Prins, B.; Zhong, G. Y.; Schwaninger, H.; Simon, C.; Buckler, E.; Myles, S. Genomics Assisted Ancestry Deconvolution in Grape. *PLoS ONE* **2013,** *8* (11), e80791. https://doi.org/10.1371/journal.pone.0080791

Saxena, R. K.; Edwards, D.; Varshney, R. K. Structural Variations in Plant Genomes. *Brief. Funct. Genom. Proteom.* **2014,** *51* (6), 1052–1059. https://doi.org/10.1093/bfgp/elu016

Schatz, M. C.; Maron, L. G.; Stein, J. C. et al. Whole Genome de novo Assemblies of Three Divergent Strains of Rice, *Oryza sativa*, Document Novel Gene Space of *aus* and *indica*. *Genome Biol.* **2014,** *15*, 506. https://doi.org/10.1186/s13059–014–0506-z

Schatz, M. C.; Maron, L. G.; Stein, J. C.; Hernandez Wences, A.; Gurtowski, J.; Biggers, E.; Lee, H.; Kramer, M.; Antoniou, E.; Ghiban, E.; Wright, M. H.; Chia, J. ming, Ware, D.; McCouch, S. R.; McCombie, W. R. Whole Genome de novo Assemblies of Three Divergent Strains of Rice, *Oryza sativa*, Document Novel Gene Space of aus and indica. *Genome Biol.* **2014,** *15* (11), 1–16. https://doi.org/10.1186/s13059–014–0506-z

Scheben, A.; Wolter, F.; Batley, J.; Puchta, H.; Edwards, D. Towards CRISPR/Cas Crops–Bringing Together Genomics and Genome Editing. *New Phytol.* **2017,** *216* (3), 682–698.

Scott, J. W. & Gardner, R. G. Breeding for Resistance to Fungal Pathogens. In: *Genetic Improvement of Solanaceous Crops: Tomato*; Razdan, M. K., Mattoo, A. K., Eds., Vol. 2; Science Publishers: Enfield, New Hampshire, **2007**; pp 421–456.

Sedbrook, J. C.; Carroll, K. L.; Hung, K. F.; Masson, P. H.; Somerville, C. R. The *Arabidopsis* SKU5 Gene Encodes an Extracellular Glycosyl Phosphatidylinositol–Anchored Glycoprotein Involved in Directional Root Growth. *Plant Cell.* **2002,** *14* (7), 1635–1648.

Septiningsih, E. M.; Trijatmiko, K. R.; Moeljopawiro, S.; McCouch, S. R. Identification of Quantitative Trait Loci for Grain Quality in an Advanced Backcross Population Derived from the *Oryza sativa* Variety IR64 and the Wild Relative *O. rufipogon. Theor. Appl. Genet.* **2003,** *107* (8), 1433–https://doi.org/10.1007/s00122–003–1376-z

Shan, Q.; Wang, Y.; Li, J.; Zhang, Y.; Chen, K.; Liang, Z. et al. Targeted Genome Modification of Crop Plants Using a CRISPR-Cas System. *Nat Biotechnol.* **2013,** *31* (8), 686–688.

Sharma, T. R.; Rai, A. K.; Gupta, S. K.; Vijayan, J.; Devanna, B. N.; Ray, S. Rice Blast Management Through Host-Plant Resistance: Retrospect and Prospects. *Agric Res.* **2012,** *25* (347-2016-16905), 37–48. https://doi.org/10.1007/s40003–011–0003–5

Sherman-Broyles, S.; Bombarely, A.; Powell, A. F.; Doyle, J. L.; Egan, A. N.; Coate, J. E.; et al. The Wild Side of a Major Crop: Soybean's Perennial Cousins from Down Under. *Am. J. Bot.* **2014,** *101*, 1651–1665. DOI: 10.3732/ajb.1400121

Shinde, B.; Gurve, S.; Gonde, A.; Hole, U. Studies on Resistance of Cotton Genotypes Against Jassids (Amrasca biguttula biguttula Ishida). *Bioinfolet-A Q. J. Life Sci.* **2014,** *11*, 758–762. DOI: 10.20546/ijcmas.2017.604.312

Smýkal, P.; Nelson, M. N.; Berger, J. D.; Von Wettberg, E. J.B. The Impact of Genetic Changes During Crop Domestication. *Agronomy* **2018,** *8* (7), 1–22. https://doi.org/10.3390/agronomy8070119

Soyk, S.; Müller, N. A.; Park, S. J.; Schmalenbach, I.; Jiang, K.; Hayama, R. et al. Variation in the Flowering Gene Self Pruning 5G Promotes Day-Neutrality and Early Yield in Tomato. *Nat. Genet.* **2017,** *49* (1), 162–168.

Spindel, J.; Begum, H.; Akdemir, D.; Virk, P.; Collard, B.; Redoña, E.; Atlin, G.; Jannink, J. L.; McCouch, S. R. Genomic Selection and Association Mapping in Rice (*Oryza sativa*), Effect of Trait Genetic Architecture, Training Population Composition, Marker Number and Statistical Model on Accuracy of Rice Genomic Selection in Elite, Tropical Rice Breeding Lines. *PLoS Genet.* **2015,** *11* (2), e1004982. https://doi.org/10.1371/journal.pgen.1004982

Stein, J. C.; Yu, Y.; Copetti, D.; Zwickl, D. J.; Zhang, L.; Zhang, C.; Chougule, K.; Gao, D.; Iwata, A.; Goicoechea, J. L.; Wei, S.; Wang, J.; Liao, Y.; Wang, M.; Jacquemin, J.; Becker, C.; Kudrna, D.; Zhang, J.; Londono, C. E.M. et al. Genomes of 13 Domesticated and Wild Rice Relatives Highlight Genetic Conservation, Turnover and Innovation Across the Genus *Oryza. Nat. Genet.* **2018,** *50* (2), 285–296. https://doi.org/10.1038/s41588–018–0040–0

Sun, C.; Hu, Z.; Zheng, T.; Lu, K.; Zhao, Y.; Wang, W.; Shi, J.; Wang, C.; Lu, J.; Zhang, D.; Li, Z.; Wei, C. RPAN: Rice Pan-Genome Browser for ~3000 Rice Genomes. *Nucleic Acids Res.* **2017,** *45* (2), 597–605. https://doi.org/10.1093/nar/gkw958

Syfert, M. M.; Castañeda-Álvarez, N. P.; Khoury, C. K.; Särkinen, T.; Sosa, C. C.; Achicanoy, H. A.; Bernau, V.; Prohens, J.; Daunay, M. C.; Knapp, S. Crop Wild Relatives of the Brinjal Eggplant (*Solanum melongena*), Poorly Represented in Genebanks and Many Species at Risk of Extinction. *Am. J. Bot.* **2016,** *103* (4), 635–651. https://doi.org/10.3732/ajb.1500539

Tanksley, S. D.; McCouch, S. R. Seed Banks and Molecular Maps: Unlocking Genetic Potential from the Wild. Science.**1997,** *277* (5329), 1063–1066. https://doi.org/10.1126/science.277.5329.1063

Tanksley, S. D.; Nelson, J. C. Advanced Backcross QTL Analysis: A Method for the Simultaneous Discovery and Transfer of Valuable QTLs from Unadapted Germplasm into Elite Breeding Lines. *Theor. Appl. Genet.* **1996,** *2* (8), 221–232. https://doi.org/10.1007/BF00223376

Tettelin, H.; Masignani, V.; Cieslewicz, M. J.; Donati, C.; Medini, D.; Ward, N. L.; Angiuoli, S. V.; Crabtree, J.; Jones, A. L.; Durkin, A. S.; DeBoy, R. T.; Davidsen, T. M.; Mora, M.;

Scarselli, M.; Margarit Y Ros, I.; Peterson, J. D.; Hauser, C. R.; Sundaram, J. P.; Nelson, W. C.; … Fraser, C. M. Genome Analysis of Multiple Pathogenic Isolates of *Streptococcus agalactiae*: Implications for the Microbial "Pan-Genome." *Proc. Natl. Acad. Sci. USA* **2005,** *102* (39), 13950–13955. https://doi.org/10.1073/pnas.0506758102

van Zonneveld, M.; Rakha, M.; Chou, Y. Y.; Chang, C. H.; Yen, J. Y.; Schafleitner, R. et al. Mapping Patterns of Abiotic and Biotic Stress Resilience Uncovers Conservation Gaps and Breeding Potential of Vigna Wild Relatives. *Sci Rep.* **2020,** *10* (1), 1–11.

Varshney, R. K.; Graner, A.; Sorrells, M. E. Genomics-Assisted Breeding for Crop Improvement. *Trends Plant Sci.* **2005,** *10* (12), 621–630. https://doi.org/10.1016/j.tplants.2005.10.004

Varshney, R. K.; Saxena, R. K.; Upadhyaya, H. D.; Khan, A. W.; Yu, Y.; Kim, C.; Rathore, A.; Kim, D.; Kim, J.; An, S.; Kumar, V.; Anuradha, G.; Yamini, K. N.; Zhang, W.; Muniswamy, S.; Kim, J. S.; Penmetsa, R. V.; Von Wettberg, E.; Datta, S. K. Whole-Genome Resequencing of 292 Pigeonpea Accessions Identifies Genomic Regions Associated with Domestication and Agronomic Traits. *Nat. Genet.* **2017,** *49* (7), 1082–1088. https://doi.org/10.1038/ng.3872

Vasil, I. K. Molecular Genetic Improvement of Cereals: Transgenic Wheat (*Triticum aestivum* L.). *Plant Cell Rep.* **2007,** *26* (8), 1133–1154. https://doi.org/10.1007/s00299–007–0338–3

Vavilov, N. I. Centers of Origin of Cultivated Plants. Origin and Geography of Cultivated Plants. 1926.

Von Wettberg, E. J.B.; Chang, P. L.; Başdemir, F.; Carrasquila-Garcia, N.; Korbu, L. B.; Moenga, S. M.; Bedada, G.; Greenlon, A.; Moriuchi, K. S.; Singh, V.; Cordeiro, M. A.; Noujdina, N. V.; Dinegde, K. N.; Shah Sani, S. G.A.; Getahun, T.; Vance, L.; Bergmann, E.; Lindsay, D.; Mamo, B. E. et al. Ecology and Genomics of an Important Crop Wild Relative as a Prelude to Agricultural Innovation. *Nat. Commun.* **2018,** *9* (1), 1–13. https://doi.org/10.1038/s41467–018–02867-z

Wang, J.-F.; Ho, F.-I.; Truong, H. T.H. et al. Identification of Major QTLs Associated with Stable Resistance of Tomato Cultivar 'Hawaii 7996' to *Ralstonia solanacearum. Euphytica* **2013a,** *190* (2), 241–252.

Wang, Y.; Yang, W.; Zhang, W.; Han, Q.; Feng, M.; Shen, H. Mapping of a Heat-Stable Gene for Resistance to Southern Root-Knot Nematode in *Solanum lycopersicum. Plant Mol. Biol. Rep.* **2013b,** 31 (2), 352–362.

Wang, W.; Mauleon, R.; Hu, Z.; Chebotarov, D.; Tai, S.; Wu, Z.; Li, M.; Zheng, T.; Fuentes, R. R.; Zhang, F.; Mansueto, L.; Copetti, D.; Sanciangco, M.; Palis, K. C.; Xu, J.; Sun, C.; Fu, B.; Zhang, H.; Gao, Y. et al. Genomic Variation in 3,010 Diverse Accessions of Asian Cultivated Rice. *Nature* **2018,** *557* (7703), 43–49. https://doi.org/10.1038/s41586–018–0063–9

Wang, Y.; Xiong, G.; Hu, J.; Jiang, L.; Yu, H.; Xu, J.; Fang, Y.; Zeng, L.; Xu, E.; Xu, J.; Ye, W.; Meng, X.; Liu, R.; Chen, H.; Jing, Y.; Wang, Y.; Zhu, X.; Li, J.; Qian, Q. Copy Number Variation at the GL7 Locus Contributes to Grain Size Diversity in Rice. *Nat. Genet.* **2015,** *47* (8), 944–948. https://doi.org/10.1038/ng.3346

Warschefsky, E.; Varma Penmetsa, R.; Cook, D. R.; Von Wettberg, E. J.B. Back to the Wilds: Tapping Evolutionary Adaptations for Resilient Crops Through Systematic Hybridization with Crop Wild Relatives. *Am. J. Bot.* **2014,** *101* (10), 1791–1800. https://doi.org/10.3732/ajb.1400116

Wei, W. H.; Qin, R.; Song, Y. C.; Guo, L. Q.; Gu, M. G. Comparative Analyses of Disease Resistant and Nonresistant Lines from Maize × *Zea diploperennis* by GISH. *Bot. Bull. Acad. Sin.* **2001,** *42*, 109–114. https://ejournal.sinica.edu.tw/bbas/content/2001/2/bot422–04.html

Wendel, J. F.; Cronn, R. C. Polyploidy and the Evolutionary History of Cotton. *Adv. Agron.* **2003,** *78* (13986), 78004–78008.

Winter, S. M.; Shelp, B. J.; Anderson, T. R.; Welacky, T. W.; Rajcan, I. QTL Associated with Horizontal Resistance to Soybean Cyst Nematode in *Glycine soja* PI464925B. *Theor. Appl. Genet.* **2007,** *114*, 461–472. DOI: 10.1007/s00122-006-0446-4

Witek, K.; Jupe, F.; Witek, A. I.; Baker, D.; Clark, M. D.; Jones, J. D.G. Accelerated Cloning of a Potato Late Blight-Resistance Gene Using RenSeq and SMRT Sequencing. *Nat. Biotechnol.* **2016,** *34* (6), 656–660. https://doi.org/10.1038/nbt.3540

Wu, Z.; Fang, D.; Yang, R.; Gao, F.; An, X.; Zhuo, X.; Li, Y.; Yi, C.; Zhang, T.; Liang, C.; Cui, P.; Cheng, Z.; Luo, Q. De novo Genome Assembly of Oryza granulata Reveals Rapid Genome Expansion and Adaptive Evolution. *Commun. Biol.* **2018,** *1* (1), 1–10. https://doi.org/10.1038/s42003–018–0089–4

Xu, X.; Liu, X.; Ge, S.; Jensen, J. D.; Hu, F.; Li, X.; Dong, Y.; Gutenkunst, R. N.; Fang, L.; Huang, L.; Li, J.; He, W.; Zhang, G.; Zheng, X.; Zhang, F.; Li, Y.; Yu, C.; Kristiansen, K.; Zhang, X. et al. Resequencing 50 Accessions of Cultivated and Wild Rice Yields Markers for Identifying Agronomically Important Genes. *Nat. Biotechnol.* **2012,** *30* (1), 105–111. https://doi.org/10.1038/nbt.2050

Yu, N.; Diers, B. W. Fine Mapping of the SCN Resistance QTL cqSCN-006 and cqSCN-007 from *Glycine soja* PI 46*Euphytica* **2017,** *213*, 54. DOI: 10.1007/s10681-016-1791-2

Yuan, W.; Guo, S.; Gao, J.; Zhong, M.; Yan, G.; Wu, W. et al. General Control Nonderepressible 2 (GCN2) Kinase Inhibits target of Rapamycin Complex 1 in response to Amino Acid Starvation in Saccharomyces Cerevisiae. *J. Biol. Chem.* **2017,** *292* (7), 2660–2669.

Yuan, Y.; Bayer, P. E.; Batley, J.; Edwards, D. Improvements in Genomic Technologies: Application to Crop Genomics. *Trends Biotechnol.* **2017,** *35* (6), 547–558. https://doi.org/10.1016/j.tibtech.2017.02.009

Zha, X.; Luo, X.; Qian, X.; He, G.; Yang, M.; Li, Y.; Yang, J. Over-Expression of the Rice LRK1 Gene Improves Quantitative Yield Components. *Plant Biotechnol. J.* **2009,** *7* (7), 611–620. https://doi.org/10.1111/j.1467-7652.2009.00428.x

Zhang, Q.; Wang, C.; Zhao, K.; Yang, W.; Qiao, F.; Zhou, Y. et al. Development of Near-Isogenic Line CBB23 with a New Resistance Gene to Bacterial Blight in Rice and Its Application. *Chin. J. Rice Sci.* **2002,** *16*, 206–210.

Zhang, X.; Pérez-Rodríguez, P.; Burgueño, J.; Olsen, M.; Buckler, E.; Atlin, G.; Prasanna, B. M.; Vargas, M.; Vicente, F. S.; Crossa, J. Rapid Cycling Genomic Selection in a Multiparental Tropical Maize Population. *G3: Genes Genom. Genet.* **2017,** *7* (7), 2315–https://doi.org/10.1534/g3.117.043141

Zhao, F. A.; Fang, W.; Xie, D.; Zhao, Y.; Tang, Z.; Li, W.; et al. Proteomic Identification of Differentially Expressed Proteins in *Gossypium thurberi* Inoculated with Cotton *Verticillium dahliae*. *Plant Sci.* **2012,** 176–184. DOI: 10.1016/j.plantsci.2011.10.007

Zhao, Q.; Feng, Q.; Lu, H.; Li, Y.; Wang, A.; Tian, Q.; Zhan, Q.; Lu, Y.; Zhang, L.; Huang, T.; Wang, Y.; Fan, D.; Zhao, Y.; Wang, Z.; Zhou, C.; Chen, J.; Zhu, C.; Li, W.; Weng, Q. et al. Pan-Genome Analysis Highlights the Extent of Genomic Variation in Cultivated and Wild Rice. *Nat. Genet.* **2018,** *50* (2), 278–284. https://doi.org/10.1038/s41588–018–0041-z

Zsögön, A.; Čermák, T.; Naves, E. R.; Notini, M. M.; Edel, K. H.; Weinl, S.; Freschi, L.; Voytas, D. F.; Kudla, J.; Peres, L. E.P. De novo Domestication of Wild Tomato Using Genome Editing. *Nat. Biotechnol.* **2018,** *36* (12), 1211–https://doi.org/10.1038/nbt.4272

CHAPTER 15

Breeding and Molecular Mechanism of Drought Stress in Maize

AMARJEET KUMAR[1], ALKA BHARATI[2,3], AJAY KUMAR CHANDRA[3], SNEHA ADHIKARI[4], ANJALI JOSHI[5], LOVE KUMAR[6], VISHAL DINKAR[7], ASHUTOSH SINGH[8], AND H. G. KENCHARADDI[9]

[1]*Department of Genetics and Plant Breeding, MTTC & VTC, Selesih, CAU, Imphal, India*

[2]*ICAR-Central Agroforestry Research Institute, Jhansi, India*

[3]*ICAR-Indian Agricultural Research Institute, New Delhi, India*

[4]*ICAR-IIWBR, Regional Station Flowerdale, Shimla, India*

[5]*Genetics and Tree Improvement Division, AFRI, Jodhpur, Rajasthan, India*

[6]*Centre for Protected Cultivation Technology, IARI, New Delhi, India*

[7]*ICAR-Central Institute of Temperate Horticulture, Srinagar, J&K, India*

[8]*Centre for Advanced Studies on Climate Change, RPCAU, Pusa, Samastipur, India*

[9]*College of Agriculture, CAU (I), Kyrdemkulai, Meghalaya, India Portugal*

ABSTRACT

Maize belonging to the Gramineae family is an important cereal to fulfil the food, feed, and nutritional requirement across the world. The growing population has an ever-increasing demand for high production and productivity. The effect of climate change has drawn the attention of researchers toward need of climate-resilient varieties in the present scenario. The erratic

Climate-Resilient Agriculture: A Molecular Perspective. Ashutosh Singh, Saurabh Pandey (Eds.)

rainfall and depletion of soil water is an alarming condition across the globe. Therefore, it is a demand of the time to develop a climate-proof genotype to fulfil the need of high production in the changing climatic condition. Plant breeders keep their eyes on genetic resources like maize germplasm wild relatives with biotic- and abiotic-resistant genes. The most frequently wild relative of maize is teosinte used as a genetic resource for biotic and abiotic stresses through wide hybridization. Drought is a complex multidimensional trait. Hence, genetic and physiological dissection is needed to understand the actual mechanism of drought stress. Plants have developed evolutionary adaptive mechanisms such as escape, avoidance, and tolerance against abiotic stresses by the accumulation of phytohormones osmolytes, and proline. The reproductive growth stage is the most sensitive stage of maize due to drought. The several key genes have been identified, that is, *ZmDREB2.7*, *ZmVPP1*, *ZmNAC111, ZmTIP1,* etc., in maize-regulating drought tolerance. Understanding the cross-talk mechanism among different pathways responding to drought stress is crucial. However, a collaboration of multidisciplinary approaches, that is, conventional breeding, molecular breeding including marker-assisted selection (MAS) and quantitative trait loci (QTL), transgenic approach, and physiology, can make a synergistic understanding of the response of maize under moisture stress and to improve the breeding efficiency.

15.1 INTRODUCTION

Maize (*Zea mays* L.) is a highly cross-pollinated crop due to monoecious plant habits and protandrous nature. Maize has the highest genetic potential for yield as compared to other cereals since it belongs to the C4-cycle crop; therefore, it has the capacity to capture light energy more efficiently to produce high food grains per unit area (Kumar et al., 2020). Therefore, it is called the "Queen of cereals." At the global level, a number of crops are grown, but only four crops, that is, sugar cane (21%, 1.9 billion tonnes), maize (12%, 1.1 billion tonnes), rice (8%, 0.8 billion tonnes), and wheat (8%, 0.8 billion tonnes) accounted for 49% of world production in 2019 and (FAOSTAT, 2021). But among the cereal crops, the highest production was found in maize, followed by wheat and rice. In India, the total area of cultivation of maize in 2021–2022 was 9900K ha, whereas the production and productivity were 32,500K MT and 3.3 MT/ha, respectively (USDA, 2021–2022).

Irrigation is an essential component of agricultural production, which also played a key role in the green revolution. During the 1960s, the irrigation-equipped land area was very less. In 2000, the cultivated area having irrigation facilities was 289 mha, whereas it reached 342 mha in 2019 (FAOSTAT, 2021). The largest area of cultivation having irrigation facilities is present in China (75 mha) followed by India (70 mha), far ahead of the USA (27 million ha). The availability of irrigation facilities also provides stability in crop production, and productivity ensures food security. At the same time, erratic rainfall and global climate change have a great impact on the production of maize. Pearl millet and sorghum are considered more drought-tolerant crops than maize. Maize is more sensitive to abiotic stresses, particularly drought as well as waterlogging, specifically at the reproductive stage during the growing season. The drought may be defined as "The inadequacy of water availability, including precipitation and soil moisture storage capacity, in quantity and distribution during the life cycle of a crop to restrict the expression of its full genetic yield potential" (Sinha,1986). In India, maize is growing throughout the year, but it is growing mainly in two seasons, that is, in Kharif, more than 85% area is cultivated, and less than 10% area is cultivated in the rabi season (Singh et al., 2005). The highest production of rabi maize is in Bihar. In both seasons maize crop faces the challenge of drought at a different critical stage of crop growth, mainly before flowering and during the reproductive stage. Kharif season faces the problem of moisture stress due to either delay or failure of monsoon, whereas, in rabi season, the maize crop does not get sufficient irrigation water at the flowering as well as at the reproductive stage, which is the critical stage for yield. Maize crop normally requires 500–800 mm water for their life span, but in some countries that receive about 300–500 mm precipitations, it is also cultivating, which is just near the critical level to get optimum yield.

The growing population and ever-increasing demand for food grain put pressure to enhance the high production and productivity. The erratic rainfall and depletion of soil water are an alarming condition. Most of the countries located in the Arabian Peninsula, Near East, and North Africa are experiencing the most severe water stress levels. In these countries, the rate of ground water withdrawal is 10 to almost 40 times per year from their available renewable freshwater resources, resulting in the rapid depletion of water. Therefore, the urgent need of the hour is to develop a climate-proof genotype to fulfil the high production in the changing climatic condition. The selection and stacking of genes/QTLs of all those traits like leaf, root, stem, ear; which makes the plant architecture climate smart and maximize the production and productivity of maize are studied. Plant breeders keep their eyes on genetic resources like maize germplasm wild relatives with biotic- and abiotic-resistant genes/QTLs suitable for wide adaptation. The exploitation of teosinte as a genetic resource for biotic and abiotic stresses is becoming a burning area of research in recent years through wide hybridization. Drought is a complex multidimensional trait. Hence, genetic and physiological dissection is needed to understand the actual mechanism of drought stress. The identification of major genes involved in bio-chemical pathways related to the stress response, the contribution of molecular biology has enormous potential. Understanding the cross-talk mechanism among different pathways responding to drought stress is crucial. In addition to these reverse and forward genetics, functional genomics and comparative genomics are assisting in developing climate-proof genotypes. However, a collaboration of multidisciplinary approaches, that is, conventional breeding, molecular breeding including marker-assisted selection (MAS) and quantitative trait loci (QTL), transgenic approach, and physiology, can make a synergistic understanding of the response of maize to water deficit and improve the breeding efficiency.

15.2 IMPACT OF DROUGHT ON MAIZE

Maize crop is highly sensitive to drought and the majority of cultivated area is under rainfed and only 15% area is under irrigated condition (Shah et al., 2020). This is a challenge for sustainable production for increasing population. Moisture stress during preflowering and reproductive stage severely affect the performance by impairing their yield contributing trait function. The most critical stage for the drought is flowering and grain filling stage and if drought coincides during these stages, yield reduction may exceed more than 50% (Classen and Shaw, 1970). The reduction in the production and productivity depending upon the growth stage, intensity, and duration of drought ranges from 30 to 90% (Pandit et al., 2018). The major growth stages of maize which is susceptible to drought are preflowering stage, flowering stage, and grain filling stage, and their contribution to yield reduction may be 25%, 50%, and 21%, respectively (Denmead and Shaw, 1960).

The indication of drought stress in maize can be assessed by screening the change in the phenotypic symptoms, that is, leaf color changes from green to green-gray, acropetal succession of leaf rolling. The important changes occur due to drought in the morphological character which includes change in the function and structure of root architecture, that is, root length, root volume, root density, number of roots, leaf morphology, that is, leaf rolling, leaf firing, flowering, that is, increase in the days to flowering, increase ASI (Anthesis-silking interval) and early maturity. Drought stress also affects the physiological and biochemical mechanisms such as accumulation of proline, change in ABA concentration, osmotic imbalance, accumulation of anti-oxidants, water channel proteins, transcription factors, signal transduction pathways. It also affects the total biomass and other traits associated with the plant reproduction overall lead to the significant yield reduction. Apart from these, drought stress also reduces the absorption of PAR (photosynthetically active radiation), RUE (radiation use efficiency) by leaf and also reduces the harvest index. Simultaneously, due to stress stomata starts closing reduction in photosynthesis sharply and finally slowing down the overall growth of maize (Edmeades, 2013). Major studies are based on moisture stress at flowering and grain filling stages, but it also influences the maize crop at different growth stages, that is, reduction in seed germination percentage (Muhammad et al., 2013), reduction in the seedling establishment (Liu et al., 2015), and overall growth and developments (Flint-Garcia et al., 2005).

15.3 EFFECT OF DROUGHT ON REPRODUCTIVE STAGE

Pollination is the process of transfer of pollen to the stigma of the flower. Pollination phenomena are crucial for determining the yield and the requirement of water at this stage is critical. Maize is a monoecious as well as geitonogamieous plant having male flower, that is, tassel at the top of the plant and female flower, that is, silk at the 3rd or 4th node of plant. Maize crop is protandrous in nature. In maize an ample number of pollens, that is, ranging from present-day hybrids 2–5 million to wild species like teosinte and land races 14–50 million across the species are produced in the tassel under normal condition. Among the tassel branches, first-time pollen production starts in the central primary branch followed by secondary branches. Breeders are trying to divert the photosynthetic assimilates toward the female part (silk) from the tassel so that smart tassel can be developed by reducing the number of pollens per tassel up to optimum which will be sufficient to pollinate a cob. Field investigations concluded that the reduction in kernel setting is only possible when at least 80% reduction of total pollen production occurs. Shedding of pollens depends on type of varieties and environmental condition, and it starts occurring in the morning time after sunrise when temperature rises and it continues up to 14 days under normal condition. The pollen mortality and sterility occur due to dehydration under drought stress. It also affects the pollen shedding, viability, shape, and their dispersal. In the process of gametogenesis, mitosis and meiotic division occurs and finally results in pollen development. Normal functioning of pollen, that is, germination and pollen tube growth require starch accumulation in the pollen (Jin et al., 2013). Under the drought stress this starch and some osmolytes act as a protectant of pollen viability, if the drought intensity is high; at the same time, the accumulation of starch reduced during pollen development.

As protandrous nature of crop, there is asynchrony of pollen shedding and silk emergence occurs. Under normal condition, this asynchrony interval is varied from 1 to 3 days. The most common phenomena under drought stress are delay in silking. Silk delay occurs due to the reduction of water potential under stress condition, since for silk growth and development high water potential is needed. This leads to the delay in silk emergence and results in the widening of ASI and hence loss of synchrony of pollen shedding and silking reduced the proper kernel setting. Under severe drought, silk delay is too long so that pollen shedding becomes over before silk emergence. Drought also affects the ovary or under high intensity zygote and embryo abortion occurs. Drought stress widens the ASI and it varied from 3 to 4 days (Herrero and Johnson, 1981). The degree of silk delay can determine the complete barren or partial filling of kernel. Since whole length of silk acts as stigma, therefore, silk receptivity is the characteristic feature which means that it allows pollens to germinate, pollen tube growth to the stylar region. After the emergence of silk, its receptivity remains maintained up to 7 days under normal condition but due to drought stress the silk elongation becomes slow (Basetti and Westgate, 1993). Apart from this pollen tube growth and fertilization also retarded under drought stress. Single maize plant requires 8–9 mm water per day during reproductive stage, and the water requirement 2 weeks before pollination and 2 weeks after pollination is the most crucial (Kumar et al., 2020). The most sensitivities stage for water stress is the reproductive stage (grain filling and soft dough) and pre-tasselling stage and crop maturity is considered less sensitive (Shaw, 1977; Aslam et al., 2015). Water stress at the pre-flowering stage of maize causes reduction in the tassel growth and pollen formation, anther exertion, delay in the silking, and finally leads to increases in the interval between anthesis and silking (Maazou et al., 2016). Therefore, the widening of anthesis–silking interval (ASI) under water stress as compared to the well water condition can be used as an indicator of drought. There is a correlation between ASI and grain yield and hence it resulted in reduction in the yield (Sari-Gorla et al., 1999). Therefore, synchrony in the anthesis and silking is essential to increase in the kernel setting to get optimum yield. The number of kernels in the ear is less if drought stress occurs particularly at the grain filling stage as compared to well water condition (Edmeades et al., 2000). Depending on the intensity of stress, a large number of variations are observed for a number of kernels setting because it determines at the time of flowering. If intensity is more there is a possibility of partial to complete abortion of ear and finally it leads to the barrenness. Lack of tip filling, that is, kernel filling at the top of cob abort is an important trait observable under drought stress.

15.4 MECHANISM OF DROUGHT RESISTANCE

Drought stress is a complex, unpredictable, uncertain, heterogenous, erratic phenomenon which impairs the production and productivity of crop plant. In the severe intensity of drought plants can even die. Therefore, plants have developed their adaptive mechanism against abiotic stresses particularly drought stress. Drought resistance is a multifunctional trait since several components and morpho-physiological, biochemical, as well as genomic processes are interacting either in root or in shoot to combat the drought stress by the plants. These mechanisms are classified into the following categories.

15.4.1 DROUGHT ESCAPE

The ability of susceptible varieties completes their life cycle before drought period and it leads to early maturity. This adaptive mechanism of varieties is suitable for late season drought stress. Therefore, the success of these kinds of strategies depends on the varietal maturity and the planting date. Drought escape strategy minimizes the coincidence of drought period and the crop growth. Early maturity varieties have low genetic yield potential as compared to medium and late maturing varieties.

15.4.2 DROUGHT AVOIDANCE

The ability of a variety maintains high water potential or maintains a relatively high tissue hydration level under the moisture stress condition. This strategy can be done by either reducing the rate of transpiration through closing stomatal, leaf rolling, leaf firing, canopy temperature, deposition of cuticle wax, leaf pubescence, osmotic adjustment, enhancing ABA concentration, etc., most commonly called as water savers plant or by increasing water uptake by increasing root length, density, and increasing number of root hairs; most commonly called as water spenders. The plants showing avoidance mechanism exhibit combination of features of both the types. Some genotypes have the ability to avoid the moisture stress either by reduction in transpiration or by increasing the water uptake efficiency.

15.4.3 LEAF ROLLING AND STOMATAL CLOUSER

Leaf rolling occurs due to loss of turgidity or water potential and delaying the leaf rolling is reflecting that the variety is maintaining the turgidity and osmotic balance in the cell, a special genotypic characteristic of variety under drought. The early leaf rolling under moisture stress gives indication of sensitivity to drought. The water potential under drought condition ranges between –1.5 and –2.5 MPa (Riboldi et al., 2016). Leaf rolling reduces the incidence of solar radiation, transpiration, dehydration, loss of photosynthetic assimilates (Saglam et al., 2014; Riboldi et al., 2015). Leaf rolling and leaf senescence is negatively correlated with grain yield in maize (Kadioglu et al., 2012). The reduction in the leaf transpiration leads to dryness of tip of the leaf causing leaf firing, an important trait for the determination of level of drought along with leaf rolling.

The Stomata is at the center of physiology of a leaf. It is a medium of gaseous exchange between cell and atmosphere. The major role of stomata under moisture stress condition is reducing the transpiration by clouser of stomata. But, the clouser of stomata affects several processes like reduction in CO2 uptake, reduction in the photosynthetic assimilates and rate (Benešová et al., 2012; Effendi et al., 2019). In the beginning stage of moisture stress the effect of stomata clouser on the rate of transpiration is more than the CO_2 assimilation, but at later stage both the processes are reduced.

15.4.4 ROOT ARCHITECTURE

Root architecture, that is, root length, density, and biomass are the most crucial traits for the assessment of drought avoidance for breeding purposes, under drought stress, thick, single, and deep root system is very much efficient for the uptake of soil water. Root is the first organ to get the signal of water scarcity in the root zone. Thereafter, the synthesis of ABA (Abscisic acid) starts and plays a crucial role in the root shoot singling. The genetic improvement of crop adaptation to moisture stress, root system is considered the critical trait (Vadez, 2014). Drought stress affects the root structure by increasing the cell wall extensibility by enhancing the structural component, that is, xyloglucan endotransglucosylases/hydrolases (Li et al., 2015).

15.4.5 DROUGHT TOLERANCE

The capability of plants to maintain their normal growth and development under drought condition is known as drought tolerance. This adaptive strategy of plant involves the physiological and biochemical changes like osmotic adjustment, accumulations of solutes or stress responsive proteins, that is, proline, synthesis of phytohormones, maintenance of cell membrane integrity, translocation of reserve photosynthates, oxidative defence mechanism enables plant to withstand under drought condition. Hence, the overall yield of crop plant performs well under stress condition. The effect of drought stress leads to increasing the solute concentration in the soil environment, resulting in flow of water from plant cells to the soil environment. These changes cause the higher solute concentration in the cells and hence water potential is lowered down resulting in changes in the several physiological and biochemical processes like disrupting cell membrane integrity and photosynthesis.

15.4.6 OSMOTIC ADJUSTMENT

The cellular change occurs under the water-deficit condition to increase the compatible solute concentration in the cell so that water influx maintains the turgidity of cell by reducing osmotic potential; this phenomenon is described as osmotic adjustment. This is the evolutionary adaptive strategy of plant under moisture stress condition to synthesize the osmoprotectant. The osmoprotectants are classified into two classes, that is, (1) nitrogenous compounds which include proline, polyols, polyamines, and glycine betaine and (2) hydroxy compounds which include polyhydric alcohols, sucrose, and oligosaccharides synthesized in response to drought stress. When drought stress is over these synthesized osmotic protecting agents subsequently degraded. Osmotic adjustment needs reduction of leaf water potential to extract water from soil through the accumulation of compatible solutes and this accumulation is proportionate to the reduction in leaf water potential (Blum, 2017). The inorganic solutes like K^+, Na^+, Ca^{2+}, Mg^{2+}, Cl^-, and NO^{3-} are more favorable to maintain the osmotic adjustment. Photosynthetic assimilates help in the osmotic adjustment in the beginning of drought stress, but photosynthesis is not much affected. The high intensity of drought also reduces the photosynthetic efficiency by stomatal closure, reduction in chlorophyll synthesis, and reduction in the enzymatic efficiency. Membrane integrity and stability is maintained by these osmoprotectant and hydroxyl group makes interaction with phospholipids and proteins present in the membrane create either hydrophilic or hydrophobic environment.

15.4.7 PHYTOHORMONES

Phytohormones are signaling chemical molecules also called plant hormones or PGR (plant growth regulators) responsible for overall growth and development of the plant. These phytohormones are present in small amount in the cell and responsible for modulating the physiological, biochemical, and molecular processes for the adaptation and acclimatization under drought stress. Phytohormones mainly responsible for drought tolerance in crop plant are auxin (IAA), cytokinin, ABA (abscisic acid), ethylene, gibberellins (GA), salicylic acid, jasmonates, and brassinosteroids. Auxin is an important phytohormone responsible for cell division, root formation, apical dominance, and other cellular functions. There are two main pathways of auxin synthesis, the first one is tryptophan-dependent pathway and the second one is tryptophan-independent pathway (Normanly, 2010). In maize, the synthesis of auxin regulated by gene AA01 having orthologues zmA0-1 (Sekimoto et al., 1997), gene YUCCA having *spi1* orthologues (Gallavotti et al., 2008), and gene *ZmNIT1, 2* (Park et al., 2003) synthesizes Aldehyde oxidase (cytoplasm), YUCCA-like flavin monooxygenase and Nitrilase, respectively, in different pathways. The expression of auxin under drought stress enhanced and promoted the lateral roots formation. Drought tolerance is enhanced by auxin through favorable modulation of several mechanisms, that is, root function, ROS biochemical process, and ABA responsive gene, that is, DREB2A, DREB2B (Shi et al., 2014). ABA is the most important phytohormone for stress conditions also called stress hormone. When plant is exposed to drought, the signal is generated for the synthesis of ABA in the root and after synthesis it is transported to the leaves for triggering the adaptive mechanism like closure of stomata, osmotic adjustment, and ROS signaling (Qi et al., 2018; Wilkinson and Davies, 2010). It has been reported by Spollen et al. (2000) that under low water potential in the soil the accumulation of ABA restricts the growth elongation of primary root of maize by inhibition of ethylene production. In this condition the adaptive behavior of maize root makes it thinner for searching the new source of water in the soil, whereas the synthesis of ethylene promotes the lateral root growth. It has been also reported that the ABA synthesis and accumulation change the pattern of root growth, and it helps the maize root to elongate and search the water resources in the root surrounding environment which will help the plant to overcome the drought stress (Duan et al., 2013). Zhang et al. (2006) reported the effect of amount of ABA accumulation on pollen germination and pollen tube elongation, when amount raised to 0.5 μM, it favors the pollen germination and pollent tube elongation but after further increase it is significantly reducing both the parameters.

15.4.8 PROLINE ACCUMULATION

In response to abiotic stresses, accumulation of proline in the plant cells plays a crucial role in protecting the plant by balancing all types of required osmolytes. In maize, also under drought stress, proline accumulation occurs in the root and leaf and the amount of proline accumulation becomes doubled in the leaf when maize plant is exposed to the drought stress for more than 1 week. The role of proline is not only for the maintenance of osmotic balance and turgidity of cells, but at the same time it helps to protect from free radicles by detoxifying ROS and subcellular structures of the cell. The precursor molecule of proline synthesis is either glutamate or ornithine. Synthesis of proline required energy molecules, that is, two molecules of NADPH and also it releases energy molecule, that is, FADH2 and NADH in the mitochondria. The redox homeostasis and metabolism in the cell are maintained due to the glutamate to proline pathway between cytosol of cell and the mitochondria (Hare and Cress, 1997). Since proline is a osmoregulator, therefore, it has been also observed that external application of proline also imparts drought stress tolerance in the maize by increasing capacity of water uptake as well as by maintaining the membrane stability (Shahid et al., 2014).

15.4.9 ROS DEFENCE MECHANISM

The abiotic stresses responsible for the generation of reactive oxygen species which is accountable for destroying the function of protein break the nucleic acids and lead to lipid peroxidation. In the normal condition, the amount of reactive oxygen species remains at the low level due to the antioxidant defence mechanism. The antioxidant system is a complex system comprised of antioxidant compounds such as glutathione, vitamin C (ascorbic acid), and phenolic compounds, flavonoids, carotenoids, α-tocopherol, the osmolyte proline, and antioxidant enzymes such as CAT (catalase), PPO (polyphenol-oxidase), SOD (superoxide dismutase), POD (peroxidase), GR (glutathione reductase), DHAR (dehydroascorbate reductase), MDHAR (monodehydro-ascorbate reductase), and APOX (ascorbate peroxidase) and nonenzymatic antioxidants, that is, AsA, DHA, GSH, and GSSG reduce the oxidative damage under drought conditions (Ding et al., 2018). Plant abiotic stress tolerance is known to be influenced by antioxidant enzymes. In addition, they play a role in reducing the effects of various reactive oxygen species (ROS) produced due to a photostat imbalance. The activity of antioxidant enzymes also enhances under drought stress in plants (Xu et al., 2010). The drought stress also induces CAT and SOD, which act as antioxidant systems and ROS scavengers. It has been reported that the free radicals generated due to drought stress target the lipids and proteins of the membrane. These enzymatic and nonenzymatic antioxidants directly or indirectly impart drought tolerance in maize crop. The photosynthesis and antioxidants synthesis are dependent on the availability of ABA in the leaf of maize, and it is essential to maintain the photosynthesis and antioxidant defence mechanism under drought condition in maize.

15.5 MOLECULAR MECHANISM

Under the stress conditions plants have their own developed adaptive mechanism to cope up with any kind of stress and several stress-regulated genes are induced to protect the life of crop plant. When the maize crop faces the water scarcity at the time of pollen formation, certain genes like galactinol and vacuole invertase genes are upregulated in the pollen leading to the osmoprotection and maintenance of viability of the pollen (Taji et al., 2002). The gene ZmCIPK8 was isolated from maize and it was cloned and characterized and found that it helps in regulating the drought stress in maize by mRNA accumulation of ZmCIPK8 genes in leaves and roots analyzed by qRT-PCR (Tai et al., 2013). Drought is a complex trait and therefore several genes, genomic regions, transcription factors are involved in drought tolerance of maize. Some of important genes are here described which is playing a critical role in the drought tolerance of maize.

15.5.1 ZMDREB2.7 (DEHYDRATION RESPONSIVE ELEMENT BINDING)

The conclusion of several research experiments conducted throughout the world by many scientists is that there are two major signaling pathways of abiotic stresses, that is, ABA-dependent pathway and ABA- independent pathway that is regulating the induction stress responsive genes. The transcription factor DREB thought that it is regulating the induction of gene in the ABA-independent signaling pathway. The *ZmDREB2.7* gene exhibited the sequence polymorphic variation in the promoter region/5' untranslated region which is associated with the degree of drought tolerance in the maize at seedling stage, but the protein coding sequence is not contributing to the drought tolerance (Liu et al., 2013). The transgenic *Arabidopsis* was developed for *ZmDREB2.7 gene and it was found that this gene overexpressed and imparted drought tolerance.* It has been also identified the favorable alleles of *ZmDREB2.7* in the drought-tolerant cultivar of maize and that can be used as genetic resources in the further breeding program to develop high-yielding varieties with the high level of drought tolerance.

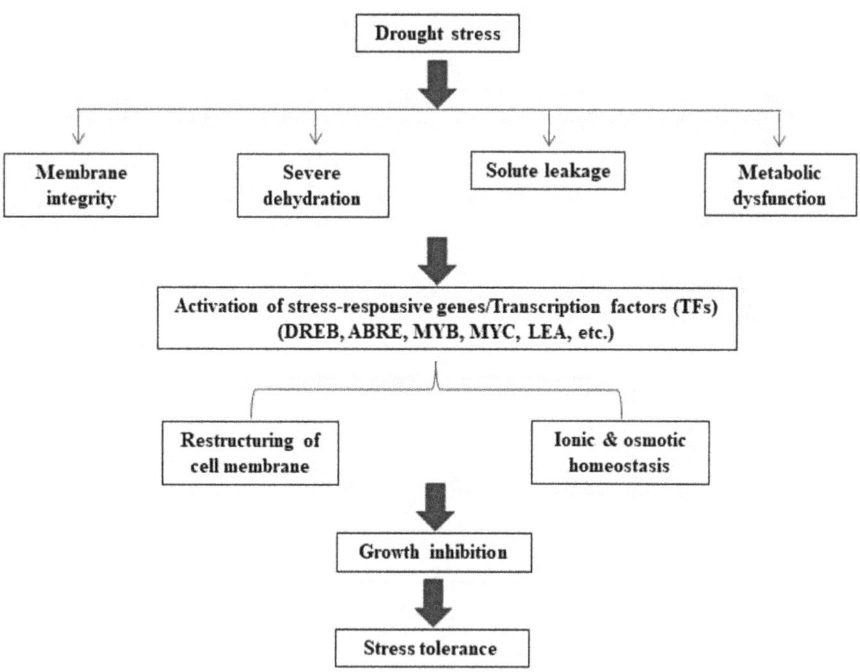

FIGURE 15.1 Effects of drought stress on the cellular functions.

15.5.2 *ZMVPP1*

The vacular proton-pumping pyrophosphatase (ZmVPP1) gene is responsible for conserving energy through hydrolyzing PPi to form Pi under the drought stress situation by maintaining the proton motive force in the cell. In maize, the gene *ZmVPP1* encodes a vacuolar-type H^+ pyrophosphatase enzyme. Wang et al. (2016) conducted the experiment to know the function of the gene and he conducted the experiment for the GWAS study at the seedling stage for drought tolerance in maize and found mainly two variations at the promotor region, that is, indels at –397 (366 bp) and –126 (5 bp) associated with the drought tolerance. The 366 bp insertion of zmVPP1 allele's possess three MYB binding cis elements and it plays a critical role in the induction of drought responsive genes and ultimately improves the drought tolerance (Wang et al., 2016). It was also observed that in the transgenic inbred maize lines the ZmVPP1 gene overexpressing under drought stress condition.

15.5.3 *ZMNAC111*

After domestication of maize, the insertion of 82 bp MITE (miniature inverted-repeat transposable element) in the promotor region of the gene *ZmNAC111* led to the development of drought tolerance in maize. It has also been observed that in the transgenic maize, enhancing the expression of *ZmNAC111* gene improves the drought tolerance at the seedling stage. This gene also induces and upregulates the drought-responsive genes in the maize under water scarcity.

15.5.4 *ZMTIP1*

The plasma membrane and tonoplast have aquaporins proteins which are controlling the movement of water across the membrane. The gene encoding the aquaporin proteins is expressed in the specialized tissues and controlling the solute and water movement. The gene *ZmTIP1* (tonoplast intrinsic protein), which encodes a functional *S*-acyltransferase, is responsible for controlling the root hairs length and the degree of drought tolerance in maize crop at the seedling stage (Barrieu et al., 1998). It has also been reported that the tonoplast water channel is mainly expressed in the epidermal and endodermal cells of root, parenchyma cells of root and stems. The high expression of *ZmTIP1* gene in the cells enhanced the water flow across the membrane under drought condition to maintain the osmotic equilibrium between cytosol and vaculor content. The drought-tolerant cultivar having *ZmTIP1* allele possesses a high number of TFB (transcription factor binding) elements in the promotor region. For the improvement of maize production under drought conditions, we can select and use the desirable alleles of *ZmTIP1 in the breeding program.*

15.5.5 ZMTCP (TEOSINTE-BRANCHED 1/CYCLOIDEA/ PROLIFERATING)

The zmTCP gene family is playing an important role in the drought tolerance in maize crop. In a study a total of 46 zmTCP have been identified in the maize genome, out of which zmTCP32 and zmTCP42 were found associated with the drought tolerance in the maize at the seedling stage. The expression analysis suggested that the RNA level of zmTCP32 and zmTCP42 is induced by drought, ABA, and polyethylene glycol treatments. These alleles can be utilized as the genetic resource to develop more cultivars with drought tolerance and high yield of maize.

15.5.6 ZMSDD1

The maize *ZmSDD1* gene expression was reported in the leaves, shoots, and roots. Liu et al. (2015) demonstrated that the *ZmSDD1* gene improved the drought stress tolerance in the transgenic maize by enhancing the WUE and photosynthetic rate. The WUE and photosynthetic rate improvement occurred due to higher Co_2 consumption rate, increased PEPCase activities, and decreased Co_2 compensation point.

15.5.7 ZMARGOS1

The *ZmARGOS1* (Auxin Regulated *Gene* involved in Organ Size) genes are responsible for reduction in the ethylene sensitivity. The Zm-ARGOS1 is localized to the endoplasmic reticulum (ER) and golgi membrane. The ETI2 (ETHYLENE-INSENSITIVE2) and RTE1(REVERSION-TO-ETHYLENE SENSITIVITY1) are ethylene signaling protein and ethylene receptors are present in the ER (endoplasmic reticulum) and golgi membrane (Shi et al., 2015). It is a negative regulator of the ethylene response and enhancing drought tolerance of transgenic maize in overexpressed condition.

TABLE 15.1 List of Genes Confer Drought Tolerance in Maize.

S. No.	Gene	Function	Reference
1	ZmDREB2.7	Overexpression of gene under drought condition, variation in the promotor region, drought tolerance at seedling stage	Liu et al. (2013)
2	*ZmNAC111*	Seedling drought tolerance, increasing expression in transgenic maize, ABA-dependent regulation, modulated stomatal closure, and drought-responsive gene expression	Mao et al. (2015)
3	*ZmVPP1*	Encoding vacuolar type H(+) pyrophosphatase, increased photosynthetic efficiency, increased root development	Wang et al. (2016)
4	*ZmTIP1*	Encodes a functional *S*-acyltransferase, enhanced expression, elongation of root hair length	Zhang et al. (2020)
5	*ZmTCP32* and *ZmTCP42*	Overexpression under drought stress, enhanced drought tolerance	Ding et al. (2019)
6	*ZmNCED3*, *ZmRAB18*, *ZmRD29B*, *ZmRD17* , *Z mPP2C*	Upregulation of ABA	-
7	GRMZM2G137961	Associated with ASI, encoding tyrosine phosphatase (PTPase), cell cycle control, cell growth, proliferation, differentiation, and transformation	Xue, et al. (2013)
8	GRMZM2G119079	Associated with ASI, encodes Acyl-CoA N-acyltransferase, involved in Krebs cycle, fatty acid metabolism, highly expressed in the mature tassel	Xue, et al. (2013)
9	ZmCIPK8	ABA and H_2O_2 signaling, two domains: N-terminal protein kinase domain and C-terminal regulatory domain with a CBL-interacting NAF/FISL motif, present on chromosome no 3	Tai et al. (2013)
10	*zmARGOS1* genes	Negative regulators of the ethylene response, enhancing drought tolerance when overexpressed in transgenic maize plants	Shi et al. (2015)
11	ZmMYB3R	Increased activity of catalase (CAT), peroxidase (POD), and superoxide dismutase (SOD) enzyme, enhanced sensitivity to ABA, regulation of the stomatal aperture, increased drought tolerance	Wu et al. (2019)
12	ZmWRKY40	Localization in the nucleus of mesophyll protoplasts, overexpression reduced ROS content and enhanced the activities of POD and CAT under drought	Wang et al. (2018)
13	Zm00001d017240	Associated with Coleoptile length and Grx-like gene, overexpression in maize leads to an increase in the stability of the plasma membrane by proline metabolism and ROS detoxification enzyme activity ultimately improves the drought tolerance	Rida et al. (2021)
14	ZmVP14	Role in the ABA synthesis under drought	Cura et al. (2019)

15.6 STRATEGIES FOR IMPROVEMENTS OF MAIZE FOR DROUGHT TOLERANCE

The improvement of crop production and productivity, broadly two main breeding approaches, that is, empirical breeding approach and analytical breeding approach is used by a plant breeder. The empirical breeding approach leading to the improvement of yield is achieved by a plant breeder through selecting the genotype directly based on the yield performance or yield component. Plant breeder evaluated the genotypes in the naturally drought environment (the area where occurrence of drought is a characteristic feature) and developed the variety or population for the target environment, most popularly known as target population environment (TPE). The concept of TPE was given by Fischer et al. (2003). In this approach yield is the prime focus and since it is a complex trait and their heritability is very low, therefore, the improvement process is slow.

In case of an analytical breeding approach, the main focus of a plant breeder is the indirect selection of secondary traits, that is, morphological, physiochemical, or molecular related traits with grain yield to improve the yield of a genotype. In this case, the traits should have high narrow sense heritability, so that the response of selection will be high leading to rapid improvements. The following secondary traits can be used as the selection parameter in maize for drought stress tolerance.

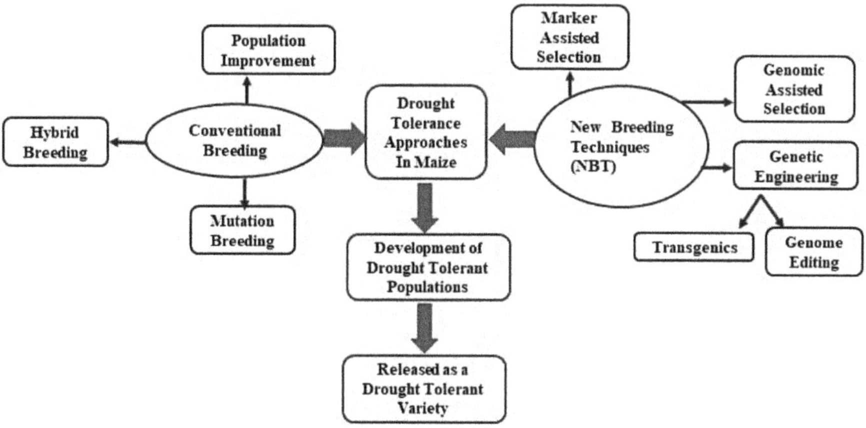

FIGURE 15.2 Breeding approaches for drought-tolerant improvement in maize.

15.6.1 ANTHESIS-SILKING INTERVAL (ASI)

Under drought stress at the reproductive stage, the Anthesis-silking-interval (ASI) in maize delayed. In the Indian subcontinent ASI may be delayed from 4 to 16 days and this is due to increase in the pollen shedding time and silking time (Dass et al., 2001). To screen the drought-tolerant genotype, ASI can be considered one of important parameters in maize. There is a correlation between longer ASI and the reduction in grain yield under drought stress. Some of the lines have been identified, that is, HKI-209, HKI-335, HKI-551, HKI-1332, HKI-1025 and some hybrids like 1035-11536x295, 1332x3-4, 335x1035-11, and 209x3-4 with shorter ASI under stress condition. It has been also reported that ASI is positively correlated with root traits, that is, root biomass and horizontal and vertical expansion of roots. The method of determining the ASI is as (1) count the total number of plants in the plot, (2) data on daily basis regarding number of plants with tassel flowering (male) or silking (female), (3) Anthesis days = the total number of days when more than 50% of all plants have tassel flowering, (4) Silking days = total number of days when more than 50% of all plants have silking, (5) ASI = Silking days – anthesis days.

TABLE 15.2 Correlation of Secondary Traits and Drought Tolerance in Maize.

S. No.	Traits	Indicative of drought tolerance
1.	Grain yield	=> increase
2.	Ears per plant	=> increase
3.	Anthesis-silking interval (ASI)	=> decrease
4.	Leaf senescence	=> decrease
5.	Leaf rolling	=> decrease
6.	Tassel size	=> decrease

15.6.2 SCORING OF LEAF ROLLING

In maize, leaf rolling can be considered one of the crucial parameters to screen the drought-tolerant genotypes. Leaf rolling can be scored on the scale 1–9, score 1 means there is no leaf rolling (under control condition) and score 9 means maximum leaf rolling under drought stress condition. Under drought condition, if the leaf rolling is nearer to leaf score 1 and if nearer to leaf rolling score 9 they are considered drought-tolerant maize line and drought-sensitive maize lines, respectively. Leaf-rolling, also known as survival rate, is one of the most common physiological indicators used to determine drought tolerance at the seedling stage. Leaf-rolling starts at a leaf water potential of − 1 MPa and reaches a maximum of − 2 MPa.

TABLE 15.3 List of Secondary Traits Used for Selection of Drought Tolerance in Maize.

S. No.	Drought tolerance mechanism	Secondary traits
1	Dehydration avoidance	Leaf rolling, a combination of leaf rolling and leaf firing, canopy temperature, leaf attributes, leaf water retention, root characteristics
2	Dehydration tolerance	Plant phenology, grain filling by translocated stem reserve, cellular membrane stability, seed germination

15.6.3 LEAF SENESCENCE AND CHLOROPHYLL CONTENT UNDER DROUGHT

Leaf senescence and chlorophyll content is highly correlated and the depletion of chlorophyll content starts before flowering and hence, initiates leaf senescence and this senescence becomes rapid at the time of grain filling in maize (Borrás et al., 2020). Therefore, under drought condition if the leaf senescence and depletion of chlorophyll is slow that can be used as a parameter for screening of drought-tolerant genotype. Due to slow rate of leaf senescence, tolerant lines either resist decreasing of chlorophyll content during stress conditions or show very little reduction. The rate of leaf senescence is more in inbreds as compared to hybrids. The composition of leaf, that is, the chlorophyll content, antioxidant system, abscisic acid (ABA) concentration during the leaf senescence changed under drought condition (Ren et al., 2020).

15.6.4 INDICES

a. Drought Tolerance Index for Yield

$$\textit{Drought tolerance index (DI)} = \frac{Y_S \times [Y_S/Y_P]}{\bar{Y}_S}$$

In the above formulas, Ys, Yp, and $\bar{Y}s$, $\bar{Y}p$ represent yield in stress and non-stress conditions for each genotype, and yield mean in stress and non-stress conditions for all genotypes, respectively (Lan, 1998).

b. Yield Index (YI)

The high index value reflects that the genotype is suitable for drought conditions

$$YI = \frac{Ys}{\bar{Y}p}$$

Y_S and Y_P are stress and optimal (potential) yield of a given genotype, respectively. $\bar{Y}_S$ and $\bar{Y}_P$ are average yield of all genotypes under stress and optimal conditions, respectively (Gavuzzi et al., 1997).

c. Stress Tolerance Index (STI)

The high STI values reflect that the maize lines is drought tolerant

$$STI = \frac{(Ys)(Yp)}{\bar{Y}p}$$

Y_S and Y_P are stress and optimal (potential) yield of a given genotype, respectively. $\bar{Y}_S$ and $\bar{Y}_P$ are average yield of all genotypes under stress and optimal conditions, respectively (Fernandez, 1992). Maize: Increase ear/plant, reduced anthesis-silking interval, and reduced tassel size (Emeades et al., 1992).

15.7 CONVENTIONAL BREEDING

The conventional method of improvement in maize is with the help of hybridization between desired, selected, and tested contrasting parents for the targeted traits. The basic need of improvement is (1) the variability should present in the available genetic resource, (2) high heritability of the targeted traits, (3) precise phenotyping in the stress environment, (4) precise selection criteria as well as the selection parameter for secondary traits. If the variability is not available in the cultivated varieties of the crop, then we have to search the variability in their wild forms and relatives. In case of wild species, one major problem used to come is the cross compatibility. In this situation we can use bridge species for transferring the desired traits conventionally. If it is cross compatible, in that condition we will go for the prebreeding concept and then utilize desired traits for improvement. The most common conventional breeding strategies are population improvement as well as hybrid varieties in the cross-pollinated crops like maize. The most economical breeding strategies for improving the drought tolerance in maize are exploitation of drought-tolerant germplasm in the breeding program to combat the drought stress at the global level (Tables 15.4 and 15.5).

TABLE 15.4 Wild Species of Maize Exploited as a Source of Drought Tolerance.

Zea mays subsp. *mexicana*	Drought tolerance	Gonzaalez et al. (2018)
Eastern gamagrass	Drought tolerance	Mammadov et al. (2018)
Zea mays parviglumis	Drought tolerance	Kumar et al. (2020); Adhikari et al. (2021a), (2021b); Sahoo et al. (2021)

TABLE 15.5 Identified Landraces for Drought Tolerance.

Land races	Locations	Traits	Reference
Tuxpeno	Global collection	Drought tolerant	Prasanna (2012)
L25, L14, L1, and L3		Drought tolerant	Andjelkovic et al. (2014)
TZm-1167,TZm-1162, TZm-1472, TZm-1508, and TZm-1506	West Africa	Drought tolerant	Nelimor et al. (2020)
GalTrini, SITexas	Mexico	Drought tolerant	Hernández et al. (2021)

In the African countries, initiation of CIMMYT led to the production of drought-tolerant hybrid seed of maize at the commercial level and also drought-tolerant open pollinated varieties, that is, ZM523 along with that farmer in Congo also produced seeds of drought-tolerant OPV ZM623 (CIMMYT, 2012). Several drought-tolerant OPV as well as hybrid varieties have been developed by different scientists across the globe through the conventional method.

Mutation breeding strategies can be utilized for the generation of variability if variability is not available in the desired traits. With the help of mutation breeding, we can develop drought-tolerant mutant in maize. Mutation-assisted breeding along with the new breeding techniques can be used as an effective tool to develop drought tolerance maize cultivars.

15.8 NEW BREEDING TECHNIQUES (NBT)

The improvement of maize through conventional breeding has major constraint, that is, time consuming and long breeding cycle due to complexity of the drought trait. This complexity is low heritability, degree of the trait, and higher genotype x environment interaction (Asrat, 2021). Therefore, NBT is essential for the rapid improvement of maize crop for desired traits. NBT includes marker-assisted selection, genomic selection, genome editing, transgenics, etc. These NBT can increase the reliability of selection since it is independent of environment. Marker-assisted selection (MAS) can be utilized for several purposes in the breeding programs like to know the genetic divergence of the parents, classification of heterotic groups and selection of inbred lines as a parent. With the help MAB (marker-assisted backcrossing) desired genes can be transferred into recipient lines. In CIMMYT, six QTL were identified on chromosomes 1, 2, 5, 6, 8, and 10, accounting together for approximately 50% of the phenotypic variability of ASI in maize. A major QTL has been detected which regulates the leaf ABA concentration in maize. Five QTLs for short ASI were introgressed through MABC from a drought donor to drought susceptible line in maize (Ribaut and Ragot, 2006). Under drought condition in maize, a major constitutive QTL (*Root-ABA1*) has been identified and correlated with total dry root mass, diameter of root, and crown root branching (Giuliani et al., 2005b). Under drought condition, due to upregulation of aquaporin genes *PIP1* and *RWC-3* in maize improve the water uptake of plant and also increase the overall root size (Giuliani et al., 2005a).

Quan et al. (2004) reported that in maize, glycine betaine is induced under abiotic stresses particularly in drought. The *choline dehydrogenase*, an enzyme encoded by *betA* gene isolated from *E. coli* responsible for the biosynthesis of glycine betaine from choline, has been transformed in the maize inbred line DH4866. The higher level of glycine betaine has been accumulated in the transgenic maize and reflected greater level of drought tolerance than normal plants (non-transgenic) at the seedling and germination stages. It has been also noticed that the grain yield of transformed plants was showing considerably higher than normal plants after drought treatment. Therefore, he concluded that due to higher level of expression and accumulation of glycine betaine in transgenic maize made cell membrane integrity protected as compared to normal plants in conditions of drought stress. In a most recent study, maize ZmSDD1 cloned and its functions associated with the increase the WUE and the photosynthetic rate under drought conditions in maize (Liu et al., 2013). The genome-editing techniques CRISPR-Cas9 in maize enhance the expression level of ARGOS8 gene to develop drought tolerance and also the promotor of ARGOS8 gene has been changed into GOS2 and finally new mutant having high grain yield under drought stress has been developed (Shi et al., 2017). The expression of ZmNF-YBs transcription factor (orthologous maize) increased under the drought stress in maize and also increased the drought tolerance by improving several traits (chlorophyll content, stomatal conductance, leaf temperature, reduced wilting, maintenance of photosynthesis). Under stress condition, the relative water and chlorophyll contents and chlorophyll fluorescence increases due to the overexpression of phytochrome-interacting factor 3 (PIF3) lead to increasing the tolerance to dehydration in ZmPIFs transgenic plants.

15.9 FUTURE STRATEGIES

In maize breeding program for drought tolerance, we have made quite good journey of improvement right from the beginning of searching the genetic resources for drought tolerance to the development of drought-tolerant varieties and the genetic dissection of drought tolerance. We have also achieved new height in the field of molecular breeding for identifications of number of QTLs for secondary traits associated with the drought tolerance. But, in the growing demand of food for the increasing population still we have to search new drought-tolerant sources as well as to develop new technology to ease the breeding program. Gene stacking and pyramiding could be used for the improvement of different desired traits for drought tolerance in maize crops. A multidisciplinary approach involving genetics, biochemistry, biotechnology, physiology plant breeding, and crop science is essential to assess the complicated and integrated responses of plants to drought.

KEYWORDS

- **drought**
- **maize**
- **MAS**
- **QTLs**
- **transgenic**
- **physiology**

REFERENCES

Achievements for Drought Tolerant Maize Breeding in Southern Africa. CIMMYT [Online] Retrieved June 1, 2022. https://www.cimmyt.org/news/achievements-for-drought-tolerant-maize-breeding-in-southern-africa/.

Adhikari, S.; Joshi, A.; Kumar, A.; Singh, N. K. Diversification of Maize (Zea mays L.) Through Teosinte (Zea mays subsp. parviglumis Iltis & Doebley) Allelic. *Genet. Res. Crop Evolut.* **2021,** *68* (7), 2983–2995.

Anđelković, V.; Kravić, N.; Babić, V.; Ignjatović-Mićić, D.; Dumanović, Z.; Vančetović, J. Estimation of Drought Tolerance Among Maize Landraces from Mini-Core Collection. *Genetika* (2014, *46* (3), 775–788.

Anjum, S. A.; Ashraf, U.; Tanveer, M.; Khan, I.; Hussain, S.; Shahzad, B.; Zohaib, A.; Abbas, F.; Saleem, M. F.; Ali, I.; Wang, L. C. Drought Induced Changes in Growth, Osmolyte Accumulation and Antioxidant Metabolism of Three Maize Hybrids. *Front. Plant Sci.* **2017,** *8*, 69. DOI: 10.3389/fpls.2017.00069

Area, Yield and Production. Ipad.fas.usda.gov. [Online] 2022. Retrieved 11 June 2022. https://ipad.fas.usda.gov/cropexplorer/util/new_get_psd_data.aspx?regionid=sasia&cntryid=IN&cntryname=India.

Asrat, Z. The Improvement of Maize (Zea mays L.) for Drought Stress Tolerance. *Int. J. Adv. Res. Biol. Sci.* **2021,** *8* (7), 90–102.

Barrieu, F.; Chaumont, F.; Chrispeels, M. J. High expression of the tonoplast aquaporin ZmTIP1 in epidermal and conducting tissues of maize. *Plant Physiol.* (1998, *117* (4), 1153–1163.

Bassetti, P.; Westgate, M. E. Emergence, Elongation, and Senescence of Maize Silks. *Crop Sci.* **1993,** *33* (2), 271–275.

Benešová, M.; Hola, D.; Fischer, L.; Jedelský, P. L.; Hnilička, F.; Wilhelmová, N.; ... Hniličková, H. The Physiology and Proteomics of Drought Tolerance in Maize: Early Stomatal Closure as a Cause of Lower Tolerance to Short-Term Dehydration?. *PLoS One* **2012,** 7(6), e38017.

Blum, A. Osmotic Adjustment is a Prime Drought Stress Adaptive Engine in Support of Plant Production. *Plant, Cell Environ.* **2017,** *40* (1), 4–10.

Borrás, L.; Maddonni, G. A.; Otegui, M. E. Leaf Senescence in Maize Hybrids: Plant Population, Row Spacing and Kernel Set Effects. *Field Crops Res.* **2003,** *82* (1), 13–26.

Brugière, N.; Zhang, W.; Xu, Q.; Scolaro, E. J.; Lu, C.; Kahsay, R. Y.; ... & Habben, J. E. Overexpression of RING Domain E3 Ligase ZmXerico1 Confers Drought Tolerance Through Regulation of ABA Homeostasis. *Plant Physiol.* **2017,** *175* (3), 1350–1369.

Curá, J. A.; Franz, D. R.; Filosofía, J. E.; Balestrasse, K. B.; Burgueño, L. E. Inoculation with Azospirillum sp. and Herbaspirillum sp. Bacteria Increases the Tolerance of Maize to Drought Stress. *Microorganisms* **2017,** *5* (3), 41.

Dass, S.; Arora, P.; Kumari, M.; Pal, D. Morphological Traits Determining Drought Tolerance in Maize (Zea mays L.). *Indian J. Agric. Res.* **2001,** *35* (3), 190–193.

Denmead, O. T.; Shaw, R. H. The Effects of Soil Moisture Stress at Different Stages of Growth on the Development and Yield of Corn 1. *Agronomy J.* **1960,** *52* (5), 272–274.

Ding, S.; Cai, Z.; Du, H.; Wang, H. Genome-Wide Analysis of TCP Family Genes in Zea mays L. Identified a Role for ZmTCP42 in Drought Tolerance. *Int. J. Mol. Sci.* **2019,** *20* (11), 2762.

Ding, F.; Wang, R. Amelioration of Postharvest Chilling Stress by Trehalose in Pepper. *Sci. Hortic.* **2018,** *232*, 52–56.

Duan, L.; Dietrich, D.; Ng, C. H.; Chan, P. M. Y.; Bhalerao, R.; Bennett, M. J.; Dinneny, J. R. Endodermal ABA Signaling Promotes Lateral Root Quiescence During Salt Stress in Arabidopsis Seedlings. *Plant Cell* **2013,** *25* (1), 324–341.

Edmeades, G. O. Progress in Achieving and Delivering Drought Tolerance in Maize-an Update; ISAAA: Ithaca, NY, 2013; p 130.

Effendi, R.; Priyanto, S. B.; Aqil, M.; Azrai, M. In *Drought Adaptation Level of Maize Genotypes Based on Leaf Rolling, Temperature, Relative Moisture Content, and Grain Yield Parameters*, IOP Conference Series: Earth and Environmental Science; IOP Publishing, 2019; vol *270* (1), p 012016.

Flint-Garcia, S. A.; Thuillet, A. C.; Yu, J.; Pressoir, G.; Romero, S. M.; Mitchell, S. E.; ... & Buckler, E. S. Maize Association Population: A High-Resolution Platform for Quantitative Trait Locus Dissection. *Plant J.* **2005,** *44* (6), 1054–1064.

Gallavotti, A.; Barazesh, S.; Malcomber, S.; Hall, D.; Jackson, D.; Schmidt, R. J.; McSteen, P. Sparse Inflorescence1 Encodes a Monocot-Specific YUCCA-Like Gene Required for Vegetative and Reproductive Development in Maize. *Proc. Natl. Acad. Sci.* **2008,** *105* (39), 15196–15201.

Giuliani, S.; Clarke, J.; Kreps, J.; Sanguineti, M. C.; Salvi, S.; Landi, P.; et al. In *Microarray Analysis of Backcrossed-Derived Lines Differing for Root-ABA1, A Major QTL Controlling Root Characteristics and ABA Concentration in Maize*, Proceedings of the International Congress in the Wake of the Double Helix: From the Green Revolution to the Gene Revolution; Tuberosa, R., Phillips, R. L., Gale, M., Eds.; Avenue Media: Bologna, 2005a; pp 463–490.

Giuliani, S.; Sanguineti, M. C.; Tuberosa, R.; Bellotti, M.; Salvi, S.; Landi, P. Root-ABA1, A Major Constitutive QTL, Affects Maize Root Architecture and Leaf ABA Concentration at Different Water Regimes. *J. Exp. Bot.* **2005b,** *56* (422), 3061–3070.

GonzaAlez, J. J. S.; Corral, J. A. R.; García, G. M.; Ojeda, G. R.; Ldlc, L.; Holland, J. B. Ecogeography of teosinte. *PloS One* **2018,** *13* (2), e0192676.

Hare, P. D.; Cress, W. A.; Van Staden, J. Dissecting the Roles of Osmolyte Accumulation During Stress. *Plant Cell Environ.* **1998,** *21* (6), 535–553.

Hernández, V. A. G.; Cruz, E. L.; Onofre, L. E. M.; Varela, A. S.; Espinosa, M. A. G.; García, F. Z. Maize (Zea mays L.) Landraces Classified by Drought Stress Tolerance at the Seedling Stage. *Emir. J. Food Agric.* **2021,** 29–36.

Herrero, M. P.; Johnson, R. R. Drought Stress and its Effects on Maize Reproductive Systems 1. *Crop Sci.* **1981,** *21* (1), 105–110.

Ipad.fas.usda.gov. Area, Yield and Production; 2022 [Online]. <https://ipad.fas.usda.gov/cropexplorer/util/new_get_psd_data.aspx?regionid=sasia&cntryid=IN&cntryname=India> (accessed June 11, 2022).

Jia, T. J., Jing-jing, L. I., Wang, L. F., Cao, Y. Y., MA, J.; Wang, H.; Zhang, D. F.; LIH. Y. Evaluation of Drought Tolerance in ZmVPP1-Overexpressing Transgenic Inbred Maize Lines and their Hybrids. *J. Integr. Agric.* **2020,** *19* (9), 2177–2187.

Jin, Y.; Yang, H.; Wei, Z.; Ma, H.; Ge, X. Rice Male Development Under Drought Stress: Phenotypic Changes and Stage-Dependent Transcriptomic Reprogramming. *Mol. Plant* **2013,** *6* (5), 1630–1645.

Kadioglu, A.; Terzi, R.; Saruhan, N.; Saglam, A. Current Advances in the Investigation of Leaf Rolling Caused by Biotic and Abiotic Stress Factors. *Plant Sci.* **2012,** *182*, 42–48.

Kumar, A.; Singh, N. K.; Jeena, A. S.; Jaiswal, J. P.; Verma, S. S. Evaluation of Teosinte Derived Maize Lines for Drought Tolerance. *Indian J. Plant Genet. Res.* **2020,** *33* (1), 60–67.

Li, R.; Zeng, Y.; Xu, J.; Wang, Q.; Wu, F.; Cao, M.; ... & Lu, Y. Genetic Variation for Maize Root Architecture in Response to Drought Stress at the Seedling Stage. *Breed. Sci.* **2015,** *65* (4), 298–307.

Li, Z.; Zhang, X.; Zhao, Y.; Li, Y.; Zhang, G.; Peng, Z.; Zhang, J. Enhancing Auxin Accumulation in Maize Root Tips Improves Root Growth and Dwarfs Plant Height. *Plant Biotechnol. J.* **2018,** *16* (1), 86–99.

Liu, M.; Li, M.; Liu, K.; Sui, N. Effects of Drought Stress on Seed Germination and Seedling Growth of Different Maize Varieties. *J. Agric. Sci.* **2015,** *7* (5), 231.

Liu, Y.; Qin, L.; Han, L.; Xiang, Y.; Zhao, D. Overexpression of Maize SDD1 (ZmSDD1) Improves Drought Resistance in Zea mays L. by Reducing Stomatal Density. *Plant Cell Tissue Organ Culture* **2015,** *122* (1), 147–159.

Liu, S.; Wang, X.; Wang, H.; Xin, H.; Yang, X.; Yan, J.; ... & Qin, F. Genome-Wide Analysis of ZmDREB Genes and their Association with Natural Variation in Drought Tolerance at Seedling Stage of Zea mays L. *PLoS Genet.* **2013,** *9* (9), e1003790.

Luo, L.; Xia, H.; Lu, B. R. Crop Breeding for Drought Resistance. *Front. Plant Sci.* **2019,** *10*, 314.

Maazou, A. R. S.; Tu, J.; Qiu, J.; Liu, Z. Breeding for Drought Tolerance in Maize (Zea mays L.). *Am. J. Plant Sci.* **2016,** *7* (14), 1858–1870. DOI: //dx.doi.org/10.4236/ajps.2016.714172

Mammadov, J.; Buyyarapu, R.; Guttikonda, S. K.; Parliament, K.; Abdurakhmonov, I. Y.; Kumpatla, S. P. Wild Relatives of Maize, Rice, Cotton, and Soybean: Treasure Troves for Tolerance to Biotic and Abiotic Stresses. *Front. Plant Sci.* **2018,** *9*, 886.

Mao, H.; Wang, H.; Liu, S.; Li, Z.; Yang, X.; Yan, J.; ... & Qin, F. A Transposable Element in a NAC Gene is Associated with Drought Tolerance in Maize Seedlings. *Nat. Commun.* **2015,** *6* (1), 1–13.

Muhammad, A.; Amjad, F.; Ihsan, K.; Qurban, A.; Muhammad, A.; Muhammad, K.; Ahsan, M. Inheritance of Various Yield Contributing Traits in Maize (Zea mays L.) at Low Moisture Condition. *Afr. J. Agric. Res.* **2013,** *8* (4), 413–420.

Nelimor, C.; Badu-Apraku, B.; Tetteh, A. Y.; Garcia-Oliveira, A. L.; N'guetta, A. S. P. Assessing the Potential of Extra-Early Maturing Landraces for Improving Tolerance to Drought, Heat, and Both Combined Stresses in Maize. *Agronomy* **2020,** *10* (3), 318.

Normanly, J. Approaching Cellular and Molecular Resolution of Auxin Biosynthesis and Metabolism. *Cold Spring Harb. Perspect. Biol.* **2010,** *2* (1), a001594.

Pandit, M.; Sah, R. P.; Chakraborty, M.; Prasad, K.; Chakraborty, M. K.; Tudu, V.; ... & Rana, M. Gene Action and Combining Ability for Dual Purpose Traits in Maize (Zea mays L.) Under Water Deficit Stress Prevailing in Eastern India. *Range Manag. Agrofor.* **2018,** *39* (1), 29–37.

Park, W. J.; Kriechbaumer, V.; Muller, A.; Piotrowski, M.; Meeley, R. B.; Gierl, A.; Glawischnig, E. The Nitrilase ZmNIT2 Converts Indole-3-Acetonitrile to Indole-3-Acetic Acid. *Plant Physiol.* **2003,** *133* (2), 794–802.

Prasanna, B. M. Diversity in Global Maize Germplasm: Characterization and Utilization. *J. Biosci.* 2012, *37* (5), 843–855.

Quan, R.; Shang, M.; Zhang, H.; Zhao, Y.; Zhang, J. Engineering of Enhanced Glycine Betaine Synthesis Improves Drought Tolerance in Maize. *Plant Biotechnol. J.* **2004,** *2* (6), 477–486.

Ren, B. Z.; Juan, H. U.; Zhang, J. W.; Dong, S. T.; Peng, L. I. U.; Bin, Z. H. A. O. Effects of Urea Mixed with Nitrapyrin on Leaf Photosynthetic and Senescence Characteristics of Summer Maize (Zea mays L.) Waterlogged in the Field. *J. Integr. Agric.* **2020,** *19* (6), 1586–1595.

Ribaut, J. M.; Ragot, M. Marker-Assisted Selection to Improve Drought Adaptation in Maize: The Backcross Approach, Perspectives, Limitations, and Alternatives. *J. Exp. Bot.* **2007,** *58* (2), 351–360.

Riboldi, L. B.; Oliveira, R. F.; Angelocci, L. R. Leaf Turgor Pressure in Maize Plants Under Water Stress. *Aust. J. Crop Sci.* **2016,** *10* (6), 878–886.

Rida, S.; Maafi, O.; López-Malvar, A.; Revilla, P.; Riache, M.; Djemel, A. Genetics of Germination and Seedling Traits Under Drought Stress in a MAGIC Population of Maize. *Plants* **2021,** *10* (9), 1786.

Saglam, A.; Kadioglu, A.; Demiralay, M.; Terzi, R. Leaf Rolling Reduces Photosynthetic Loss in Maize Under Severe Drought. *Acta Botanica Croatica* **(2014,** *73* (2), 315–323.

Sah, R. P.; Chakraborty, M.; Prasad, K.; Pandit, M.; Tudu, V. K.; Chakravarty, M. K.; ... & Moharana, D. Impact of Water Deficit Stress in Maize: Phenology and Yield Components. *Sci. Rep.* **2020,** *10* (1), 1–15.

Sahoo, S.; Adhikari, S.; Joshi, A.; Singh, N. K. Use of Wild Progenitor Teosinte in Maize (Zea mays subsp. mays) Improvement: Present Status and Future Prospects. *Trop. Plant Biol.* **2021,** *14* (2), 156–179.

Sari-Gorla, M.; Krajewski, P.; Di Fonzo, N.; Villa, M.; Frova, C. Genetic Analysis of Drought Tolerance in Maize by Molecular Markers. II. Plant Height and Flowering. *Theor. Appl. Genet.* **1999,** *99* (1), 289–295.

Sekimoto, H.; Seo, M.; Dohmae, N.; Takio, K.; Kamiya, Y.; Koshiba, T. Cloning and Molecular Characterization of Plant Aldehyde Oxidase. *J. Biol. Chem.* **1997,** *272* (24), 15280–15285.

Shi, J.; Gao, H.; Wang, H.; Lafitte, H. R.; Archibald, R. L.; Yang, M.; ... & Habben, J. E. ARGOS 8 Variants Generated by CRISPR-Cas9 Improve Maize Grain Yield Under Field Drought Stress Conditions. *Plant Biotechnol. J.* **2017,** *15* (2), 207–216.

Spollen, W. G.; LeNoble, M. E.; Samuels, T. D.; Bernstein, N.; Sharp, R. E. Abscisic Acid Accumulation Maintains Maize Primary Root Elongation at Low Water Potentials by Restricting Ethylene Production. *Plant Physiol.* **2000,** *122* (3), 967–976.

Tai, F.; Wang, Q.; Yuan, Z.; Yuan, Z.; Li, H.; Wang, W. Characterization of Five CIPK Genes Expressions in Maize Under Water Stress. *Acta Physiol. Plant.* **2013,** *35* (5), 1555–1564.

Vadez, V. Root Hydraulics: The Forgotten Side of Roots in Drought Adaptation. *Field Crops Res.* **2014,** *165*, 15–24.

Wang, B.; Li, Z.; Ran, Q.; Li, P.; Peng, Z.; Zhang, J. ZmNF-YB16 Overexpression Improves Drought Resistance and Yield by Enhancing Photosynthesis and the Antioxidant Capacity of Maize Plants. *Front. Plant Sci.* **2018,** *9*, 709.

Wang, C. T.; Ru, J. N.; Liu, Y. W.; Yang, J. F.; Li, M.; Xu, Z. S.; Fu, J. D. The maize WRKY Transcription Factor ZmWRKY40 Confers Drought Resistance in Transgenic Arabidopsis. *Int. J. Mol. Sci.* **2018,** *19* (9), 2580.

Wang, X.; Wang, H.; Liu, S.; Ferjani, A.; Li, J.; Yan, J.; ... & Qin, F. Genetic Variation in ZmVPP1 Contributes to Drought Tolerance in Maize Seedlings. *Nat. Genet.* **2016,** *48* (10), 1233–1241.

Wu, J.; Jiang, Y.; Liang, Y.; Chen, L.; Chen, W.; Cheng, B. Expression of the Maize MYB Transcription Factor ZmMYB3R Enhances Drought and Salt Stress Tolerance in Transgenic Plants. *Plant Physiol. Biochem.* **2019,** *137*, 179–188.

Wu, G.; Zhao, Y.; Shen, R.; Wang, B.; Xie, Y.; Ma, X.; ... & Wang, H. Characterization of Maize Phytochrome-Interacting Factors in Light Signaling and Photomorphogenesis. *Plant Physiol.* **2019,** *181* (2), 789–803.

Xu, S. C.; Li, Y. P.; Jin, H. U.; Guan, Y. J.; Zheng, Y. Y.; Zhu, S. J. Responses of Antioxidant Enzymes to Chilling Stress in Tobacco Seedlings. *Agric. Sci. China* **2010,** *9* (11), 1594–1601.

Xue, Y.; Warburton, M. L.; Sawkins, M.; Zhang, X.; Setter, T.; Xu, Y.; ... & Yan, J. Genome-wide Association Analysis for Nine Agronomic Traits in Maize Under Well-Watered and Water-Stressed Conditions. *Theor. Appl. Genet.* **2013,** *126* (10), 2587–2596.

Zhang, X.; Mi, Y.; Mao, H.; Liu, S.; Chen, L.; Qin, F. Genetic Variation in ZmTIP1 Contributes to Root Hair Elongation and Drought Tolerance in Maize. *Plant Biotechnol. J.* **2020,** *18* (5), 1271–1283.

CHAPTER 16

Next-Generation Breeding for Heat Stress Improvement in Maize

SURESH H. ANTRE[1], SAUMYA KUMARI[1], A. E. SUNIL SUBRAMANYA[1], and ASHUTOSH SINGH[2]

[1]*University of Agricultural Sciences, GKVK Campus, Bangalore, India*

[2]*Centre for Advance Studies on Climate Change, RPCAU, Pusa, Samastipur, India*

ABSTRACT

The researchers have massive challenges in upcoming years to improve yield and productivity of food crops due an ever-increasing population, dwindling agricultural area, and changing climates. Heat stress has been becoming a one of the major problems especially in tropical rain-fed environments worldwide for sustainable production of food crops including maize. The plant breeding techniques in last two decades have dramatically changed and replaced with advanced genomics and breeding tools. Hear stress tolerance is a complex trait with the combination of several quantitative traits. The high-throughput genotyping and phenotyping techniques with advanced sequencing tools have made understanding of genetics of heat tolerance possible. Speed breeding is another effective strategy that focuses on reducing the crop generation time and boosting the improvement programs. CRISPR/Cas based genome editing is the cutting-edge methodology used to improve the trait expression by precise altering of genomic regions. In the present chapter, successful integration of next generation breeding platforms and multidisciplinary approaches for maize improvement under heat stress has been discussed which opens exciting routes to develop stress resilient crops towards global food security.

Climate-Resilient Agriculture: A Molecular Perspective. Ashutosh Singh, Saurabh Pandey (Eds.)

16.1 INTRODUCTION

The current pattern of climate crisis is potent threat to achieve food security to burgeoning global population by significantly increasing global surface mean temperature, changing precipitation patterns which in turn reducing the crop productivity. Maize (*Zea mays* L.) is considered as an essential cereal with a significant appeal for not only as food, feed but also as fodder across the world. It is one such a crop that is widely cultivated around the globe and second most valuable crop in terms of acreage. Globally it was cultivated on 193.10 million hectares with 1147.7 million metric tonnes production and 5.75 metric tonnes per hectare productivity (FAOSTAT, 2020). It is the third most crucial cereal crop in terms of production after wheat and rice in India covering 9.02 million hectares area with 27.71 million metric tonnes production and 3.07 metric tonnes per hectare productivity (India Agri Stat., 2020). It has been predicted that, the worldwide mean temperature may rise by 2–5°C or more by the year 2050 (www.ipcc.ch). Tropical and subtropical climates which have coverage of 160 million hectares will be most vulnerable due to climate change that affects the temperature pattern and water availability. The rain-fed system is the prevailing production system especially in tropical and subtropical regions of the world (Edmeades, 2008) with a global yield loss of about 25%. It has been observed that out of total 19 million hectares cultivated area in tropical Asia, 80% area is rain-fed and drought-prone. Thus, yield losses are anticipated to increase owing to the erratic rainfall pattern in near future (Mhike et al., 2012).

Heat stress negatively affect the growth stages of crop depending upon the region where the crop is growing. At seed filling stage it causes the reduction in storage granules and proteins number and endosperms number, which further leads to lower starch and protein build-up in kernels of maize (Monjardino et al., 2005). Four important stages of maize *viz.* germination, photosynthesis, vegetative, and reproductive (pollen viability and fertilization) are affected by high temperature stress where threshold temperature recoded for each stage 15–40°C, 33–38°C, 38°C, and 36–40°C, respectively (Govindaraj et al., 2018). Additionally, Mahmood et al. (2010) observed a substantial reduction in forage quality in maize due to heat stress. Heat stress substantially lowers the maize yield through increased kernel abortion, decreasing photosynthetic efficiency, and reduced starch storage (Cantarero et al., 1999; Wilhelm et al., 1999; Edreira and Otegui, 2013). Reproductive system found to be a more sensitive to heat stress specifically male system (Schoper et al., 1987). Increase temperatures also aggravate drought and induces significant yield losses (Mallya et al., 2013; Lobell et al., 2014). For example, in United States it was reported that combined drought and heat stress reduced yields by 27% in 2012 (Rippey, 2015). Tolerance to heat stress is a complex phenomenon influenced by interaction of morphological, physiological and molecular factors and different levels. Several plant breeding methods has been utilized in last few decades to improve heat stress tolerance in maize with limited success. To fulfill the future food requirements, researchers are extensively in search of one or other novel alternative technologies which could help in climate resilient smart breeding to accelerate crop improvement programmes. Moreover, in recent years, the advanced genomics tools have opened a way for faster development of stress tolerance cultivars. The conventional breeding approach as well current next generation high throughput tools for improving heat stress tolerance, particularly in maize are discussed in this chapter.

16.2 HEAT STRESS TOLERANCE AND SCREENING METHODOLOGIES IN MAIZE

Heat stress tolerance is a complex quantitative traits and mode of inheritance of traits associated with heat stress has prime importance for successful breeding strategies. Hence, effective selection strategies for heat tolerance requires for evaluating genotypes with replicated trials in multiple environments (Bai, 2003). Under high temperature stress, non-additive gene action has been reported for yield and contributing characters in maize (Akbar et al., 2008). Other traits like seed vigor, germination percentage, growth rate of plant, relative water content, rolling of leaf, anthesis silking duration, pollen size and viability, silk receptivity, shelling percentage, ear number per plant, leaf senescence, plant maturity days, and economic yield a dominance gene action has been found to be predominant (Tassawar et al., 2012). Akbar et al. (2008) described non-additive gene action for yield and contributing characters under high temperature stress in maize.

Environmental interactions and other factors are the crucial challenges for heat tolerance screening at field level. However, presence of several other relevant traits allows successful selection of tolerant genotypes under the field conditions (Yadav et al., 2011). Controlled environments, on the other hand, restrict natural selection for the components interacting with the heat stress tolerance which limits the possiblity for wider applicability of trait screening under field conditions (Fritsche-Neto et al., 2012). Assessment of heat tolerance is important and can be done by a range of approaches including visual assessment, measurements, assays and evaluating in hotspot site which is in details reviewed by (Govindaraj et al., 2018) and included (1) Thermo-stability test for cell membrane, (2) Chlorophyll fluorescence measurement, (3) Membrane lipid saturation estimation, (4) Canopy temperature depression (CTD), (5) Visual assessment and morphological methods, (6) Selection in hot production environments, (7) Pollen-based selection and screening of genotypes. These screening methods at different stages are found effective for selection of genotypes for heat tolerance in various crops including maize (Setimela et al., 2005; Singh et al., 2017; Singh et al., 2020). However, these screening techniques need to be validated under field conditions for their applicability in crop improvement. This can be done by effective utilization of high throughput phenotyping platforms. It is also recommended that, besides using screening method, it is also necessary to develop potent thermo-tolerance markers that can be perhaps utilized for future applications in maize breeding for heat stress tolerance.

16.3 APPROACHES FOR IMPROVEMENT IN HEAT STRESS TOLERANCE

16.3.1 MARKER BASED BREEDING APPROACH

Germplasm resources play an important role in any crop breeding program. Several heat stress tolerant related gene and their inheritance pattern have been identified through various conventional breeding approaches (Wahid et al., 2007; Farooq et al., 2011). One of the major challenges in developing an improved heat stress tolerant genotypes is the governance of traits by small effect of QTLs or several epistatic QTLs (Bita and Gerats, 2013). Hence, effective use of the DNA markers in identifying the major regions or QTLs or candidate genes in the genome responsible for heat stress tolerance in a germplasm provides a basic starting material for breeding program which further led to development stress tolerant cultivars (Fig. 16.1). The recent advances in genotyping tools and marker discovery are being used to identify precise location of QTLs in various crops (Jagadish et al., 2010; Paliwal et al., 2012; Pinto et al., 2010). In case of maize, the QTLs were identified for heat stress related traits such as pollen germination, pollen tube growth and cellular membrane stability (Ottaviano et al., 1991; Frova et al., 1994). Further, marker-assisted selection (MAS) or genomic selection (GS) strategies can be utilized to introgress such a candidate QTLs into elite lines for heat stress tolerance (Tester and Langridge, 2010). Thus QTL linked markers allow the breeding of stress-tolerant crops either by combining two or more QTLs for stress tolerance. Another approach is "physiological" trait-based breeding, where parent material usually selected from the high temperature regions. The aim in both approaches is to identify and combine the target loci or QTL/s with high heritability that sustain yield not only at normal but also at high temperature conditions.

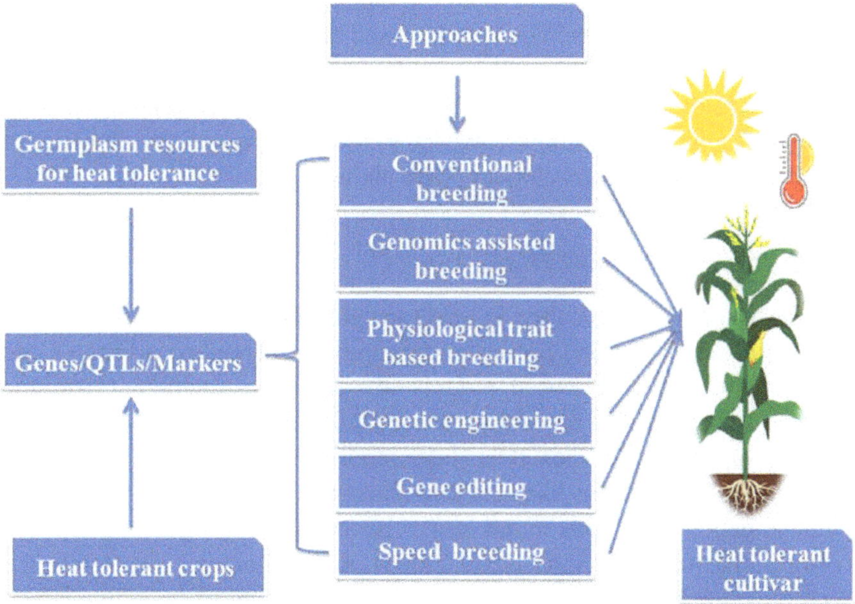

FIGURE 16.1 An integrated approach of developing heat stress tolerant cultivars in maize.

16.3.2 GENOME WIDE ASSOCIATION MAPPING STUDIES

Genome-wide association study (GWAS) is a strong and powerful method to link the complex phenotypic traits back to their fundamental genetics. This approach is an efficient to understand the genetic basis of complex trait such as heat tolerance using naturally occurring genetic diversity. The GWAS approach provides a high mapping resolution as compare to bi-parental mapping populations to detect marker-trait association for the trait of interest (Liu et al., 2016). GWAS approach uses the existing genetic variability between divergent sub-populations with available SNP information and LD decay values (Cooper et al., 2019; Shikha et al., 2021; Mazaheri et al., 2019) to identify marker trait association (Xiao et al., 2017). Multitude germplasm screening and GWAS used for stress tolerant inbred lines identification in maize (Rodriguez et al., 2010; Farooqi et al., 2016). Six significant SNP haplotype on chromosome 4 associated with grain yield under combined heat and drought stress conditions has been identified through GWAS (Yuan et al., 2019). They also identified significant hot-spots on chromosome 8 for grain yield under heat stress and on chromosome 4 for combined heat and drought stress conditions. In another study, 12 substantial SNP associations under heat stress for grain yield were identified using a panel of 662 double haploid lines with tropical origin in sub-tropical maize on chromosomes 1, 3, 6, 7, and 10 (Longmei et al., 2021). The recent advances and evidences shown that, next generation sequencing platform and SNP genotyping provides the inputs to characterize genetic diversity at a high resolution and allows breeder to select for beneficial diversity to advance new stress tolerant cultivars with higher yield.

16.3.3 GENOMIC SELECTION

The use of GWAS has resulted in development of potential approach called as genomic selection (GS) for designing innovative breeding programs and developing new marker-based models for genetic evaluations. Genomic selection speeds up the breeding cycle by allowing for rapid selection of preferred superior genotypes. It has been reported that marker assisted selection for multifaceted traits like heat tolerance due to genotype × environment and gene-gene interactions found less effective which resulted in a reduced breeding efficiency (Collins et al., 2008). GS technique used to predict breeding value from the data generated from screening of huge number of molecular markers designated for great genome coverage. On the other hand, MAS strategies uses molecular markers with substantial association through a trait of interest (Bita and Gerats, 2013). Simulation experiment shown that GS can improve the correlations between predicted and true breeding values across numerous generations without re-phenotyping. As a result, GS may lead to lowered analysis cost and improves the rates of genetic gain (Habier et al., 2009; Heffner et al., 2009).

16.3.4 GENETIC ENGINEERING APPROACHES FOR IMPROVED STRESS TOLERANCE

Substantial progress is being carried out in different plant science domains like genomics, molecular biology, plant physiology, metabolomics, and biotechnology. Hence, via bringing together information from these different disciplines, some of the efficient strategies might be designed which could advance heat tolerance in the previously existing genotypes and incorporate it in the susceptible ones. The transgenic technology has made it possible to manipulate native genes or introduce gene (s) into the genome of an organism that are not found in the crop gene pools, and it is becoming a popular method for developing crops that are tolerant to both abiotic and biotic stresses (Ashraf et al., 2008). Therefore, to get stress tolerant plants, efforts can be made to engineering signalling pathways related to abiotic stress and characterization and cloning of transcription factors which are found to regulate the gene expression (Ma et al., 2016). It was also seen that stress tolerance gene can negatively affect the yield. Thus, transformation strategies play a crucial role where transgene expression could be restricted to a particular tissue *via* tissue specific or stress inducible promoter with the minimal effect on yield potential of crop (Nakashima et al., 2009). Moreover, multiple stresses tolerance at various stages of crop development has also been addressed by gene pyramiding through genetic engineering (Datta et al., 2002).

It has been proved from current genetic progress that heat tolerance can be improved utilizing conventional protocols and transgenic approaches. It has been thereby easier to identify the function of candidate genes through genome-wide expression analysis for heat stress responses along with their regulatory and tolerance mechanism with the advancement in functional genomics (Vij and Tyagi, 2007; Sreenivasulu et al., 2007; Urano et al., 2010). Development of tolerance against heat stress at various plant developmental stages in different tissues, and study of components of tolerance regulated by distinct gene sets are essential (Howarth, 2005; Bohnert et al., 2006). The genetic engineering and genome editing has been so far mainly focusing on engineering genes that encodes heat tolerance governed by transcription factors, heat shock proteins (HSPs), osmolytes, chaperones, plant growth regulators, and antioxidants, etc., (Ashraf, 2010; Grover et al., 2013). These factors mainly regulate the plant growth, development, and are also involved in response to stress conditions. Thus, transgenic modifications of them surely support to comprehend multigene regulation pathways and their relationship to stress tolerance. The transcriptomics and proteomics studies have revealed numerous heat responsive genes and proteins in maize. Genetic engineering of heat shock factors (HSF), heat shock proteins (HSPs) related genes and/or and their modification have huge potential by altering their expression levels to improve heat stress tolerance in crops. Heat tolerance can also be altered by increasing the levels of osmolytes, antioxidant enzymes, and by altering membrane fluidity. Hence, identification of genotypes for the genes showing heat tolerance related traits could be useful for future gene selection, cloning or breeding purposes (Jagadish et al., 2010; Mangrauthia et al., 2016).

In this direction, several efforts to develop heat-tolerant transgenic lines by over-expression or by down-regulation for particular trait sequestered from diverse crop species with maize. For example, over-expression of *OsMYB55* gene provides heat tolerance (Casaretto et al., 2016). It has been found that, a substantial number of genes showing responses to different abiotic stresses *viz.,* high temperature, dehydration and oxidative stress were also enhanced during the overexpression. Recent developments in plant tissue culture, genetic transformation, and expression analysis have enhanced our knowledge about biochemical and genetic basis of heat stress-tolerance. Moreover, modulation of the expression of the identified genes, proteins, enzymes, or compounds via genetic engineering produced improved heat stress tolerance in diverse plant species (Zang et al., 2001; Rontein et al., 2002). Hence, multiple genes could now possibly transferred which additively increase plant stress tolerance with the current transformation technology. Identification of heat tolerance components their identification at gene level and cloning may permit transformation of plants with multiple genes and production of highly stress-tolerant transgenic plants.

16.4 GENOME EDITING

The advances in next generation sequencing have increased the accessibility of genome sequences for large number of crops, allowing novel genome editing tools to be used for breeding the crops for any desired traits. Advancement in genome editing technologies has caused revolution in agriculture sector. Genome editing has helped breeders to precisely knock in/knock out or knock down any desirable sequence in the genome and thus, became the most effective tools for crop improvement in this decade. There are different genome editing tools available like Zinc Finger Nucleases (ZFNs), Transcription activator-like effector nucleases (TALENs), mega nucleases, and CRISPR-Cas9 system which helps in site-directed mutagenesis. Among all these clustered regularly interspaced short palindromic repeat (CRISPR) genome editing tool is most frequently used nowadays due to its simplicity of use. It involves a duplex RNA structure (CRISPR RNA (cr RNA): Trans activating CRISPR RNA (Tracr RNA)) which guides Cas9 nucleases to target DNA. It involves a simple designing and cloning method, where guide RNA can be syntheised, attached with the same Cas9 endonuclease and further would recognize intended sequence in the genome and Cas9 enzyme cuts at the target DNA. Such genome editing tool with site-specific nucleases has resulted in the successful illustration of precise gene editing in both plant and animal systems. Site-specific nucleases creates a double stranded breaks (DSB) in the target DNA which are then repaired through cell's own repair machinery either by non-homologous end joining (NHEJ) or homology-directed recombination (HDR) pathways which results into insertion/deletion (INDELS) and substitution mutations in the target region(s) in the genomes correspondingly (Jinek et al., 2012). Employing a CRISPR project comprises of following steps- (1) Synthesizing a single gRNA (sgRNA) having a spacer and scaffold region, (2) Cloning the sgRNA attached to Cas9 protein into a suitable binary vector, (3) introducing cloned construct into desired host/cell lines through suitable transformation methods, (4) screening, (5) Confirmation and validation of edited lines.

CRISPR-Cas9 system has become a powerful genome engineering tool which is being exploited for crop improvement with precise gene-editing of targeted traits. It has been successfully practiced in several major crops. Heat tolerance being a complex trait, targeting multiple genes simultaneously is a big challenge so this revolution of CRISPR/Cas can be used to modify multiple genes to enhance heat stress tolerance. The CRISPR/Cas9 technology is continuously opening novel directions in plant research. Recent advancement in efficient delivery methods of CRISPR/Cas9 resulted improved endonuclease system with advance features, like minimum off-targets, improved precise nuclease activity, broader target range, and numerous PAM sites (Koonin et al., 2017). This system employed as precise and effective gene editing tool for multiplexed genome editing, which involves the simultaneous targeting of several genes (Malenica et al., 2021).

16.5 SPEED BREEDING

The vision of climate resilient smart breeding requires connecting all the activity starting from wild germplasm to future ready crop through the layers of next generation breeding (genomic selection, GWAS, haplotype breeding, etc.) coupled with Omics (genomics, proteomics, metabolomics, and phonemics), rapid generation advancement (speed breeding) as well as modern computer science things like Artificial intelligence (AI). Recently, in association with high throughput genotyping and phenotyping technologies, a new technique has been developed in this direction named as "speed breeding," a rapid generation-advance technology which can shorten the breeding cycle and accelerate crop research. Speed breeding methods comprises controlled growth condition with high density planting, management of light (continuous light) and temperatures (optimal temperatures), and use of different selection approaches to select key traits. Speed Breeding can be supported out in various ways, for example by extending the daily exposure to light combined with early seed harvest from the plant and thus by reducing the generation time. This increases the photosynthesis activity, and further leads to faster flowering and early seed development. Different next generation plant breeding approaches can be coupled with speed breeding to cut short the lengthy breeding cycle and also for the efficient use of resources (Hickey et al., 2017; Samineni et al., 2019; Watson et al., 2018 and Varshney et al., 2021).

The basic principle behind speed breeding is modulating the optimum light (quality/intensity) to extend crop photoperiod using supplementary lighting with metal halide or sodium vapor lamps (SVL) and light-emitting diode (LED) lights. As a result, it mimics daily light and dark cycle, and plants are exposed to prolonged photoperiod of about 22 h by combining different light sources. These thereby are helpful in controlling temperature to accelerate the crop plant photosynthesis and flowering coupled with early seed harvesting to fasten the plant life cycle. Speed breeding is understood to have revolutionized the agriculture sector and may be used to hasten the crop breeding programs (Hickey et al., 2019).

In recent years, application of speed-breeding demonstrated to increase the number of crops per year, hastening the progress of breeding studies, and genomics in different crops. Speed-breeding can be coupled with other breeding methodologies which can become a real boon for breeders in changing the time game to get any variety or product by accelerating breeding procedures. It can be integrated with other modern crop breeding technologies like high-throughput genotyping, precision phenotyping, genomic selection (GS), marker-assisted selection (MAS), and genome editing (Pandey et al., 2022; Kumar et al., 2021). It has been found to ideally suit for backcrossing, where the goal is to integrate a comparatively simple inherited trait into a new variety. To considerably enhance the efficacy for breeding for heat tolerance in maize, speed breeding can be coupled with marker-assisted breeding (MAS and genomic selection) and then followed by high throughput phenotyping (Fig. 16.2). It is interesting to find that integration of speed breeding with gene editing specifically modify the plant genes for various abiotic and biotic stresses. As a result, combing speeding technology with existing crop breeding methods to accelerate crop breeding programs will provide an exciting path forward for crop improvement especially under abiotic stresses.

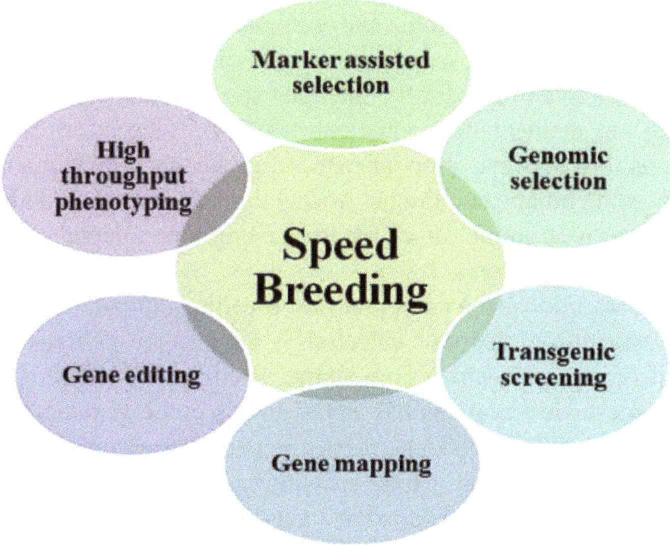

FIGURE 16.2 Speed breeding can be coupled with other breeding methodologies.

16.6 ARTIFICIAL INTELLIGENCE

Artificial intelligence (AI) has recently received attention in agriculture as a means of addressing issues such as plant stresses, herbicide resistance, soil health, and crop yield by constituting the intelligent prediction models (Muraya et al., 2017). AI mainly involves machine learning (ML) algorithm models similar to support vector machine (SVM), artificial neural network (ANN), random forest (RF), advanced hi-tech technology such as internet of things (IoT), and deep neural network (DNN). It is a fascinating hi-tech system with many agricultural uses that opens up new opportunities in the area of digital agriculture (Montesinos-Lopez et al., 2018). Artificial Intelligence-Machine Learning involves different stages which helps breeders to access enormous multi-omics and environmental data so that they can deploy the system to attain an anticipated phenotype in the perspective of maximum advantage for genotype. Different stages are as follows (1) Data capture which involves-Data sources and data collection, (2) Data storage, (3) Data pre-processing, (4) Data analysis, (5) Data interpretation.

The Next-Gen AI has enormous ability to automate data analysis with wider crop applications. The first step towards identification and breeding of stress tolerant crops is to define the symptoms/problem associated with high temperature in this context and then target specific solution by best utilizing new algorithms, data collection, and computational power. Combination and definition of the target space is vital with direction for an algorithm to produce a predictive model for the intended space. These statistical models are nothing but mathematical equations which provide information about the relationships between crop growth, crop health, yield, management technology, and climatic conditions. Artificial intelligence-based algorithms can be thus, be used in the future for forecasting heat-tolerance and stress-tolerance indices in maize.

Genomic selection is the main focus for breeding climate resilient plants. The basic assertion of genomic selection is based upon heritable (genetic) component of any trait which is outcome of collective effect of all inter-linked genome-wide variations like SNPs, indels in the genome. Therefore, models for estimating the effects of each genetic variant on the phenotype are the crucial component while accounting for environmental effects. These statistical model is developed in a genotyped and phenotyped training population for testing. The AI-based models are appropriate for agricultural, hydrological, environmental, and other approach to the way. The methods used in AI are based on statistical-learning-based approaches, probability, categorization, logic, and several types of search engine optimizations along with mathematical programming. Overall, AI based methods in GS trying to map the genomic data layer (e.g., SNPs) to different phenotypes (e.g., heat tolerance) (Harfouche et al., 2019). The developed statistical model is further tested to an independent breeding population for their phenotype prediction based upon available genotypic data which can save much time for phenotyping. These models with higher prediction accuracy can translate into increased yields in shorter breeding cycles.

16.7 HIGH THROUGHPUT PHENOTYPING

Plant phenotyping is an important method to inquire the plants relationship with its environment. Plant phenomics along with genetics and modern high-throughput platforms provide insights about plant stress data and enables high resolution imaging and screening of multiple plants (Roitsch et al., 2019). The term "phenome" refers to the plants whole phenotypic profile and a phenotype is result of genomic expression in relation to their environmental conditions (Houle et al., 2010). One of the most promising technologies, high-throughput phenotyping combines with engineering, plant science, maths, computation, and information science with externally advanced AI and ML tools tries to uncover various stress related phenotypes of genotypes essential to produce the ideal crop cultivars (Furbank et al., 2019). Three approaches make up high-throughput phenotyping: first, the target trait controlling a distinctive stress response is selected for quick, accurate, repeatable, and precise data collection. The next step is based on the imaging and estimation of data, whereas, in the subsequent step computational analysis used to predict a biological response based upon the available data (Esposito et al., 2020).

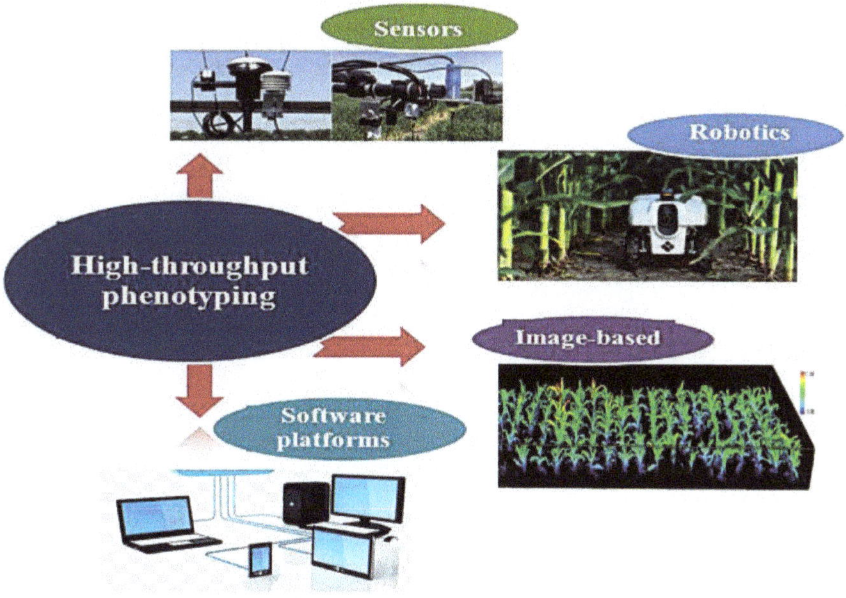

FIGURE 16.3 Tools involved in high throughput phenotyping.

Crop phenomics has undergone a revolution because of various advanced implements like sensors, robotics, software platforms, and imaging tools (Fig. 16.3). Unmanned aerial vehicles (UAVs), are presently used on a massive scale and includes phenotyping towers, drones, satellites, phenomobiles, environmental sensor networks, air crafts, autonomous ground vehicles, helicopters, field scanning platforms, and zeppelins are the few most popular plant phenotyping platforms (Liebisch et al., 2015). Image-based phenotyping technologies can swiftly and specifically measure plant stress resistance traits, grain yield, and quality thereby have a higher probability to speed up crop breeding. The quantitative measurement of heat resistance indicators in combination with high-throughput phenotyping platforms will provide a chance to measure both novel and traditional heat resistant traits and related genes.

In conclusion, heat stress in maize affects negatively to both vegetative and reproductive stages. Breeding for heat stress tolerance requires germplasm resources that to identify genomic regions/QTLs/gene(s) confers heat tolerance at field level. Further, improvement in heat stress tolerance achieved through conventional plant breeding techniques with limited success. Next generation breeding tools based on high throughput sequencing technologies along with speed breeding and CRISPR has fasten the improvement programs. Recent applications of high throughput phenotyping and artificial intelligence based on machine learning and integrated breeding approach have open up the next era for breeding which help to improve crop yield not only under heat stress but in other biotic and abiotic stress in various crops.

KEYWORDS

- **maize**
- **heat stress**
- **breeding tools**
- **genome editing**
- **genomics**
- **CRISPR/Cas**
- **speed breeding**

REFERENCES

Ashraf, M. Inducing Drought Tolerance in Plants: Recent Advances. *Biotechnol. Adv.* **2010,** *28*, 169–183.

Ashraf, M.; Athar, H. R.; Harris, P. J. C.;Kwon, T. R. Some Prospective Strategies for Improving Crop Salt Tolerance. *Adv. Agron.* **2008,** *97*, 45–110.

Bai, J. Genetic Variation of Heat Tolerance and Correlation with Other Agronomic Traits in a Maize (Zea mays L.) Recombinant Inbred Line Population. Doctoral Dissertation, Texas Tech University, 2003, pp 1–91.

Bita, C. E.; Gerats, T. Plant Tolerance to High Temperature in a Changing Environment: Scientific Fundamentals and Production of Heat Stress-Tolerant Crops. *Front. Plant Sci.* 2013, *4* (273), 1–18.

Bohnert, H. J.; Gong, Q.; Li, P.;Ma, S. Unraveling Abiotic Stress Tolerance Mechanisms—Getting Genomics Going. *Curr. Opin. Plant Biol.* **2006,** *9*, 180–188.

Cantarero, M.; Cirilo, A.;Andrade, F. Night Temperature at Silking Affects Set in Maize. *Crop Sci.* **1999,** *39*, 703–710.

Casaretto, J. A.; El-kereamy, A.; Zeng, B.; et al. Expression of OsMYB55 in Maize Activates Stress-Responsive Genes and Enhances Heat and Drought Tolerance. *BMC Genomic* **2016,** *17*, 312.

Collins, N. C.; Tardieu, F.; Tuberosa, R. Quantitative Trait Loci and Crop Performance Under Abiotic Stress: Where do we Stand? *Plant Physiol.* **2008,** *147*, 469–486.

Datta, K.; Baisakh, N.; Maung Thet, K.; Tu, J.; Datta, S. Pyramiding Transgenes for Multiple Resistance in Rice Against Bacterial Blight, Yellow Stem Borer and Sheath Blight. *Theor. Appl. Genet.* **2002,** *106*, 1–8.

Edmeades, G. Drought Tolerance in Maize: An Emerging Reality; Glob. Status Commer.39, 2008.

Edreira, J. R.; Otegui, M. E. Heat Stress in Temperate and Tropical Maize Hybrids: A Novel Approach for Assessing Sources of Kernel Loss in Field Conditions. *Field Crops Res.* **2013,** *142*, 58–67.

Esposito, S.; Carputo, D.; Cardi, T.; Tripodi, P. Applications and Trends of Machine Learning in Genomics and Phenomics for Next-Generation Breeding. *Plants* **2020,** *9*, 34.

Farooq, M.; Bramley, H.; Palta, J. A.; Siddique, K. H. M. Heat Stress in Wheat During Reproductive and Grain Filling Phases. *Crit. Rev. Plant Sci.* **2011,** *30*, 1–17.

Frova, C.; Sari-Gorla, M. Quantitative Expression of Maize HSPs: Genetic Dissection and Association with Thermo Tolerance. *Theor. Appl. Genet.* 1993, *86*, 213–220.

Frova, C.; Sari-Gorla, M. Quantitative Trait Loci (QTLs) for Pollen Thermo Tolerance Detected in Maize. *Mol. Genet. Genom.* **1994,** *245*, 424–430.

Furbank, R. T.; Jimenez-Berni, J. A.; George-Jaeggli, B.; Potgieter, A. B.; Deery, D. M. Field Crop Phenomics: Enabling Breeding for Radiation use Efficiency and Biomass in Cereal Crops. *New Phytol.* **2019,** *223*, 1714–1727.

Govindaraj, M.; Pattanashetti, S. K.; Kanatti, N. P. A. A. Breeding Cultivars for Heat Stress Tolerance in Staple Food Crops. In *Next Generation Plant Breeding*; Çiftçi, Y. Ö., Ed.; IntechOpen, 2018; vol *76480* (3).

Grover, A.; Mittal, D.; Negi, M.;Lavania, D. Generating High Temperature Tolerant Transgenic Plants: Achievements and Challenges. *Plant Sci.* **2013,** *205–206*, 38–47.

Habier, D.; Fernando, R. L.; Dekkers, J. C. M. Genomic Selection Using Low-Density Marker Panels. *Genetics* **2009,** *182*, 343–353.

Harfouche, A.; Jacobson, D. A.; Kainer, D.; Romero, J. C.; Harfouche, A. H.; Mugnozza, G. S.; Moshelion, M.; Tuskan, G. A.; Keurentjes, J. J. B.; Altman, A. Accelerating Climate Resilient Plant Breeding by Applying Next-Generation Artificial Intelligence. *Trends Biotechnol.* **2019,** *37* (11), 1317–1235.

Heffner, E. L.; Sorrells, M. E.; Jannink, J. L. Genomic Selection for Crop Improvement. *Crop Sci.* **2009,** *49*, 1–12.
Hickey, L. T.; German, S. E.; Pereyra, S. A.; Diaz, J. E.; Ziems, L. A.; Fowler, R. A.; Platz, G. J.; Franckowiak, J. D.; Dieters, M. J. Speed Breeding for Multiple Disease Resistance in Barley. *Euphytica* **2017,** *213*, 64.
Hickey, L. T.; Hafeez, A. N.; Robinson, H.; Jackson, S. A.; Leal-Bertioli, S. C.; Tester, M.; et al. Breeding Crops to Feed 10 Billion. *Nat. Biotechnol.* **2019,** *37*, 744–754.
Houle, D.; Govindaraju, D. R.; Omholt, S. Phenomics: The Next Challenge. *Nat. Rev. Genet.* 2010, 11: 855–866.
Jagadish, S. V. K.; Cairns, J.; Lafitte, R.; Wheeler, T. R.; Price, A. H.; Craufurd, P. Q. Genetic Analysis of Heat Tolerance at Anthesis in Rice. *Crop Sci.* **2010,** *50*, 1633–1641.
Jagadish, S.; Muthurajan, R.; Oane, R.; Wheeler, T. R.; Heuer, S.; Bennett, J.; Craufurd, P. Q. Physiological and Proteomic Approaches to Address Heat Tolerance During Anthesis in Rice (Oryza sativa L.). *J. Exp. Bot.* **2010,** *61*, 143–156.
Jinek, M.; Chylinski, K.; Fonfara, I.; Hauer, M.; Doudna, J. A.; Charpentier, E. A Programmable Dual-RNA–Guided DNA Endonuclease in Adaptive Bacterial Immunity. *Science* **2012,** *337*, 816–821.
Koonin, E. V.; Makarova, K. S.; Zhang, F. Diversity, Classification and Evolution of CRISPR-Cas Systems. *Curr. Opin. Microbiol.* **2017,** *37*, 67–78.
Kumar, K.; Gupta, M.; Singh, A.; Aggarwal, C.; Choudhary, M.; Parihar, C. M.; Singh, I.; Yadava, P. Frontier Technologies in Maize Improvement. In *Maize Research in India: Retrospect and Prospect*; New India Publishing Agency: New Delhi, India, 2021; pp 541–563.
Liebisch, F.; Kirchgessner, N.; Schneider, D.; Walter, A.; Hund, A. Remote, Aerial Phenotyping of Maize Traits with a Mobile Multi-Sensor Approach. *Plant Methods* **2015,** *11*, 9.
Lobell, D. B.; Roberts, M. J.; Schlenker, W.; Braun, N.; Little, B, B.; Rejesus, R. M.; Hammer, G., L. Greater Sensitivity to Drought Accompanies Maize Yield Increase in the U.S. Midwest. *Science* **2014,** *344* (6183), 516–519.
Longmei, N.; et al. Genome Wide Association Mapping for Heat Tolerance in Sub-tropical Maize. *BMC Genom.* 2021, *22*, 154.
Ma, X.; Zhu, Q.; Chen, Y.; Liu, G. CRISPR/Cas9 Platforms for Genome Editing in Plants: Developments and Applications. *Mol. Plant* 2016, *9* (7), 961–974.
Mahmood, S.; Wahid, A.; Javed, F.; Basra, S. M. Heat Stress Effects on Forage Quality Characteristics of Maize (Zea mays) Cultivars. *Int. J. Agri. Biol.* **2010,** *12*, 701–706.
Malenica, N.; Dunić, J. A.; Vukadinović, L.; Cesar, V.; Šimić, D. Genetic Approaches to Enhance Multiple Stress Tolerance in Maize. *Genes* **2021,** *12* (11), 1760.
Mallya, G.; Zhao, L.; Song, X.; Niyogi, D.; Govindaraju, R. 2012 Midwest Drought in the United States. *J. Hydrol. Eng.* **2013,** *18*, 737–745.
Mangrauthia, S. K.; Agarwal, S.; Sailaja, B.; Sarla, N.; Voleti, S. Transcriptome Analysis of Oryza sativa (Rice) Seed Germination at High Temperature Shows Dynamics of Genome Expression Associated with Hormones Signaling and Abiotic Stress Pathways. *Tropic Plant Biol.* **2016,** *9*, 215–228.
Mhike, X.; Okori, P.; Magorokosho, C.; Ndlela, T. Validation of the Use of Secondary Traits and Selection Indices for Drought Tolerance in Tropical Maize (Zea mays L.). *Afr. J. Plant Sci.* **2012,** *6*, 96–102.
Monjardino, P.; Smith, A. G.; Jones, R. J. Heat Stress Effects on Protein Accumulation of Maize Endosperm. *Crop Sci.* **2005,** *45*, 1203–1210.

Montesinos-López, O. A.; Martín-Vallejo, J.; Crossa, J.; Gianola, D.; Hernández- Suárez, C. M.; Montesinos-López, A.; et al. New Deep Learning Genomic-Based Prediction Model for Multiple Traits with Binary, Ordinal, and Continuous Phenotypes. *Genes Genomes Genet.* **2019,** *9*, 1545–1556.

Muraya, M. M.; Chu, J.; Zhao, Y.; Junker, A.; Klukas, C.; Reif, J. C.; et al. Genetic Variation of Growth Dynamics in Maize (Zea mays L.) Revealed Through Automated Non-invasive Phenotyping. *Plant J.* **2017,** *89*, 366–380.

Nakashima, K.; Ito, Y.; Yamaguchi-Shinozaki, K. Transcriptional Regulatory Networks in Response to Abiotic Stresses in Arabidopsis and Grasses. *Plant Physiol.* **2009,** *149*, 188–195.

Ottaviano, E.; Sari-Gorla, M.; Pé, E.; Frova, C. Molecular Markers (RFLPs and HSPs) for the Genetic Dissection of Thermo Tolerance in Maize. *Theor. Appl. Genet.* **1991,** *81*, 713–719.

Pandey, S.; Singh, A.; Parida, S. K.; Prasad, M. Combining Speed Breeding with Traditional and Genomics-Assisted Breeding for Crop Improvement. *Plant Breed.* **2022**.

Paliwal, R.; Roder, M. S.; Kumar, U.; Srivastava, J. P.; Joshi, A. K. QTL Mapping of Terminal Heat Tolerance in Hexaploid Wheat (T. aestivum L.). *Theor. Appl. Genet.* **2012,** *125*, 561–575.

Pinto, R. S.; Reynolds, M. P.; Mathews, K. L.; McIntyre, C. L.; Olivares-Villegas, J. J.; Chapman, S. C. Heat and Drought Adaptive QTL in a Wheat Population Designed to Minimize Confounding Agronomic Effects. *Theor. Appl. Genet.* **2010,** *121*, 1001–1021.

Rippey, B. R. The U.S. Drought of 2012. *Weather Clim. Extremes* **2015,** *10*, 57–64.

Roitsch, T.; Cabrera-Bosquet, L.; Fournier, A.; Ghamkhar, K.; Jiménez-Berni, J.; Pinto, F.; et al. New Sensors and Data-Driven Approaches-A Path to Next Generation Phenomics. *Plant Sci.* **2019,** *282*, 2–10.

Rontein, D.; Basset, G.; Hanson, A. D. Metabolic Engineering of Osmoprotectant Accumulation in Plants. *Metab. Eng.* **2002,** *4*, 49–56.

Samineni, S.; Sen, M.; Sajja, S. B.; Gaur, P. M. Rapid Generation Advance (RGA) in Chickpea to Produce up to Seven Generations Per Year and Enable Speed Breeding. *Crop J.* **2019,** *8* (1), 164–169.

Schoper, J. B.; Lambert, R. J.; Vasilas, B. L.;Westgate, M. E. Plant Factors Controlling Seed Set in Maize the Influence of Silk, Pollen, and Ear-Leaf Water Status and Tassel Heat Treatment at Pollination. *Plant Physiol.* **1987,** *83*, 121–125.

Setimela, P. S.; Andrews, D. J.; Partridge, J.; Eskridge, K. M. Screening Sorghum Seedlings for Heat Tolerance Using a Laboratory Method. *Eur. J. Agron.* **2005,** *23* (2), 103–107.

Shikha, K.; Shahi, J. P.; Vinayan, M. T.; Zaidi, P. H.; Singh, A. K.; Sinha, B. Genome-Wide Association Mapping in Maize: Status and Prospects. *3 Biotech.* **2021,** *11* (5), 244.

Sreenivasulu, N.; Sopory, S. K.; Kavi Kishore, P. V. Deciphering the Regulatory Mechanisms of Abiotic Stress Tolerance in Plants by Genomic Approaches. *Gene* **2007,** *388*, 1–13.

Tassawar, H.; Iftikhar, A.; Zulfiqar, A. Study on Gene Action and Combining Abilities for Thermotolerant Ablilities of Corn (Zea mays L.). *Int. J. Plant Prod.* **2012,** *1* (1), 1–12.

Tester, M.; Langridge, P. Breeding Technologies to Increase Crop Production in a Changing World. *Science* **2010,** *327*, 818–822.

Urano, K.; Kurihara, Y.; Seki, M.; Shinozaki, K. 'Omics' Analyses of Regulatory Networks in Plant Abiotic Stress Responses. *Curr. Opin. Plant Biol.* **2010,** *13*, 132–138.

Vij, S.; Tyagi, A. K. Emerging Trends in the Functional Genomics of the Abiotic Stress Response in Crop Plants. *Plant Biotechnol. J.* **2007,** *5*, 361–380.

Wahid, A.; Gelani, S.; Ashraf, M.; Foolad, M. R. Heat Tolerance in Plants: An Overview. *Environ. Exp. Bot.* **2007,** *61*, 199–223.

Watson, A.; Ghosh, S.; Williams, M. J.; Cuddy, W. S.; Simmonds, J.; Hickey, L. T. Speed Breeding is a Powerful Tool to Accelerate Crop Research and Breeding. *Nat. Plants* **2018,** *4* (1), 23–29.

Wilhelm, E.; Mullen, R.; Keeling, P.; Singletary, G. Heat Stress During Grain Filling in Maize: Effects on Kernel Growth and Metabolism. *Crop Sci.* **1999,** *39*, 1733–1741.

Yadav, S. S.; Redden, R.; Hatfield, J. L; Lotze-Campen, H.; Hall, A. Crop Adaptation to Climate Change; Wiley, New York, 2011.

Yuan, Y.; Cairns, J. E.; Babu, R.; Gowda, M.; Makumbi, D.; Magorokosho, C.; Zhang, A.; Liu, Y.; Wang, N.; Hao, Z.; San Vicente, F. Genome-Wide Association Mapping and Genomic Prediction Analyses Reveal the Genetic Architecture of Grain Yield and Flowering Time Under Drought and Heat Stress Conditions in Maize. *Front. Plant Sci.* **2019,** *9*, 1919.

Zhang, H. X.; Hodson, J. N.; Williams, J. P.; Blumwald. E. Engineering Salt-Tolerant Brassica Plants: Caracterization of Yield and Seed Oil Quality in Transgenic Plants with Increased Vacuolar Sodium Accumulation. *PNAS* **2001,** *98*, 6896–6901.

Index

D

H

I

K

L

M

For Product Safety Concerns and Information please contact our EU
representative GPSR@taylorandfrancis.com
Taylor & Francis Verlag GmbH, Kaufingerstraße 24, 80331 München, Germany

www.ingramcontent.com/pod-product-compliance
Ingram Content Group UK Ltd.
Pitfield, Milton Keynes, MK11 3LW, UK
UKHW021513250926
475702UK00043B/269

* 9 7 8 1 7 7 4 9 1 5 0 6 6 *